An Introduction to Sparse Stochastic Processes

Providing a novel approach to sparsity, this comprehensive book presents the theory of stochastic processes that are ruled by linear stochastic differential equations and that admit a parsimonious representation in a matched wavelet-like basis.

Two key themes are the statistical property of infinite divisibility, which leads to two distinct types of behavior – Gaussian and sparse – and the structural link between linear stochastic processes and spline functions, which is exploited to simplify the mathematical analysis. The core of the book is devoted to investigating sparse processes, including a complete description of their transform-domain statistics. The final part develops practical signal-processing algorithms that are based on these models, with special emphasis on biomedical image reconstruction.

This is an ideal reference for graduate students and researchers with an interest in signal/image processing, compressed sensing, approximation theory, machine learning, or statistics.

MICHAEL UNSER is Professor and Director of the Biomedical Imaging Group at the École Polytechnique Fédérale de Lausanne (EPFL), Switzerland. He is a member of the Swiss Academy of Engineering Sciences, a Fellow of EURASIP, and a Fellow of the IEEE.

POUYA D. TAFTI is a data scientist currently residing in Germany, and a former member of the Biomedical Imaging Group at EPFL, where he conducted research on the theory and applications of probabilistic models for data.

"Over the last twenty years, sparse representation of images and signals became a very important topic in many applications, ranging from data compression, to biological vision, to medical imaging. The book *Sparse Stochastic Processes* by Unser and Tafti is the first work to systematically build a coherent framework for non-Gaussian processes with sparse representations by wavelets. Traditional concepts such as Karhunen-Loéve analysis of Gaussian processes are nicely complemented by the wavelet analysis of Levy Processes which is constructed here. The framework presented here has a classical feel while accommodating the innovative impulses driving research in sparsity. The book is extremely systematic and at the same time clear and accessible, and can be recommended both to engineers interested in foundations and to mathematicians interested in applications."

David Donoho, *Stanford University*

"This is a fascinating book that connects the classical theory of generalised functions (distributions) to the modern sparsity-based view on signal processing, as well as stochastic processes. Some of the early motivations given by I. Gelfand on the importance of generalised functions came from physics and, indeed, signal processing and sampling. However, this is probably the first book that successfully links the more abstract theory with modern signal processing. A great strength of the monograph is that it considers both the continuous and the discrete model. It will be of interest to mathematicians and engineers having appreciations of mathematical and stochastic views of signal processing."

Anders Hansen, *University of Cambridge*

An Introduction to Sparse Stochastic Processes

MICHAEL UNSER and POUYA D. TAFTI

École Polytechnique Fédérale de Lausanne

CAMBRIDGE
UNIVERSITY PRESS

University Printing House, Cambridge CB2 8BS, United Kingdom

Cambridge University Press is part of the University of Cambridge.

It furthers the University's mission by disseminating knowledge in the pursuit of
education, learning and research at the highest international levels of excellence.

www.cambridge.org
Information on this title: www.cambridge.org/9781107058545

First published 2014

Printed in the United Kingdom by Clays, St Ives plc

A catalog record for this publication is available from the British Library

Library of Congress Cataloging in Publication Data
Unser, Michael A., author.
An introduction to sparse stochastic processes / Michael Unser and Pouya Tafti, École polytechnique fédérale,
Lausanne.
    pages   cm
Includes bibliographical references and index.
ISBN 978-1-107-05854-5 (Hardback)
1. Stochastic differential equations.   2. Random fields.   3. Gaussian processes.   I. Tafti, Pouya, author.
II. Title.
QA274.23.U57 2014
519.2′3–dc23   2014003923

ISBN 978-1-107-05854-5 Hardback

To Gisela, Lucia, Klaus, and Murray

Contents

Preface

In the years since 2000, there has been a significant shift in paradigm in signal processing, statistics, and applied mathematics that revolves around the concept of sparsity and the search for "sparse" representations of signals. Early signs of this (r)evolution go back to the discovery of wavelets, which have now superseded classical Fourier techniques in a number of applications. The other manifestation of this trend is the emergence of data-processing schemes that minimize an ℓ_1 norm as opposed to the squared ℓ_2 norm associated with the traditional linear methods. A highly popular research topic that capitalizes on those ideas is compressed sensing. It is the quest for a statistical framework that would support this change of paradigm that led us to the writing of this book.

The cornerstone of our formulation is the classical innovation model, which is equivalent to the specification of stochastic processes as solutions of linear stochastic differential equations (SDE). The non-standard twist here is that we allow for non-Gaussian driving terms (white Lévy noise) which, as we shall see, has a dramatic effect on the type of signal being generated. A fundamental property, hinted in the title of the book, is that the non-Gaussian solutions of such SDEs admit a sparse representation in an adapted wavelet-like basis. While a sizable part of the present material is an outgrowth of our own research, it is founded on the work of Lévy (1930) and Gelfand (arguably, the second most famous Soviet mathematician after Kolmogorov), who derived general functional tools and results that are hardly known by practitioners but, as we argue in the book, are extremely relevant to the issue of sparsity. The other important source of inspiration is spline theory and the observation that splines and stochastic processes are ruled by the same differential equations. This is the reason why we opted for the innovation approach which facilitates the transposition of analytical techniques from one field to the other. While the formulation requires advanced mathematics that are carefully explained in the book, the underlying model has a strong engineering appeal since it constitutes the natural extension of the traditional filtered-white-noise interpretation of a Gaussian stationary process.

The book assumes that the reader has a good understanding of linear systems (ordinary differential equations, convolution), Hilbert spaces, generalized functions (i.e., inner products, Dirac impulses, linear operators), the Fourier transform, basic statistical signal processing, and (multivariate) statistics (probability density and characteristic functions). By contrast, there is no requirement for prior knowledge of splines, stochastic differential equations, or advanced functional analysis (function spaces, Bochner's theorem, operator theory, singular integrals) since these topics are treated in a self-contained fashion.

Several people have had a crucial role in the genesis of this book. The idea of defining sparse stochastic processes originated during the preparation of a talk for Martin Vetterli's 50th birthday (which coincided with the anniversary of the launching of Sputnik) in an attempt to build a bridge between his signals with a finite rate of innovation and splines. We thank him for his long-time friendship and for convincing us to undertake this writing project. We are grateful to our former collaborator, Thierry Blu, for his precious help in the elucidation of the functional link between splines and stochastic processes. We are extremely thankful to Arash Amini, Julien Fageot, Pedram Pad, Qiyu Sun, and John-Paul Ward for many helpful discussions and their contributions to mathematical results. We are indebted to Emrah Bostan, Ulugbek Kamilov, Hagai Kirshner, Masih Nilchian, and Cédric Vonesch for turning the theory into practice and for running the signal- and image-processing experiments described in Chapters 10 and 11. We are most grateful to Philippe Thévenaz for his intelligent editorial advice and his spotting of multiple errors and inconsistencies, while we take full responsibility for the remaining ones. We also thank Phil Meyler, Sarah Marsh and Gaja Poggiogalli from Cambridge University Press, as well as John King for his careful copy-editing.

The authors also acknowledge very helpful and stimulating discussions with Ben Adcock, Emmanuel Candès, Volkan Cevher, Robert Dalang, Mike Davies, Christine De Mol, David Donoho, Pier-Luigi Dragotti, Michael Elad, Yonina Eldar, Jalal Fadili, Mario Figueiredo, Vivek Goyal, Rémy Gribonval, Anders Hansen, Nick Kingsbury, Gitta Kutyniok, Stamatis Lefkimmiatis, Gabriel Peyré, Robert Novak, Jean-Luc Stark, and Dimitri Van De Ville, as well as a number of other researchers involved in the field.

The European Research Commission (ERC) and the Swiss National Science Foundation provided partial support throughout the writing of the book.

Notation

Abbreviations

ADMM	Alternating-direction method of multipliers
AL	Augmented Lagrangian
AR	Autoregressive
ARMA	Autoregressive moving average
AWGN	Additive white Gaussian noise
BIBO	Bounded input, bounded output
CAR	Continuous-time autoregressive
CARMA	Continuous-time autoregressive moving average
CCS	Consistent cycle spinning
DCT	Discrete cosine transform
fBm	Fractional Brownian motion
FBP	Filtered backprojection
FFT	Fast Fourier transform
FIR	Finite impulse response
FISTA	Fast iterative shrinkage/thresholding algorithm
ICA	Independent-component analysis
id	Infinitely divisible
i.i.d.	Independent identically distributed
IIR	Infinite impulse response
ISTA	Iterative shrinkage/thresholding algorithm
JPEG	Joint Photographic Experts Group
KLT	Karhunen–Loève transform
LMMSE	Linear minimum-mean-square error
LPC	Linear predictive coding
LSI	Linear shift-invariant
MAP	Maximum a posteriori
MMSE	Minimum-mean-square error
MRI	Magnetice resonance imaging
PCA	Principal-component analysis
pdf	Probability density function
PSF	Point-spread function
ROI	Region of interest

SαS	Symmetric-alpha-stable
SDE	Stochastic differential equation
SNR	Signal-to-noise ratio
WSS	Wide-sense stationary

Sets

$\mathbb{N}, \mathbb{Z}^+$	Non-negative integers, including 0
$\mathbb{Z}$	Integers
$\mathbb{R}$	Real numbers
$\mathbb{R}^+$	Non-negative real numbers
$\mathbb{C}$	Complex numbers
$\mathbb{R}^d$	d-dimensional Euclidean space
$\mathbb{Z}^d$	d-dimensional integers

Various notation

j	Imaginary unit such that $j^2 = -1$
$\lceil x \rceil$	Ceiling: smallest integer at least as large as x
$\lfloor x \rfloor$	Floor: largest integer not exceeding x
$(x_1 : x_n)$	n-tuple $(x_1, x_2, \ldots, x_n)$
$\|f\|$	Norm of the function f (see Section 3.1.2)
$\|f\|_{L_p}$	L_p-norm of the function f (in the sense of Lebesgue)
$\|a\|_{\ell_p}$	ℓ_p-norm of the sequence a
$\langle \varphi, s \rangle$	Scalar (or duality) product
$\langle f, g \rangle_{L_2}$	L_2 inner product
$f^\vee$	Reversed signal: $f^\vee(r) = f(-r)$
$(f * g)(r)$	Continuous-domain convolution
$(a * b)[n]$	Discrete-domain convolution
$\widehat{\varphi}(\omega)$	Fourier transform of φ: $\int_{\mathbb{R}^d} \varphi(r) e^{-j\langle \omega, r \rangle}\, dr$
$\widehat{f} = \mathscr{F}\{f\}$	Fourier transform of f (classical or generalized)
$f = \mathscr{F}^{-1}\{\widehat{f}\}$	Inverse Fourier transform of $\widehat{f}$
$\overline{\mathscr{F}}\{f\}(\omega) = \mathscr{F}\{f\}(-\omega)$	Conjugate Fourier transform of f

Signals, functions, and kernels

$f, f(\cdot)$, or $f(r)$	Continuous-domain signal: function $\mathbb{R}^d \to \mathbb{R}$
φ	Generic test function in $\mathscr{S}(\mathbb{R}^d)$
$\psi_L = L^*\phi$	Operator-like wavelet with smoothing kernel ϕ
$s, \langle \varphi, s \rangle$	Generalized function $\mathscr{S}(\mathbb{R}^d) \to \mathbb{R}$
μ_h	Measure associated with h: $\langle \varphi, h \rangle = \int_{\mathbb{R}^d} \varphi(r) \mu_h(dr)$
δ	Dirac impulse: $\langle \varphi, \delta \rangle = \varphi(0)$
$\delta(\cdot - r_0)$	Shifted Dirac impulse
β_L	Generalized B-spline associated with the operator L
φ_{int}	Spline interpolation kernel

$\beta_+^n = \beta_{\mathrm{D}^{n+1}}$	Causal polynomial B-spline of degree n
$x_+^n = \max(0, x)^n$	One-sided power function
β_α	First-order exponential B-spline with pole $\alpha \in \mathbb{C}$
$\beta_{(\alpha_1:\alpha_N)}$	Nth-order exponential B-spline: $\beta_{\alpha_1} * \cdots * \beta_{\alpha_N}$
$a, a[\cdot]$, or $a[\boldsymbol{n}]$	Discrete-domain signal: sequence $\mathbb{Z}^d \to \mathbb{R}$
$\delta[\boldsymbol{n}]$	Discrete Kronecker impulse

Spaces

$\mathcal{X}, \mathcal{Y}$	Generic vector spaces (normed or nuclear)
$L_2(\mathbb{R}^d)$	Finite-energy functions $\int_{\mathbb{R}^d} |f(\boldsymbol{r})|^2 \, \mathrm{d}\boldsymbol{r} < \infty$
$L_p(\mathbb{R}^d)$	Functions such that $\int_{\mathbb{R}^d} |f(\boldsymbol{r})|^p \, \mathrm{d}\boldsymbol{r} < \infty$
$L_{p,\alpha}(\mathbb{R}^d)$	Functions such that $\int_{\mathbb{R}^d} |f(\boldsymbol{r})(1 + |\boldsymbol{r}|)^\alpha|^p \, \mathrm{d}\boldsymbol{r} < \infty$
$\mathcal{D}(\mathbb{R}^d)$	Smooth and compactly supported test functions
$\mathcal{D}'(\mathbb{R}^d)$	Distributions or generalized functions over $\mathbb{R}^d$
$\mathcal{S}(\mathbb{R}^d)$	Smooth and rapidly decreasing test functions
$\mathcal{S}'(\mathbb{R}^d)$	Tempered distributions (generalized functions)
$\mathcal{R}(\mathbb{R}^d)$	Bounded functions with rapid decay
$\ell_2(\mathbb{Z}^d)$	Finite-energy sequences $\sum_{\boldsymbol{k} \in \mathbb{Z}^d} |a[\boldsymbol{k}]|^2 < \infty$
$\ell_p(\mathbb{Z}^d)$	Sequences such that $\sum_{\boldsymbol{k} \in \mathbb{Z}^d} |a[\boldsymbol{k}]|^p < \infty$

Operators

Id	Identity
$\mathrm{D} = \frac{\mathrm{d}}{\mathrm{d}t}$	Derivative
D_{d}	Finite difference (discrete derivative)
D^N	Nth-order derivative
$\partial^{\boldsymbol{n}}$	Partial derivative of order $\boldsymbol{n} = (n_1, \ldots, n_d)$
L	Whitening operator (LSI)
$\widehat{L}(\boldsymbol{\omega})$	Frequency response of L (Fourier multiplier)
ρ_{L}	Green's function of L
L^*	Adjoint of L such that $\langle \varphi_1, \mathrm{L}\varphi_2 \rangle = \langle \mathrm{L}^*\varphi_1, \varphi_2 \rangle$
L^{-1}	Right inverse of L such that $\mathrm{L}\mathrm{L}^{-1} = \mathrm{Id}$
$h(\boldsymbol{r}_1, \boldsymbol{r}_2)$	Generalized impulse response of L^{-1}
L^{-1*}	Left inverse of L^* such that $(\mathrm{L}^{-1*})\mathrm{L}^* = \mathrm{Id}$
L_{d}	Discrete counterpart of L
$\mathcal{N}_{\mathrm{L}}$	Null space of L
P_α	First-order differential operator: $\mathrm{D} - \alpha \mathrm{Id}, \alpha \in \mathbb{C}$
$\mathrm{P}_{(\alpha_1:\alpha_N)}$	Differential operator of order N: $\mathrm{P}_{\alpha_1} \circ \cdots \circ \mathrm{P}_{\alpha_N}$
Δ_α	First-order weighted difference
$\Delta_{(\alpha_1:\alpha_N)}$	Nth-order weighted differences: $\Delta_{\alpha_1} \circ \cdots \circ \Delta_{\alpha_N}$
∂_τ^γ	Fractional derivative of order $\gamma \in \mathbb{R}^+$ and phase τ
$(-\Delta)^{\frac{\gamma}{2}}$	Fractional Laplacian of order $\gamma \in \mathbb{R}^+$
$\mathrm{I}_p^{\gamma*}$	L_p-stable left inverse of $(-\Delta)^{\frac{\gamma}{2}}$

Probability

X, Y	Generic scalar random variables
$\mathscr{P}_X$	Probability measure on $\mathbb{R}$ of X
$p_X(x)$	Probability density function (univariate)
$\Phi_X(x)$	Potential function: $-\log p_X(x)$
$\mathrm{prox}_{\Phi_X}(x, \lambda)$	Proximal operator
$p_{\mathrm{id}}(x)$	Infinitely divisible probability law
$\mathbb{E}\{\cdot\}$	Expected value operator
m_n	nth-order moment: $\mathbb{E}\{X^n\}$
κ_n	nth-order cumulant
$\widehat{p}_X(\omega)$	Characteristic function of X: $\mathbb{E}\{e^{j\omega X}\}$
$f(\omega)$	Lévy exponent: $\log \widehat{p}_{\mathrm{id}}(\omega)$
$v(a)$	Lévy density
$p_{(X_1:X_N)}(\boldsymbol{x})$	Multivariate probability density function
$\widehat{p}_{(X_1:X_N)}(\boldsymbol{\omega})$	Multivariate characteristic function
$m_{\boldsymbol{n}}$	Moment with multi-index $\boldsymbol{n} = (n_1, \ldots, n_N)$
$\kappa_{\boldsymbol{n}}$	Cumulant with multi-index $\boldsymbol{n}$
$H_{(X_1:X_N)}$	Differential entropy
$I(X_1, \ldots, X_N)$	Mutual information
$D(p\|q)$	Kullback–Leibler divergence

Generalized stochastic processes

w	Continuous-domain white noise (innovation)
$\langle \varphi, w \rangle$	Generic scalar observation of innovation process
$f_\varphi(\omega)$	Modified Lévy exponent: $\log \widehat{p}_{\langle \varphi, w \rangle}(\omega)$
$v_\varphi(a)$	Modified Lévy density
s	Generalized stochastic process: $\mathrm{L}^{-1}w$
u	Generalized increment process: $\mathrm{L}_{\mathrm{d}}s = \beta_{\mathrm{L}} * w$
W	1-D Lévy process with $\mathrm{D}W = w$
B_H	Fractional Brownian motion with Hurst index H
$\widehat{\mathscr{P}}_s(\varphi)$	Characteristic functional: $\mathbb{E}\{e^{j\langle \varphi, s \rangle}\}$
$\mathscr{B}_s(\varphi_1, \varphi_2)$	Correlation functional: $\mathbb{E}\{\langle \varphi_1, s \rangle \langle \varphi_2, s \rangle\}$
$R_s(\boldsymbol{r}_1, \boldsymbol{r}_2)$	Autocorrelation function: $\mathbb{E}\{s(\boldsymbol{r}_1)\overline{s(\boldsymbol{r}_2)}\}$

1   Introduction

1.1   Sparsity: Occam's razor of modern signal processing?

The hypotheses of Gaussianity and stationarity play a central role in the standard statistical formulation of signal processing. They fully justify the use of the Fourier transform as the optimal signal representation and naturally lead to the derivation of optimal linear filtering algorithms for a large variety of statistical estimation tasks. This classical view of signal processing is elegant and reassuring, but it is not at the forefront of research anymore.

Starting with the discovery of the wavelet transform in the late 1980s [Dau88, Mal89], researchers in signal processing have progressively moved away from the Fourier transform and have uncovered powerful alternatives. Consequently, they have ceased modeling signals as Gaussian stationary processes and have adopted a more deterministic, approximation-theoretic point of view. The key developments that are presently reshaping the field, and which are central to the theory presented in this book, are summarized below.

- *Novel transforms and dictionaries for the representation of signals.* New redundant and non-redundant representations of signals (wavelets, local cosine, curvelets) have emerged since the mid 1990s and have led to better algorithms for data compression, data processing, and feature extraction. The most prominent example is the wavelet based JPEG-2000 standard for image compression [CSE00], which outperforms the widely-used JPEG method based on the DCT (discrete cosine transform). Another illustration is wavelet-domain image denoising, which provides a good alternative to more traditional linear filtering [Don95]. The various dictionaries of basis functions that have been proposed so far are tailored to specific types of signals; there does not appear to be one that fits all.
- *Sparsity as a new paradigm for signal processing.* At the origin of this new trend is the key observation that many naturally occurring signals and images – in particular, the ones that are piecewise-smooth – can be accurately reconstructed from a "sparse" wavelet expansion that involves many fewer terms than the original number of samples [Mal98]. The concept of sparsity has been systematized and extended to other transforms, including redundant representations (a.k.a. frames); it is at the heart of recent developments in signal processing. Sparse signals are easy to compress and to denoise by simple pointwise processing (e.g., shrinkage) in the transformed domain. Sparsity provides an equally powerful framework for dealing

with more difficult, ill-posed signal-reconstruction problems [CW08, BDE09]. Promoting sparse solutions in linear models is also of interest in statistics: a popular regression shrinkage estimator is LASSO, which imposes an upper bound on the ℓ_1-norm of the model coefficients [Tib96].

- *New sampling strategies with fewer measurements.* The theory of compressed sensing deals with the problem of the reconstruction of a signal from a minimal, but suitably chosen, set of measurements [Don06, CW08, BDE09]. The strategy there is as follows: among the multitude of solutions that are consistent with the measurements, one should favor the "sparsest" one. In practice, one replaces the underlying ℓ_0-norm minimization problem, which is NP hard, by a convex ℓ_1-norm minimization which is computationally much more tractable. Remarkably, researchers have shown that this simplification does yield the correct solution under suitable conditions (e.g., restricted isometry) [CW08]. Similarly, it has been demonstrated that signals with a finite rate of innovation (the prototypical example being a stream of Dirac impulses with unknown locations and amplitudes) can be recovered from a set of uniform measurements at twice the "innovation rate" [VMB02], rather than twice the bandwidth, as would otherwise be dictated by Shannon's classical sampling theorem.

- *Superiority of non-linear signal-reconstruction algorithms.* There is increasing empirical evidence that non-linear variational methods (non-quadratic or sparsity-driven regularization) outperform the classical (linear) algorithms (direct or iterative) that are being used routinely for solving bioimaging reconstruction problems [CBFAB97, FN03]. So far, this has been demonstrated for the problem of image deconvolution and for the reconstruction of non-Cartesian MRI [LDP07]. The considerable research effort in this area has also resulted in the development of novel algorithms (ISTA, FISTA) for solving convex optimization problems that were previously considered out of numerical reach [FN03, DDDM04, BT09b].

1.2   Sparse stochastic models: the step beyond Gaussianity

While the recent developments listed above are truly remarkable and have resulted in significant algorithmic advances, the overall picture and understanding is still far from being complete. One limiting factor is that the current formulations of compressed sensing and sparse-signal recovery are fundamentally deterministic. By drawing on the analogy with the classical linear theory of signal processing, where there is an equivalence between quadratic energy-minimization techniques and minimum-mean-square-error (MMSE) estimation under the Gaussian hypothesis, there are good chances that further progress is achievable by adopting a complementary statistical-modeling point of view. [1] The crucial ingredient that is required to guide such an investigation is a sparse counterpart to the classical family of Gaussian stationary processes (GSP). This

---

[1] It is instructive to recall the fundamental role of statistical modeling in the development of traditional signal processing. The standard tools of the trade are the Fourier transform, Shannon-type sampling, linear filtering, and quadratic energy-minimization techniques. These methods are widely used in practice: they are powerful, easy to deploy, and mathematically convenient. The important conceptual point is that they

book focuses on the formulation of such a statistical framework, which may be aptly qualified as the next step after Gaussianity under the functional constraint of linearity.

In light of the elements presented in the introduction, the basic requirements for a comprehensive theory of sparse stochastic processes are as follows:

- *Backward compatibility*. There is a large body of literature and methods based on the modeling of signals as realizations of GSP. We would like the corresponding identification, linear filtering, and reconstruction algorithms to remain applicable, even though they obviously become suboptimal when the Gaussian hypothesis is violated. This calls for an extended formulation that provides the same control of the correlation structure of the signals (second-order moments, Fourier spectrum) as the classical theory does.
- *Continuous-domain formulation*. The proper interpretation of qualifying terms such as "piecewise-smooth," "translation-invariant," "scale-invariant," "rotation-invariant" calls for continuous-domain models of signals that are compatible with the conventional (finite-dimensional) notion of sparsity. Likewise, if we intend to optimize or possibly redesign the signal-acquisition system as in generalized sampling and compressed sensing, the very least is to have a model that characterizes the information content prior to sampling.
- *Predictive power*. Among other things, the theory should be able to explain why wavelet representations can outperform the older Fourier-related types of decompositions, including the KLT, which is optimal from the classical perspective of variance concentration.
- *Ease of use*. To have practical usefulness, the framework should allow for the derivation of the (joint) probability distributions of the signal in any transformed domain. This calls for a linear formulation with the caveat that it needs to accommodate non-Gaussian distributions. In that respect, the best thing beyond Gaussianity is

are justifiable based on the theory of Gaussian stationary processes (GSP). Specifically, one can invoke the following optimality results:

- The Fourier transform as well as several of its real-valued variants (e.g., DCT) are asymptotically equivalent to the Karhunen–Loève transform (KLT) for the whole class of GSP. This supports the use of sinusoidal transforms for data compression, data processing, and feature extraction. The underlying notion of optimality here is energy compaction, which implies decorrelation. Note that the decorrelation is equivalent to independence in the Gaussian case only.
- *Optimal filters*. Given a series of linear measurements of a signal corrupted by noise, one can readily specify its optimal reconstruction (LMMSE estimator) under the general Gaussian hypothesis. The corresponding algorithm (Wiener filter) is linear and entirely determined by the covariance structure of the signal and noise. There is also a direct connection with variational reconstruction techniques since the Wiener solution can also be formulated as a quadratic energy-minimization problem (Gaussian MAP estimator) (see Section 10.2.2).
- *Optimal sampling/interpolation strategies*. While this part of the story is less known, one can also invoke estimation-theoretic arguments to justify a Shannon-type, constant-rate sampling, which ensures a minimum loss of information for a large class of predominantly lowpass GSP [PM62, Uns93]. This is not totally surprising since the basis functions of the KLT are inherently bandlimited. One can also derive minimum-mean-square-error interpolators for GSP in general. The optimal signal-reconstruction algorithm takes the form of a hybrid Wiener filter whose input is discrete (signal samples) and whose output is a continuously defined signal that can be represented in terms of generalized B-spline basis functions [UB05b].

infinite divisibility, which is a general property of random variables that is preserved under arbitrary linear combinations.

- *Stochastic justification and refinement of current reconstruction algorithms.* A convincing argument for adopting a new theory is that it must be compatible with the state of the art, while it also ought to suggest new directions of research. In the present context, it is important to be able to establish the connection with deterministic recovery techniques such as ℓ_1-norm minimization.

The good news is that the foundations for such a theory exist and can be traced back to the pioneering work of Paul Lévy, who defined a broad family of "additive" stochastic processes, now called *Lévy processes*. *Brownian motion* (a.k.a. the Wiener process) is the only Gaussian member of this family, and, as we shall demonstrate, the only representative that does not exhibit any degree of sparsity. The theory that is developed in this book constitutes the full linear, multidimensional extension of those ideas where the essence of Paul Lévy's construction is embodied in the definition of *Lévy innovations* (or white Lévy noise), which can be interpreted as the derivative of a Lévy process in the sense of distributions (a.k.a. generalized functions). The Lévy innovations are then linearly transformed to generate a whole variety of processes whose spectral characteristics are controlled by a linear mixing operator, while their sparsity is governed by the innovations. The latter can also be viewed as the driving term of some corresponding linear *stochastic differential equation* (SDE).

Another way of describing the extent of this generalization is to consider the representation of a general continuous-domain Gaussian process by a stochastic Wiener integral,

$$s(t) = \int_{\mathbb{R}} h(t, \tau) \, dW(\tau), \tag{1.1}$$

where $h(t, \tau)$ is the kernel – that is, the infinite-dimensional analog of the matrix representation of a transformation in $\mathbb{R}^n$ – of a general, L_2-stable linear operator. W is a random measure which is such that

$$W(t) = \int_0^t dW(\tau)$$

is the Wiener process, where the latter equation constitutes a special case of (1.1) with $h(t, \tau) = \mathbb{1}_{\{t > \tau \geq 0\}}$. Here, $\mathbb{1}_{\Omega}$ denotes the indicator function of the set Ω. If $h(t, \tau) = h(t - \tau)$ is a convolution kernel, then (1.1) defines the whole class of Gaussian stationary processes.

The essence of the present formulation is to replace the Wiener measure by a more general non-Gaussian, multidimensional Lévy measure. The catch, however, is that we shall not work with measures but rather with generalized functions and generalized stochastic processes. These are easier to manipulate in the Fourier domain and better suited for specifying general linear transformations. In other words, we shall rewrite (1.1) as

$$s(t) = \int_{\mathbb{R}} h(t, \tau) w(\tau) \, d\tau, \tag{1.2}$$

where the entity w (the continuous-domain innovation) needs to be given a proper mathematical interpretation. The main advantage of working with innovations is that they provide a very direct link with the theory of linear systems, which allows for the use of standard engineering notions such as the impulse and frequency responses of a system.

1.3 From splines to stochastic processes, or when Schoenberg meets Lévy

We shall start our journey by making an interesting connection between splines, which are deterministic objects with some inherent sparsity, and Lévy processes with a special focus on the compound-Poisson process, which constitutes the archetype of a sparse stochastic process. The key observation is that both categories of signals – namely, deterministic and random – are ruled by the same differential equation. They can be generated via the proper integration of an "innovation" signal that carries all the necessary information. The fun is that the underlying differential system is only marginally stable, which requires the design of a special antiderivative operator. We then use the close relationship between splines and wavelets to gain insight into the ability of wavelets to provide sparse representations of such signals. Specifically, we shall see that most non-Gaussian Lévy processes admit a better M-term representation in the Haar wavelet basis than in the classical Karhunen–Loève transform (KLT), which is usually believed to be optimal for data compression. The explanation for this counter-intuitive result is that we are breaking some of the assumptions that are implicit in the proof of optimality of the KLT.

1.3.1 Splines and Legos revisited

Splines constitute a general framework for converting series of data points (or samples) into continuously defined signals or functions. By extension, they also provide a powerful mechanism for translating tasks that are specified in the continuous domain into efficient numerical algorithms (discretization).

The cardinal setting corresponds to the configuration where the sampling grid is on the integers. Given a sequence of sample values $f[k], k \in \mathbb{Z}$, the basic cardinal interpolation problem is to construct a continuously defined signal $f(t), t \in \mathbb{R}$ that satisfies the interpolation condition $f(t)|_{t=k} = f[k]$, for all $k \in \mathbb{Z}$. Since the general problem is obviously ill-posed, the solution is constrained to live in a suitable reconstruction subspace (e.g., a particular space of cardinal splines) whose degrees of freedom are in one-to-one correspondence with the data points. The most basic concretization of those ideas is the construction of the piecewise-constant interpolant

$$f_1(t) = \sum_{k \in \mathbb{Z}} f[k] \beta_+^0 (t - k), \tag{1.3}$$

which involves rectangular basis functions (informally described as "Legos") that are shifted replicates of the causal [2] B-spline of degree zero:

$$\beta_+^0(t) = \begin{cases} 1, & \text{for } 0 \leq t < 1 \\ 0, & \text{otherwise.} \end{cases} \tag{1.4}$$

---

[2] A function $f_+(t)$ is said to be causal if $f_+(t) = 0$, for all $t < 0$.

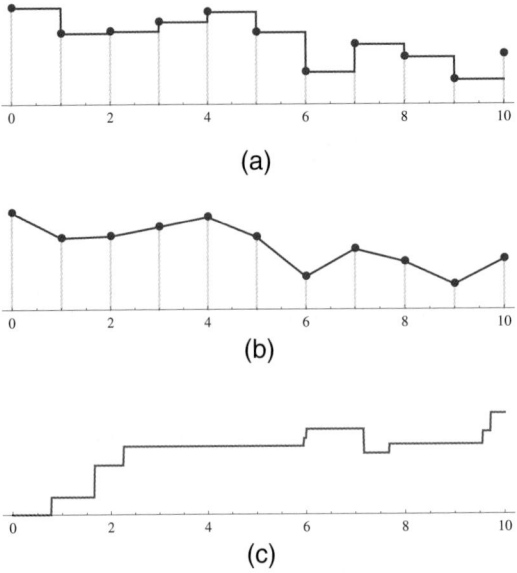

Figure 1.1 Examples of spline signals. (a) Cardinal spline interpolant of degree zero (piecewise-constant). (b) Cardinal spline interpolant of degree one (piecewise-linear). (c) Non-uniform D-spline or compound-Poisson process, depending on the interpretation (deterministic vs. stochastic).

Observe that the basis functions $\{\beta_+^0(t - k)\}_{k \in \mathbb{Z}}$ are non-overlapping and orthonormal, and that their linear span defines the space of cardinal polynomial splines of degree zero. Moreover, since $\beta_+^0(t)$ takes the value one at the origin and vanishes at all other integers, the expansion coefficients in (1.3) coincide with the original samples of the signal. Equation (1.3) is nothing but a mathematical representation of the sample-and-hold method of interpolation which yields the type of "Lego-like" signal shown in Figure 1.1a.

A defining property of piecewise-constant signals is that they exhibit "sparse" first-order derivatives that are zero almost everywhere, except at the points of transition where differentiation is only meaningful in the sense of distributions. In the case of the cardinal spline specified by (1.3), we have that

$$Df_1(t) = \sum_{k \in \mathbb{Z}} a_1[k]\delta(t - k), \tag{1.5}$$

where the weights of the integer-shifted Dirac impulses $\delta(\cdot - k)$ are given by the corresponding jump size of the function: $a_1[k] = f[k] - f[k - 1]$. The main point is that the application of the operator $D = \frac{d}{dt}$ uncovers the spline discontinuities (a.k.a. knots) which are located on the integer grid: its effect is that of a mathematical A-to-D conversion since the right-hand side of (1.5) corresponds to the continuous-domain representation of a discrete signal commonly used in the theory of linear systems. In the nomenclature of splines, we say that $f_1(t)$ is a *cardinal* D-spline, [3] which is a special case

---

[3] Other brands of splines are defined in the same fashion by replacing the derivative D by some other differential operator generically denoted by L.

$$\beta_+^0(t) = \mathbb{1}_+(t) - \mathbb{1}_+(t-1)$$

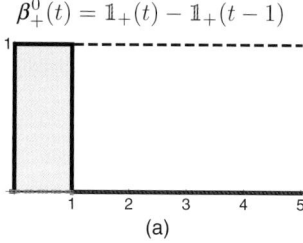 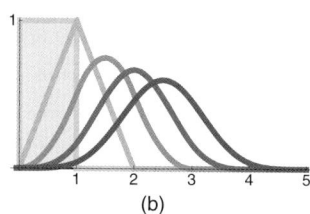

(a)   (b)

Figure 1.2 Causal polynomial B-splines. (a) Construction of the B-spline of degree zero starting from the causal Green's function of D. (b) B-splines of degree $n = 0, \ldots, 4$ (light to dark), which become more bell-shaped (and beautiful) as n increases.

of a general *non-uniform* D-spline where the knots can be located arbitrarily (see Figure 1.1c).

The next fundamental observation is that the expansion coefficients in (1.5) are obtained via a finite-difference scheme which is the discrete counterpart of differentiation. To get some further insight, we define the finite-difference operator

$$D_d f(t) = f(t) - f(t-1).$$

The latter turns out to be a smoothed version of the derivative

$$D_d f(t) = (\beta_+^0 * Df)(t),$$

where the smoothing kernel is precisely the B-spline generator for the expansion (1.3). An equivalent manifestation of this property can be found in the relation

$$\beta_+^0(t) = D_d D^{-1} \delta(t) = D_d \mathbb{1}_+(t), \tag{1.6}$$

where the unit step $\mathbb{1}_+(t) = \mathbb{1}_{[0,+\infty)}(t)$ (a.k.a. the Heaviside function) is the causal Green's function [4] of the derivative operator. This formula is illustrated in Figure 1.2a. Its Fourier-domain counterpart is

$$\widehat{\beta}_+^0(\omega) = \int_{\mathbb{R}} \beta_+^0(t) e^{-j\omega t}\, dt = \frac{1 - e^{-j\omega}}{j\omega}, \tag{1.7}$$

which is recognized as being the ratio of the frequency responses of the operators D_d and D, respectively.

Thus, the basic Lego component, β_+^0, is much more than a mere building block: it is also a kernel that characterizes the approximation that is made when replacing a continuous-domain derivative by its discrete counterpart. This idea (and its generalization for other operators) will prove to be one of the key ingredients in our formulation of sparse stochastic processes.

---

[4] We say that $\rho(t)$ is the causal Green's function of the shift-invariant operator L if ρ is causal and satisfies $L\rho = \delta$. This can also be written as $L^{-1}\delta = \rho$, meaning that ρ is the causal impulse response of the shift-invariant inverse operator L^{-1}.

1.3.2   Higher-degree polynomial splines

A slightly more sophisticated model is to select a piecewise-linear reconstruction which admits the similar B-spline expansion

$$f_2(t) = \sum_{k \in \mathbb{Z}} f[k+1] \beta_+^1(t - k), \tag{1.8}$$

where

$$\beta_+^1(t) = (\beta_+^0 * \beta_+^0)(t) = \begin{cases} t, & \text{for } 0 \le t < 1 \\ 2 - t, & \text{for } 1 \le t < 2 \\ 0, & \text{otherwise} \end{cases} \tag{1.9}$$

is the causal B-spline of degree one, a triangular function centered at $t = 1$. Note that the use of a causal generator is compensated by the unit shifting of the coefficients in (1.8), which is equivalent to recentering the basis functions on the sampling locations. The main advantage of f_2 in (1.8) over f_1 in (1.3) is that the underlying function is now continuous, as illustrated in Figure 1b.

In an analogous manner, one can construct higher-degree spline interpolants that are piecewise polynomials of degree n by considering B-spline atoms of degree n obtained from the $(n+1)$-fold convolution of $\beta_+^0(t)$ (see Figure 1.2b). The generic version of such a higher-order spline model is

$$f_{n+1}(t) = \sum_{k \in \mathbb{Z}} c[k] \beta_+^n(t - k), \tag{1.10}$$

with

$$\beta_+^n(t) = \big(\underbrace{\beta_+^0 * \beta_+^0 * \cdots * \beta_+^0}_{n+1} \big)(t).$$

The catch, though, is that, for $n > 1$, the expansion coefficients $c[k]$ in (1.10) are not identical to the sample values $f[k]$ anymore. Yet, they are in a one-to-one correspondence with them and can be determined efficiently by solving a linear system of equations that has a convenient band-diagonal Toeplitz structure [Uns99].

The higher-order counterparts of relations (1.7) and (1.6) are

$$\widehat{\beta_+^n}(\omega) = \left(\frac{1 - e^{-j\omega}}{j\omega} \right)^{n+1}$$

and

$$\begin{aligned} \beta_+^n(t) &= D_d^{n+1} D^{-(n+1)} \delta(t) \\ &= \frac{D_d^{n+1}(t)_+^n}{n!} \\ &= \sum_{k=0}^{n+1} (-1)^k \binom{n+1}{k} \frac{(t-k)_+^n}{n!} \end{aligned} \tag{1.11}$$

with $(t)_+ = \max(0, t)$. The latter explicit time-domain formula follows from the fact that the impulse response of the $(n + 1)$-fold integrator (or, equivalently, the causal Green's function of D^{n+1}) is the one-sided power function $D^{-(n+1)}\delta(t) = \frac{t_+^n}{n!}$. This elegant formula is due to Schoenberg, the father of splines [Sch46]. He also proved that the polynomial B-spline of degree n is the shortest cardinal D^{n+1}-spline and that its integer translates form a Riesz basis of such polynomial splines. In particular, he showed that the B-spline representation (1.10) is unique and stable, in the sense that

$$\|f_n\|_{L_2}^2 = \int_{\mathbb{R}} |f_n(t)|^2 \, dt \leq \|c\|_{\ell_2}^2 = \sum_{k \in \mathbb{Z}} |c[k]|^2.$$

Note that the inequality above becomes an equality for $n = 0$ since the squared L_2-norm of the corresponding piecewise-constant function is easily converted into a sum. This also follows from Parseval's identity because the B-spline basis $\{\beta_+^0(\cdot - k)\}_{k \in \mathbb{Z}}$ is orthonormal.

One last feature is that polynomial splines of degree n are inherently smooth, in the sense that they are n-times differentiable everywhere with bounded derivatives – that is, Hölder continuous of order n. In the cardinal setting, this follows from the property that

$$D^n \beta_+^n(t) = D^n D_d^{n+1} D^{-(n+1)} \delta(t)$$

$$= D_d^n D_d D^{-1} \delta(t) = D_d^n \beta_+^0(t),$$

which indicates that the nth-order derivative of a B-spline of degree n is piecewise-constant and bounded.

1.3.3 Random splines, innovations, and Lévy processes

To make the link with Lévy processes, we now express the random counterpart of (1.5) as

$$Ds(t) = \sum_n A_n \delta(t - t_n) = w(t), \tag{1.12}$$

where the locations t_n of the Dirac impulses are uniformly distributed over the real line (Poisson distribution with rate parameter λ) and the weights A_n are independent identically distributed (i.i.d.) with amplitude distribution $p_A(a)$. For simplicity, we are also assuming that p_A is symmetric with finite variance $\sigma_A^2 = \int_{\mathbb{R}} a^2 p_A(a) da$. We shall refer to w as the *innovation* of the signal s since it contains all the parameters that are necessary for its description. Clearly, s is a signal with a finite rate of innovation, a term that was coined by Vetterli *et al.* [VMB02].

The idea now is to reconstruct s from its innovation w by integrating (1.12). This requires the specification of some boundary condition to fix the integration constant. Since the constraint in the definition of Lévy processes is $s(0) = 0$ (with probability one), we first need to find a suitable antiderivative operator, which we shall denote by

D_0^{-1}. In the event when the input function is Lebesgue integrable, the relevant operator is readily specified as

$$D_0^{-1}\varphi(t) = \int_{-\infty}^{t}\varphi(\tau)\,d\tau - \int_{-\infty}^{0}\varphi(\tau)\,d\tau = \begin{cases} \int_{0}^{t}\varphi(\tau)\,d\tau, & \text{for } t \geq 0 \\[2mm] -\int_{t}^{0}\varphi(\tau)\,d\tau, & \text{for } t < 0. \end{cases}$$

It is the corrected version (subtraction of the proper signal-dependent constant) of the conventional shift-invariant integrator D^{-1} for which the integral runs from $-\infty$ to t. The Fourier counterpart of this definition is

$$D_0^{-1}\varphi(t) = \int_{\mathbb{R}} \frac{e^{j\omega t} - 1}{j\omega}\widehat{\varphi}(\omega)\frac{d\omega}{2\pi},$$

which can be extended, by duality, to a much larger class of generalized functions (see Chapter 5). This is feasible because the latter expression is a regularized version of an integral that would otherwise be singular, since the division by $j\omega$ is tempered by a proper correction in the numerator: $e^{j\omega t} - 1 = j\omega t + O(\omega^2)$. It is important to note that D_0^{-1} is scale-invariant (in the sense that it commutes with scaling), but not shift-invariant, unlike D^{-1}. Our reason for selecting D_0^{-1} over D^{-1} is actually more fundamental than just imposing the "right" boundary conditions. It is guided by stability considerations: D_0^{-1} is a valid right inverse of D in the sense that $DD_0^{-1} = \text{Id}$ over a large class of generalized functions, while the use of the shift-invariant inverse D^{-1} is much more constrained. Other than that, both operators share most of their global properties. In particular, since the finite-difference operator has the convenient property of annihilating the constants that are in the null space of D, we see that

$$\beta_+^0(t) = D_d D_0^{-1}\delta(t) = D_d D^{-1}\delta(t). \tag{1.13}$$

Having the proper inverse operator at our disposal, we can apply it to formally solve the stochastic differential equation (1.12). This yields the explicit representation of the sparse stochastic process:

$$s(t) = D_0^{-1}w(t) = \sum_n A_n D_0^{-1}\{\delta(\cdot - t_n)\}(t)$$

$$= \sum_n A_n\big(\mathbb{1}_+(t - t_n) - \mathbb{1}_+(-t_n)\big), \tag{1.14}$$

where the second term $\mathbb{1}_+(-t_n)$ in the last parenthesis ensures that $s(0) = 0$. Clearly, the signal defined by (1.14) is piecewise-constant (random spline of degree zero) and its construction is compatible with the classical definition of a compound-Poisson process, which is a special type of Lévy process. A representative example is shown in Figure 1.1c.

It can be shown that the innovation w specified by (1.12), made of random impulses, is a special type of continuous-domain white noise with the property that

$$\mathbb{E}\{w(t)w(t')\} = R_w(t - t') = \sigma_w^2\delta(t - t'), \tag{1.15}$$

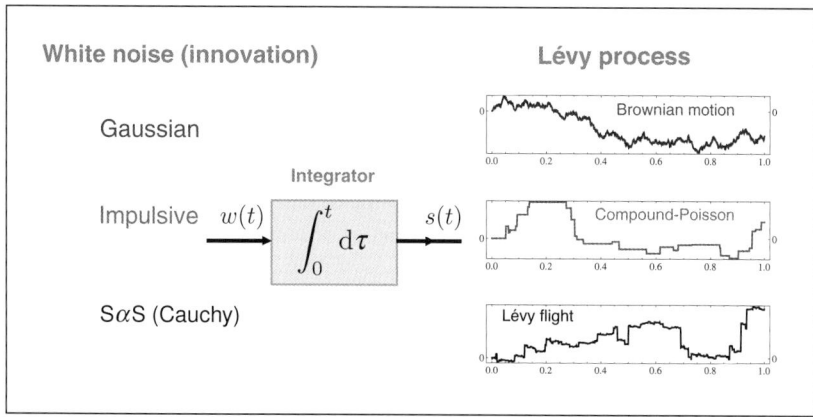

Figure 1.3 Synthesis of different brands of Lévy processes by integration of a corresponding continuous-domain white noise. The alpha-stable excitation in the bottom example is such that the increments of the Lévy process have a symmetric Cauchy distribution.

where $\sigma_w^2 = \lambda \sigma_A^2$ is the product of the Poisson rate parameter λ and the variance σ_A^2 of the amplitude distribution. More generally, we can determine the *correlation functional* of the innovation, which is given by

$$\mathbb{E}\{\langle \varphi_1, w \rangle \langle \varphi_2, w \rangle\} = \sigma_w^2 \langle \varphi_1, \varphi_2 \rangle \qquad (1.16)$$

for any real-valued functions $\varphi_1, \varphi_2 \in L_2(\mathbb{R})$ and $\langle \varphi_1, \varphi_2 \rangle = \int_{\mathbb{R}} \varphi_1(t) \varphi_2(t) \, dt$.

This suggests that we can apply the same operator-based synthesis to other types of continuous-domain white noise, as illustrated in Figure 1.3. In doing so, we are able to generate the whole family of Lévy processes. In the case where w is a white Gaussian noise, the resulting signal is a Brownian motion which has the property of being continuous almost everywhere. A more extreme case arises when w is an alpha-stable noise which yields a stable Lévy process whose sample path has a few really large jumps and is rougher than a Brownian motion.

In the classical literature on stochastic processes, Lévy processes are usually defined in terms of their increments, which are i.i.d. and infinitely divisible random variables (see Chapter 7). Here, we shall consider the so-called increment process $u(t) = s(t) - s(t - 1)$, which has a number of remarkable properties. The key observation is that u, in its continuous-domain version, is the convolution of a white noise (innovation) with the B-spline kernel β_+^0. Indeed, the relation (1.13) leads to

$$u(t) = D_d s(t) = D_d D_0^{-1} w(t) = (\beta_+^0 * w)(t). \qquad (1.17)$$

This implies, among other things, that u is stationary, while the original Lévy process s is not (since D_0^{-1} is not shift-invariant). It also suggests that the samples of the increment process u are independent if they are taken at a distance of 1 or more apart, the limit corresponding to the support of the rectangular convolution kernel β_+^0. When the autocorrelation function $R_w(\tau)$ of the driving noise is well defined and given by (1.15), we can easily determine the autocorrelation of u as

$$R_u(\tau) = \mathbb{E}\{u(t+\tau)u(t)\} = \left(\beta_+^0 * (\beta_+^0)^{\vee} * R_w\right)(\tau) = \sigma_w^2 \beta_+^1(\tau - 1), \tag{1.18}$$

where $(\beta_+^0)^{\vee}(t) = \beta_+^0(-t)$. It is proportional to the autocorrelation of a rectangle, which is a triangular function (centered B-spline of degree one).

Of special interest to us are the samples of u on the integer grid, which are characterized for $k \in \mathbb{Z}$ as

$$u[k] = s(k) - s(k-1) = \langle \mathbb{1}_{(k-1,k]}, w \rangle = \langle \beta_+^0(k - \cdot), w \rangle.$$

The relation on the right-hand side can be used to show that the $u[k]$ are i.i.d. because w is white, stationary, and the supports of the analysis functions $\beta_+^0(k-t)$ are non-overlapping. We shall refer to $\{u[k]\}_{k \in \mathbb{Z}}$ as the *discrete innovation* of s. Its determination involves the sampling of s at the integers and a discrete differentiation (finite differences), in direct analogy with the generation of the continuous-domain innovation $w(t) = Ds(t)$.

The discrete innovation sequence $u[\cdot]$ will play a fundamental role in signal processing because it constitutes a convenient tool for extracting the statistics and characterizing the samples of a stochastic process. It is probably the best practical way of presenting the information because

- we never have access to the full signal $s(t)$, which is a continuously defined entity, and
- we cannot implement the whitening operator (derivative) exactly, not to mention that the continuous-domain innovation $w(t)$ does not admit an interpretation as an ordinary function of t. For instance, Brownian motion is not differentiable anywhere in the classical sense.

This points to the fact that the continuous-domain innovation model is a theoretical construct. Its primary purpose is to facilitate the determination of the joint probability distributions of any series of linear measurements of a wide class of sparse stochastic processes, including the discrete version of the innovation which has the property of being maximally decoupled.

1.3.4   Wavelet analysis of Lévy processes and M-term approximations

Our purpose so far has been to link splines and Lévy processes to the derivative operator D. We shall now exploit this connection in the context of wavelet analysis. To that end, we consider the Haar basis $\{\psi_{i,k}\}_{i \in \mathbb{Z}, k \in \mathbb{Z}}$, which is generated by the Haar wavelet

$$\psi_{\text{Haar}}(t) = \begin{cases} 1, & \text{for } 0 \leq t < \frac{1}{2} \\ -1, & \text{for } \frac{1}{2} \leq t < 1 \\ 0, & \text{otherwise.} \end{cases} \tag{1.19}$$

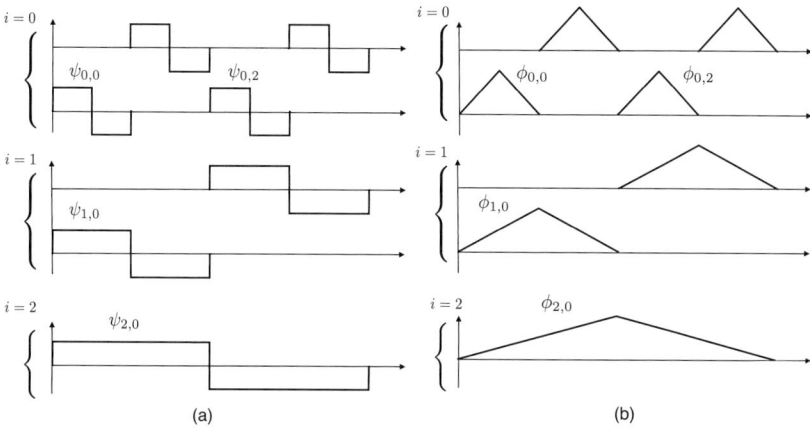

Figure 1.4 Dual pair of multiresolution bases where the first kind of functions (wavelets) are the derivatives of the second (hierarchical basis functions). (a) (Unnormalized) Haar wavelet basis. (b) Faber–Schauder basis (a.k.a. Franklin system).

The basis functions, which are orthonormal, are given by

$$\psi_{i,k}(t) = 2^{-i/2} \psi_{\mathrm{Haar}} \left(\frac{t - 2^i k}{2^i} \right), \tag{1.20}$$

where i and k are the scale (dilation of ψ_{Haar} by 2^i) and location (translation of $\psi_{i,0}$ by $2^i k$) indices, respectively. A closely related system is the Faber–Schauder basis $\{\phi_{i,k}(\cdot)\}_{i \in \mathbb{Z}, k \in \mathbb{Z}}$, which is made up of B-splines of degree one in a wavelet-like configuration (see Figure 1.4).

Specifically, the hierarchical triangle basis functions are given by

$$\phi_{i,k}(t) = \beta_+^1 \left(\frac{t - 2^i k}{2^{i-1}} \right). \tag{1.21}$$

While these functions are orthogonal within any given scale (because they are non-overlapping), they fail to be so across scales. Yet, they form a Schauder basis, which is a somewhat weaker property than being a Riesz basis of $L_2(\mathbb{R})$.

The fundamental observation for our purpose is that the Haar system can be obtained by differentiating the Faber–Schauder one, up to some amplitude factor. Specifically, we have the relations

$$\psi_{i,k} = 2^{i/2-1} \mathrm{D}\phi_{i,k} \tag{1.22}$$

$$\mathrm{D}_0^{-1} \psi_{i,k} = 2^{i/2-1} \phi_{i,k}. \tag{1.23}$$

Let us now apply (1.22) to the formal determination of the wavelet coefficients of the Lévy process $s = \mathrm{D}_0^{-1} w$. The crucial manipulation, which will be justified rigorously within the framework of generalized stochastic processes (see Chapter 3), is $\langle s, \mathrm{D}\phi_{i,k} \rangle = \langle \mathrm{D}^* s, \phi_{i,k} \rangle = -\langle w, \phi_{i,k} \rangle$, where we have used the adjoint relation $\mathrm{D}^* = -\mathrm{D}$

and the right-inverse property of D_0^{-1}. This allows us to express the wavelet coefficients as

$$Y_{i,k} = \langle s, \psi_{i,k} \rangle = -2^{i/2-1} \langle w, \phi_{i,k} \rangle,$$

which, up to some scaling factors, amounts to a Faber–Schauder analysis of the innovation $w = Ds$. Since the triangle functions $\phi_{i,k}$ are non-overlapping within a given scale and the innovation is independent at every point, we immediately deduce that the corresponding wavelet coefficients are also independent. However, the decoupling is not perfect across scales due to the parent-to-child overlap of the triangle functions. The residual correlation can be determined from the correlation functional (1.16) of the noise, according to

$$\mathbb{E}\{Y_{i,k} Y_{i',k'}\} = 2^{(i+i')/2-2} \mathbb{E}\left\{ \langle w, \phi_{i,k} \rangle \langle w, \phi_{i',k'} \rangle \right\} \propto \langle \phi_{i,k}, \phi_{i',k'} \rangle.$$

Since the triangle functions are non-negative, the residual correlation is zero if and only if $\phi_{i,k}$ and $\phi_{i',k'}$ are non-overlapping, in which case the wavelet coefficients are independent as well. We can also predict that the wavelet transform of a compound-Poisson process will be sparse (i.e., with many vanishing coefficients) because the random Dirac impulses of the innovation will intersect only a few Faber–Schauder functions, an effect that becomes more and more pronounced as the scale gets finer. The level of sparsity can therefore be expected to be directly dependent upon λ (the density of impulses per unit length).

To quantify this behavior, we applied Haar wavelets to the compression of sampled realizations of Lévy processes and compared the results with those of the "optimal" textbook solution for transform coding. In the case of a Lévy process with finite variance, the KLT can be determined analytically from the knowledge of the covariance function $\mathbb{E}\{s(t)s(t')\} = C(|t| + |t'| - |t - t'|)$, where C is an appropriate constant. The KLT is also known to converge to the discrete cosine transform (DCT) as the size of the signal increases. The present compression task is to reconstruct a series of 4096-point signals from their M largest transform coefficients, which is the minimum-error selection rule dictated by Parseval's relation. Figure 1.5 displays the graph of the relative quadratic M-term approximation errors for the three types of Lévy processes shown in Figure 1.3. We also considered the identity transform as baseline, and the DCT as well, whose results were found to be indistinguishable from those of the KLT. We observe that the KLT performs best in the Gaussian scenario, as expected. It is also slightly better than wavelets at large compression ratios for the compound-Poisson process (piecewise-constant signal with Gaussian-distributed jumps). In the latter case, however, the situation changes dramatically as M increases since one is able to reconstruct the signal perfectly from a fraction of the wavelet coefficients, by reason of the sparse behavior explained above. The advantage of wavelets over the KLT/DCT is striking for the Lévy flight (symmetric-alpha-stable, or SαS, distribution with $\alpha = 1$). While these findings are surprising at first, they do not contradict the classical theory which tells us that the KLT has the minimum basis-restriction error for the given class of processes. The twist here is that the selection of the M largest transform coefficients amounts to some adaptive reordering of the basis functions, which is not accounted for in the derivation

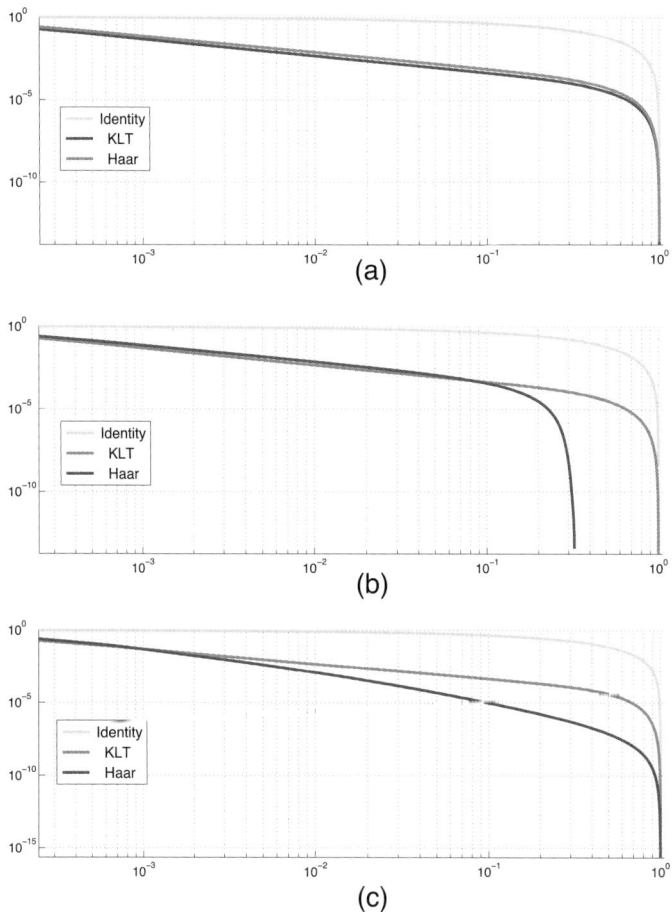

Figure 1.5 Haar wavelets vs. KLT: M-term approximation errors for different brands of Lévy processes. (a) Gaussian (Brownian motion). (b) Compound-Poisson with Gaussian jump distribution and $e^{-\lambda} = 0.9$. (c) Alpha-stable (symmetric Cauchy). The results are averages over 1000 realizations.

of the KLT. The other point is that the KLT solution is not defined for the third type of $S\alpha S$ process, whose theoretical covariances are unbounded – this does not prevent us from applying the Gaussian solution/DCT to a finite-length realization whose ℓ_2-norm is finite (almost surely). This simple experiment with various stochastic models corroborates the results obtained with image compression where the superiority of wavelets over the DCT (e.g., JPEG2000 vs. JPEG) is well established.

1.3.5   Lévy's wavelet-based synthesis of Brownian motion

We close this introductory chapter by making the connection with a multiresolution scheme that Paul Lévy developed in the 1930s to characterize the properties of Brownian motion. To do so, we adopt a point of view that is the dual of the one in Section

1.3.4: it essentially amounts to interchanging the analysis and synthesis functions. As first step, we expand the innovation w in the orthonormal Haar basis and obtain

$$w = \sum_{i\in\mathbb{Z}}\sum_{k\in\mathbb{Z}} Z_{i,k}\psi_{i,k} \text{ with } Z_{i,k} = \langle w, \psi_{i,k}\rangle.$$

This is acceptable [5] under the finite-variance hypothesis on w. Since the Haar basis is orthogonal, the coefficients $Z_{i,k}$ in the above expansion are fully decorrelated, but not necessarily independent, unless the white noise is Gaussian or the corresponding basis functions do not overlap. We then construct the Lévy process $s = D_0^{-1}w$ by integrating the wavelet expansion of the innovation, which yields

$$s(t) = \sum_{i\in\mathbb{Z}}\sum_{k\in\mathbb{Z}} Z_{i,k}D_0^{-1}\psi_{i,k}(t)$$

$$= \sum_{i\in\mathbb{Z}}\sum_{k\in\mathbb{Z}} 2^{i/2-1}Z_{i,k}\phi_{i,k}(t). \tag{1.24}$$

The representation (1.24) is of special interest when the noise is Gaussian, in which case the coefficients $Z_{i,k}$ are i.i.d. and follow a standardized Gaussian distribution. The formula then maps into Lévy's recursive mid-point method of synthesizing Brownian motion, which Yves Meyer singles out as the first use of wavelets to be found in the literature (see [JMR01, pp. 20–24]). The Faber–Schauder expansion (1.24) stands out as a localized, practical alternative to Wiener's original construction of Brownian motion, which involves a sum of harmonic cosines (KLT-type expansion).

1.4   Historical notes: Paul Lévy and his legacy

Paul Lévy is a highly original thinker who ended up being one of the most influential figures of modern probability theory [Tay75, Loè73]. Among his many contributions to the field are the introduction of the characteristic function as an analytical tool, the characterization of the limit of sums of independent random variables with unbounded variance (stable distributions), and the investigation of infinitely divisible laws which ultimately led to the specification of the complete family of additive – or Lévy – processes. In this latter respect, Michel Loève singles out his 1934 report on the integration/summation of independent random components [Lév34] as one of the most important probability papers ever published. There, Lévy is bold enough to make the transition from a discrete to a continuous indexing in a running sum. This results in the construction of a random function that is a generalization of Brownian motion and one of the earliest instances of a non-Gaussian stochastic process. If one leaves the

---

[5] The convergence in the sense of distributions is ensured since the wavelet coefficients of a rapidly decaying test function φ are rapidly decaying as well.

mathematical technicalities aside, this is very much in the spirit of (1.2), except for the presence of the more general weighting kernel h.

During his tenure as professor at the prestigious École Polytechnique in Paris from 1920 to 1959, Paul Lévy only supervised four Ph.D. students.[6] Every one of them turned out to be a brilliant scientist whose work is intertwined with the material presented in this book.

The first, Wolfgang Döblin (Ph.D. 1938 at age 23; co-advised by Maurice Fréchet), was a German jew who acquired French citizenship in 1936. Döblin was an extraordinarily gifted mathematician. His short career ended tragically on the front of World War II when he took his own life to avoid being captured by the German troops entering France. Yet, during the year he served as a French soldier, he was able to make fundamental contributions to the theory of Markov processes and stochastic integration that predate the work of Itô; these were discovered in 2000 in a sealed envelope deposited at the French Academy of Sciences – see [BY02] as well as [Pet05] for a romanced account of Döblin's life.

Lévy's second student, Michel Loève (Ph.D. 1941; co-advised by Maurice Fréchet), is a prominent name in modern probability theory. He was a famous professor at Berkeley who is best known for the development of the spectral representation of second-order stationary processes (the Karhunen–Loève transform).

The third student was Benoit B. Mandelbrot (Ph.D 1952), who is universally recognized as the inventor of fractals. In his early work, Mandelbrot introduced the use of non-Gaussian random walks – that is, Lévy processes – into financial statistics. In particular, he showed that the rate of change of prices in markets was much better characterized by alpha-stable distributions (which are heavy-tailed) than by Gaussians [Man63]. Interestingly, it is also statistics, albeit Gaussian statistics, that led him to the discovery of fractals when he characterized the class of self-similar processes known as fractional Brownian motion (fBm). While fBm corresponds to a fractional-order integration of white Gaussian noise, the construction is somewhat technical for it involves the resolution of a singular integral.[7] The relevance to the present study is that an important subclass of sparse processes is made up by the non-Gaussian cousins of fBms and their multidimensional extension (see Section 7.5).

Lévy's fourth and last student, Georges Matheron (Ph.D. 1958), was the founding father of the field of geostatistics. Being interested in the prediction of ore concentration, he developed a statistical method for the interpolation of random fields from non-uniform samples [Mat63]. His method, called *kriging*, uses the prior knowledge that the field is a Brownian motion and determines the interpolant that minimizes the mean-square estimation error. Interestingly, the solution, which is specified by a space-dependent regression equation, happens to be a spline function whose type is determined

---

[6] Source: Mathematics Genealogy Project at http://genealogy.math.ndsu.nodak.edu/.

[7] Retrospectively, we cannot help observing the striking parallel between the stochastic integral that defines fBm (see Equation (7.48)) and the Lévy–Khintchine representation of alpha-stable laws (an area in which Mandelbrot was obviously an expert), which involves the same kind of singularity (see the Lévy density $v(a) \propto 1/a^{1+\alpha}$ in the last line of Table 4.1).

by the correlation structure (variogram) of the underlying field. There is also an intimate link between kriging and data-approximation methods based on radial basis functions and/or reproducing-kernel Hilbert spaces [Mye92] – in particular, thin-plate splines that are associated with the Laplace operator [Wah90]. While the Gaussian hypothesis is implicit to Matheron's work, it is arguably the earliest link established between splines and stochastic processes.

2   Roadmap to the book

The writing of this book was motivated by our desire to formalize and extend the ideas presented in Section 1.3 to a class of differential operators much broader than the derivative D. Concretely, this translates into the investigation of the family of stochastic processes specified by the general innovation model that is summarized in Figure 2.1. The corresponding generator of random signals (upper part of the diagram) has two fundamental components: (1) a continuous-domain noise excitation w, which may be thought of as being composed of a continuum of independent identically distributed (i.i.d.) random atoms (innovations), and (2) a deterministic mixing procedure (formally described by L^{-1}) which couples the random contributions and imposes the correlation structure of the output. The concise description of the model is $Ls = w$, where L is the whitening operator. The term "innovation" refers to the fact that w represents the unpredictable part of the process. When the inverse operator L^{-1} is linear shift-invariant (LSI), the signal generator reduces to a simple convolutional system which is characterized by its impulse response (or, equivalently, its frequency response). Innovation modeling has a long tradition in statistical communication theory and signal processing; it is the basis for the interpretation of a Gaussian stationary process as a filtered version of a white Gaussian noise [Kai70, Pap91].

In the present context, the underlying objects are continuously defined. The innovation model then results from defining a stochastic process (or random field when the index variable r is a vector in $\mathbb{R}^d$) as the solution of a stochastic differential equation (SDE) driven by a particular brand of noise. The non-standard aspect here is that we are considering the innovation model in its greatest generality, allowing for non-Gaussian inputs and differential systems that are not necessarily stable. We shall argue that these extensions are essential for making this type of modeling compatible with the latest developments in signal processing pertaining to the use of wavelets and sparsity-promoting reconstruction algorithms. Specifically, we shall see that it is possible to generate a wide variety of sparse processes by replacing the traditional Gaussian input by some more general brand of (Lévy) noise, within the limits of mathematical admissibility. We shall also demonstrate that such processes admit a sparse representation in a wavelet basis under the assumption that L is scale-invariant. The difficulty there is that scale-invariant SDEs are inherently unstable (due to the presence of poles at the origin); yet, we shall see that they can still result in a proper specification of fractal-type processes, albeit not within the usual framework of stationary processes. The non-trivial aspect of these generalizations is that they

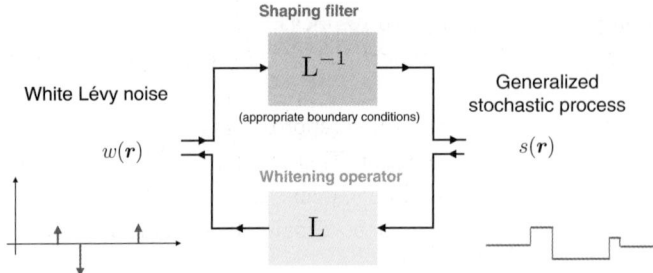

Figure 2.1 Innovation model of a generalized stochastic process. The process is generated by application of the (linear) inverse operator L^{-1} to a continuous-domain white-noise process w. The generation mechanism is general in the sense that it applies to the complete family of Lévy noises, including Gaussian noise as the most basic (non-sparse) excitation. The output process s is stationary if and only if L^{-1} is shift-invariant.

necessitate the resolution of instabilities – in the form of singular integrals. This is required not only at the system level, to allow for non-stationary processes, but also at the stochastic level because the most interesting sparsity patterns are associated with unbounded Lévy measures.

Before proceeding with the statistical characterization of sparse stochastic processes, we shall highlight the central role of the operator L and make a connection with spline theory and the construction of signal-adapted wavelet bases.

2.1   On the implications of the innovation model

To motivate our approach, we start with an informal discussion, leaving the technicalities aside. The stochastic process s in Figure 2.1 is constructed by applying the (integral) operator L^{-1} to some continuous-domain white noise w. In most cases of interest, L^{-1} has an infinitely supported impulse response which introduces long-range dependencies. If we are aiming at a concise statistical characterization of s, it is essential that we somehow invert this integration process, the ideal being to apply the operator L which would give back the innovation signal w that is fully decoupled. Unfortunately, this is not feasible in practice because we do not have access to the signal $s(r)$ over the entire domain $r \in \mathbb{R}^d$, but only to its sampled values on a lattice or, more generally, to a series of coefficients in some appropriate basis. Our analysis options are essentially two-fold, as described in Sections 2.1.1 and 2.1.2.

2.1.1   Linear combination of sampled values

Given the sampled values $s(k), k \in \mathbb{Z}^d$, the best we can aim at is to implement a discrete version of the operator L, which is denoted by L_d. In effect, L_d will act on the sampled version of the signal as a digital filter. The corresponding continuous-domain description of its impulse response is

$$L_d \delta(r) = \sum_{k \in \mathbb{Z}^d} d_L[k] \delta(r - k)$$

with some appropriate weights d_L. To fix ideas, L_d may correspond to the numerical version of the operator provided by the finite-difference method of approximating derivatives.

The interest is now to characterize the (approximate) decoupling effect of this discrete version of the whitening operator. This is quite feasible when the continuous-domain composition of the operators L_d and L^{-1} is shift-invariant with impulse response $\beta_L(r)$ which is assumed to be absolutely integrable (BIBO stability). In that case, one readily finds that

$$u(r) = L_d s(r) = (\beta_L * w)(r),\qquad(2.1)$$

where

$$\beta_L(r) = L_d L^{-1}\delta(r).\qquad(2.2)$$

This suggests that the decoupling effect will be the strongest when the convolution kernel β_L is the most localized (minimum support) and closest to an impulse. [1] We call β_L the *generalized B-spline* associated with the operator L. For a given operator L, the challenge will be to design the most localized kernel β_L, which is the way of approaching the discretization problem that best matches our statistical objectives. The good news is that this is a standard problem in spline theory, meaning that we can take advantage of the large body of techniques available in this area, even though they have hardly been applied to the stochastic setting so far.

2.1.2   Wavelet analysis

The second option is to analyze the signal s using wavelet-like functions $\{\psi_i(\cdot - r_k)\}$. For that purpose, we assume that we have at our disposal some real-valued "L-compatible" *generalized wavelets* which, at a given resolution level i, are such that

$$\psi_i(r) = L^*\phi_i(r).\qquad(2.3)$$

Here, L^* is the adjoint operator of L and ϕ_i is some smoothing kernel with good localization properties. The interpretation is that the wavelet transform provides some kind of multiresolution version of the operator L with the effective width of the kernels ϕ_i increasing in direct proportion to the scale; typically, $\phi_i(r) \propto \phi_0(r/2^i)$. Then, the wavelet analysis of the stochastic process s reduces to

$$\langle s, \psi_i(\cdot - r_0)\rangle = \langle s, L^*\phi_i(\cdot - r_0)\rangle$$
$$= \langle Ls, \phi_i(\cdot - r_0)\rangle$$
$$= \langle w, \phi_i(\cdot - r_0)\rangle = (\phi_i^\vee * w)(r_0),\qquad(2.4)$$

where $\phi_i^\vee(r) = \phi_i(-r)$ is the reversed version of ϕ_i. The remarkable aspect is that the effect is essentially the same as in (2.1), so that it makes good sense to develop a common framework to analyze white noise.

---

[1] One may be tempted to pretend that β_L is a Dirac impulse, which amounts to neglecting all discretization effects. Unfortunately, this is incorrect and most likely to result in false statistical conclusions. In fact, we shall see that the localization deteriorates as the order of the operator increases, inducing higher (Markov) orders of dependencies.

This is all nice in principle as long as one can construct "L-compatible" wavelet bases. For instance, if L is a pure nth-order derivative operator – or by extension, a scale-invariant differential operator – then the above reasoning is directly applicable to conventional wavelet bases. Indeed, these are known to behave like multiscale versions of derivatives due to their vanishing-moment property [Mey90, Dau92, Mal09]. In prior work, we have linked this behavior, as well as a number of other fundamental wavelet properties, to the polynomial B-spline convolutional factor that is necessarily present in every wavelet that generates a multiresolution basis of $L_2(\mathbb{R})$ [UB03]. What is not so widely known is that the spline connection extends to a much broader variety of operators – not necessarily scale-invariant – and that it also provides a general recipe for constructing wavelet-like basis functions that are matched to some given operator L. This has been demonstrated in one dimension (1-D) for the entire family of ordinary differential operators [KU06]. The only significant difference with the conventional theory of wavelets is that the smoothing kernels ϕ_i are not necessarily rescaled versions of each other.

Note that the "L-compatible" property is relatively robust. For instance, if $L = L'L_0$, then an "L-compatible" wavelet is also L'-compatible with $\phi_i' = L_0\phi_i$. The design challenge in the context of stochastic modeling is thus to come up with a suitable wavelet basis such that ϕ_i in (2.3) is most localized – possibly, of compact support.

2.2   Organization of the book

The reasoning of Section 2.1 is appealing because of its conceptual simplicity and generality. Yet, the precise formulation of the theory requires some special care because the underlying stochastic objects are infinite-dimensional and possibly highly singular. For instance, we are faced with a major difficulty at the outset because the continuous-domain input of our model (the innovation w) does not admit a conventional interpretation as a function of the domain variable r. This entity can only be probed indirectly by forming scalar products with test functions in accordance with Laurent Schwartz' theory of distributions, so that the use of advanced mathematics is unavoidable.

For the benefit of readers who would be unfamiliar with concepts used in this book, we provide the relevant mathematical background in Chapter 3, which also serves the purpose of introducing the notation. The first part is devoted to the definition of the relevant function spaces, with special emphasis on generalized functions (a.k.a. tempered distributions), which play a central role in our formulation. The second part reviews the classical, finite-dimensional tools of probability theory and shows how some concepts (e.g., characteristic function, Bochner's theorem) are extendable to the infinite-dimensional setting within the framework of Gelfand's theory of generalized stochastic processes [GV64].

Chapter 4 is devoted to the mathematical specification of the innovation model. Since the theory gravitates around the notion of *Lévy exponents*, we start with a systematic investigation of such functions, denoted by $f(\omega)$, which are fundamental to the (classical) study of *infinitely divisible* probability laws. In particular, we discuss their canonical representation given by the Lévy–Khintchine formula. In Section 4.4, we make use of the powerful Minlos–Bochner theorem to transfer those representations to the infinite-dimensional setting. The fundamental result of our theory is that every admissible continuous-domain innovation for the model in Figure 2.1 belongs to the so-called family of white Lévy noises. This implies that an innovation process is completely characterized by its Lévy exponent $f(\omega)$. We conclude the chapter with the presentation of mathematical criteria for the existence of solutions of Lévy-driven SDEs (stochastic differential equations) and provide the functional tools for the complete statistical characterization of these processes. Interestingly, the classical Gaussian processes are covered by the formulation (by setting $f(\omega) = -\frac{1}{2}\omega^2$), but they turn out to be the only non-sparse members of the family.

Besides the random excitation w, the second fundamental component of the innovation model in Figure 2.1 is the inverse L^{-1} of the whitening operator L. It must fulfill some continuity/boundedness condition in order to yield a proper solution of the underlying SDE. The construction of such inverses (shaping filters) is the topic of Chapter 5, which presents a systematic catalog of the solutions that are currently available, including recent constructs for scale-invariant/unstable SDEs.

In Chapter 6, we review the tools that are available from the theory of splines in relation to the specification of the analysis kernels in Equations (2.1) and (2.3). The techniques are quite generic and applicable to any operator L that admits a proper inverse L^{-1}. Moreover, by writing a generalized B-spline as $\beta_L = L_d L^{-1}\delta$, one can appreciate that the construction of a B-spline for some operator L implicitly provides the solution of two innovation-related problems at once: (1) the formal inversion of the operator L (for solving the SDE) and (2) the proper discretization of L through a finite-difference scheme. The leading thread in our formulation is that these two tasks should not be dissociated – this is achieved formally via the identification of β_L, which actually results in simplified and streamlined mathematics. Remarkably, these generalized B-splines are also the key for constructing wavelet-like basis functions that are "L-compatible."

In Chapter 7, we apply our framework to the functional specification of a variety of generalized stochastic processes, including the classical family of Gaussian stationary processes and their sparse counterparts. We also characterize non-stationary processes that are solutions of unstable SDEs. In particular, we describe higher-order extensions of Lévy processes, as well as a whole variety of fractal-type processes.

In Chapter 8, we rely on our functional characterization to obtain a maximally decoupled representation of sparse stochastic processes by application of the discretized version of the whitening operator or by suitable wavelet expansion. Based on the characteristic form of these processes, we are able to deduce the transform-domain statistics and to precisely assess residual dependencies. These ideas are illustrated with examples of sparse processes for which operator-like wavelets outperform the classical KLT (or DCT) and result in an independent component analysis.

An implicit property of the innovation model is that the statistical distribution of the inner product between a sparse stochastic process and a particular basis function (e.g., wavelet) is uniquely characterized by a "modified" Lévy exponent. Our main point in Chapter 9 is to use this result to show that the sparsity of the input noise is transferred to the transformed domain. Apart from a shaping effect that can be quantified, the resulting probability density function remains within the same family of infinitely divisible laws.

In the final part of the book, we illustrate the use of these stochastic models (and the corresponding analytical tools) with the formulation of algorithms for the recovery of signals and images from incomplete, noisy measurements. In Chapter 10, we develop a general framework for the discretization of linear inverse problems in a B-spline basis, which is analogous to the *finite-element method* for solving PDEs. The central element is the "projection" of the continuous-domain stochastic model onto the (finite-dimensional) reconstruction space in order to specify the prior statistical distribution of the signal. This naturally yields the *maximum a posteriori* solution to the signal-reconstruction problem. The framework is illustrated with the derivation of practical algorithms for magnetic resonance imaging, deconvolution microscopy, and tomographic reconstruction. Remarkably, the resulting MAP estimators are compatible with the non-quadratic regularization schemes (e.g., ℓ_1-minimization, LASSO, and/or non-convex ℓ_p relaxation) that are currently in favor in imaging. To get a handle on the quality of the reconstruction, we then rely on the innovation model to investigate the extent to which one is able to "optimally" denoise sparse signals. In particular, we demonstrate the feasibility of MMSE reconstruction when the signal belongs to the class of Lévy processes, which provides us with a gold standard against which to compare other algorithms.

In Chapter 11, we present alternative wavelet-based reconstruction methods that are typically faster than the fixed-scale techniques of Chapter 10. These methods capitalize on the orthonormality of the wavelet basis, which provides direct control of the norm of the signal. We show that the underlying optimization task is amenable to iterative thresholding algorithms (ISTA or FISTA) which are simple to deploy and well suited for large-scale problems. We also investigate the effect of cycle spinning, which is a fundamental ingredient for making wavelets competitive in terms of image quality. Our closing topic is the use of statistical modeling for the improvement of standard wavelet-based denoising – in particular, the optimization of the wavelet-domain thresholding functions and the search of a consensus solution across multiple wavelet expansions in order to minimize the global estimation error.

3 Mathematical context and background

In this chapter we summarize some of the mathematical preliminaries for the remaining chapters. These concern the function spaces used in the book, duality, generalized functions, probability theory, and generalized random processes. Each of these topics is discussed in a separate section.

For the most part, the theory of function spaces and generalized functions can be seen as an infinite-dimensional generalization of linear algebra (function spaces generalize $\mathbb{R}^n$, and continuous linear operators generalize matrices). Similarly, the theory of generalized random processes involves the generalization of the idea of a finite random vector in $\mathbb{R}^n$ to an element of an infinite-dimensional space of generalized functions.

To give a taste of what is to come, we briefly compare finite- and infinite-dimensional theories in Tables 3.1 and 3.2. The idea, in a nutshell, is to replace vectors by (generalized) functions. Formally, this extension amounts to replacing some finite sums (in the finite-dimensional formulation) by integrals. Yet, in order for this to be mathematically sound, one needs to properly define the underlying objects as elements of some infinite-dimensional vector space, to specify the underlying notion(s) of convergence (which is not an issue in $\mathbb{R}^n$) while ensuring that some basic continuity conditions are met.

Fundamental to our formulation is the material on infinite-dimensional probability theory from Section 3.4.4 to the end of the chapter. The mastery of those notions requires a good understanding of function spaces and generalized functions, which are covered in the first part of the chapter. The impatient reader who is not directly concerned with the full mathematical details may skip what follows the tables at first reading and consult the relevant sections later as needed.

3.1   Some classes of function spaces

By the term "function" we shall mean elements of various *function spaces*. At a minimum, a function space is a set $\mathscr{X}$ along with some criteria for determining, first, whether or not a given "function" $\varphi = \varphi(r)$ belongs to $\mathscr{X}$ (in mathematical notation, $\varphi \in \mathscr{X}$) and, second, given $\varphi, \varphi_0 \in \mathscr{X}$, whether or not φ and φ_0 describe the same object in $\mathscr{X}$ (in mathematical notation, $\varphi = \varphi_0$). Most often, in addition to these, the space $\mathscr{X}$ has additional *structure* (see below).

In this book we shall deal largely with two types of function spaces: complete normed spaces such as Lebesgue L_p spaces, and nuclear spaces such as the Schwartz space $\mathscr{S}$

Table 3.1 Comparison of notions of linear algebra with those of functional analysis and the theory of distributions (generalized functions). See Sections 3.1–3.3 for an explanation.

Finite-dimensional theory (linear algebra)	Infinite-dimensional theory (functional analysis)
Euclidean space $\mathbb{R}^N$, complexification $\mathbb{C}^N$	Function spaces such as the Lebesgue space $L_p(\mathbb{R}^d)$ and the space of tempered distributions $\mathscr{S}'(\mathbb{R}^d)$, among others
Vector $x = (x_1,\ldots,x_N)$ in $\mathbb{R}^N$ or $\mathbb{C}^N$	Function $f(r)$ in $\mathscr{S}'(\mathbb{R}^d)$, $L_p(\mathbb{R}^d)$, etc.
bilinear scalar product $\langle x, y\rangle = \sum_{n=1}^{N} x_n y_n$	$\langle \varphi, g\rangle = \int \varphi(r) g(r)\, dr$ $\varphi \in \mathscr{S}(\mathbb{R}^d)$ (test function), $g \in \mathscr{S}'(\mathbb{R}^d)$ (generalized function), or $\varphi \in L_p(\mathbb{R}^d)$, $g \in L_q(\mathbb{R}^d)$ with $\frac{1}{p} + \frac{1}{q} = 1$, for instance
Equality: $x = y \iff x_n = y_n$ $\iff \quad \langle u, x\rangle = \langle u, y\rangle, \ \forall u \in \mathbb{R}^N$ $\iff \quad \|x - y\|^2 = 0$	Various notions of equality (depends on the space), such as weak equality of distributions: $f = g \in \mathscr{S}'(\mathbb{R}^d) \iff \langle \varphi, f\rangle = \langle \varphi, g\rangle$ for all $\varphi \in \mathscr{S}(\mathbb{R}^d)$, almost-everywhere equality: $f = g \in L_p(\mathbb{R}^d) \iff$ $\int_{\mathbb{R}^d} |f(r) - g(r)|^p\, dr = 0$
Linear operators $\mathbb{R}^N \to \mathbb{R}^M$ $y = \mathbf{A}x \Rightarrow y_m = \sum_{n=1}^{N} a_{mn} x_n$	Continuous linear operators $\mathscr{S}(\mathbb{R}^d) \to \mathscr{S}'(\mathbb{R}^d)$ $g = A\varphi \Rightarrow g(r) = \int_{\mathbb{R}^d} a(r,s)\varphi(s)\, ds$ for some $a \in \mathscr{S}'(\mathbb{R}^d \times \mathbb{R}^d)$ (Schwartz' kernel theorem)
Transpose $\langle x, \mathbf{A}y\rangle = \langle \mathbf{A}^{\mathsf{T}}x, y\rangle$	Adjoint $\langle \varphi, Ag\rangle = \langle A^*\varphi, g\rangle$

Table 3.2 Comparison of notions of finite-dimensional statistical calculus with the theory of generalized stochastic processes. See Section 3.4 for an explanation.

Finite-dimensional	Infinite-dimensional
Random variable X in $\mathbb{R}^N$	Generalized stochastic process s in $\mathscr{S}'$
Probability measure $\mathscr{P}_X$ on $\mathbb{R}^N$ $\mathscr{P}_X(E) = \text{Prob}(X \in E) = \int_E p_X(x)\, dx$ (p_X is a generalized [i.e., hybrid] pdf) for suitable subsets $E \subset \mathbb{R}^N$	Probability measure $\mathscr{P}_s$ on $\mathscr{S}'$ $\mathscr{P}_s(E) = \text{Prob}(s \in E) = \int_E \mathscr{P}_s(dg)$ for suitable subsets $E \subset \mathscr{S}'$
Characteristic function $\widehat{\mathscr{P}}_X(\omega) = \mathbb{E}\{e^{j\langle \omega, X\rangle}\} = \int_{\mathbb{R}^N} e^{j\langle \omega, x\rangle} p_X(x)\, dx$, $\omega \in \mathbb{R}^N$	Characteristic functional $\widehat{\mathscr{P}}_s(\varphi) = \mathbb{E}\{e^{j\langle \varphi, s\rangle}\} = \int_{\mathscr{S}'} e^{j\langle \varphi, g\rangle} \mathscr{P}_s(dg)$, $\varphi \in \mathscr{S}$

and the space $\mathcal{D}$ of compactly supported test functions, as well as their duals $\mathcal{S}'$ and $\mathcal{D}'$, which are spaces of generalized functions. These two categories of spaces (complete-normed and nuclear) cannot overlap, except in finite dimensions. Since the function spaces that are of interest to us are infinite-dimensional (they do not have a finite vector-space basis), the two categories are mutually exclusive.

The structure of each of the aforementioned spaces has two aspects. First, as a vector space over the real numbers or its complexification, the space has an *algebraic* structure. Second, with regard to the notions of convergence and taking of limits, the space has a *topological* structure. The algebraic structure lends meaning to the idea of a *linear* operator on the space, while the topological structure gives rise to the concept of a *continuous* operator or map, as we shall see shortly.

All the spaces considered here have a similar algebraic structure. They are either vector spaces over $\mathbb{R}$, meaning that, for any φ, φ_0 in the space and any $a \in \mathbb{R}$, the operations of addition $\varphi \mapsto \varphi + \varphi_0$ and multiplication by scalars $\varphi \mapsto a\varphi$ are defined and map the space (denoted henceforth by $\mathcal{X}$) into itself. Or, we may take the complexification of a real vector space $\mathcal{X}$, composed of elements of the form $\varphi = \varphi_r + \mathrm{j}\varphi_i$, with $\varphi_r, \varphi_i \in \mathcal{X}$ and j denoting the imaginary unit. The complexification is then a vector space over $\mathbb{C}$. In the remainder of the book, we shall denote a real vector space and its complexification by the same symbol. The distinction, when important, will be clear from the context.

For the spaces with which we are concerned in this book, the topological structure is completely specified by providing a criterion for the convergence of sequences.[1] By this we mean that, for any given sequence (φ_i) in $\mathcal{X}$ and any $\varphi \in \mathcal{X}$, we are equipped with the knowledge of whether or not φ is the limit of (φ_i). A *topological space* is a set $\mathcal{X}$ with topological structure. For normed spaces, the said criterion is given in terms of a *norm,* while in nuclear spaces it is given in terms of a *family of seminorms,* as we shall discuss below. But before that, let us first define linear and continuous operators.

An operator is a mapping from one vector space to another; that is, a rule that associates an output function $A\{\varphi\} \in \mathcal{Y}$ (also written as $A\varphi$) with each input $\varphi \in \mathcal{X}$.

DEFINITION 3.1 (Linear operator) An operator $A : \mathcal{X} \to \mathcal{Y}$ where $\mathcal{X}$ and $\mathcal{Y}$ are vector spaces is called *linear* if, for any $\varphi, \varphi_0 \in \mathcal{X}$ and $a, b \in \mathbb{R}$ (or $\mathbb{C}$),

$$A\{a\varphi + b\varphi_0\} = aA\{\varphi\} + bA\{\varphi_0\}. \tag{3.1}$$

DEFINITION 3.2 (Continuous operator) Let $\mathcal{X}, \mathcal{Y}$ be topological spaces. An operator $A : \mathcal{X} \to \mathcal{Y}$ is called *sequentially continuous* (with respect to the topologies of $\mathcal{X}$ and $\mathcal{Y}$) if, for any convergent sequence (φ_i) in $\mathcal{X}$ with limit $\varphi \in \mathcal{X}$, the sequence $(A\{\varphi_i\})$ converges to $A\{\varphi\}$ in $\mathcal{Y}$, that is,

$$\lim_i A\{\varphi_i\} = A\{\lim_i \varphi_i\}.$$

The above definition of continuity coincides with the stricter topological definition for spaces we are interested in.

---

[1] This is in contrast with those topological spaces where one needs to consider generalizations of the notion of a sequence involving partially ordered sets (the so-called nets or filters). Spaces in which a knowledge of sequences suffices are called *sequential.*

We shall assume that the topological structure of our vector spaces is such that the operations of addition and multiplication by scalars in $\mathbb{R}$ (or $\mathbb{C}$) are continuous. With this compatibility condition, our object is called a *topological vector space*.

Having defined the two types of structure (algebraic and topological) and their relation with operators in abstract terms, let us now show concretely how the topological structure is defined for some important classes of spaces.

3.1.1   About the notation: mathematics vs. engineering

So far, we have considered a function in abstract terms as an element of a vector space: $\varphi \in \mathcal{X}$. The more conventional view is that of a map $\varphi : \mathbb{R}^d \to \mathbb{R}$ (or $\mathbb{C}$) that associates a value $\varphi(r)$ with each point $r = (r_1, \ldots, r_d) \in \mathbb{R}^d$. Following the standard convention in engineering, we shall therefore also use the notation $\varphi(r)$ [instead of $\varphi(\cdot)$ or φ] to represent the function using r as our generic d-dimensional index variable, the norm of which is denoted by $|r|^2 = \sum_{i=1}^d |r_i|^2$. This is to be contrasted with the point values (or samples) of φ which will be denoted using subscripted index variables; i.e., $\varphi(r_k)$ stands for the value of φ at $r = r_k$. Likewise, $\varphi(r - r_0) = \varphi(\cdot - r_0)$ refers to the function φ shifted by r_0.

A word of caution is in order here. While the engineering notation has the advantage of being explicit, it can also be felt as being abusive because the point values of φ are not necessarily well defined, especially when the function presents discontinuities, not to mention the case of generalized functions that do not have a pointwise interpretation. $\varphi(r)$ should therefore be treated as an alternative notation for φ that reminds us of the domain of the function, and not be interpreted literally.

3.1.2   Normed spaces

A *norm* on $\mathcal{X}$ is a map $\mathcal{X} \to \mathbb{R}$, usually denoted as $\varphi \mapsto \|\varphi\|$ (with indices used if needed to distinguish between different norms), which fulfills the following properties for all $a \in \mathbb{R}$ (or $\mathbb{C}$) and $\varphi, \varphi_0 \in \mathcal{X}$.

- $\|\varphi\| \geq 0$ (*non-negativity*)
- $\|a\varphi\| = |a| \, \|\varphi\|$ (*positive homogeneity*)
- $\|\varphi + \varphi_0\| \leq \|\varphi\| + \|\varphi_0\|$ (*triangular inequality*)
- $\|\varphi\| = 0$ implies $\varphi = 0$ (*separation of points*).

By relaxing the last requirement we obtain a *seminorm*.

A *normed space* is a vector space $\mathcal{X}$ equipped with a norm.

A sequence (φ_i) in a normed space $\mathcal{X}$ is said to *converge to* φ (in the topology of $\mathcal{X}$), in symbols

$$\lim_i \varphi_i = \varphi,$$

if and only if

$$\lim_i \|\varphi - \varphi_i\| = 0.$$

Let (φ_i) be a sequence in $\mathscr{X}$ such that for any $\epsilon > 0$ there exists an $N \in \mathbb{N}$ with

$$\|\varphi_i - \varphi_j\| < \epsilon \quad \text{for all } i, j \geq N.$$

Such a sequence is called a *Cauchy* sequence. A normed space $\mathscr{X}$ is *complete* if it does not have any holes, in the sense that, for every Cauchy sequence in $\mathscr{X}$, there exists a $\varphi \in \mathscr{X}$ such that $\lim_i \varphi_i = \varphi$ (in other words if every Cauchy sequence has a limit in $\mathscr{X}$). A normed space that is not complete can be completed by introducing new points corresponding to the limits of equivalent Cauchy sequences. For example, the real line is the completion of the set of rational numbers with respect to the absolute-value norm.

Examples

Important examples of complete normed spaces are the Lebesgue spaces. The Lebesgue spaces $L_p(\mathbb{R}^d)$, $1 \leq p \leq \infty$, are composed of functions whose $L_p(\mathbb{R}^d)$ norm, denoted as $\|\cdot\|_p$, is finite, where

$$\|\varphi(r)\|_{L_p} = \begin{cases} \left(\int_{\mathbb{R}^d} |\varphi(r)|^p \, dr\right)^{\frac{1}{p}} & \text{for } 1 \leq p < \infty \\ \text{ess sup}_{r \in \mathbb{R}^d} |\varphi(r)| & \text{for } p = \infty \end{cases}$$

and where two functions that are equal almost everywhere are considered to be equivalent.

We may also define weighted L_p spaces by replacing the shift-invariant Lebesgue measure (dr) by a weighted measure $w(r)dr$ in the above definitions. In that case, $w(r)$ is assumed to be a measurable function that is (strictly) positive almost everywhere. In particular, for $w(r) = 1 + |r|^\alpha$ with $\alpha \geq 0$ (or, equivalently, $w(r) = (1 + |r|)^\alpha$), we denote the associated norms as $\|\cdot\|_{p,\alpha}$, and the corresponding normed spaces as $L_{p,\alpha}(\mathbb{R}^d)$. The latter spaces are useful when characterizing the decay of functions at infinity. For example, $L_{\infty,\alpha}(\mathbb{R}^d)$ is the space of functions that are bounded by a constant multiple of $\frac{1}{1+|r|^\alpha}$ almost everywhere (a.e.).

Some remarkable inclusion properties of $L_{p,\alpha}(\mathbb{R}^d)$, $1 \leq p \leq \infty$, $\alpha > 0$, are

- $\alpha > \alpha_0$ implies $L_{p,\alpha}(\mathbb{R}^d) \subset L_{p,\alpha_0}(\mathbb{R}^d)$.
- $L_{\infty,\frac{d}{p}+\epsilon}(\mathbb{R}^d) \subset L_p(\mathbb{R}^d)$ for any $\epsilon > 0$.

Finally, we define the *space of rapidly decaying functions*, $\mathscr{R}(\mathbb{R}^d)$, as the intersection of all $L_{\infty,\alpha}(\mathbb{R}^d)$ spaces, $\alpha > 0$, or, equivalently, as the intersection of all $L_{\infty,\alpha}(\mathbb{R}^d)$ with $\alpha \in \mathbb{N}$. In other words, $\mathscr{R}(\mathbb{R}^d)$ contains all bounded functions that essentially decay faster than $1/|r|^\alpha$ at infinity for all $\alpha \in \mathbb{R}^+$. A sequence (f_i) converges in (the topology of) $\mathscr{R}(\mathbb{R}^d)$ if and only if it converges in all $L_{\infty,\alpha}(\mathbb{R}^d)$ spaces.

The causal exponential $\rho_\alpha(r) = \mathbb{1}_{[0,\infty)}(r)e^{\alpha r}$ with $\text{Re}(\alpha) < 0$ that is central to linear systems theory is a prototypical example of a function included in $\mathscr{R}(\mathbb{R})$.

3.1.3   Nuclear spaces

Defining nuclear spaces is neither easy nor particularly intuitive. Fortunately, for our purpose in this book, knowing the definition is not necessary. We shall simply assert that certain function spaces are nuclear, in order to use certain results that are true for nuclear spaces (specifically, the Minlos–Bochner theorem; see below). For the sake of completeness, a general definition of nuclear spaces is given at the end of this section, but this definition may safely be skipped without compromising the presentation.

Specifically, it will be sufficient for us to know that the spaces $\mathscr{D}(\mathbb{R}^d)$ and $\mathscr{S}(\mathbb{R}^d)$, which we shall shortly define, are nuclear, as are the Cartesian products and powers of nuclear spaces, and their closed subspaces.

To define these spaces, we need to identify their members, as well as the criterion of convergence for sequences in the space.

The space $\mathscr{D}(\mathbb{R}^d)$

The space of *compactly supported smooth test functions* is denoted by $\mathscr{D}(\mathbb{R}^d)$. It consists of infinitely differentiable functions with compact support in $\mathbb{R}^d$. To define its topology, we provide the following criterion for convergence in $\mathscr{D}(\mathbb{R}^d)$:

A sequence (φ_i) of functions in $\mathscr{D}(\mathbb{R}^d)$ is said to converge (in the topology of $\mathscr{D}(\mathbb{R}^d)$) if

(1) there exists a compact (here, meaning closed and bounded) subset K of $\mathbb{R}^d$ such that all φ_i are supported inside K; and
(2) the sequence (φ_i) converges in all of the seminorms

$$\|\varphi\|_{K,n} = \sup_{r \in K} |\partial^n \varphi(r)| \quad \text{for all } n \in \mathbb{N}^d.$$

Here, $n = (n_1, \ldots, n_d) \in \mathbb{N}^d$ is what is called a *multi-index,* and ∂^n is shorthand for the partial derivative $\partial_{r_1}^{n_1} \cdots \partial_{r_d}^{n_d}$. We take advantage of the present opportunity also to introduce two other notations: $|n|$ for $\sum_{i=1}^{d} |n_i|$ and r^n for the product $\prod_{i=1}^{d} r_i^{n_i}$.

The space $\mathscr{D}(\mathbb{R}^d)$ is nuclear (for a proof, see for instance [GV64]).

The Schwartz space $\mathscr{S}(\mathbb{R}^d)$

The *Schwartz space* or the space of so-called *smooth and rapidly decaying test functions,* denoted as $\mathscr{S}(\mathbb{R}^d)$, consists of infinitely differentiable functions φ on $\mathbb{R}^d$ for which all of the seminorms defined below are finite:

$$\|\varphi\|_{m,n} = \sup_{r \in \mathbb{R}^d} |r^m \partial^n \varphi(r)| \quad \text{for all } m, n \in \mathbb{N}^d.$$

In other words, $\mathscr{S}(\mathbb{R}^d)$ is populated by functions that, together with all of their derivatives, decay faster than the inverse of any polynomial at infinity.

The topology of $\mathscr{S}(\mathbb{R}^d)$ is defined by positing that a sequence (φ_i) converges in $\mathscr{S}(\mathbb{R}^d)$ if and only if it converges in all of the above seminorms.

The Schwartz space has the remarkable property that its complexification is invariant under the Fourier transform. In other words, the Fourier transform, defined by the integral

$$\varphi(r) \mapsto \widehat{\varphi}(\omega) = \mathscr{F}\{\varphi\}(\omega) = \int_{\mathbb{R}^d} e^{-j\langle r, \omega \rangle} \varphi(r) \, dr$$

and inverted by

$$\widehat{\varphi}(\boldsymbol{\omega}) \mapsto \varphi(\boldsymbol{r}) = \mathscr{F}^{-1}\{\widehat{\varphi}\}(\boldsymbol{r}) = \int_{\mathbb{R}^d} e^{j\langle r, \omega\rangle} \widehat{\varphi}(\boldsymbol{\omega}) \frac{d\boldsymbol{\omega}}{(2\pi)^d},$$

is a continuous map from $\mathscr{S}(\mathbb{R}^d)$ into itself. Our convention here is to use $\boldsymbol{\omega} \in \mathbb{R}^d$ as the generic Fourier-domain index variable.

In addition, both $\mathscr{S}(\mathbb{R}^d)$ and $\mathscr{D}(\mathbb{R}^d)$ are closed and continuous under differentiation of any order and multiplication by polynomials. Lastly, they are included in $\mathscr{R}(\mathbb{R}^d)$ and hence in all the Lebesgue spaces, $L_p(\mathbb{R}^d)$, which do not require any smoothness.

General definition of nuclear spaces

Defining a nuclear space requires us to define *nuclear operators*. These are operators that can be approximated by operators of *finite rank* in a certain sense (an operator between vector spaces is of finite rank if its range is finite-dimensional).

We first recall the notation $\ell_p(\mathbb{N})$, $1 \le p < \infty$, for the space of p-summable sequences; that is, sequences $\boldsymbol{c} = (c_i)_{i \in \mathbb{N}}$ for which

$$\sum_{i \in \mathbb{N}} |c_i|^p$$

is finite. We also denote by $\ell_\infty(\mathbb{N})$ the space of all bounded sequences.

In a complete normed space $\mathscr{Y}$, let $(\psi_i)_{i \in \mathbb{N}}$ be a sequence with bounded norm (i.e., $\|\psi_i\| \le M$ for some $M \in \mathbb{R}$ and all $i \in \mathbb{N}$). We then denote by M_ψ the linear operator $\ell_1(\mathbb{N}) \to \mathscr{Y}$ which maps a sequence $\boldsymbol{c} = (c_i)_{i \in \mathbb{N}}$ in ℓ_1 to the weighted sum

$$\sum_{i \in \mathbb{N}} c_i \psi_i$$

in $\mathscr{Y}$ (the sum converges in norm by the triangular inequality).

An operator $A : \mathscr{X} \to \mathscr{Y}$, where $\mathscr{X}, \mathscr{Y}$ are complete normed spaces, is called *nuclear* if there exists a continuous linear operator

$$\widetilde{A} : \mathscr{X} \to \ell_\infty : \varphi \mapsto (a_i(\varphi)),$$

an operator

$$\Lambda : \ell_\infty \to \ell_1 : (c_i) \mapsto (\lambda_i c_i)$$

where $\sum_i |\lambda_i| < \infty$, and a bounded sequence $\boldsymbol{\psi} = (\psi_i)$ in $\mathscr{Y}$, such that we can write

$$A = M_\psi \, \Lambda \, \widetilde{A}.$$

This is equivalent to the following decomposition of A into a sum of rank 1 operators:

$$A : \varphi \mapsto \sum_{i \in \mathbb{N}} \lambda_i a_i(\varphi) \psi_i.$$

The continuous linear operator $\mathscr{X} \to \mathscr{Y} : \varphi \mapsto \lambda_i a_i(\varphi)\psi_i$ is of rank 1 because it maps $\mathscr{X}$ into the 1-D subspace of $\mathscr{Y}$ spanned by ψ_i; compare (ψ_i) with a basis and $(a_i(\varphi))$ with the coefficients of $A\varphi$ in this basis.

More generally, given an arbitrary topological vector space $\mathscr{X}$ and a complete normed space $\mathscr{Y}$, the operator A : $\mathscr{X} \to \mathscr{Y}$ is said to be *nuclear* if there exists a complete normed space $\mathscr{X}_1$, a nuclear operator $A_1 : \mathscr{X}_1 \to \mathscr{Y}$, and a continuous operator B : $\mathscr{X} \to \mathscr{X}_1$, such that

$$A = A_1 B.$$

Finally, $\mathscr{X}$ is a *nuclear space* if any continuous linear map $\mathscr{X} \to \mathscr{Y}$, where $\mathscr{Y}$ is a complete normed space, is nuclear.

3.2   Dual spaces and adjoint operators

Given a space $\mathscr{X}$, a *functional* on $\mathscr{X}$ is a map f that takes $\mathscr{X}$ to the scalar field $\mathbb{R}$ (or $\mathbb{C}$). In other words, f takes a function $\varphi \in \mathscr{X}$ as argument and returns the number $f(\varphi)$.

When $\mathscr{X}$ is a vector space, we may consider *linear* functionals on it, where linearity has the same meaning as in Definition 3.1. When f is a linear functional, it is customary to use the notation $\langle \varphi, f \rangle$ in place of $f(\varphi)$.

The set of all linear functionals on $\mathscr{X}$, denoted as $\mathscr{X}^*$, can be given the structure of a vector space in the obvious way by the identity

$$\langle \varphi, af + bf_0 \rangle = a \langle \varphi, f \rangle + b \langle \varphi, f_0 \rangle,$$

where $\varphi \in \mathscr{X}, f, f_0 \in \mathscr{X}^*$, and $a, b \in \mathbb{R}$ (or $\mathbb{C}$) are arbitrary. The resulting vector space $\mathscr{X}^*$ is called the *algebraic dual* of $\mathscr{X}$.

The map from $\mathscr{X} \times \mathscr{X}^*$ to $\mathbb{R}$ (or $\mathbb{C}$) that takes the pair (φ, f) to their so-called *scalar product* $\langle \varphi, f \rangle$ is then *bilinear* in the sense that it is linear in each of the arguments φ and f. Note that the reasoning about linear functionals works both ways, so that we can also switch the order of the pairing. This translates into the formal commutativity rule $\langle f, \varphi \rangle = \langle \varphi, f \rangle$ with a dual interpretation of the two sides of the equality.

Given vector spaces $\mathscr{X}, \mathscr{Y}$ with algebraic duals $\mathscr{X}^*, \mathscr{Y}^*$ and a linear operator A : $\mathscr{X} \to \mathscr{Y}$, the *adjoint* or *transpose* of A, denoted as A^*, is the linear operator $\mathscr{Y}^* \to \mathscr{X}^*$ defined by

$$A^* f = f \circ A$$

for any linear functional $f : \mathscr{Y} \to \mathbb{R}$ (or $\mathbb{C}$) in $\mathscr{Y}^*$, where $\circ$ denotes composition. The motivation behind the above definition is to have the identity

$$\langle A\varphi, f \rangle = \langle \varphi, A^* f \rangle \tag{3.2}$$

hold for all $\varphi \in \mathscr{X}$ and $f \in \mathscr{Y}^*$.

If $\mathscr{X}$ is a topological vector space, it is of interest to consider the subspace of $\mathscr{X}^*$ composed of those linear functionals on $\mathscr{X}$ that are continuous with respect to the topology of $\mathscr{X}$. This subspace is denoted as $\mathscr{X}'$ and called the *topological* or *continuous dual* of $\mathscr{X}$. Note that, unlike $\mathscr{X}^*$, the continuous dual generally depends on the topology of $\mathscr{X}$. In other words, the same vector space $\mathscr{X}$ with different topologies will generally have different continuous duals.

As a general rule, in this book we shall adopt some standard topologies and only work with the corresponding continuous dual space, which we shall simply call the dual. Also, henceforth, we shall assume the scalar product $\langle \cdot, \cdot \rangle$ to be restricted to $\mathcal{X} \times \mathcal{X}'$. There, the space $\mathcal{X}$ may vary but is necessarily paired with its continuous dual.

Following the restrictions of the previous paragraph, we sometimes say that the adjoint of $A : \mathcal{X} \to \mathcal{Y}$ *exists,* meaning that the algebraic adjoint $A^* : \mathcal{Y}^* \to \mathcal{X}^*$, when restricted to $\mathcal{Y}'$, maps into $\mathcal{X}'$, so that we can write

$$\langle A\varphi, f \rangle = \langle \varphi, A^* f \rangle,$$

where the scalar products on the two sides are now restricted to $\mathcal{Y} \times \mathcal{Y}'$ and $\mathcal{X} \times \mathcal{X}'$, respectively.

One can define different topologies on $\mathcal{X}'$ by providing various criteria for convergence. The only one we shall need to deal with is the *weak-∗ topology,* which indicates (for a sequential space $\mathcal{X}$) that (f_i) converges to f in $\mathcal{X}'$ if and only if

$$\lim_i \langle \varphi, f_i \rangle = \langle \varphi, f \rangle \quad \text{for all } \varphi \in \mathcal{X}.$$

This is precisely the topology of *pointwise convergence* for all "points" $\varphi \in \mathcal{X}$.

We shall now mention some examples.

3.2.1 The dual of L_p spaces

The dual of the Lebesgue space $L_p(\mathbb{R}^d)$, $1 \leq p < \infty$, can be identified with the space $L_{p'}(\mathbb{R}^d)$ with $1 < p' \leq \infty$ satisfying $1/p + 1/p' = 1$, by defining

$$\langle \varphi, f \rangle = \int_{\mathbb{R}^d} \varphi(r) f(r) \, dr \tag{3.3}$$

for $\varphi \in L_p(\mathbb{R}^d)$ and $f \in L_{p'}(\mathbb{R}^d)$. In particular, $L_2(\mathbb{R}^d)$, which is the only Hilbert space of the family, is its own dual.

To see that linear functionals described by the above formula with $f \in L_{p'}$ are continuous on L_p, we can rely on Hölder's inequality, which states that

$$|\langle \varphi, f \rangle| \leq \int_{\mathbb{R}^d} |\varphi(r) f(r)| \, dr \leq \|\varphi\|_{L_p} \|f\|_{L_{p'}} \tag{3.4}$$

for $1 \leq p, p' \leq \infty$ and $1/p + 1/p' = 1$. The special case of (3.4) for $p = 2$ yields the Cauchy–Schwarz inequality.

3.2.2 The duals of $\mathscr{D}$ and $\mathscr{S}$

In this section, we give the mathematical definition of the duals of the nuclear spaces $\mathscr{D}$ and $\mathscr{S}$. A physical interpretation of these definitions is postponed until Section 3.3.

The dual of $\mathscr{D}(\mathbb{R}^d)$, denoted as $\mathscr{D}'(\mathbb{R}^d)$, is the so-called *space of distributions* over $\mathbb{R}^d$ (although we shall use the term "distribution" more generally to mean any generalized

function in the sense of Section 3.3). Ordinary locally integrable functions[2] (in particular, all L_p functions and all continuous functions) can be identified with elements of
$\mathscr{D}'(\mathbb{R}^d)$ by using (3.3). By this we mean that any locally integrable function f defines a
continuous linear functional on $\mathscr{D}(\mathbb{R}^d)$ where, for $\varphi \in \mathscr{D}(\mathbb{R}^d)$, $\langle \varphi, f \rangle$ is given by (3.3).
However, not all elements of $\mathscr{D}'(\mathbb{R}^d)$ can be characterized in this way. For instance,
the *Dirac functional* δ (a.k.a. Dirac impulse), which maps $\varphi \in \mathscr{D}(\mathbb{R}^d)$ to the value
$\langle \varphi, \delta \rangle = \varphi(\mathbf{0})$, belongs in $\mathscr{D}'(\mathbb{R}^d)$ but cannot be written as an integral à la (3.3). Even
in this and similar cases, we may sometimes write $\int_{\mathbb{R}^d} \varphi(r) f(r) \, dr$, keeping in mind
that the integral is no longer a true (i.e., Lebesgue) integral, but simply an alternative
notation for $\langle \varphi, f \rangle$.

In similar fashion, the dual of $\mathscr{S}(\mathbb{R}^d)$, denoted as $\mathscr{S}'(\mathbb{R}^d)$, is defined and called the
space of tempered (or *Schwartz*) *distributions*. Since $\mathscr{D} \subset \mathscr{S}$ and any sequence that
converges in the topology of $\mathscr{D}$ also converges in $\mathscr{S}$, it follows that $\mathscr{S}'(\mathbb{R}^d)$ is (can
be identified with) a smaller space (i.e., a subspace) of $\mathscr{D}'(\mathbb{R}^d)$. In particular, not every
locally integrable function belongs in $\mathscr{S}'$. For example, locally integrable functions
of exponential growth have no place in $\mathscr{S}'$ as their scalar product with Schwartz test
functions via (3.3) is not in general finite (much less continuous). Once again, $\mathscr{S}'(\mathbb{R}^d)$
contains objects that are not functions on $\mathbb{R}^d$ in the true sense of the word. For example,
δ also belongs in $\mathscr{S}'(\mathbb{R}^d)$.

3.2.3   Distinction between Hermitian and duality products

We use the notation $\langle f, g \rangle_{L_2} = \int_{\mathbb{R}^d} f(r) \overline{g(r)} \, dr$ to represent the usual (Hermitian-
symmetric) L_2 inner product. The latter is defined for $f, g \in L_2(\mathbb{R}^d)$ (the Hilbert space
of complex finite-energy functions); it is equivalent to Schwartz' duality product only
when the second argument is real-valued (due to the presence of complex conjugation).
The corresponding Hermitian adjoint of an operator A is denoted by A^H; it is defined as
$\langle A^H f, g \rangle_{L_2} = \langle f, Ag \rangle_{L_2} = \langle f, \overline{Ag} \rangle$ which implies that $A^H = A^*$. The distinction between
these types of adjoints is only relevant when considering signal expansions or analyses
in terms of complex basis functions.

The classical Fourier transform is defined as

$$\hat{f}(\omega) = \mathscr{F}\{f\}(\omega) = \int_{\mathbb{R}^d} f(r) e^{-j\langle r, \omega \rangle} \, dr$$

for any $f \in L_1(\mathbb{R}^d)$. This definition admits a unique extension, $\mathscr{F} : L_2(\mathbb{R}^d) \to L_2(\mathbb{R}^d)$,
which is an isometry map (Plancherel's theorem). The fact that the Fourier transform
preserves the L_2 norm of a function (up to a normalization factor) is a direct consequence
of Parseval's relation,

$$\langle f, g \rangle_{L_2} = \frac{1}{(2\pi)^d} \langle \hat{f}, \hat{g} \rangle_{L_2},$$

whose duality product equivalent is $\langle f, \hat{g} \rangle = \langle \hat{f}, g \rangle$.

---

[2] A function on $\mathbb{R}^d$ is called locally integrable if its integral over any closed bounded set is finite.

3.3   Generalized functions

3.3.1   Intuition and definition

We begin with some considerations regarding the modeling of physical phenomena. Let us suppose that the object of our study is some physical quantity f that varies in relation to some parameter $r \in \mathbb{R}^d$ representing space and/or time. We assume that our way of obtaining information about f is by making measurements that are localized in space–time using sensors $(\varphi, \psi, \ldots)$. We shall denote the measurement of f procured by φ as $\langle \varphi, f \rangle$.[3] Let us suppose that our sensors form a vector space, in the sense that for any two sensors φ, ψ and any two scalars $a, b \in \mathbb{R}$ (or $\mathbb{C}$), there is a real or virtual sensor $a\varphi + b\psi$ such that

$$\langle a\varphi + b\psi, f \rangle = a\langle \varphi, f \rangle + b\langle \psi, f \rangle.$$

In addition, we may reasonably suppose that the phenomenon under observation has some form of continuity, meaning that

$$\lim_i \langle \varphi_i, f \rangle = \langle \varphi, f \rangle,$$

where (φ_i) is a sequence of sensors that tend to φ in a certain sense. We denote the set of all sensors by $\mathcal{X}$. In the light of the above notions of linear combinations and limits defined in $\mathcal{X}$, mathematically the space of sensors then has the structure of a topological vector space.

Given the above properties and the definitions of the previous sections, we conclude that f represents an element of the continuous dual $\mathcal{X}'$ of $\mathcal{X}$. Given that our sensors, as previously noted, are assumed to be localized in $\mathbb{R}^d$, we may model them as compactly supported or rapidly decaying functions on $\mathbb{R}^d$, denoted by the same symbols $(\varphi, \psi, \ldots)$ and, in the case where f also corresponds to a function on $\mathbb{R}^d$, relate the observation $\langle \varphi, f \rangle$ to the functional form of φ and f by the identity

$$\langle \varphi, f \rangle = \int_{\mathbb{R}^d} \varphi(r) f(r) \, dr.$$

We exclude from consideration those functions f for which the above integral is undefined or infinite for some $\varphi \in \mathcal{X}$.

However, we are not limited to taking f to be a true function of $r \in \mathbb{R}^d$. By requiring our sensor or *test* functions to be smooth, we can permit f to become singular; that is, to depend on the value of φ and/or of its derivatives at isolated points/curves inside $\mathbb{R}^d$. An example of a singular *generalized function* f, which we have already noted, is the Dirac distribution (or impulse) δ that measures the value of φ at the single point $r = \mathbf{0}$ (i.e., $\langle \varphi, \delta \rangle = \varphi(\mathbf{0})$).

Mathematically, we define *generalized functions* as members of the continuous dual $\mathcal{X}'$ of a nuclear space $\mathcal{X}$ of functions, such as $\mathscr{D}(\mathbb{R}^d)$ or $\mathscr{S}(\mathbb{R}^d)$.

---

[3] The connection with previous sections should already be apparent from this choice of notation.

Implicit to the manipulation of generalized functions is the notion of weak equality (or equality in the sense of distributions). Concretely, this means that one should interpret the statement $f = g$ with $f, g \in \mathcal{X}'$ as

$$\langle \varphi, f \rangle = \langle \varphi, g \rangle \quad \text{for all} \quad \varphi \in \mathcal{X}.$$

3.3.2   Operations on generalized functions

Following (3.2), any continuous linear operator $\mathcal{D} \to \mathcal{D}$ or $\mathcal{S} \to \mathcal{S}$ can be transposed to define a continuous linear operator $\mathcal{D}' \to \mathcal{D}'$ or $\mathcal{S}' \to \mathcal{S}'$. In particular, since $\mathcal{D}(\mathbb{R}^d)$ and $\mathcal{S}(\mathbb{R}^d)$ are closed under differentiation, we can define derivatives of distributions.

First, note that, formally,

$$\langle \partial^n \varphi, f \rangle = \langle \varphi, \partial^{n*} f \rangle.$$

Now, using integration by parts in (3.3), for φ, f in $\mathcal{D}(\mathbb{R}^d)$ or $\mathcal{S}(\mathbb{R}^d)$ we see that $\partial^{n*} = (-1)^{|n|} \partial^n$. In other words, we can write

$$\langle \varphi, \partial^n f \rangle = (-1)^{|n|} \langle \partial^n \varphi, f \rangle. \tag{3.5}$$

The idea is then to use (3.5) as the defining formula in order to extend the action of the derivative operator ∂^n for any $f \in \mathcal{D}'(\mathbb{R}^d)$ or $\mathcal{S}'(\mathbb{R}^d)$.

Formulas for scaling, shifting (translation), rotation, and other geometric transformations of distributions are obtained in a similar manner. For instance, the translation by r_0 of a generalized function f is defined via the identity

$$\langle \varphi, f(\cdot - r_0) \rangle = \langle \varphi(\cdot + r_0), f \rangle.$$

More generally, we give the following definition.

DEFINITION 3.3 (Dual extension principle)   Given operators $U, U^* : \mathcal{S}(\mathbb{R}^d) \to \mathcal{S}(\mathbb{R}^d)$ that form an adjoint pair on $\mathcal{S}(\mathbb{R}^d) \times \mathcal{S}(\mathbb{R}^d)$, we extend their action to $\mathcal{S}'(\mathbb{R}^d) \to \mathcal{S}'(\mathbb{R}^d)$ by defining Uf and U^*f so as to have

$$\langle \varphi, Uf \rangle = \langle U^* \varphi, f \rangle$$
$$\langle \varphi, U^* f \rangle = \langle U \varphi, f \rangle$$

for all f. A similar definition gives the extension of adjoint pairs $\mathcal{D}(\mathbb{R}^d) \to \mathcal{D}(\mathbb{R}^d)$ to operators $\mathcal{D}'(\mathbb{R}^d) \to \mathcal{D}'(\mathbb{R}^d)$.

Examples of operators $\mathcal{S}(\mathbb{R}^d) \to \mathcal{S}(\mathbb{R}^d)$ that can be extended in the above fashion include derivatives, rotations, scaling, translation, time-reversal, and multiplication by smooth functions of slow growth in the space–time domain. The other fundamental operation is the Fourier transform, which is treated in Section 3.3.3.

3.3.3   The Fourier transform of generalized functions

We have already noted that the Fourier transform $\mathscr{F}$ is a reversible operator that maps the (complexified) space $\mathscr{S}(\mathbb{R}^d)$ into itself. The additional relevant property is that $\mathscr{F}$ is self-adjoint: $\langle \varphi, \mathscr{F}\psi \rangle = \langle \mathscr{F}\varphi, \psi \rangle$, for all $\varphi, \psi \in \mathscr{S}(\mathbb{R}^d)$. This helps us specify the generalized Fourier transform of distributions in accordance with the general extension principle in Definition 3.3.

DEFINITION 3.4   The *generalized Fourier transform* of a distribution $f \in \mathscr{S}'(\mathbb{R}^d)$ is the distribution $\widehat{f} = \mathscr{F}\{f\} \in \mathscr{S}'(\mathbb{R}^d)$ that satisfies

$$\langle \varphi, \widehat{f} \rangle = \langle \widehat{\varphi}, f \rangle$$

for all $\varphi \in \mathscr{S}$, where $\widehat{\varphi} = \mathscr{F}\{\varphi\}$ is the classical Fourier transform of φ given by the integral

$$\widehat{\varphi}(\boldsymbol{\omega}) = \int_{\mathbb{R}^d} e^{-j\langle r, \omega \rangle} \varphi(r) \, dr.$$

For example, since we have

$$\int_{\mathbb{R}^d} \varphi(r) \, dr = \langle \varphi, 1 \rangle = \widehat{\varphi}(0) = \langle \widehat{\varphi}, \delta \rangle,$$

we conclude that the (generalized) Fourier transform of δ is the constant function 1.

The fundamental property of the generalized Fourier transform is that it maps $\mathscr{S}'(\mathbb{R}^d)$ into itself and that it is invertible with $\mathscr{F}^{-1} = \frac{1}{(2\pi)^d}\overline{\mathscr{F}}$ where $\overline{\mathscr{F}}\{f\} = \mathscr{F}\{f^\vee\}$. This quasi-self-reversibility – also expressed by the first row of Table 3.3 – implies that any operation on generalized functions that is admissible in the space–time domain has its counterpart in the Fourier domain, and vice versa. For instance, multiplication with a smooth function in the Fourier domain corresponds to a convolution in the signal

Table 3.3 Basic properties of the (generalized) Fourier transform.

Temporal or spatial domain	Fourier domain
$\widehat{f}(r) = \mathscr{F}\{f\}(r)$	$(2\pi)^d f(-\boldsymbol{\omega})$
$f^\vee(r) = f(-r)$	$\widehat{f}(-\boldsymbol{\omega}) = \widehat{f}^\vee(\boldsymbol{\omega})$
$\overline{f(r)}$	$\overline{\widehat{f}(-\boldsymbol{\omega})}$
$f(\mathbf{A}^{\mathrm{T}} r)$	$\frac{1}{|\det \mathbf{A}|}\widehat{f}(\mathbf{A}^{-1}\boldsymbol{\omega})$
$f(r - r_0)$	$e^{-j\langle r_0, \omega \rangle}\widehat{f}(\boldsymbol{\omega})$
$e^{j\langle r, \omega_0 \rangle}f(r)$	$\widehat{f}(\boldsymbol{\omega} - \boldsymbol{\omega}_0)$
$\partial^n f(r)$	$(j\boldsymbol{\omega})^n \widehat{f}(\boldsymbol{\omega})$
$r^n f(r)$	$j^{|n|}\partial^n \widehat{f}(\boldsymbol{\omega})$
$(g * f)(r)$	$\widehat{g}(\boldsymbol{\omega})\widehat{f}(\boldsymbol{\omega})$
$g(r)f(r)$	$(2\pi)^{-d}(\widehat{g} * \widehat{f})(\boldsymbol{\omega})$

domain. Consequently, the familiar functional identities concerning the classical Fourier transform, such as the formulas for change of variables and differentiation, among others, also hold true for this generalization. These are summarized in Table 3.3.

In addition, the reader can find in Appendix A a table of Fourier transforms of some important singular generalized functions in one and several variables.

3.3.4   The kernel theorem

The kernel theorem provides a characterization of continuous operators $\mathscr{X} \to \mathscr{X}'$ (with respect to the nuclear topology on $\mathscr{X}$ and the weak-$*$ topology on $\mathscr{X}'$). We shall state a version of the theorem for $\mathscr{X} = \mathscr{S}(\mathbb{R}^d)$, which is the one we shall use. The version for $\mathscr{D}(\mathbb{R}^d)$ is obtained by replacing the symbol $\mathscr{S}$ with $\mathscr{D}$ everywhere in the statement of the theorem.

THEOREM 3.1 (Schwartz' kernel theorem: first form)   *Every continuous linear operator* $A : \mathscr{S}(\mathbb{R}^d) \to \mathscr{S}'(\mathbb{R}^d)$ *can be written in the form*

$$\varphi(r) \mapsto A\{\varphi\}(r) = \int_{\mathbb{R}^d} \varphi(s)a(r,s)\, ds, \qquad (3.6)$$

where $a(\cdot, \cdot)$ *is a generalized function in* $\mathscr{S}'(\mathbb{R}^d \times \mathbb{R}^d)$.

We can interpret the above formula as some sort of continuous-domain matrix-vector product, where r, s play the role of the row and column indices, respectively (see the list of analogies in Table 3.1). This characterization of continuous linear operators as infinite-dimensional matrix-vector products partly justifies our earlier statement that nuclear spaces "resemble" finite-dimensional spaces in fundamental ways.

The kernel $a \in \mathscr{S}'(\mathbb{R}^d \times \mathbb{R}^d)$ associated with the linear operator A can be identified as

$$a(\cdot, s_0) = A\{\delta(\cdot - s_0)\}, \qquad (3.7)$$

which corresponds to making the formal substitution $\varphi = \delta(\cdot - s_0)$ in (3.6). One can therefore view $a(\cdot, s_0)$ as the generalized *impulse response* of A.

An equivalent statement of Theorem 3.1 is as follows.

THEOREM 3.2 (Schwartz kernel theorem: second form)   *Every continuous bilinear form* $l : \mathscr{S}(\mathbb{R}^d) \times \mathscr{S}(\mathbb{R}^d) \to \mathbb{R}$ *(or* $\mathbb{C}$*) can be written as*

$$l(\varphi_1, \varphi_2) = \int_{\mathbb{R}^d \times \mathbb{R}^d} \varphi_1(r)\varphi_2(s)a(r,s)\, ds\, dr, \qquad (3.8)$$

where the kernel a *is some generalized function in* $\mathscr{S}'(\mathbb{R}^d \times \mathbb{R}^d)$.

One may argue that the signal-domain notation that is used in both (3.6) and (3.8) is somewhat abusive since $A\{\varphi\}$ and a do not necessarily have an interpretation as classical functions (see statement on the notation in Section 3.1.1). The purists therefore prefer to denote (3.8) as

$$l(\varphi_1, \varphi_2) = \langle \varphi_1 \otimes \varphi_2, a \rangle \qquad (3.9)$$

with $(\varphi_1 \otimes \varphi_2)(r,s) = \varphi_1(r)\varphi_2(s)$ for all $\varphi_1, \varphi_2 \in \mathscr{S}(\mathbb{R}^d)$.

The connection among the representations (3.6)–(3.9) is clarified by relating the continuous bilinear form l to the underlying continuous linear operator $A : \mathscr{S}(\mathbb{R}^d) \to \mathscr{S}'(\mathbb{R}^d)$ by means of the identity

$$l(\varphi_1, \varphi) = \langle \varphi_1, A\varphi \rangle,$$

where $A\varphi \in \mathscr{S}'(\mathbb{R}^d)$ is the generalized function specified by (3.6) or, equivalently, by the inner "integral" (duality product) with respect to s in (3.8).

3.3.5 Linear shift-invariant operators and convolutions

Let S_{r_0} denote the shift operator $\varphi \mapsto \varphi(\cdot - r_0)$. We call an operator U *shift-invariant* if $US_{r_0} = S_{r_0}U$ for all $r_0 \in \mathbb{R}^d$.

As a corollary of the kernel theorem, we have the following characterization of *linear shift-invariant (LSI)* operators $\mathscr{S} \to \mathscr{S}'$ (and a similar characterization for $\mathscr{D} \to \mathscr{D}'$).

COROLLARY 3.3 *Every continuous linear shift-invariant operator $\mathscr{S}(\mathbb{R}^d) \to \mathscr{S}'(\mathbb{R}^d)$ can be written as a convolution*

$$\varphi(r) \mapsto (\varphi * h)(r) = \int_{\mathbb{R}^d} \varphi(s) h(r - s) \, ds$$

with some generalized function $h \in \mathscr{S}'(\mathbb{R}^d)$.

The idea there is that the kernel (or generalized impulse response) in (3.6) is a function of the relative displacement only: $a(r, s) = h(r - s)$ (shift-invariance property). Moreover, in this case we have the convolution-multiplication formula

$$\mathscr{F}\{h * \varphi\} = \widehat{\varphi}\widehat{h}. \tag{3.10}$$

Note that the convolution of a test function and a distribution is in general a distribution. The latter is smooth (and therefore equivalent to an ordinary function), but not necessarily rapidly decaying. However, $\varphi * h$ will once again belong continuously to $\mathscr{S}$ if $\widehat{h}$, the Fourier transform of h, is a *smooth* (infinitely differentiable) function with at most polynomial growth at infinity because the smoothness of $\widehat{h}$ translates into h having rapid decay in the spatio-temporal domain, and vice versa. In particular, we note that the condition is met when $h \in \mathscr{R}(\mathbb{R}^d)$ (since $r^n h(r) \in L_1(\mathbb{R}^d)$ for any $n \in \mathbb{N}^d$). A classical situation in dimension $d = 1$ where the decay is guaranteed to be exponential is when the Fourier transform of h is a rational transfer function of the form

$$\widehat{h}(\omega) = C_0 \frac{\prod_{m=1}^{M} (j\omega - z_m)}{\prod_{n=1}^{N} (j\omega - p_n)}$$

with no purely imaginary pole (i.e., with $\mathrm{Re}(p_n) \neq 0$, $1 \leq n \leq N$).[4]

Since any sequence that converges in some L_p space, with $1 \leq p \leq \infty$, also converges in $\mathscr{S}'$, the kernel theorem implies that any continuous linear operator $\mathscr{S}(\mathbb{R}^d) \to L_p(\mathbb{R}^d)$ can be written in the form specified by (3.6).

In defining the convolution of two distributions, some caution should be exerted. To be consistent with the previous definitions, we can view convolutions as continuous

---

[4] For M or $N = 0$, we shall take the corresponding product to be equal to 1.

LSI operators. The convolution of two distributions will then correspond to the composition of two LSI operators. To fix ideas, let us take two distributions f and h, with corresponding operators A_f and A_h. We then wish to identify $f * h$ with the composition $A_f A_h$. However, note that, by the kernel theorem, A_f and A_h are initially defined $\mathscr{S} \to \mathscr{S}'$. Since the codomain of A_h (the space $\mathscr{S}'$) does not match the domain of A_f (the space $\mathscr{S}$), this composition is a priori undefined.

There are two principal situations where we can get around the above limitation. The first is when the range of A_h is limited to $\mathscr{S} \subset \mathscr{S}'$ (i.e., A_h maps $\mathscr{S}$ to itself instead of the much larger $\mathscr{S}'$). This is the case for the distributions with a smooth Fourier transform that we discussed previously.

The second situation where we may define the convolution of f and h is when the range of A_h can be restricted to some space $\mathscr{X}$ (i.e., $A_g : \mathscr{S} \to \mathscr{X}$) and, furthermore, A_f has a continuous extension to $\mathscr{X}$; that is, we can extend it as $A_f : \mathscr{X} \to \mathscr{S}'$.

An important example of the second situation is when the distributions in question belong to the spaces $L_p(\mathbb{R}^d)$ and $L_q(\mathbb{R}^d)$ with $1 \leq p, q \leq \infty$ and $1/p + 1/q \leq 1$. In this case, their convolution is well defined and can be identified with a function in $L_r(\mathbb{R}^d)$, $1 \leq r \leq \infty$, with

$$1 + \frac{1}{r} = \frac{1}{p} + \frac{1}{q}.$$

Moreover, for $f \in L_p(\mathbb{R}^d)$ and $h \in L_q(\mathbb{R}^d)$, we have

$$\|f * h\|_{L_r} \leq \|f\|_{L_p} \|h\|_{L_q}.$$

This result is *Young's inequality for convolutions*. An important special case of this identity, most useful in derivations, is obtained for $q = 1$ and $p = r$:

$$\|h * f\|_{L_p} \leq \|h\|_{L_1} \|f\|_{L_p}. \tag{3.11}$$

The latter formula indicates that $L_p(\mathbb{R}^d)$ spaces are "stable" under convolution with elements of $L_1(\mathbb{R}^d)$ (*stable filters*).

3.3.6   Convolution operators on $L_p(\mathbb{R}^d)$

While the condition $h \in L_1(\mathbb{R}^d)$ in (3.11) is very useful in practice and plays a central role in the classical theory of linear systems, it does not cover the entire range of bounded convolution operators on $L_p(\mathbb{R}^d)$. Here we shall be more precise and characterize the complete class of such operators for the cases $p = 1, 2, +\infty$. In harmonic analysis, these operators are commonly referred to as L_p *Fourier multipliers*, using (3.10) as the starting point for their definition.

DEFINITION 3.5 (Fourier multiplier)   An operator $T : L_p(\mathbb{R}^d) \to L_p(\mathbb{R}^d)$ is called an L_p *Fourier multiplier* if it is continuous on $L_p(\mathbb{R}^d)$ and can be represented as $Tf = \mathscr{F}^{-1}\{\hat{f}H\}$. The function $H : \mathbb{R}^d \to \mathbb{C}$ is the frequency response of the underlying filter.

The first observation is that the definition guarantees linearity and shift-invariance. Moreover, since $\mathscr{S}(\mathbb{R}^d) \subset L_p(\mathbb{R}^d) \subset \mathscr{S}'(\mathbb{R}^d)$, the multiplier operator can be written as a convolution $Tf = h * f$ (see Corollary 3.3) where $h \in \mathscr{S}'(\mathbb{R}^d)$ is the impulse response of the operator T: $h = \mathscr{F}^{-1}\{H\} = T\delta$. Conversely, we also have that $H = \hat{h} = \mathscr{F}\{h\}$.

Since we are dealing with a linear operator on a normed vector space, we can rely on the equivalence between continuity (in accordance with Definition 3.2) and the boundedness of the operator.

DEFINITION 3.6 (Operator norm)   The *norm* of the linear operator T : $L_p(\mathbb{R}^d) \to L_p(\mathbb{R}^d)$ is given by

$$\|T\|_{L_p} = \sup_{f \in L_p(\mathbb{R}^d) \setminus \{0\}} \frac{\|Tf\|_{L_p}}{\|f\|_{L_p}}.$$

The operator is said to be bounded if its norm is finite.

In practice, it is often sufficient to work out bounds for the extreme cases (e.g., $p = 1, +\infty$) and to then invoke the Riesz–Thorin interpolation theorem to extend the results to the p values in between.

THEOREM 3.4 (Riesz–Thorin)   *Let T be a linear operator that is bounded on $L_{p_1}(\mathbb{R}^d)$ as well as on $L_{p_2}(\mathbb{R}^d)$, with $1 \leq p_1 \leq p_2$. Then, T is also bounded for any $p \in [p_1, p_2]$ in the sense that there exist constants $C_p = \|T\|_{L_p}$ with $\min(C_{p_1}, C_{p_2}) \leq C_p \leq \max(C_{p_1}, C_{p_2})$ such that*

$$\|Tf\|_{L_p} \leq C_p \|f\|_{L_p}$$

for all $f \in L_p(\mathbb{R}^d)$.

The next theorem summarizes the main results that are available on the characterization of convolution operators on $L_p(\mathbb{R}^d)$.

THEOREM 3.5 (Characterization of L_p Fourier multipliers)   *Let T be a Fourier-multiplier operator with frequency response $H : \mathbb{R}^d \to \mathbb{C}$ and (generalized) impulse response $h = \mathscr{F}^{-1}\{H\} = T\{\delta\}$. Then, the following statements apply:*

(1) The operator T is an L_1 Fourier multiplier if and only if there exists a finite complex-valued Borel measure denoted by μ_h such that $H(\omega) = \int_{\mathbb{R}^d} e^{-j\langle \omega, r \rangle} \mu_h(dr)$.
(2) The operator T is an L_∞ Fourier multiplier if and only if H is the Fourier transform of a finite complex-valued Borel measure, as stated in (1).
(3) The operator T is an L_2 Fourier multiplier if and only if $H = \hat{h} \in L_\infty(\mathbb{R}^d)$.

The corresponding operator norms are

$$\|T\|_{L_1} = \|T\|_{L_\infty} = \|\mu_h\|_{TV}$$

$$\|T\|_{L_2} = \frac{1}{(2\pi)^{d/2}} \|H\|_{L_\infty},$$

where $\|\mu_h\|_{TV}$ is the total variation (TV) of the underlying measure. Finally, T is an L_p Fourier multiplier for the whole range $1 \leq p \leq +\infty$ if the condition on H in (1) or (2) *is met with*

$$\|T\|_{L_p} \leq \|\mu_h\|_{TV} = \sup_{\|\varphi\|_{L_\infty} \leq 1} \langle \varphi, h \rangle. \tag{3.12}$$

We note that the above theorem is an extension of (3.11) since being a finite Borel measure is less restrictive a condition than $h \in L_1(\mathbb{R}^d)$. To see this, we invoke Lebesgue's decomposition theorem, stating that a finite measure μ_h admits a unique decomposition as

$$\mu_h = \mu_{ac} + \mu_{sing},$$

where μ_{ac} is an absolutely continuous measure and μ_{sing} a singular measure whose mass is concentrated on a set whose Lebesgue measure is zero. If $\mu_{sing} = 0$, then there exists a unique function $h \in L_1(\mathbb{R}^d)$ – the Radon–Nikodym derivative of μ_h with respect to the Lebesgue measure – such that

$$\int_{\mathbb{R}^d} \varphi(r)\mu_h(dr) = \int_{\mathbb{R}^d} \varphi(r)h(r)\,dr = \langle \varphi, h \rangle.$$

We then recall Hölder's inequality (3.4) with $(p, p') = (\infty, 1)$,

$$|\langle \varphi, h \rangle| \leq \|\varphi\|_{L_\infty} \|h\|_{L_1},$$

to see that the total-variation norm defined by (3.12) reduces to the L_1 norm: $\|\mu_h\|_{TV} = \|h\|_{L_1}$. Under those circumstances, there is an equivalence between (3.11) and (3.12).

More generally, when $\mu_{sing} \neq 0$ we can make the same kind of association between μ_h and a generalized function h which is no longer in $L_1(\mathbb{R}^d)$. The typical case is when μ_{sing} is a discrete measure which results in a generalized function $h_{sing} = \sum_k h_k \delta(\cdot - r_k)$ that is a sum of Dirac impulses. The total variation of μ_h is then given by $\|\mu_h\|_{TV} = \|h_{ac}\|_{L_1} + \sum_k |h_k|$.

Statement (3) in Theorem 3.5 is a consequence of Parseval's identity. It is consistent with the intuition that a "stable" filter should have a bounded frequency response, as a minimal requirement. The class of convolution kernels that satisfy this condition are sometimes called *pseudo-measures*. These are more general entities than measures because the Fourier transform of a finite measure is necessarily uniformly continuous in addition to being bounded.

The last result in Theorem 3.5 is obtained by interpolation between Statements (1) and (2) using the Riesz–Thorin theorem. The extent to which the TV condition $\|\mu_h\|_{TV} < \infty$ can be relaxed for $p \neq 1, 2, \infty$ is not yet settled and considered to be a difficult mathematical problem. A borderline example of a one-dimensional (1-D) convolution operator that is bounded for $1 < p < \infty$ (see Theorem 3.6 below), but fails to meet the *necessary and sufficient* TV condition for $p = 1, \infty$, is the Hilbert transform. Its frequency response is $H_{Hilbert}(\omega) = -j\,sign(\omega)$, which is bounded since $|H_{Hilbert}(\omega)| = 1$ (all-pass filter), but which is not uniformly continuous because of the jump at $\omega = 0$. Its impulse response is the generalized function $h(r) = \frac{1}{\pi r}$, which is not included in $L_1(\mathbb{R})$ for two reasons: the singularity at $r = 0$ and the lack of sufficient decay at infinity.

The case of the Hilbert transform is covered by Mikhlin's multiplier theorem, which provides a sufficient condition on the frequency response of a filter for L_p-stability.

THEOREM 3.6 (Mikhlin)   *A Fourier-multiplier operator is bounded in $L_p(\mathbb{R}^d)$ for $1 < p < \infty$ if its frequency response $H : \mathbb{R}^d \to \mathbb{C}$ satisfies the differential estimate*

$$\left| \boldsymbol{\omega}^n \partial^n H(\boldsymbol{\omega}) \right| \leq C_n \quad \textit{for all } |\boldsymbol{n}| \leq (d/2) + 1.$$

Mikhlin's condition, which can absorb some degree of discontinuity at the origin, is easy to check in practice. It is stronger than the minimal boundedness requirement for $p = 2$.

3.4 Probability theory

3.4.1 Probability measures

Probability measures are mathematical constructs that permit us to assign numbers (probabilities) between 0 (almost impossible) to 1 (almost sure) to *events*. An event is modeled by a subset A of the *universal set* Ω_X of all outcomes of a certain experiment X, which is assumed to be known. The symbol $\mathcal{P}_X(A)$ then gives the probability that some element of A occurs as the outcome of experiment X. Note that, in general, we may assign probabilities only to *some* subsets of Ω_X. We shall denote the collection of all subsets of Ω_X for which $\mathcal{P}_X$ is defined as $\mathfrak{S}_X$.

The probability measure $\mathcal{P}_X$ then corresponds to a function $\mathfrak{S}_X \to [0, 1]$. The triple $(\Omega_X, \mathfrak{S}_X, \mathcal{P}_X)$ is called a *probability space.*

Frequently, the collection $\mathfrak{S}_X$ contains open and closed sets, as well as their countable unions and intersections, collectively known as *Borel sets*. In this case we call $\mathcal{P}_X$ a *Borel probability measure.*

An important application of the notion of probability is in computing the "average" value of some (real- or complex-valued) quantity f that depends on the outcome in Ω_X. This quantity, the computation of which we shall discuss shortly, is called the *expected value* of f, and is denoted as $\mathbb{E}\{f(X)\}$.

An important context for probabilistic computations is when the outcome of X can be encoded as a finite-dimensional numerical sequence, which implies that we can identify Ω_X with $\mathbb{R}^n$ (or a subset thereof). In this case, within the proper mathematical setting, we can find a (generalized) function p_X, called the *probability distribution*[5] or *density function (pdf)* of X, such that

$$\mathcal{P}_X(A) = \int_A p_X(\boldsymbol{x}) \, \mathrm{d}\boldsymbol{x}$$

for suitable subsets A of $\mathbb{R}^n$.[6]

---

[5] Probability distributions should not be confused with the distributions in the sense of Schwartz (i.e., generalized functions) that were introduced in Section 3.3. It is important to distinguish the two usages, in part because, as we describe here, in finite dimensions a connection can be made between probability distributions and positive generalized functions.

[6] In classical probability theory, a pdf is defined as the Radon–Nikodym derivative of a probability measure with respect to some other measure, typically the Lebesgue measure (as we shall assume). This requires the probability measure to be absolutely continuous with respect to the latter measure. The definition

More generally, the expected value of $f : X \to \mathbb{C}$ is here given by

$$\mathbb{E}\{f(X)\} = \int_{\mathbb{R}^n} f(x) p_X(x) \, dx. \tag{3.13}$$

We say "more generally" because $\mathscr{P}_X(A)$ can be seen as the expected value of the indicator function $\mathbb{1}_A(X)$. Since the integral of complex-valued f can be written as the sum of its real and imaginary parts, without loss of generality we shall consider only real-valued functions where convenient.

When the outcome of the experiment is a vector with infinitely many coordinates (for instance a function $\mathbb{R} \to \mathbb{R}$), it is typically not possible to characterize probabilities with probability distributions. It is nevertheless still possible to define probability measures on subsets of Ω_X, and also to define the integral (average value) of many a function $f : \Omega_X \to \mathbb{R}$. In effect, a definition of the integral of f *with respect to probability measure* $\mathscr{P}_X$ is obtained using a limit of "simple" functions (finite weighted sums of indicator functions) that approximate f. For this general definition of the integral we use the notation

$$\mathbb{E}\{f(X)\} = \int_{\Omega_X} f(x) \, \mathscr{P}_X(dx),$$

which we may also use, in addition to (3.13), in the case of a finite-dimensional Ω_X.

In general, given a function $f : \Omega_X \to \Omega_Y$ that defines a new outcome $y \in \Omega_Y$ for every outcome $x \in \Omega_X$ of experiment X, one can see the result of applying f to the outcome of X as a new experiment Y. The probability of an event $B \subset \Omega_Y$ is the same as the combined probability of all outcomes of X that generate an outcome in B. Thus, mathematically,

$$\mathscr{P}_Y(B) = \mathscr{P}_X(f^{-1}(B)) = \mathscr{P}_X \circ f^{-1}(B),$$

where the inverse image $f^{-1}(B)$ is defined as

$$f^{-1}(B) = \{x \in \Omega_X : f(x) \in B\}.$$

$\mathscr{P}_Y = \mathscr{P}_X(f^{-1} \cdot)$ is called the *push-forward* of $\mathscr{P}_X$ through f.

3.4.2   Joint probabilities and independence

When two experiments X and Y with probabilities $\mathscr{P}_X$ and $\mathscr{P}_Y$ are considered simultaneously, one can imagine a joint probability space $(\Omega_{X,Y}, \mathfrak{S}_{X,Y}, \mathscr{P}_{X,Y})$ that *supports* both X and Y, in the sense that there exist functions $f : \Omega_{X,Y} \to \Omega_X$ and $g : \Omega_{X,Y} \to \Omega_Y$ such that

$$\mathscr{P}_X(A) = \mathscr{P}_{X,Y}(f^{-1}(A)) \quad \text{and} \quad \mathscr{P}_Y(B) = \mathscr{P}_{X,Y}(g^{-1}(B))$$

for all $A \in \mathfrak{S}_X$ and $B \in \mathfrak{S}_Y$.

of the generalized pdf given here is more permissive, and also includes measures that are singular with respect to the Lebesgue measure (for instance the Dirac measure of a point, for which the generalized pdf is a Dirac distribution). This generalization relies on identifying measures on the Euclidean space with positive linear functionals.

The functions f, g above are assumed to be fixed, and the joint event that A occurs for X and B for Y is given by

$$f^{-1}(A) \cap g^{-1}(B).$$

If the outcome of X has no bearing on the outcome of Y and vice versa, then X and Y are said to be independent. In terms of probabilities, this translates into the probability factorization rule

$$\mathscr{P}_{X,Y}(f^{-1}(A) \cap g^{-1}(B)) = \mathscr{P}_X(A) \cdot \mathscr{P}_Y(B) = \mathscr{P}_{X,Y}(f^{-1}(A)) \cdot \mathscr{P}_{X,Y}(g^{-1}(B)).$$

The above ideas can be extended to any finite collection of experiments $X_1, \ldots, X_M$ (and even to infinite ones, with appropriate precautions and adaptations).

3.4.3   Characteristic functions in finite dimensions

In finite dimensions, given a probability measure $\mathscr{P}_X$ on $\Omega_X = \mathbb{R}^n$, for any vector $\boldsymbol{\omega} \in \mathbb{R}^n$ we can compute the expected value (integral) of the bounded function $\boldsymbol{x} \mapsto e^{j\langle \boldsymbol{\omega}, \boldsymbol{x} \rangle}$. This permits us to define a complex-valued function on $\mathbb{R}^n$ by the formula

$$\widehat{p}_X(\boldsymbol{\omega}) = \mathbb{E}\{e^{j\langle \boldsymbol{\omega}, \boldsymbol{x} \rangle}\} = \int_{\mathbb{R}^n} e^{j\langle \boldsymbol{\omega}, \boldsymbol{x} \rangle} p_X(\boldsymbol{x}) \, d\boldsymbol{x} = \overline{\mathscr{F}}\{p_X\}(\boldsymbol{\omega}), \qquad (3.14)$$

which corresponds to a slightly different definition of the Fourier transform of the (generalized) probability distribution p_X. The convention in probability theory is to define the forward Fourier transform with a positive sign for $j\langle \boldsymbol{\omega}, \boldsymbol{x} \rangle$, which is the opposite of the convention used in analysis.

One can prove that $\widehat{p}_X$, as defined above, is always continuous at $\boldsymbol{0}$ with $\widehat{p}_X(\boldsymbol{0}) = 1$, and that it is *positive definite* (see Definition B.1 in Appendix B).

Remarkably, the converse of the above fact is also true. We record the latter result, which is due to Bochner, together with the former observation, as Theorem 3.7.

THEOREM 3.7 (Bochner)   *Let $\widehat{p}_X : \mathbb{R}^n \to \mathbb{C}$ be a function that is positive definite, fulfills $\widehat{p}_X(\boldsymbol{0}) = 1$, and is continuous at $\boldsymbol{0}$. Then, there exists a unique Borel probability measure $\mathscr{P}_X$ on $\mathbb{R}^n$, such that*

$$\widehat{p}_X(\boldsymbol{\omega}) = \int_{\mathbb{R}^n} e^{j\langle \boldsymbol{\omega}, \boldsymbol{x} \rangle} \mathscr{P}_X(d\boldsymbol{x}) = \mathbb{E}\{e^{j\langle \boldsymbol{\omega}, \boldsymbol{x} \rangle}\}.$$

Conversely, the function specified by (3.14) with $p_X(\boldsymbol{r}) \geq 0$ and $\int_{\mathbb{R}^n} p_X(\boldsymbol{r}) \, d\boldsymbol{r} = 1$ is positive definite, uniformly continuous, and such that $|\widehat{p}_X(\boldsymbol{\omega})| \leq \widehat{p}_X(\boldsymbol{0}) = 1$.

The interesting twist (which is due to Lévy) is that the positive definiteness of $\widehat{p}_X$ and its continuity at $\boldsymbol{0}$ implies continuity everywhere (as well as boundedness).

Since, by the above theorem, $\widehat{p}_X$ uniquely identifies $\mathscr{P}_X$, it is called the *characteristic function* of probability measure $\mathscr{P}_X$ (recall that the probability measure $\mathscr{P}_X$ is related to the density p_X by $\mathscr{P}_X(E) = \int_E p_X(\boldsymbol{x}) \, d\boldsymbol{x}$ for sets E in the σ-algebra over $\mathbb{R}^n$).

The next theorem characterizes weak convergence of measures on $\mathbb{R}^n$ in terms of their characteristic functions.

THEOREM 3.8 (Lévy's continuity theorem)   *Let $(\mathscr{P}_{X_i})$ be a sequence of probability measures on $\mathbb{R}^n$ with respective sequence of characteristic functions $(\widehat{p}_{X_i})$. If there exists a function $\widehat{p}_X$ such that*

$$\lim_i \widehat{p}_{X_i}(\boldsymbol{\omega}) = \widehat{p}_X(\boldsymbol{\omega})$$

pointwise on $\mathbb{R}^n$, and if, in addition, $\widehat{p}_X$ is continuous at $\mathbf{0}$, then $\widehat{p}_X$ is the characteristic function of a probability measure $\mathscr{P}_X$ on $\mathbb{R}^n$. Moreover, $\mathscr{P}_{X_i}$ converges weakly to $\mathscr{P}_X$, in symbols

$$\mathscr{P}_{X_i} \xrightarrow{w} \mathscr{P}_X,$$

meaning, for any continuous function $f : \mathbb{R}^n \to \mathbb{R}$,

$$\lim_i \mathbb{E}_{X_i}\{f\} = \mathbb{E}_X\{f\}.$$

The reciprocal of the above theorem is also true; namely, if $\mathscr{P}_{X_i} \xrightarrow{w} \mathscr{P}_X$, then $\widehat{p}_{X_i}(\boldsymbol{\omega}) \to \widehat{p}_X(\boldsymbol{\omega})$ pointwise.

3.4.4   Characteristic functionals in infinite dimensions

Given a probability measure $\mathscr{P}_X$ on the continuous dual $\mathscr{X}'$ of some test function space $\mathscr{X}$, one can define an analog of the finite-dimensional characteristic function, dubbed the *characteristic functional of $\mathscr{P}_X$* and denoted as $\widehat{\mathscr{P}}_X$, by means of the identity

$$\widehat{\mathscr{P}}_X(\varphi) = \mathbb{E}\{e^{j\langle \varphi, X \rangle}\}. \tag{3.15}$$

Comparing the above definition with (3.14), one notes that $\mathbb{R}^n$, as the domain of the characteristic function $\widehat{p}_X$, is now replaced by the space $\mathscr{X}$ of test functions.

As was the case in finite dimensions, the characteristic functional fulfills two important conditions:

- Positive definiteness: $\widehat{\mathscr{P}}_X$ is positive definite, in the sense that, for any N (test) functions $\varphi_1, \ldots, \varphi_N$, for any N, the $N \times N$ matrix with entries $p_{ij} = \widehat{\mathscr{P}}_X(\varphi_i - \varphi_j)$ is non-negative definite.
- Normalization: $\widehat{\mathscr{P}}_X(0) = 1$.

In view of the finite-dimensional result (Bochner's theorem), it is natural to ask if a condition in terms of continuity can be given also in the infinite-dimensional case, so that any functional $\widehat{\mathscr{P}}_X$ fulfilling this continuity condition, in addition to the above two, uniquely identifies a probability measure on $\mathscr{X}'$. In the case where $\mathscr{X}$ is a nuclear space (and, in particular, for $\mathscr{X} = \mathscr{S}(\mathbb{R}^d)$ or $\mathscr{D}(\mathbb{R}^d)$; see Section 3.1.3), such a condition is given by the Minlos–Bochner theorem.

THEOREM 3.9 (Minlos–Bochner)   *Let $\mathscr{X}$ be a nuclear space and let $\widehat{\mathscr{P}}_X : \mathscr{X} \to \mathbb{C}$ be a functional that is positive definite in the sense discussed above, fulfills $\widehat{\mathscr{P}}_X(0) = 1$, and is continuous $\mathscr{X} \to \mathbb{C}$. Then, there exists a unique probability measure $\mathscr{P}_X$ on $\mathscr{X}'$ (the continuous dual of $\mathscr{X}$), such that*

$$\widehat{\mathscr{P}}_X(\varphi) = \int_{\mathscr{X}'} e^{j\langle \varphi, x \rangle} \mathscr{P}_X(\mathrm{d}x) = \mathbb{E}\{e^{j\langle \varphi, X \rangle}\}.$$

Conversely, the characteristic functional associated with some probability measure $\mathscr{P}_X$ on $\mathscr{X}'$ is positive definite, continuous over $\mathscr{X}$, and such that $\widehat{\mathscr{P}}_X(0) = 1$.

The practical implication of this result is that one can rely on characteristic functionals to indirectly specify infinite-dimensional measures (most importantly, probabilities of stochastic processes) – which are difficult to pin down otherwise. Operationally, the characteristic functional $\widehat{\mathscr{P}}_X(\varphi)$ is nothing but a mathematical rule (e.g., $\widehat{\mathscr{P}}_X(\varphi) = e^{-\frac{1}{2}\|\varphi\|_2^2}$) that returns a value in $\mathbb{C}$ for any given function $\varphi \in \mathscr{S}$. The truly powerful aspect is that this rule condenses *all* the information about the statistical distribution of some underlying infinite-dimensional random object X. When working with characteristic functionals, we shall see that computing probabilities and deriving various properties of the said processes are all reduced to analytical derivations.

3.5   Generalized random processes and fields

In this section, we present an introduction to the theory of generalized random processes, which is concerned with defining probabilities on function spaces, that is, infinite-dimensional vector spaces with some notion of limit and convergence. We have made the point before that the theory of generalized functions is a natural extension of finite-dimensional linear algebra. The same kind of parallel can be drawn between the theory of generalized stochastic processes and conventional probability calculus (which deals with finite-dimensional random vector variables). Therefore, before getting into more detailed explanations, it is instructive to have a look back at Table 3.2, which provides a side-by-side summary of the primary probabilistic concepts that have been introduced so far. The reader is then referred to Table 3.4, which presents a comparison of finite- and infinite-dimensional "innovation models." To give the basic idea, in finite dimensions, an "innovation" is a vector in $\mathbb{R}^n$ of independent identically distributed (i.i.d.) random variables. An "innovation model" is obtained by transforming such a vector by means of a linear operator (a matrix), which embodies the structure of dependencies of the model. In infinite dimensions, the notion of an i.i.d. vector is replaced by that of a random process with independent values at every point (which we shall call an "innovation process"). The transformation is achieved by applying a continuous linear operator which constitutes the generalization of a matrix. The characterization of such models is made possible by their characteristic functionals, which, as we saw in Section 3.4.4, are the infinite-dimensional equivalents of characteristic functions of random variables.

3.5.1   Generalized random processes as collections of random variables

A *generalized stochastic process* [7] is essentially a randomization of the idea of a generalized function (Section 3.3), in much the same way as an ordinary stochastic process is a randomization of the concept of a function.

---

[7] We shall use the terms random/stochastic "process" and "field" almost interchangeably. The distinction, in general, lies in the fact that, for a random process, the parameter is typically interpreted as time, while, for a field, the parameter is typically multidimensional and interpreted as a spatial or spatio-temporal location.

Table 3.4 Comparison of innovation models in finite- and infinite-dimensional settings. See Sections 4.3–4.5 for a detailed explanation.

Finite-dimensional	Infinite-dimensional
Standard Gaussian i.i.d. vector	Standard Gaussian white noise w
$W = (W_1, \ldots, W_N)$ $\widehat{p}_W(\boldsymbol{\omega}) = \mathrm{e}^{-\frac{1}{2}\lvert\boldsymbol{\omega}\rvert^2}, \boldsymbol{\omega} \in \mathbb{R}^N$	$\widehat{\mathscr{P}}_w(\varphi) = \mathrm{e}^{-\frac{1}{2}\lVert\varphi\rVert_2^2}, \varphi \in \mathscr{S}$
Multivariate Gaussian vector X	Gaussian generalized process s
$X = \mathbf{A}W$ $\widehat{p}_X(\boldsymbol{\omega}) = \mathrm{e}^{-\frac{1}{2}\lvert\mathbf{A}^{\mathsf{T}}\boldsymbol{\omega}\rvert^2}$	$s = \mathrm{A}w$ (for continuous $\mathrm{A} : \mathscr{S}' \to \mathscr{S}'$) $\widehat{\mathscr{P}}_s(\varphi) = \mathrm{e}^{-\frac{1}{2}\lVert\mathrm{A}^*\varphi\rVert^2}$
General i.i.d. vector $W = (W_1, \ldots, W_N)$ with exponent f $\widehat{p}_W(\boldsymbol{\omega}) = \mathrm{e}^{\sum_{n=1}^{N} f(\omega_n)}$	General white noise w with Lévy exponent f $\widehat{\mathscr{P}}_w(\varphi) = \mathrm{e}^{\int_{\mathbb{R}^d} f(\varphi(r))\,\mathrm{d}r}$
Linear transformation of general i.i.d. random vector W (innovation model)	Linear transformation of general white noise s (innovation model)
$X = \mathbf{A}W$ $\widehat{p}_X(\boldsymbol{\omega}) = \widehat{p}_W(\mathbf{A}^{\mathsf{T}}\boldsymbol{\omega})$	$s = \mathrm{A}w$ $\widehat{\mathscr{P}}_s(\varphi) = \widehat{\mathscr{P}}_w(\mathrm{A}^*\varphi)$

At a minimum, the definition of a generalized stochastic process s should permit us to associate probabilistic models with observations made using test functions. In other words, with any test function φ in some suitable test-function space $\mathscr{X}$ is associated a random variable $s(\varphi)$, also denoted as $\langle\varphi, s\rangle$. This is to be contrasted with an observation $s(t)$ at time t, which would be modeled by a random variable in the case of an ordinary stochastic process. We shall denote the probability measure of the random variable $\langle\varphi, s\rangle$ as $\mathscr{P}_{s,\varphi}$. Similarly, to any finite collection of observations $\langle\varphi_n, s\rangle$, $1 \leq n \leq N, N \in \mathbb{N}$, corresponds a joint probability measure $\mathscr{P}_{s,\varphi_1:\varphi_N}$ on $\mathbb{R}^N$ (we shall only consider real-valued processes here, and therefore assume the observations to be real-valued).

Moreover, finite families of observations $\langle\varphi_n, s\rangle$, $1 \leq n \leq N$, and $\langle\psi_m, s\rangle$, $1 \leq m \leq M$, need to be *consistent* or *compatible* to ensure that all computations of the probability of an event involving finite observations yield the same value for the probability. In modeling physical phenomena, it is also reasonable to assume some weak form of continuity in the probability of $\langle\varphi, s\rangle$ as a function of φ.

Mathematically, these requirements are fulfilled by the kind of probabilistic model induced by a *cylinder-set probability measure*. In other words, a cylinder-set probability measure provides a consistent probabilistic description for all finite sets of observations of some phenomenon s using test functions $\varphi \in \mathscr{X}$. Furthermore, a cylinder-set probability measure can always be specified via its *characteristic functional* $\widehat{\mathscr{P}}_s(\varphi) = \mathbb{E}\{\mathrm{e}^{\mathrm{j}\langle\varphi, s\rangle}\}$, which makes it amenable to analytic computations.

The only conceptual limitation of such a probability model is that, at least a priori, it does not permit us to associate the *sample paths* of the process with (generalized) functions. Put differently, in this framework, we are not allowed to interpret s as a random entity belonging to the dual $\mathscr{X}'$ of $\mathscr{X}$, since we have not yet defined a proper probability measure on $\mathscr{X}'$.[8] Doing so involves some additional steps.

3.5.2 Generalized random processes as random generalized functions

Fortunately, the above existence and interpretation problem is fully resolved by taking $\mathscr{X}$ to be a nuclear space, thanks to the Minlos–Bochner theorem (Theorem 3.9). This allows for the extension of the underlying cylinder-set probability measure to a proper (by which here we mean countably additive) probability measure on $\mathscr{X}'$ (the topological dual of $\mathscr{X}$).

In this case, the joint probabilities $\mathscr{P}_{s,\varphi_1:\varphi_N}$, $\varphi_1,\ldots,\varphi_N \in \mathscr{X}$, $N \in \mathbb{N}$, corresponding to the random variables $\langle \varphi_n, s \rangle$ for all possible choices of test functions, collectively define a probability measure $\mathscr{P}_s$ on the infinite-dimensional dual space $\mathscr{X}'$. This means that we can view s as an element drawn randomly from $\mathscr{X}'$ according to the *probability law* $\mathscr{P}_s$.

In particular, if we take $\mathscr{X}$ to be either $\mathscr{S}(\mathbb{R}^d)$ or $\mathscr{D}(\mathbb{R}^d)$, then our generalized random process/field will have realizations that are distributions in $\mathscr{S}'(\mathbb{R}^d)$ or $\mathscr{D}'(\mathbb{R}^d)$, respectively. We can then also think of $\langle \varphi, s \rangle$ as the measurement of this random object s by means of some sensor (test function) φ in $\mathscr{S}$ or $\mathscr{D}$.

Since we shall rely on this fact throughout the book, we reiterate once more that a complete probabilistic characterization of s as a probability measure on the space $\mathscr{X}'$ (dual to some nuclear space $\mathscr{X}$) is provided by its characteristic functional. The truly powerful aspect of the Minlos–Bochner theorem is that the implication goes both ways: *any* continuous, positive definite functional $\widehat{\mathscr{P}}_s : \mathscr{X} \to \mathbb{C}$ with proper normalization identifies a unique probability measure $\mathscr{P}_s$ on $\mathscr{X}'$. Therefore, to define a generalized random process s with realizations in $\mathscr{X}'$, it suffices to produce a functional $\widehat{\mathscr{P}}_s : \mathscr{X} \to \mathbb{C}$ with the noted properties.

3.5.3 Determination of statistics from the characteristic functional

The characteristic functional of the generalized random process s contains complete information about its probabilistic properties, and can be used to compute all probabilities, and to derive or verify the probabilistic properties related to s.

Most importantly, it can yield the Nth-order joint probability density of any set of linear observations of s by suitable N-dimensional inverse Fourier transformation. This follows from a straightforward manipulation in the domain of the (joint) characteristic function and is recorded for further reference.

PROPOSITION 3.10 *Let $\mathbf{y} = (Y_1,\ldots,Y_N)$ with $Y_n = \langle \varphi_n, s \rangle$, where $\varphi_1,\ldots,\varphi_N \in \mathscr{X}$, be a set of linear measurements of the generalized stochastic process s with*

---

[8] In fact, $\mathscr{X}'$ may very well be too small to support such a description (while the *algebraic dual*, $\mathscr{X}^*$, can support the measure – by Kolmogorov's extension theorem – but is too large for many practical purposes). An important example is that of white Gaussian noise, which one may conceive of as associating a Gaussian random variable with variance $\|\varphi\|_2^2$ to any test function $\varphi \in L_2$. However, the "energy" of white Gaussian noise is clearly infinite. Therefore it cannot be modeled as a randomly chosen function in $(L_2)' = L_2$.

characteristic functional $\widehat{\mathscr{P}}_s(\varphi) = \mathbb{E}\{e^{j\langle\varphi,s\rangle}\}$ that is continuous over the function space $\mathscr{X}$. Then,

$$\widehat{p}_{(Y_1:Y_N)}(\boldsymbol{\omega}) = \widehat{\mathscr{P}}_{s,\varphi_1:\varphi_N}(\boldsymbol{\omega}) = \widehat{\mathscr{P}}_s\left(\sum_{n=1}^N \omega_n\varphi_n\right)$$

and the joint pdf of $\mathbf{y}$ is given by

$$p_{(Y_1:Y_N)}(\mathbf{y}) = \mathscr{F}^{-1}\{\widehat{p}_{(Y_1:Y_N)}\}(\mathbf{y}) = \int_{\mathbb{R}^N} \widehat{\mathscr{P}}_s\left(\sum_{n=1}^N \omega_n\varphi_n\right)e^{-j\langle\mathbf{y},\boldsymbol{\omega}\rangle}\frac{d\boldsymbol{\omega}}{(2\pi)^N},$$

where the observation functions $\varphi_n \in \mathscr{X}$ are fixed and $\boldsymbol{\omega} = (\omega_1,\ldots,\omega_N)$ plays the role of the N-dimensional Fourier variable.

Proof The continuity assumption over the function space $\mathscr{X}$ (which need not be nuclear) ensures that the manipulation is legitimate. Starting from the definition of the characteristic function of $\mathbf{y} = (Y_1,\ldots,Y_N)$, we have

$$\widehat{p}_{(Y_1:Y_N)}(\boldsymbol{\omega}) = \mathbb{E}\left\{\exp\left(j\langle\boldsymbol{\omega},\mathbf{y}\rangle\right)\right\}$$
$$= \mathbb{E}\left\{\exp\left(j\sum_{n=1}^N \omega_n\langle\varphi_n,s\rangle\right)\right\}$$
$$= \mathbb{E}\left\{\exp\left(j\langle\sum_{n=1}^N \omega_n\varphi_n,s\rangle\right)\right\} \qquad \text{(by linearity of duality product)}$$
$$= \widehat{\mathscr{P}}_s\left(\sum_{n=1}^N \omega_n\varphi_n\right). \qquad \text{(by definition of } \widehat{\mathscr{P}}_s(\varphi))$$

The density $p_{(Y_1:Y_N)}$ is then obtained by inverse (conjugate) Fourier transformation. □

Similarly, the formalism allows one to retrieve all first- and second-order moments of the generalized stochastic process s. To that end, one considers the mean and correlation functionals defined and computed as

$$\mathscr{M}_s(\varphi) = \mathbb{E}\{\langle\varphi,s\rangle\} = (-j)\frac{d}{d\omega}\widehat{\mathscr{P}}_{s,\varphi}(\omega)\Big|_{\omega=0}$$
$$= (-j)\frac{d}{d\omega}\widehat{\mathscr{P}}_s(\omega\varphi)\Big|_{\omega=0}$$
$$\mathscr{B}_s(\varphi_1,\varphi_2) = \mathbb{E}\{\langle\varphi_1,s\rangle\langle\varphi_2,s\rangle\} = (-j)^2\frac{\partial^2}{\partial\omega_1\partial\omega_2}\widehat{\mathscr{P}}_{s,\varphi_1,\varphi_2}(\omega_1,\omega_2)\Big|_{\omega_1,\omega_2=0}$$
$$= (-j)^2\frac{\partial^2}{\partial\omega_1\partial\omega_2}\widehat{\mathscr{P}}_s(\omega_1\varphi_1+\omega_2\varphi_2)\Big|_{\omega_1,\omega_2=0}.$$

When the space of test functions is nuclear ($\mathscr{X} = \mathscr{S}(\mathbb{R}^d)$ or $\mathscr{D}(\mathbb{R}^d)$) and the above quantities are well defined, we can find generalized functions m_s (the generalized *mean*) and R_s (the generalized *autocorrelation* function) such that

$$\mathscr{M}_s(\varphi) = \int_{\mathbb{R}^d} \varphi(r) m_s(r) \, dr \tag{3.16}$$

$$\mathscr{B}_s(\varphi_1, \varphi_2) = \int_{\mathbb{R}^d} \varphi_1(r) \varphi_2(s) R_s(r, s) \, dr. \tag{3.17}$$

The first identity is simply a consequence of $\mathscr{M}_s$ being a continuous linear functional on $\mathscr{X}$, while the second is an application of Schwartz' kernel theorem (Theorem 3.2).

3.5.4 Operations on generalized stochastic processes

In constructing stochastic models, it is of interest to separate the essential randomness of the models (the "innovation") from their deterministic structure. Our way of approaching this objective is by encoding the random part in a characteristic functional $\widehat{\mathscr{P}}_w$ and the deterministic structure of dependencies in an operator U (or, equivalently, in its adjoint U*). In the following paragraphs, we first review the mathematics of this construction, before we come back to, and clarify, the said interpretation. The concepts presented here in an abstract form are illustrated and made intuitive in the remainder of the book.

Given a continuous linear operator $U : \mathscr{X} \to \mathscr{Y}$ with continuous adjoint $U^* : \mathscr{Y}' \to \mathscr{X}'$, where $\mathscr{X}, \mathscr{Y}$ need not be nuclear, and a functional

$$\widehat{\mathscr{P}}_w : \mathscr{Y} \to \mathbb{C}$$

that satisfies the three conditions of Theorem 3.9 (continuity, positive definiteness, and normalization), we obtain a new functional

$$\widehat{\mathscr{P}}_s : \mathscr{X} \to \mathbb{C}$$

fulfilling the same properties by composing $\widehat{\mathscr{P}}_w$ and U as per

$$\widehat{\mathscr{P}}_s(\varphi) = \widehat{\mathscr{P}}_w(U\varphi) \quad \text{for all } \varphi \in \mathscr{X}. \tag{3.18}$$

Writing

$$\widehat{\mathscr{P}}_s(\omega\varphi) = \mathbb{E}\{e^{j\omega\langle\varphi,s\rangle}\} = \widehat{p}_{\langle\varphi,s\rangle}(\omega)$$

and

$$\widehat{\mathscr{P}}_w(\omega U\varphi) = \mathbb{E}\{e^{j\omega\langle U\varphi,w\rangle}\} = \widehat{p}_{\langle U\varphi,w\rangle}(\omega)$$

for generalized processes s and w, we deduce that the random variables $\langle\varphi, s\rangle$ and $\langle U\varphi, w\rangle$ have the same characteristic functions and therefore follow

$$\langle\varphi, s\rangle = \langle U\varphi, w\rangle \quad \text{in probability law.}$$

The manipulation that led to Proposition 3.10 shows that a similar relation exists, more generally, for any finite collection of observations $\langle\varphi_n, s\rangle$ and $\langle U\varphi_n, w\rangle$, $1 \leq n \leq N$, $N \in \mathbb{N}$.

Therefore, symbolically at least, by the definition of the adjoint $U^* : \mathscr{Y}' \to \mathscr{X}'$ of U, we may write

$$\langle \varphi, s \rangle = \langle \varphi, U^* w \rangle.$$

This seems to indicate that, in a sense, the random model s, which we have defined using (3.18), can be interpreted as the application of U^* to the original random model w. However, things are complicated by the fact that, unless $\mathscr{X}$ and $\mathscr{Y}$ are nuclear spaces, we may not be able to interpret w and s as random elements of $\mathscr{Y}'$ and $\mathscr{X}'$, respectively. Therefore the application of $U^* : \mathscr{Y}' \to \mathscr{X}'$ to s should be understood to be merely a formal construction.

On the other hand, by requiring $\mathscr{X}$ to be nuclear and $\mathscr{Y}$ to be either nuclear or completely normed, we see immediately that $\widehat{\mathscr{P}}_s : \mathscr{X} \to \mathbb{C}$ fulfills the requirements of the Minlos–Bochner theorem, and thereby defines a generalized random process with realizations in $\mathscr{X}'$.

The previous discussion suggests the following approach to defining generalized random processes: take a continuous, positive definite functional $\widehat{\mathscr{P}}_w : \mathscr{Y} \to \mathbb{C}$ on some (nuclear or completely normed) space $\mathscr{Y}$. Then, for any continuous operator U defined from a nuclear space $\mathscr{X}$ into $\mathscr{Y}$, the composition

$$\widehat{\mathscr{P}}_s = \widehat{\mathscr{P}}_w(U\cdot)$$

is the characteristic functional of a generalized random process s with realizations in $\mathscr{X}'$.

In subsequent chapters, we shall mostly focus on the situation where $U = L^{-1*}$ and $U^* = L^{-1}$ for some given (whitening) operator L that admits a continuous inverse in the suitable topology, the typical choice of spaces being $\mathscr{X} = \mathscr{S}(\mathbb{R}^d)$ and $\mathscr{Y} = L_p(\mathbb{R}^d)$. The underlying hypothesis is that one is able to invert the linear operator U and to recover w from s, which is formally written as $w = Ls$; that is,

$$\langle \varphi, w \rangle = \langle \varphi, Ls \rangle, \quad \text{for all } \varphi \in \mathscr{Y}.$$

The above ideas are summarized in Figure 3.1.

3.5.5   Innovation processes

In a certain sense, the most fundamental class of generalized random processes we can use to play the role of w in the construction of Section 3.5.4 are those with *independent values at every point* in $\mathbb{R}^d$ [GV64, Chapter 4, pp. 273–288]. The reason is that we can then isolate the spatio-temporal dependency of the probabilistic model in the *mixing operator* (U^* in Figure 3.1), and attribute randomness to independent contributions (*innovations*) at geometrically distinct points in the domain. We call such a construction an *innovation model*.

Let us attempt to make the notion of independence at every point more precise in the context of generalized stochastic processes, where the objects of study are, more accurately, not pointwise observations, but rather observations made through scalar products with test functions. To qualify a generalized process s as having independent values

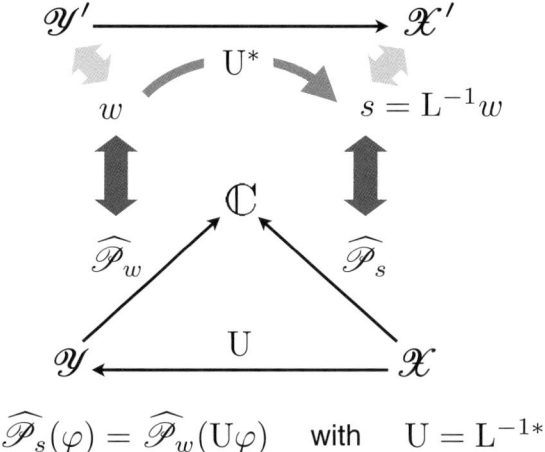

$$\widehat{\mathscr{P}_s}(\varphi) = \widehat{\mathscr{P}_w}(\mathrm{U}\varphi) \quad \text{with} \quad \mathrm{U} = \mathrm{L}^{-1*}$$

Figure 3.1 Definition of linear transformation of generalized stochastic processes using characteristic functionals. In this book, we shall focus on innovation models where w is a white noise process. The operator $\mathrm{L} = \mathrm{U}^{-1*}$ (if it exists) is called the *whitening operator* of s, since $\mathrm{L}s = w$.

at every point, we therefore require that the random variables $\langle \varphi_1, w \rangle$ and $\langle \varphi_2, w \rangle$ be independent whenever the test functions φ_1 and φ_2 have disjoint supports.

Since the joint characteristic function of independent random variables factorizes (is separable), we can formulate the above property in terms of the characteristic functional $\widehat{\mathscr{P}_w}$ of w as

$$\widehat{\mathscr{P}_w}(\varphi_1 + \varphi_2) = \widehat{\mathscr{P}_w}(\varphi_1)\,\widehat{\mathscr{P}_w}(\varphi_2).$$

An important class of characteristic functionals fulfilling this requirement are those that can be written in the form

$$\widehat{\mathscr{P}_w}(\varphi) = \mathrm{e}^{\int_{\mathbb{R}^d} f(\varphi(r))\,\mathrm{d}r}. \tag{3.19}$$

To have $\widehat{\mathscr{P}_w}(0) = 1$ (normalization), we require that $f(0) = 0$. The requirement of positive definiteness narrows down the class of admissible functions f much further, practically to those identified by the Lévy–Khintchine formula. This will be the subject of the greater part of our next chapter.

3.5.6   Example: filtered white Gaussian noise

In the above framework, we can define *white Gaussian noise* or *innovation* on $\mathbb{R}^d$ as a random element of the space of Schwartz' generalized functions, $\mathscr{S}'(\mathbb{R}^d)$, whose characteristic functional is given by

$$\widehat{\mathscr{P}_w}(\varphi) = \mathrm{e}^{-\frac{1}{2}\|\varphi\|_2^2}.$$

Note that this functional is a special instance of (3.19) with $f(\omega) = -\frac{1}{2}\omega^2$. The Gaussian appellation is justified by observing that, for any N test functions $\varphi_1, \ldots, \varphi_N$, the random variables $\langle \varphi_1, w \rangle, \ldots, \langle \varphi_N, w \rangle$ are jointly Gaussian. Indeed, we can apply Proposition 3.10 to obtain the joint characteristic function

$$\widehat{\mathscr{P}_{\varphi_1 : \varphi_N}}(\boldsymbol{\omega}) = \exp\left(-\frac{1}{2} \left\| \sum_{n=1}^{N} \omega_n \varphi_n \right\|_2^2 \right).$$

By taking the inverse Fourier transform of the above expression, we find that the random variables $\langle \varphi_n, w \rangle$, $n = 1, \ldots, N$, have a multivariate Gaussian distribution with mean $\mathbf{0}$ and covariance matrix with entries

$$C_{mn} = \langle \varphi_m, \varphi_n \rangle.$$

The independence of $\langle \varphi_1, w \rangle$ and $\langle \varphi_2, w \rangle$ is obvious whenever φ_1 and φ_2 have disjoint support. This justifies calling the process *white*. [9] In this special case, even mere orthogonality of φ_1 and φ_2 is enough for independence, since for $\varphi_1 \perp \varphi_2$ we have $C_{mn} = 0$.

From (3.16) and (3.17), we also find that w has 0 mean and "correlation function" $R_w(\boldsymbol{r}, \boldsymbol{s}) = \delta(\boldsymbol{r} - \boldsymbol{s})$, which should also be familiar. In fact, this last expression is sometimes used to formally "define" white Gaussian noise.

A *filtered* white Gaussian noise is obtained by applying a continuous convolution (i.e., LSI) operator $U^* : \mathscr{S}' \to \mathscr{S}'$ to the Gaussian innovation, in the sense described in Section 3.5.4.

Let us denote the convolution kernel of the operator $U : \mathscr{S} \to \mathscr{S}$ (the adjoint of U^*) by h. [10] The convolution kernel of $U^* : \mathscr{S}' \to \mathscr{S}'$ is then $h^\vee$. Following Section 3.5.4, we find the following characteristic functional for the filtered process $U^* w = h^\vee * w$:

$$\widehat{\mathscr{P}_{U^* w}}(\varphi) = e^{-\frac{1}{2} \| h * \varphi \|_2^2}.$$

In turn, it yields the following mean and correlation functions

$$m_{U^* w}(\boldsymbol{r}) = 0$$
$$R_{U^* w}(\boldsymbol{r}, \boldsymbol{s}) = \left(h * h^\vee \right)(\boldsymbol{r} - \boldsymbol{s}),$$

as expected.

3.6   Bibliographical pointers and historical notes

Sections 3.1 and 3.2
Recommended references on functional analysis, topological vector spaces, and duality are the books by Schaefer [Sch99] and Rudin [Rud73].

Much of the theory of nuclear spaces was developed by Grothendieck [Gro55] in his thesis work under the direction of Schwartz. For detailed information, we refer to Pietsch [Pie72].

---

[9] Our notion of *whiteness* in this book goes further than having a white spectrum. By whiteness we mean that the process is stationary and has truly independent (not merely uncorrelated) values over disjoint sets.

[10] Recall that, for the convolution to map back into $\mathscr{S}$, h needs to have a smooth Fourier transform, which implies rapid decay in the temporal or spatial domain. This is the case, in particular, for any rational transfer function that lacks purely imaginary poles.

Section 3.3

For a comprehensive treatment of generalized functions, we recommend the books of Gelfand and Shilov [GS64] and Schwartz [Sch66] (the former being more accessible while maintaining rigor). The results on Fourier multipliers are covered by Hörmander [Hör80] and Mikhlin *et al.* [MP86].

A historical precursor to the theory of generalized functions is the "operational method" of Heaviside, appearing in his collected works in the last decade of the nineteenth century [Hea71]. The introduction of the Lebesgue integral was a major step that gave a precise meaning to the concept of the almost-everywhere equivalence of functions. Dirac introduced his eponymous distribution as a convenient notation in the 1920s. Sobolev [Sob36] developed a theory of generalized functions in order to define weak solutions of partial differential equations. But it was Laurent Schwartz [Sch66] who put forth the formal and comprehensive theory of generalized functions (distributions) as we use it today (first edition published in 1950). His work was further developed and exposed by the Russian school of Gelfand *et al.*

Section 3.4

Kolmogorov is the founding father of the modern axiomatic theory of probability which is based on measure theory. We still recommend his original book [Kol56] as the main reference for the material presented here. Newer and more advanced results can be found in the encyclopedic works of Bogachev [Bog07] and Fremlin [Fre00, Fre01, Fre02, Fre03, Fre08] on measure theory.

Paul Lévy defined the characteristic function in the early 1920s and is responsible for turning the Fourier–Stieltjes apparatus into one of the most useful tools of probability theory [Lév25, Tay75]. The foundation of the finite-dimensional Fourier approach is Bochner's theorem, which appeared in 1932 [Boc32].

Interestingly, it was Kolmogorov himself who introduced the characteristic functional in 1935 as an equivalent (infinite-dimensional) Fourier-based description of a measure on a Banach space [Kol35]. This tool then lay dormant for many years. The theoretical breakthrough came when Minlos proved the equivalence between this functional and the characterization of probability measures on duals of nuclear spaces (Theorem 3.9) – as hypothesized by Gelfand [Min63, Kol59]. This powerful framework now constitutes the infinite-dimensional counterpart of the traditional Fourier approach to probability theory.

What is lesser known is that Laurent Schwartz, who also happened to be Paul Lévy's son-in-law, revisited the theory of probability measures on infinite-dimensional topological vector spaces, including developments from the French school, in the final years of his career [Sch73b, Sch81b]. These later works are highly abstract, as one may expect from their author. This makes for an interesting contrast with Paul Lévy, who had a limited interest in axioms and whose research was primarily guided by an extraordinary intuition.

Section 3.5

The concept of generalized stochastic processes, including the characterization of continuously defined white noises, was introduced by Gelfand in 1955 [Gel55]. Itô contributed to the topic by formulating the correlation theory of such processes [Itô54]; see also [Itô84]. The basic reference for the material presented here is [GV64, Chapter 3].

The first applications of the characteristic functional to the study of stochastic processes have been traced back to 1947; they are due to Le Cam [LC47] and Bochner [Boc47], who both appear to have (re)discovered the tool independently. Le Cam was concerned with the practical problem of modeling the relation between rainfall and riverflow, while Bochner was aiming at a fundamental characterization of stochastic processes. Another early promoter is Bartlett, who, in collaboration with Kendall, determined the characteristic functional of several Poisson-type processes that are relevant to biology and physics [BK51]. The framework was consolidated by Gelfand [Gel55] and Minlos in the 1960s. They provided the extension to generalized functions and also addressed the fundamental issue of the uniqueness and consistency of this infinite-dimensional description.

4 Continuous-domain innovation models

The stochastic processes that we wish to characterize are those generated by linear transformation of non-Gaussian white noise. If we were operating in the discrete domain and restricting ourselves to a finite number of dimensions, we would be able to use any sequence of i.i.d. random variables w_n as system input and rely on conventional multivariate statistics to characterize the output. This strongly suggests that the specification of the mixing matrix (L^{-1}) and the probability density function (pdf) of the innovation is sufficient to obtain a complete description of a linear stochastic process, at least in the discrete setting.

But our goal is more ambitious since we place ourselves in the context of continuously defined processes. The situation is then not quite as straightforward because: (1) we are dealing with infinite-dimensional objects, (2) it is much harder to properly define the notion of continuous-domain white noise, and (3) there are theoretical restrictions on the class of admissible innovations. While this calls for an advanced mathematical machinery, the payoff is that the continuous-domain formalism lends itself better to analytical computations, by virtue of the powerful tools of functional and harmonic analysis. Another benefit is that the non-Gaussian members of the family are necessarily sparse as a consequence of the theory which rests upon the powerful characterization and existence theorems by Lévy–Khintchine, Minlos, Bochner, and Gelfand–Vilenkin.

As in the subsequent chapters, we start by providing some intuition in the first section and then proceed with a more formal characterization. Section 4.2 is devoted to an in-depth investigation of Lévy exponents, which are intimately tied to the family of infinitely divisible distributions in the classical (scalar) theory of probability. What is non-standard here and fundamental to our argument is the link that is made between infinite divisibility and sparsity in Section 4.2.3. In Section 4.3, we apply those results to the Fourier-domain characterization of a multivariate linear model driven by an infinitely divisible noise vector, which primarily serves as preparation for the subsequent infinite-dimensional generalization. In Section 4.4, we extend the formulation to the continuous domain, which results in the proper specification of white Lévy noise w (or non-Gaussian innovations) as a generalized stochastic process (in the sense of Gelfand and Vilenkin) with independent "values" at every point. The fundamental result is that a given brand of noise (or innovations) is uniquely specified by its Lévy exponent $f(\omega)$ via its characteristic functional $\widehat{\mathscr{P}}_w(\varphi)$. Finally, in Section 4.5, we characterize the statistical effect of the mixing operator L^{-1} (general linear model) and provide mathematical conditions on f and L that ensure that the resulting process $s = \mathrm{L}^{-1}w$ is well defined mathematically.

4.1   Introduction: from Gaussian to sparse probability distributions

Intuitively, a continuous-domain white-noise process is formed by the juxtaposition of a continuum of i.i.d. random contributions. Since these atoms of randomness are infinitesimal, the realizations (a.k.a. sample paths) of such processes are highly singular (discontinuous), meaning that they do not admit a classical interpretation as (random) functions of the index variable $r \in \mathbb{R}^d$. Consequently, the random variables associated with the sample values $w(r_0)$ are undefined. The only concrete way of observing such noises is by probing them through some localized analysis window $\varphi(\cdot - r_0)$ centered around some location r_0. This produces some scalar quantity $X = \langle \varphi(\cdot - r_0), w \rangle$ which is a conventional random variable with some pdf $p_{X(\varphi)}$. Note that $p_{X(\varphi)}$ is independent of the position r_0, which reflects the fact that w is stationary. In order to get some sense of the variety of achievable random patterns, we propose to convert the continuous-domain process w into some corresponding i.i.d. sequence X_k (discrete white noise) by selecting a sequence of non-overlapping rectangular windows:

$$X_k = \langle \mathrm{rect}(\cdot - k), w \rangle.$$

The concept is illustrated in Figure 4.1. The main point that will be made clearer in what follows is that there is a one-to-one correspondence between the pdf of X_k – the so-called canonical pdf $p_{\mathrm{id}}(x) = p_{X(\mathrm{rect})}(x)$ – and the complete functional description of w via its characteristic functional, which we shall investigate in Section 4.4. What is more remarkable (and distinct from the discrete setting) is that this canonical pdf cannot be arbitrary: the theory dictates that it must be part of the family of *infinitely divisible* (id) laws (see Section 4.2).

  The prime example of an id pdf is the Gaussian distribution illustrated in Figure 4.1a. As already mentioned, it is the only non-sparse member of the family. All other distributions exhibit either a mass density at the origin (like the compound-Poisson example in Figure 4.1c with $\mathrm{Prob}(x = 0) = e^{-\lambda} = 0.75$ and Gaussian amplitude distribution) or a slower rate of decay at infinity (heavy-tail behavior). The Laplace probability law of Figure 4.1b results in the mildest possible form of sparsity – indeed, it can be proven that there is a gap between the Gaussian and the other members of the family in the sense that there is no id distribution with $p(x) = e^{-O(|x|^{1+\epsilon})}$ with $0 < \epsilon < 1$. In other words, a non-Gaussian $p_{\mathrm{id}}(x)$ is constrained to decay like $e^{-\lambda|x|}$ or slower – typically, like $O(1/|x|^r)$ with $r > 1$ (inverse polynomial/algebraic decay). The sparsest example in Figure 4.1 is provided by the Cauchy distribution $p_{\mathrm{Cauchy}}(x) = \frac{1}{\pi(x^2+1)}$, which is part of the symmetric-alpha-stable (SαS) family (here, $\alpha = 1$). The SαS distributions with $\alpha \in (0, 2)$ are notorious for their heavy-tail behavior and the fact that their moments $\mathbb{E}\{|x|^p\}$ are unbounded for $p > \alpha$.

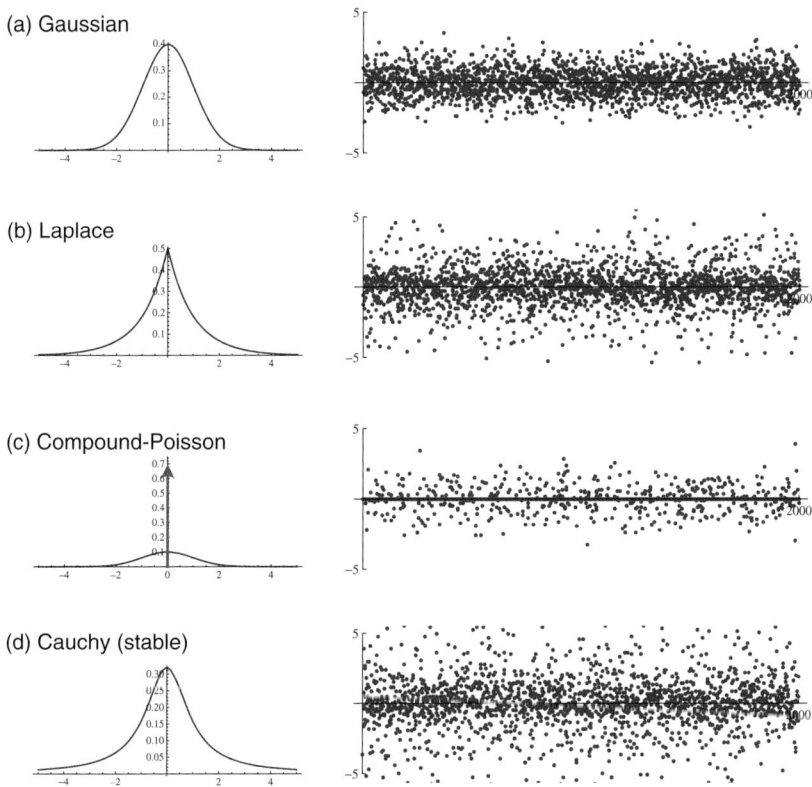

Figure 4.1 Examples of canonical, infinitely divisible probability density functions and corresponding observations of a continuous-domain white-noise process through an array of non-overlapping rectangular integration windows. (a) Gaussian distribution (not sparse). (b) Laplace distribution (moderately sparse). (c) Compound-Poisson distribution (finite rate of innovation). (d) Cauchy distribution (ultra-sparse = heavy-tailed with unbounded variance).

4.2 Lévy exponents and infinitely divisible distributions

The investigation of sparse stochastic processes requires a solid understanding of the classical notions of Lévy exponents and infinite divisibility, which constitute the pillars of our formulation. This section provides a self-contained presentation of the required mathematical background. It also brings out the link with sparsity.

DEFINITION 4.1 (Lévy exponent) A continuous, complex-valued function $f : \mathbb{R} \to \mathbb{C}$ such that $f(0) = 0$ is a valid Lévy exponent if and only if it is conditionally positive definite of order one, so that

$$\sum_{m=1}^{N} \sum_{n=1}^{N} f(\omega_m - \omega_n) \xi_m \overline{\xi}_n \geq 0$$

under the condition $\sum_{m=1}^{N} \xi_m = 0$ for every possible choice of $\omega_1, \ldots, \omega_N \in \mathbb{R}, \xi_1, \ldots, \xi_N \in \mathbb{C}$ and $N \in \mathbb{Z}^+$.

The importance of Lévy exponents in mathematical statistics is that they are tightly linked with the property of infinite divisibility.

DEFINITION 4.2 (Infinite divisibility)   A random variable X with generic pdf p_{id} is *infinitely divisible* (id) if and only if, for any $N \in \mathbb{Z}^+$, there exist i.i.d. random variables $X_1, \ldots, X_N$ such that X has the same distribution as $X_1 + \cdots + X_N$.

The foundation of the theory of such random variables is that their characteristic functions are in one-to-one correspondence with Lévy exponents. While the better-known formulation of this equivalence is provided by the Lévy–Khintchine theorem (Theorem 4.2), we prefer first to express it in functional terms, building upon the work of three giants in harmonic analysis: Lévy, Bochner, and Schoenberg.

THEOREM 4.1 (Lévy–Schoenberg)   *Let $\widehat{p}_{\mathrm{id}}(\omega) = \mathbb{E}\{e^{j\omega X}\} = \int_{\mathbb{R}} e^{j\omega x} p_{\mathrm{id}}(x)\, dx$ be the characteristic function of an infinitely divisible (id) random variable X. Then,*

$$f(\omega) = \log \widehat{p}_{\mathrm{id}}(\omega)$$

is a Lévy exponent in the sense of Definition 4.1. Conversely, if f is a valid Lévy exponent, then the inverse Fourier integral

$$p_{\mathrm{id}}(x) = \int_{\mathbb{R}} e^{f(\omega)} e^{-j\omega x} \frac{d\omega}{2\pi}$$

yields the pdf of an id random variable.

The proof is given in the supplementary material in Section 4.2.4. As for the latter implication, we observe that the condition $f(0) = 0 \Leftrightarrow \widehat{p}_{\mathrm{id}}(0) = 1$ implies that $\int_{\mathbb{R}} p_{\mathrm{id}}(x)\, dx = 1$, while the positive definiteness ensures that $p_{\mathrm{id}}(x) \geq 0$ so that it is a valid pdf.

4.2.1   Canonical Lévy–Khintchine representation

The second, more explicit statement of the announced equivalence with id distributions capitalizes on the property that Lévy exponents admit a canonical representation in terms of a Lévy measure μ_v or some equivalent density v, which is the notational choice [1] that we are favoring here.

DEFINITION 4.3 (Lévy measure/density)   A (positive) measure μ_v on $\mathbb{R}\backslash\{0\}$ is called a *Lévy measure* if it satisfies the admissibility condition

$$\int_{\mathbb{R}\backslash\{0\}} \min(a^2, 1)\mu_v(da) = \int_{\mathbb{R}\backslash\{0\}} \min(a^2, 1)v(a)\, da < \infty. \tag{4.1}$$

The corresponding density function $v : \mathbb{R} \to \mathbb{R}^+$, which is such that $\mu_v(da) = v(a)\, da$, is called the Lévy density.

---

[1] In most mathematical texts, the Lévy–Khintchine decomposition is formulated in terms of a Lévy measure μ_v rather than a density. Even though Lévy measures need not always have a density in the sense of the Radon–Nikodym derivative with respect to the Lebesgue measure (i.e., as an ordinary function), following Bourbaki we may still identify them with positive linear functionals, which we represent notationally as integrals against a "generalized" density: $\mu_v(E) = \int_E v(a)\, da$ for any set in the Borel algebra on $\mathbb{R}\backslash\{0\}$.

We observe that, as in the case of a pdf, the density v is not necessarily an ordinary function, for it may include isolated Dirac impulses (discrete part of the measure) as well as a singular component.

THEOREM 4.2 (Lévy–Khintchine) *A probability distribution p_{id} is infinitely divisible (id) if and only if its characteristic function can be written as*

$$\widehat{p}_{id}(\omega) = \int_{\mathbb{R}} p_{id}(x) e^{j\omega x}\, dx = \exp\big(f(\omega)\big), \tag{4.2}$$

with

$$f(\omega) = jb_1'\omega - \frac{b_2\omega^2}{2} + \int_{\mathbb{R}\setminus\{0\}} \big(e^{ja\omega} - 1 - ja\omega \mathbb{1}_{|a|<1}(a)\big) v(a)\, da, \tag{4.3}$$

where $b_1' \in \mathbb{R}$ and $b_2 \in \mathbb{R}^+$ are some arbitrary constants, and where v is an admissible Lévy density; $\mathbb{1}_{|a|<1}(a)$ is the indicator function of the set $\Omega = \{a \in \mathbb{R} : |a| < 1\}$ (i.e., $\mathbb{1}_\Omega(a) = 1$ if $a \in \Omega$ and $\mathbb{1}_\Omega(a) = 0$ otherwise).

The admissibility condition (4.1) guarantees that the right-hand integral in (4.3) is well defined; this follows from the bounds $|e^{ja\omega} - 1 - ja\omega| < a^2\omega^2$ and $|e^{ja\omega} - 1| < \min(|a\omega|, 2)$. An important aspect of the theory is that this allows for (non-integrable) Lévy densities with a singular behavior around the origin; for instance, $v(a) = O(1/|a|^{2+\epsilon})$ with $\epsilon \in [0, 1)$ as $a \to 0$.

The connection with Theorem 4.1 is that the Lévy–Khintchine expansion (4.3) provides a complete characterization of the conditionally positive definite functions of order one, as specified in Definition 4.1. This theme is further developed in Appendix B, which contains the proof of the above statement and also makes interesting links with theoretical results that are fundamental to machine learning and approximation theory.

In Section 4.4, we shall indicate how id distributions (or, equivalently, Lévy exponents) can be used to specify an extended family of continuous-domain white-noise processes. In that context, we shall typically require that p_{id} has a well-defined first-order absolute moment and/or that it is symmetric with respect to the origin, which leads to the following simplifications of the canonical representation.

COROLLARY 4.3 *Let p_{id} be an infinitely divisible pdf whose characteristic function is given by $\widehat{p}_{id}(\omega) = e^{f(\omega)}$. Then, depending on the properties of p_{id} (or, equivalently, on the Lévy density v), the Lévy exponent f admits the following Lévy–Khintchine-type representations:*

(1) p_{id} id symmetric $\big($i.e., $p_{id}(x) = p_{id}(-x)\big)$ if and only if

$$f(\omega) = -\frac{b_2\omega^2}{2} + \int_{\mathbb{R}\setminus\{0\}} \big(\cos(a\omega) - 1\big) v(a)\, da \tag{4.4}$$

with $v(a) = v(-a)$.

(2) p_{id} id with $\int_{\mathbb{R}} |x|\, p_{id}(x)\, dx < \infty$ if and only if

$$f(\omega) = jb_1''\omega - \frac{b_2\omega^2}{2} + \int_{\mathbb{R}\setminus\{0\}} \big(e^{ja\omega} - 1 - ja\omega\big) v(a)\, da \tag{4.5}$$

with $\int_{|a|>1} |a|\, v(a)\, da < \infty$.

(3) p_{id} *id with* $\int_{\mathbb{R}\backslash\{0\}}|a|v(a)\,\mathrm{d}a < \infty$ *if and only if*

$$f(\omega) = \mathrm{j}b_1\omega - \frac{b_2\omega^2}{2} + \int_{\mathbb{R}\backslash\{0\}}\left(\mathrm{e}^{\mathrm{j}a\omega}-1\right)v(a)\,\mathrm{d}a \tag{4.6}$$

where $b_1 \in \mathbb{R}$, $b_2 \in \mathbb{R}^+$, $b_1 = b_1'' - \int_{\mathbb{R}\backslash\{0\}}a\,v(a)\,\mathrm{d}a$ *and* $v(a) \geq 0$ *is an admissible Lévy density.*

These are obtained by direct manipulation of (4.3) with $b_1'' = b_1' + \int_{|a|\geq 1}a\,v(a)\,\mathrm{d}a$. Equation (4.4) is valid in all generality, provided that we interpret the integral as a Cauchy principal-value limit (see Appendix A) to handle potential (symmetric) singularities around the origin. The Lévy–Khintchine formulas (4.4) and (4.5) are fundamental because they give an explicit, constructive characterization of the noise functionals that are central to our formulation. From now on, we rely on Corollary 4.3 to specify admissible Lévy exponents: the parameters (b_1, b_2, v) will be referred to as the *Lévy triplet* of $f(\omega)$.

For completeness, we also mention the less classical (but equivalent) representation of the Lévy exponent as

$$f(\omega) = \mathrm{j}b_1\omega - \frac{b_2\omega^2}{2} + \widehat{v}(\omega) - \widehat{v}(0), \tag{4.7}$$

where $\widehat{v}(\omega) = \overline{\mathscr{F}\{v\}}(\omega)$ is the (conjugate) Fourier transform of v in the sense of generalized functions. The idea here is to rely on the powerful theory of generalized functions to seamlessly absorb the (potential) singularity [2] of v at $a = 0$. The interested reader can refer to Appendix A for complementary explanations.

Below is a summary of known criteria for identifying admissible Lévy exponents, some being more operational than others [GV64, pp. 275–282]. These are all consequences of Bochner's theorem, which provides a Fourier-domain equivalence between continuous, positive definite functions and probability density functions (or positive Borel measures). See Appendix B for an overview and discussion of the functional notion of positive definiteness and corresponding mathematical tools.

PROPOSITION 4.4   *The following statements on f are equivalent:*

(1) $f(\omega)$ is a continuous, conditionally positive definite function of order one.
(2) $p_{\mathrm{id}}(x) = \mathscr{F}^{-1}\{\mathrm{e}^{f(\omega)}\}(x)$ is an infinitely divisible distribution.
(3) f admits a Lévy–Khintchine representation as in Theorem 4.2.
(4) Let $p_{X_\tau}(x) = \mathscr{F}^{-1}\{\mathrm{e}^{\tau f(\omega)}\}$ for $\tau \geq 0$. Then, $\{p_{X_\tau}\}_{\tau\in\mathbb{R}^+}$ is a family of valid pdfs; that is, $p_{X_\tau}(x) \geq 0$ and $\int_{\mathbb{R}}p_{X_\tau}(x)\,\mathrm{d}x = 1$ for all $\tau \geq 0$.
(5) $\widehat{p}_{X_\tau}(\omega) = \mathrm{e}^{\tau f(\omega)}$ is a continuous, positive definite function of $\omega \in \mathbb{R}$ with $\widehat{p}_{X_\tau}(0) = 1$ for any $\tau \in [0, \infty)$.

Interestingly, it was Schoenberg (the father of splines) who first established the equivalence between Statements (1) and (5) (see proof of the direct part in Section 4.2.4).

---

[2] This corresponds to interpreting v as the generalized function associated with the finite part (p.f., i.e., Hadamard's "partie finie") of the classical Lévy measure: $\langle\varphi, v\rangle = \mathrm{p.f.}\int_{\mathbb{R}}\varphi(a)\mu_v(\mathrm{d}a)$.

The equivalence between (4) and (5) follows from Bochner's theorem (Theorem 3.7). The fact that (2) implies (4) is a side product of the proof in Section 4.2.4, while the converse implication is a direct consequence of (3). Indeed, if $f(\omega)$ has a Lévy–Khintchine representation, then the same is true for $\tau f(\omega)$, which also implies that the whole family of pdfs $\{p_{X_\tau}\}_{\tau \in \mathbb{R}^+}$ is infinitely divisible. The latter is uniquely specified by f and therefore in one-to-one correspondence with the canonical id distribution $p_{id}(x) = p_{X_\tau}(x)|_{\tau=1}$. Another important observation is that $\widehat{p}_{X_\tau}(\omega) = \left(e^{f(\omega)}\right)^\tau = \left(\widehat{p}_{id}(\omega)\right)^\tau$, so that p_{X_τ} in Statement (4) may be interpreted as the τ-fold (possibly, fractional) convolution of p_{id}.

In our work, we sometimes need to limit ourselves to some particular subset of Lévy exponents.

DEFINITION 4.4 (*p*-admissibility)   A Lévy exponent f with derivative f' is called *p-admissible* if it satisfies the inequality

$$|f(\omega)| + |\omega|\,|f'(\omega)| \leq C\,|\omega|^p \tag{4.8}$$

for some constant $C > 0$ and $0 < p \leq 2$.

PROPOSITION 4.5   *The generic Lévy exponents*

(1) $f_1(\omega) = \int_{\mathbb{R}\setminus\{0\}} \left(e^{ja\omega} - 1\right) v_1(a)\, da$ with $\int_{\mathbb{R}} |a| v_1(a)\, da < \infty$
(2) $f_2(\omega) = \int_{\mathbb{R}\setminus\{0\}} \left(\cos(a\omega) - 1\right) v_2(a)\, da$ with $\int_{\mathbb{R}} a^2 v_2(a)\, da < \infty$
(3) $f_3(\omega) = \int_{\mathbb{R}\setminus\{0\}} \left(e^{ja\omega} - 1 - ja\omega\right) v_3(a)\, da$ with $\int_{\mathbb{R}} a^2 v_3(a)\, da < \infty$

are p-admissible with $p_1 = 1$, $p_2 = 2$, *and* $p_3 = 2$, *respectively.*

Proof   The first result follows from the bounds $|e^{ja\omega} - 1| \leq |a| \cdot |\omega|$ and $|\frac{de^{ja\omega}}{d\omega}| < |a|$. The second is based on $|\cos(a\omega) - 1| \leq |a\omega|^2$ and $|\sin(a\omega)| \leq |a\omega|$. Specifically,

$$|f_2(\omega)| \leq \int_{\mathbb{R}} |a\omega|^2 v_2(a)\, da = |\omega|^2 \int_{\mathbb{R}} a^2 v_2(a)\, da$$

$$|\omega||f_2'(\omega)| = |\omega| \left| \int_{\mathbb{R}\setminus\{0\}} a \sin(a\omega) v_2(a)\, da \right|$$

$$\leq |\omega| \int_{\mathbb{R}} |a|\,|a\omega| v_2(a)\, da = |\omega|^2 \int_{\mathbb{R}} a^2 v_2(a)\, da.$$

As for the third exponent, we also use the inequality $|e^{ja\omega} - 1 - ja\omega| \leq |a\omega|^2$, which yields

$$|f_3(\omega)| \leq \int_{\mathbb{R}\setminus\{0\}} |a\omega|^2 v_3(a)\, da = |\omega|^2 \int_{\mathbb{R}} a^2 v_3(a)\, da$$

$$|\omega|\,|f_3'(\omega)| = |\omega| \left| \int_{\mathbb{R}\setminus\{0\}} ja(e^{ja\omega} - 1) v_3(a)\, da \right|$$

$$\leq |\omega| \int_{\mathbb{R}} |a|\,|a\omega| v_3(a)\, da = |\omega|^2 \int_{\mathbb{R}} a^2 v_3(a)\, da,$$

which completes the proof. □

Since the p-admissibility property is preserved through summation, this covers a large portion of the Lévy exponents specified in Corollary 4.3.

Examples

The power law $f_\alpha(\omega) = -|\omega|^\alpha$ with $0 < \alpha \le 2$ is Lévy α-admissible; it generates the SαS id distributions [Fel71]. Note that $f_\alpha(\omega)$ fails to be conditionally positive definite for $\alpha > 2$, meaning that the inverse Fourier transform of $e^{-|\omega|^\alpha}$ exhibits negative values and is therefore not a valid pdf. The upper acceptable limit is $\alpha = 2$ and corresponds to the Gaussian law. More generally, a Lévy exponent that is symmetric and twice-differentiable at the origin (which is equivalent to the variance of the corresponding id distribution being finite) is p-admissible with $p = 2$; this follows as a direct consequence of Corollary 4.3 and Proposition 4.5.

Another fundamental instance, which generates the complete family of id compound-Poisson distributions, is

$$f_{\text{Poisson}}(\omega) = \lambda \int_{\mathbb{R}\setminus\{0\}} \left(e^{j a \omega} - 1\right) p_A(a)\, da,$$

where $\lambda > 0$ is the Poisson rate and $p_A(a) \ge 0$ the amplitude pdf with $\int_{\mathbb{R}} p_A(a)\, da = 1$. In general, $f_{\text{Poisson}}(\omega)$ is p-admissible with $p = 1$ provided that $\mathbb{E}\{|A|\} = \int_{\mathbb{R}} |a| p_A(a)\, da < \infty$ (see Proposition 4.5). If, in addition, p_A is symmetric with a bounded variance, then the Poisson range of admissibility extends to $p \in [1,2]$. Further examples of symmetric id distributions are documented in Table 4.1. Their Lévy exponent is simply obtained by taking $f(\omega) = \log \widehat{p}_X(\omega)$.

The relevance of id distributions for signal processing is that any linear combination of independent id random variables is id as well. Indeed, let X_1 and X_2 be two independent id random variables with Lévy exponents f_1 and f_2, respectively; then, it is not difficult to show that $a_1 X_1 + a_2 X_2$, where a_1 and a_2 are arbitrary constants, is id with Lévy exponent $f(\omega) = f_1(a_1\omega) + f_2(a_2\omega)$.

4.2.2 Deciphering the Lévy–Khintchine formula

From a harmonic-analysis perspective, the Lévy–Khintchine representation is closely related to Bochner's theorem, which states that a positive definite function g can always be expressed as the Fourier transform of a positive finite Borel measure; i.e., $g(\omega) = \int_{\mathbb{R}} e^{j\omega a} v(a)\, da$ with $v(a) \ge 0$ and $g(0) = \int_{\mathbb{R}} v(a)\, da < \infty$. Here, the additional requirement is that $f(0) = 0$ (conditional positive definiteness of order one), which is enforced by proper subtraction of a linear correction in $j a \omega$ within the integral, the latter being partly compensated by the addition of the component $j \omega b_1$. The side benefit of this regularization is that it enlarges the set of admissible densities to those satisfying $\int_{\mathbb{R}} \min(a^2, 1) v(a)\, da < \infty$, which allows for a singular behavior around the origin. As for the linear and quadratic terms outside the integral, they map into the singular point distribution $b_1 \delta'(a) + b_2 \delta''(a)$ (weighted derivatives of Dirac impulses) that is concentrated at the origin $a = 0$ and excluded from the (classical) Lebesgue integral. For the complete details, we refer the reader to Appendix B. The proposed treatment

Table 4.1 Primary families of symmetric, infinitely divisible probability laws. The special functions $K_\alpha(x)$, $\Gamma(z)$, and $B(x,y)$ are defined in Appendix C.

Type	$p_X(x)$	Var(X)	$\hat{p}_X(\omega) = \int_\mathbb{R} p_X(x)e^{j\omega x}\,dx$	Lévy density $v(a)$
Gaussian	$\dfrac{1}{\sqrt{2\pi\sigma^2}}e^{-\frac{x^2}{2\sigma^2}}$	σ^2	$e^{-\frac{\omega^2\sigma^2}{2}}$	N/A
Compound Poisson	$e^{-\lambda}\delta(x) + (1-e^{-\lambda})p_{A,\lambda}(x)$		$e^{\lambda(\hat{p}_A(\omega)-1)}$ $\exp\left(\lambda\int_\mathbb{R}(e^{j\omega a}-1)p_A(a)\,da\right)$	$\lambda\,p_A(a)\in L_1(\mathbb{R})$
Laplace ($\lambda\in\mathbb{R}^+$)	$\dfrac{\lambda}{2}e^{-\lambda|x|}$	$\dfrac{2}{\lambda^2}$	$\dfrac{\lambda^2}{\lambda^2+\omega^2}$	$\dfrac{e^{-\lambda|a|}}{|a|}$
Symmetric gamma $r\in\mathbb{R}^+$	$\dfrac{2^{\frac{1}{2}-r}|x|^{r-\frac{1}{2}}K_{\frac{1}{2}-r}(|x|)}{\Gamma(r)\sqrt{\pi}}$	$2r$	$\left(\dfrac{1}{1+\omega^2}\right)^r$	$\dfrac{r\,e^{-|a|}}{|a|}$
Hyperbolic secant $\sigma_0\in\mathbb{R}^+$	$\dfrac{\text{sech}\left(\frac{\pi x}{2\sigma_0}\right)}{2\sigma_0}$	σ_0^2	$\text{sech}(\omega\sigma_0) = \dfrac{1}{\cosh(\omega\sigma_0)}$	$\dfrac{1}{2}\dfrac{1}{a\sinh\frac{\pi a}{2\sigma_0}}$
Meixner $r,s\in\mathbb{R}^+$	$2^{r-2}\dfrac{|\Gamma(\frac{r}{2}+j\frac{x}{2s})|^2}{s\pi\Gamma(r)}$	$s^2 r$	$\left(\dfrac{1}{\cosh(s\omega)}\right)^r$	$\dfrac{r}{2}\dfrac{1}{a\sinh\frac{\pi a}{2s}}$
Cauchy $s\in\mathbb{R}^+$	$\dfrac{1}{\pi}\dfrac{s}{s^2+x^2}$	N/A	$e^{-s|\omega|}$	$\dfrac{s}{\pi a^2}$
Sym Student $r\in\mathbb{R}^+$	$\dfrac{1}{B\left(r,\frac{1}{2}\right)}\left(\dfrac{1}{x^2+1}\right)^{r+\frac{1}{2}}$	$r\leq 1$: N/A $r>1$: $\frac{1}{2r-2}$	$\dfrac{\sqrt{\pi}2^{1-r}|\omega|^r K_{-r}(|\omega|)}{\Gamma\left(r+\frac{1}{2}\right)B\left(r,\frac{1}{4}\right)}$	unknown
SαS, $\alpha\in(0,2)$, $s\in\mathbb{R}^+$	$p_\alpha(x;\alpha,s)$	N/A	$e^{-|s\omega|^\alpha}$	$\dfrac{C_{\alpha,s}}{|a|^{1+\alpha}}$

relies on Gelfand and Vilenkin's distributional characterization of conditionally positive definiteness of order n in Theorem B.4. Despite the greater generality of the result, we find its proof more enlightening and of a less technical nature than the traditional derivation of the Lévy–Khintchine formula, which is summarized in Section 4.2.4, for completeness.

From a statistical perspective, the exponent f specified by the Lévy–Khintchine formula is the logarithm of the characteristic function of an id random variable. This means that breaking f into additive subparts is in fact equivalent to factorizing the pdf into convolutional factors. Specifically, let $\widehat{p}_X(\omega) = e^{\sum_{n=1}^N f_n(\omega)}$ be the characteristic function of a (compound) id distribution where the f_n are valid Lévy exponents. Then, $\widehat{p}_X(\omega) = \prod_{n=1}^N \widehat{p}_{X_n}(\omega)$ with $\widehat{p}_{X_n}(\omega) = e^{f_n(\omega)} = \mathbb{E}\{e^{j\omega X_n}\}$, which translates into the convolution relation

$$p_X(x) = \left(p_{X_1} * p_{X_2} * \cdots * p_{X_N}\right)(x).$$

The statistical interpretation is that $X = X_1 + \cdots + X_N$ where the X_n are independent with id pdf p_{X_n}. The *infinitely divisible* property simply translates into the fact that, for a given $f(\omega)$ and any $N > 0$, X can always be broken down into N i.i.d. components with Lévy exponent $f(\omega)/N$. Indeed, it is easy to see from the Lévy representation that the admissibility of $f(\omega)$ implies that $\tau f(\omega)$ is a valid Lévy exponent as well for any $\tau \geq 0$.

To further our understanding of id distributions, it is instructive to characterize the atoms of the Lévy–Khintchine representation. Focusing on the simplest form (4.6), we identify three types of elementary constituents, with the third type being motivated by the decomposition of a (continuous) Lévy density into a weighted "sum" of Dirac impulses: $v(a) = \int_{\mathbb{R}} v(\tau)\delta(a-\tau)\,d\tau \approx \sum_n \lambda_n \delta(a-\tau_n)$ with $\lambda_n = v(\tau_n)(\tau_n - \tau_{n-1})$:

(1) Linear term $f_1(\omega) = jb_1\omega$. This corresponds to the (degenerate) pdf of a constant $X_1 = b_1$ with $p_{X_1}(x) = \delta(x - b_1)$.

(2) Quadratic term $f_2(\omega) = \frac{-b_2\omega^2}{2}$. As already mentioned, this leads to the centered Gaussian with variance b_2 given by

$$p_{X_2}(x) = \mathcal{F}^{-1}\{e^{-b_2\omega^2/2}\}(x) = \frac{1}{\sqrt{2\pi b_2}}\exp\left(-\frac{x^2}{2b_2}\right).$$

(3) Exponential (or Poisson) term $f_3(\omega) = \lambda(e^{j\tau\omega} - 1)$, which is associated with the elementary Lévy triplet $(0, 0, \lambda\delta(a-\tau))$. Based on the Taylor-series expansion $\widehat{p}_{X_3}(\omega) = e^{\lambda(z-1)} = e^{-\lambda}\sum_{m=0}^{+\infty}\frac{(\lambda z)^m}{m!}$ with $z = e^{j\omega\tau}$, we readily obtain the pdf by (generalized) inverse Fourier transformation:

$$p_{X_3}(x) = \sum_{m=0}^{\infty}\frac{e^{-\lambda}\lambda^m}{m!}\delta(x - m\tau).$$

This formula coincides with the continuous-domain representation of a Poisson distribution [3] with Poisson parameter λ and gain factor τ; that is, $\text{Prob}(X_3 = \tau m) = \frac{e^{-\lambda}\lambda^m}{m!}$.

More generally, when $v(a) = \lambda p(a)$ where $p(a) \geq 0$ is some arbitrary pdf with $\int_{\mathbb{R}} p(a)\,da = 1$, we can make a compound-Poisson [4] interpretation with

$$f_{\text{Poisson}}(\omega) = \lambda\int_{\mathbb{R}}(e^{ja\omega} - 1)p(a)\,da = \lambda(\widehat{p}(\omega) - 1),$$

where $\widehat{p}(\omega) = \int_{\mathbb{R}} e^{ja\omega}p(a)\,da$ is the characteristic function of $p = p_A$. Using the fact that $\widehat{p}(\omega)$ is bounded, we apply the same type of Taylor-series argument and express the characteristic function as

$$e^{f_{\text{Poisson}}(\omega)} = e^{-\lambda}\sum_{m=0}^{\infty}\frac{(\lambda\widehat{p}(\omega))^m}{m!} = \widehat{p}_Y(\omega).$$

---

[3] The standard form of the discrete Poisson probability model is $\text{Prob}(N = n) = \frac{e^{-\lambda}\lambda^n}{n!}$ with $n \in \mathbb{N}$. It provides the probability of a given number of independent events (n) occurring in a fixed space–time interval when the average rate of occurrence is λ. The Poisson parameter is equal to the expected value of N, but also to its variance: $\lambda = \mathbb{E}\{N\} = \text{Var}\{N\}$.

[4] The compound-Poisson probability model has two components. The first is a random variable N that follows a Poisson distribution with parameter λ, and the second a series of i.i.d. random variables $A_1, A_2, A_3, \ldots$ with pdf p_A which are drawn at each trial of N. Then, $Y = \sum_{n=1}^{N} A_n$ is a compound-Poisson random variable with Poisson parameter λ and amplitude pdf p_A. Its mean and variance are given by $\mathbb{E}\{Y\} = \lambda\mathbb{E}\{A\}$ and $\text{Var}\{Y\} = \lambda\text{Var}(A)$, respectively.

Finally, by using the property that $\widehat{p}(\omega)^m$ is the characteristic function of the m-fold convolution of p, we get the general formula for the compound-Poisson pdf with Poisson parameter λ and amplitude distribution p:

$$p_Y(x) = e^{-\lambda}\left(\delta(x) + \frac{\lambda}{1!}p(x) + \frac{\lambda^2}{2!}(p*p)(x) + \frac{\lambda^3}{3!}(p*p*p)(x) + \cdots\right). \qquad (4.9)$$

Thus, in essence, the Lévy–Khintchine formula is a description of the Fourier transform of a distribution that is the convolution of three components: an impulse $\delta(\cdot - b_1)$ (shifting), a Gaussian of variance b_2 (smoothing), and a compound-Poisson distribution (spreading). The effect of the first term is a simple recentering of the pdf around b_1. The third compound-Poisson component is itself obtained via a suitable composition of m-fold convolutions of some primary pdf p. It is much more concentrated at the origin than the Gaussian, because of the presence of the Dirac distribution with weight $e^{-\lambda}$, but also heavier-tailed because of the spreading effect of the m-fold convolution.

The additional linear correction terms in (4.3) and (4.5) allow for a wider variety of distributions that have the common property of being limits of compound-Poisson distributions.

PROPOSITION 4.6 *Every id distribution is the weak limit of a sequence of Poisson distributions.*

Proof Let $\widehat{p}_{\mathrm{id}}$ be the characteristic function of some id distribution p_{id} and consider an arbitrary sequence $\tau_n \downarrow 0$. Then,

$$\widehat{p}_{X_n}(\omega) = \exp\left(\tau_n^{-1}\widehat{p}_{\mathrm{id}}(\omega)^{\tau_n} - 1\right)$$

is the characteristic function of a compound-Poisson distribution with parameter $\lambda = \tau_n^{-1}$ and amplitude distribution $p(x) = \overline{\mathcal{F}}^{-1}\{\widehat{p}_{\mathrm{id}}(\omega)^{\tau_n}\}(x)$. Moreover, we have that

$$\widehat{p}_{X_n}(\omega) = \exp\left(\tau_n^{-1}(e^{\tau_n \log \widehat{p}_{\mathrm{id}}(\omega)} - 1)\right) = \exp\left(\log \widehat{p}_{\mathrm{id}}(\omega) + O(\tau_n)\right)$$

for every ω as $n \to \infty$. Hence, $\widehat{p}_{X_n}(\omega) \to \exp(\log \widehat{p}_{\mathrm{id}}(\omega)) = \widehat{p}_{\mathrm{id}}(\omega)$ so that p_{X_n} converges weakly to p_{id} by Lévy's continuity theorem (Theorem 3.8). $\qquad \square$

The id pdfs for which $v \notin L_1(\mathbb{R})$ are generally smoother than the compound-Poisson ones for they do not display a singularity (Dirac impulse) at the origin, unlike (4.9). Yet, depending on the degree of concentration (or singularity) of v around the origin, they will typically exhibit a peaky behavior around the mean. While this class of distributions is responsible for the additional level of complication in the Lévy–Khintchine formula – as compared to the simpler Poisson version (4.6) – we argue that it is highly relevant for applications because of the many possibilities that it offers. Somewhat surprisingly, there are many families of id distributions with singular Lévy density that are more tractable mathematically than their compound-Poisson cousins found in Table 4.1, at least when considering their pdf.

4.2.3   Gaussian vs. sparse categorization

The family of id distributions allows for a range of behaviors that varies between the purely Gaussian and sparse extremes. In the context of Lévy processes, these are often referred to as the diffusive and jump modes. To make our point, we consider two distinct scenarios.

Finite-variance case

We first assume that the second moment $m_2 = \int_{\mathbb{R}} a^2 \, v(a) \, da$ of the Lévy density is finite, which also implies that $\int_{|a|>1} |a| v(a) \, da < \infty$ because of the admissibility condition. Hence, the corresponding Lévy–Khintchine representation is (4.5). An interesting non-Poisson example of infinitely divisible probability laws that falls into this category (with non-integrable v) is the Laplace density with Lévy triplet $\left(0, 0, v(a) = \frac{e^{-|a|}}{|a|}\right)$ and $p_{\mathrm{id}}(x) = \frac{1}{2}e^{-|x|}$ (see Figure 4.1b). This model is particularly relevant in the context of sparse signal recovery because it provides a tight connection between Lévy processes and total-variation regularization [UT11, Section VI].

Now, if the Lévy density is Lebesgue integrable, we can pull the linear correction out of the Lévy–Khintchine integral and represent f using (4.6) with $v(a) = \lambda p_A(a)$ and $\int_{\mathbb{R}} p_A(a) \, da = 1$. This implies that we can decompose X into the sum of two independent Gaussian and compound-Poisson random variables. The variances of the Gaussian and Poisson components are $\sigma^2 = b_2$ and $\lambda \mathbb{E}\{A^2\}$, respectively. The Poisson component is sparse because its probability density function exhibits the mass distribution $e^{-\lambda}\delta(x)$ at the origin shown in Figure 4.1c, meaning that the chances, for a continuous amplitude distribution, of getting zero are overwhelmingly higher than any other value, especially for smaller values of $\lambda > 0$. It is therefore justifiable to use $0 \leq e^{-\lambda} < 1$ as our Poisson sparsity index.

Infinite-variance case

We now turn our attention to the case where the second moment of the Lévy density is unbounded, which we like to label as "super-sparse." To justify this terminology, we invoke the Ramachandran–Wolfe theorem [Ram69, Wol71], which states that the pth moment $\mathbb{E}\{|x|^p\}$ with $p \in \mathbb{R}^+$ of an infinitely divisible distribution is finite if and only if $\int_{|a|>1} |a|^p \, v(a) \, da < \infty$ (see Section 9.5). For $p \geq 2$, the latter is equivalent to $\int_{\mathbb{R}} |a|^p \, v(a) \, da < \infty$ because of the Lévy admissibility condition. It follows that the cases that are not covered by the previous scenario (including the Gaussian + Poisson model) necessarily give rise to distributions whose moments of order p are unbounded for $p \geq 2$. The prototypical representatives of such heavy-tail distributions are the alpha-stable ones (see Figure 4.1d) or, by extension, the broad family of infinitely divisible probability laws that are in their domain of attraction. It has been shown that these distributions precisely fulfill the requirement for ℓ_p compressibility [AUM11], which is a stronger form of sparsity than the presence of a mass probability density at the origin.

4.2.4   Proofs of Theorems 4.1 and 4.2

For completeness, we end this section on Lévy exponents with the proofs of the two key theorems in the theory of infinitely divisible distributions. The Lévy–Schoenberg theorem is central to our formulation because it makes the link between the id property and the fundamental notion of positive definiteness. In the case of the Lévy–Khintchine theorem, we have opted for a sketch of proof which is adapted from the literature. The main intent there was to provide additional insights into the nature of the singularities of the Lévy density and their effect on the form of the exponent.

Proof of Theorem 4.1 (Lévy–Schoenberg)

Let $\widehat{p}_{id}(\omega) = \int_{\mathbb{R}} e^{j\omega x} p_{id}(x)\,dx$ be the characteristic function of an id random variable. Then, by definition, $\left(\widehat{p}_{id}(\omega)\right)^{1/n}$ is a valid characteristic function for any $n \in \mathbb{Z}^+$. Since the convolution of two pdfs is a pdf, we can also take integer powers, which results in $\left(\widehat{p}_{id}(\omega)\right)^{m/n}$ being a characteristic function. For any irrational number $\tau > 0$, we can specify a sequence of rational numbers m/n that converges to τ so that $\left(\widehat{p}_{id}(\omega)\right)^{m/n} \to \left(\widehat{p}_{id}(\omega)\right)^{\tau}$ with the limit function being continuous. This implies that $\widehat{p}_{X_\tau}(\omega) = \left(\widehat{p}_{id}(\omega)\right)^{\tau}$ is a characteristic function for any $\tau \geq 0$ by Lévy's continuity theorem (Theorem 3.8). Moreover, $\widehat{p}_{X_\tau}(x) = \left(\widehat{p}_{X_{\tau/s}}(\omega)\right)^{s}$ must be non-zero for any finite τ, owing to the fact that $\lim_{s \to \infty} \widehat{p}_{X_{\tau/s}}(\omega) = 1$. In particular, $\widehat{p}_{id}(\omega) = \widehat{p}_{X_\tau}(\omega)\big|_{\tau=1} \neq 0$ so that we can write it as $\widehat{p}_{id}(\omega) = e^{f(\omega)}$, where $f(\omega)$ is continuous with $\mathrm{Re}\left(f(\omega)\right) \leq 0$ and $f(0) = 0$. Hence, $\widehat{p}_{X_\tau}(\omega) = \left(\widehat{p}_{id}(\omega)\right)^{\tau} = e^{\tau f(\omega)} = \int_{\mathbb{R}} e^{j\omega x} p_{X_\tau}(x)\,dx$, where $p_{X_\tau}(x)$ is a valid pdf for any $\tau \in [0, \infty)$, which is Statement (4) in Proposition 4.4. By Bochner's theorem (Theorem 3.7), this is equivalent to $e^{\tau f(\omega)}$ being positive definite for any $\tau \geq 0$ with $f(\omega)$ continuous and $f(0) = 0$. The first part of Theorem 4.1 then follows as a corollary of the next fundamental result, which is due to Schoenberg.

THEOREM 4.7 (Schoenberg correspondence)   *The function $f(\omega)$ is conditionally positive definite of order one if and only if $e^{\tau f(\omega)}$ is positive definite for any $\tau > 0$.*

Proof   We only give the easier part (if statement) and refer to [Sch38, Joh66] for the complete details. The property that $e^{\tau f(\omega)}$ is positive definite for every $\tau > 0$ is expressed as

$$\sum_{m=1}^{N}\sum_{n=1}^{N} \xi_m \overline{\xi}_n e^{\tau f(\omega_m - \omega_n)} \geq 0,$$

for every possible choice of $\omega_1, \ldots, \omega_N \in \mathbb{R}$, $\xi_1, \ldots, \xi_N \in \mathbb{C}$, and $N \in \mathbb{Z}^+$. In the more restricted setup of Definition 4.1 where $\sum_{n=1}^{N} \xi_n = 0$, this can also be restated as

$$\frac{1}{\tau}\sum_{m=1}^{N}\sum_{n=1}^{N} \xi_m \overline{\xi}_n \left(e^{\tau f(\omega_m - \omega_n)} - 1\right) \geq 0.$$

The next step is to take the limit

$$\lim_{\tau \to 0} \sum_{m=1}^{N} \sum_{n=1}^{N} \xi_m \bar{\xi}_n \frac{e^{\tau f(\omega_m - \omega_n)} - 1}{\tau} = \sum_{m=1}^{N} \sum_{n=1}^{N} \xi_m \bar{\xi}_n f(\omega_m - \omega_n) \geq 0,$$

which implies that $f(\omega)$ is conditionally positive definite of order one. $\qquad\square$

This also makes the second part of Theorem 4.1 easy because $\widehat{p}_{id}(\omega) = e^{f(\omega)}$ can be factorized into a product of N identical positive definite subparts with Lévy exponent $\frac{1}{N} f(\omega)$.

Sketch of proof of Theorem 4.2 (Lévy–Khintchine)
We start from the equivalence of the id property and Statement (4) in Proposition 4.4 established above. This result is restated as

$$\frac{e^{\tau f(\omega)} - 1}{\tau} = \int_{\mathbb{R}} (e^{j\omega x} - 1) \frac{p_{X_\tau}(x)}{\tau} \, dx,$$

the limit of which as $\tau \to 0$ exists and is equal to $f(\omega)$. Next, we define the measure $K_\tau(dx) = \frac{x^2}{x^2 + 1} \frac{p_{X_\tau}(x)}{\tau} \, dx$, which is bounded for all $\tau > 0$ because $\frac{x^2}{x^2 + 1} \leq 1$ and p_{X_τ} is a valid pdf. We then express the Lévy exponent as

$$f(\omega) = \lim_{\tau \to 0} \left(\frac{e^{\tau f(\omega)} - 1}{\tau} \right) = \lim_{\tau \to 0} \int_{\mathbb{R}} (e^{j\omega x} - 1) \frac{p_{X_\tau}(x)}{\tau} \, dx$$

$$= \lim_{\tau \to 0} \int_{\mathbb{R}} \left(e^{j\omega x} - 1 - \frac{j x \omega}{1 + x^2} \right) \frac{x^2 + 1}{x^2} K_\tau(dx) + j\omega \lim_{\tau \to 0} a(\tau),$$

where

$$a(\tau) = \int_{\mathbb{R}} \frac{x}{1 + x^2} \frac{p_{X_\tau}(x)}{\tau} \, dx.$$

The technical part of the work, which is quite tedious and is not included here, is to show that the above integrals are bounded and that the two limits are well defined in the sense that $a(\tau) \to a_0$ and $K_\tau \to K$ (weakly) as $\tau \downarrow 0$ with K being a finite measure. This ultimately yields Khintchine's canonical representation,

$$f(\omega) = j\omega a_0 + \int_{\mathbb{R}} \left(e^{j\omega x} - 1 - \frac{j x \omega}{1 + x^2} \right) \frac{x^2 + 1}{x^2} K(dx),$$

where $a_0 \in \mathbb{R}$ and K is some bounded Borel measure. A potential advantage of Khintchine's representation is that the corresponding measure K is not singular. The connection with the standard Lévy–Khintchine formula is $b_2 = K(0^+) - K(0^-)$ and $v(x) \, dx = \frac{x^2 + 1}{x^2} K(dx)$ for $x \neq 0$. It is also possible to work out a relation between a and b_1, which depends upon the type of linear compensation in the canonical representation.

The above manipulation shows that the coefficients of the linear and quadratic terms of the Lévy–Khintchine formula (4.3) are primarily due to the non-integrable part of $g(x) = \lim_{\tau \downarrow 0} \frac{p_{X_\tau}(x)}{\tau} = \frac{x^2 + 1}{x^2} k(x)$ where $k(x) \, dx = K(dx)$.

By convention, the classical Lévy density v is assumed to be zero at the origin so that it differs from g by a point distribution that is concentrated at the origin. By invoking a basic theorem in distribution theory stating that a distribution entirely localized at the origin can always be expressed as a linear combination of the Dirac impulse and its derivatives, we can write that $g(x) - v(x) = b_0 \delta(x) + b_1' \delta'(x) + b_2 \delta''(x)$, where the higher-order derivatives of δ are excluded because of the admissibility condition.

For the indirect part of the proof, we start from the integral of the Lévy–Khintchine formula and consider the sequence of distributions whose exponent is

$$f_n(\omega) = \int_{|a|>1/n} \left(e^{ja\omega} - 1 - ja\omega \mathbb{1}_{|a|<1}(a) \right) v(a)\, da$$

$$= -j\omega \underbrace{\int_{1/n<|a|<1} a v(a)\, da}_{a_n} + \int_{|a|>1/n} \left(e^{ja\omega} - 1 \right) v(a)\, da.$$

Since the leading constant a_n is finite and $\int_{|a|>1/n} v(a)\, da < \infty$ for any fixed n (due to the admissibility condition on v), this corresponds to the exponent of a shifted compound-Poisson distribution whose characteristic function is $e^{f_n(\omega)}$. This allows us to deduce that $\widehat{p}_n(\omega) = e^{-b_1' j\omega - \frac{b_2}{2} \omega^2 + f_n(\omega)}$ is a valid characteristic function for any $n \in \mathbb{Z}^+$. Finally, we have the convergence of the sequence $\widehat{p}_n(\omega) \to e^{f(\omega)}$ as $n \to \infty$ where $f(\omega)$ is given by (4.3). Since $f(\omega)$ – and therefore $e^{f(\omega)}$ – is continuous around $\omega = 0$, we infer that $e^{f(\omega)}$ is a valid pdf (by Lévy's continuity theorem). The continuity of f is established by bounding the Lévy–Khintchine integral and invoking Lebesgue's dominated convergence theorem. The id part is obvious.

4.3   Finite-dimensional innovation model

To set the stage for the infinite-dimensional extension to come in Section 4.4, it is instructive to investigate the structure of a purely discrete innovation model whose input is the random vector $\mathbf{u} = (U_1, \ldots, U_N)$ of i.i.d. infinitely divisible random variables. The generic Nth-order pdf of the discrete innovation variable $\mathbf{u}$ is

$$p_{(U_1:U_N)}(u_1, \ldots, u_N) = \prod_{n=1}^{N} p_{\mathrm{id}}(u_n), \tag{4.10}$$

where $p_{\mathrm{id}}(x) = \mathscr{F}^{-1}\{e^{f(\omega)}\}(x)$ and f is the Lévy exponent of the underlying id distribution. Since $p_{(U_1:U_N)}(u_1, \ldots, u_N)$ is separable due to the independence assumption, we can write its characteristic function as the product of individual id factors:

$$\widehat{p}_{(U_1:U_N)}(\boldsymbol{\omega}) = \mathbb{E}_{(U_1:U_N)}\{e^{j\langle \boldsymbol{\omega}, \mathbf{u} \rangle}\} = \prod_{n=1}^{N} e^{f(\omega_n)}$$

$$= \exp\left(\sum_{n=1}^{N} f(\omega_n) \right), \tag{4.11}$$

where $\boldsymbol{\omega} = (\omega_1, \dots, \omega_N)$ is the frequency variable. The N-dimensional output signal $\mathbf{x} = (X_1, \dots, X_N)$ is then specified as the solution of the matrix-vector innovation equation

$$\mathbf{u} = \mathbf{L}\mathbf{x},$$

where the $N \times N$ whitening matrix $\mathbf{L}$ is assumed to be invertible. This implies that $\mathbf{x} = \mathbf{A}\mathbf{u}$ is a linear transformation of the excitation noise with $\mathbf{A} = \mathbf{L}^{-1}$. Its Nth-order characteristic function is obtained by simple (linear) change of variable:

$$\widehat{p}_{(X_1:X_N)}(\boldsymbol{\omega}) = \mathbb{E}_{(U_1:U_N)}\{e^{j\langle \boldsymbol{\omega}, \mathbf{A}\mathbf{u}\rangle}\} = \mathbb{E}_{(U_1:U_N)}\{e^{j\langle \mathbf{A}^T\boldsymbol{\omega}, \mathbf{u}\rangle}\}$$

$$= \widehat{p}_{(U_1:U_N)}(\mathbf{A}^T\boldsymbol{\omega})$$

$$= \exp\left(\sum_{n=1}^{N} f([\mathbf{A}^T\boldsymbol{\omega}]_n)\right). \tag{4.12}$$

Based on this equation, we can determine any marginal distribution by setting the appropriate frequency variables to zero. For instance, we find that the first-order pdf of X_n, the nth component of $\mathbf{x}$, is given by

$$p_{X_n}(x) = \overline{\mathscr{F}}^{-1}\left\{e^{f_n(\omega)}\right\}(x),$$

where

$$f_n(\omega) = \sum_{m=1}^{N} f(a_{nm}\omega)$$

with weighting coefficients $a_{nm} = [\mathbf{A}]_{nm} = [\mathbf{L}^{-1}]_{nm}$. The key observation here is that f_n is an admissible Lévy exponent, which implies that the underlying distribution is infinitely divisible (by Theorem 4.1), with the same being true for all the marginals and, by extension, the distribution of any linear measurement(s) of $\mathbf{x}$. While this provides a general mechanism for characterizing the probability law(s) of the discrete signal $\mathbf{x}$ within the classical framework of multivariate statistics, it is a priori difficult to perform the required computations (matrix inverse and inverse Fourier transforms) analytically, except in the Gaussian case, where the exponent is quadratic. Indeed, in this latter situation, (4.12) simplifies to $\widehat{p}_{(X_1:X_N)}(\boldsymbol{\omega}) = e^{-\frac{1}{2}\|\mathbf{A}^T\boldsymbol{\omega}\|_2^2}$, which is the Fourier transform of a multivariate Gaussian distribution with zero mean and covariance matrix $\mathbb{E}\{\mathbf{x}\mathbf{x}^T\} = \mathbf{A}\mathbf{A}^T$.

As we shall see in the next two sections, these results are transposable to the infinite-dimensional setting (see Table 3.4). While this may look like an unnecessary complication at first sight, the payoff is a theory that lends itself better to an analytical treatment using the powerful tools of harmonic analysis. The essence of the generalization is to replace the frequency variable $\boldsymbol{\omega}$ by a generic test function $\varphi \in \mathscr{S}(\mathbb{R}^d)$, the sums in (4.11) and (4.12) by Lebesgue integrals, and the matrix inverses by Green's functions which can often be specified explicitly. To make an analogy, it is conceptually and practically easier to formulate a comprehensive (deterministic) theory of linear systems by using Fourier analysis and convolution operators than by relying on linear algebra, with the same applying here. At the end of the exercise, it is still possible to come back to a finite-dimensional signal representation by projecting the continuous-domain model onto a suitable set of basis functions, as will be shown in Chapter 10.

4.4   White Lévy noises or innovations

Having gained a solid understanding of Lévy exponents, we can now move to the specification of a corresponding family of continuous-domain white-noise processes to drive the innovation model in Figure 2.1. To that end, we rely on Gelfand's theory of generalized stochastic processes [GV64], which was briefly summarized in Section 3.5. This powerful formalism allows for the complete and remarkably concise description of a generalized stochastic process by its characteristic functional. While the latter is not widely used in the standard formulation of stochastic processes, it lends itself quite naturally to the specification of generalized white-noise processes in terms of Lévy exponents, in direct analogy with what we have done before for id distributions.

DEFINITION 4.5 (White Lévy noise or innovation)   A generalized stochastic process w in $\mathcal{D}'(\mathbb{R}^d)$ is called a *white Lévy noise* (or *innovation*) if its characteristic functional is given by

$$\widehat{\mathcal{P}_w}(\varphi) = \mathbb{E}\{e^{j\langle \varphi, w \rangle}\} = \exp\left(\int_{\mathbb{R}^d} f\big(\varphi(r)\big)\,\mathrm{d}r\right), \tag{4.13}$$

where f is a valid Lévy exponent and φ is a generic test function in $\mathcal{D}(\mathbb{R}^d)$ (the space of infinitely differentiable functions of compact support).

Equation (4.13) is very similar to (4.2) and its multivariate extension (4.11). The key differences are that the frequency variable is now replaced by the generic test function $\varphi \in \mathcal{D}(\mathbb{R}^d)$ (which is a more general infinite-dimensional entity) and that the sum inside the exponential in (4.11) is replaced by an integral over the domain of φ. The fundamental point is that $\widehat{\mathcal{P}_w}(\varphi)$ is a continuous, positive definite functional on $\mathcal{D}(\mathbb{R}^d)$ with the key property that $\widehat{\mathcal{P}_w}(\varphi_1 + \varphi_2) = \widehat{\mathcal{P}_w}(\varphi_1)\widehat{\mathcal{P}_w}(\varphi_2)$ whenever φ_1 and φ_2 have non-overlapping support (i.e., $\varphi_1(r)\varphi_2(r) = 0$). The first part of the statement ensures that these generalized processes are well defined (by the Minlos–Bochner theorem), while the separability property implies that they take independent values at all points, which partially justifies the "white noise" nomenclature. Remarkably, Gelfand and Vilenkin have shown that there is also a converse implication [GV64, Theorem 6, p. 283]: Equation (4.13) specifies a continuous, positive definite functional on $\mathcal{D}(\mathbb{R}^d)$ (and hence defines an admissible white-noise process in $\mathcal{D}'(\mathbb{R}^d)$) if and only if f is a Lévy exponent. This ensures that the Lévy family constitutes the broadest possible class of acceptable white-noise inputs for our innovation model.

4.4.1   Specification of white noise in Schwartz' space $\mathscr{S}'$

In the present work, which relies heavily on convolution operators and Fourier analysis, we find it more convenient to define generalized stochastic processes with respect to test functions in the nuclear space $\mathscr{S}(\mathbb{R}^d)$, rather than the smaller space $\mathcal{D}(\mathbb{R}^d)$ used by Gelfand and Vilenkin. This requires a minimal restriction on the class of admissible Lévy densities in reference to Definition 4.3 to compensate for the lack of compact support of the functions in $\mathscr{S}(\mathbb{R}^d)$.

THEOREM 4.8 (Lévy–Schwartz admissibility)   *A white Lévy noise specified by (4.13) with $\varphi \in \mathscr{S}(\mathbb{R}^d)$ is a generalized stochastic process in $\mathscr{S}'(\mathbb{R}^d)$ provided that $f(\cdot)$ is characterized by the Lévy–Khintchine formula (4.3) with Lévy triplet (b'_1, b_2, v), where the Lévy density $v(a) \geq 0$ satisfies*

$$\int_{\mathbb{R}} \min(a^2, |a|^\epsilon) v(a)\, da < \infty \quad \text{for some } \epsilon > 0. \tag{4.14}$$

Proof   By the Minlos–Bochner theorem (Theorem 3.9), it suffices to show that $\widehat{\mathscr{P}_w}(\varphi)$ is a continuous, positive definite functional over $\mathscr{S}(\mathbb{R}^d)$ with $\widehat{\mathscr{P}_w}(0) = 1$, where the latter follows trivially from $f(0) = 0$. The positive definiteness is a direct consequence of the exponential nature of the characteristic functional and the conditional positive definiteness of the Lévy exponent $\big($see Section 4.4.3 and the paragraph following Equation (4.20)$\big)$.

The only delicate part is to prove continuity in the topology of $\mathscr{S}(\mathbb{R}^d)$. To that end, we consider a series of functions φ_n that converge to φ in $\mathscr{S}(\mathbb{R}^d)$. First, we recall that $\mathscr{S}(\mathbb{R}^d) \subset L_p(\mathbb{R}^d)$ for all $0 < p \leq +\infty$. Moreover, the convergence in $\mathscr{S}(\mathbb{R}^d)$ implies the convergence in all the L_p spaces. Indeed, if we select $k > 0$ such that $kp > 1$ and $\epsilon_0 > 0$, we have, for n sufficiently large,

$$|\varphi_n(r) - \varphi(r)| \leq \frac{\epsilon_0}{(\|r\|^2 + 1)^k},$$

which implies that

$$\|\varphi_n - \varphi\|_{L_p}^p \leq \epsilon_0^p \int \frac{dr}{(\|r\|^2 + 1)^{kp}}.$$

Since the right-hand-side integral is convergent and independent of n, we conclude that $\lim_{n \to \infty} \|\varphi_n - \varphi\|_{L_p} = 0$.

Since continuity is preserved through the composition of continuous maps and the exponential function is continuous, we only need to establish the continuity of $f(\varphi) = \log \widehat{\mathscr{P}_w}(\varphi)$. The continuity of the Gaussian part is obvious since $\int \varphi_n\, dr \to \int \varphi\, dr$ and $\|\varphi_n\|_{L_2} \to \|\varphi\|_{L_2}$. We therefore concentrate on the functional

$$G(\varphi) = \int_{\mathbb{R}^d} \int_{\mathbb{R}\setminus\{0\}} \left(e^{ja\varphi(r)} - 1 - ja\varphi(r) \mathbb{1}_{|a|\leq 1}(a) \right) v(a)\, da\, dr$$

that corresponds to the non-Gaussian part of the Lévy–Khintchine representation of f. Next, we write

$$|G(\varphi_n) - G(\varphi)| \leq \int_{\mathbb{R}^d} \int_{|a|>1} \left| e^{ja\varphi_n(r)} - e^{ja\varphi(r)} \right| v(a)\, da\, dr$$

$$+ \int_{\mathbb{R}^d} \int_{0<|a|\leq 1} \left| e^{ja\varphi_n(r)} - e^{ja\varphi(r)} - ja\big(\varphi_n(r) - \varphi(r)\big) \right| v(a)\, da\, dr$$

$$= (1) + (2).$$

To bound the first integral, we use the inequality

$$\left| e^{jx} - e^{jy} \right| = \left| e^{jy}(e^{j(x-y)} - 1) \right| \leq \min(2, |x-y|) \leq 2\left(\frac{|x-y|}{2} \right)^{\epsilon}$$

under the (non-restrictive) condition that $\epsilon \leq 1$. This yields

$$(1) \leq 2^{1-\epsilon} \int_{\mathbb{R}^d} \int_{|a|>1} |a|^{\epsilon} |\varphi_n(r) - \varphi(r)|^{\epsilon} v(a) \, da \, dr$$

$$= 2^{1-\epsilon} \left(\int_{|a|>1} |a|^{\epsilon} v(a) \, da \right) \|\varphi_n - \varphi\|_{L_\epsilon}^{\epsilon}.$$

As for the second integral, we use the bound

$$\left| e^{jx} - e^{jy} - j(x-y) \right| = \left| e^{jy}(e^{j(x-y)} - 1 - j(x-y)) + j(x-y)(e^{jy} - 1) \right|$$

$$\leq \left| e^{j(x-y)} - 1 - j(x-y) \right| + \left| (x-y)(e^{jy} - 1) \right|$$

$$\leq (x-y)^2 + |x-y| \cdot |y|.$$

Therefore,

$$(2) \leq \int_{\mathbb{R}^d} \int_{0<|a|\leq 1} a^2 \big(\varphi_n(r) - \varphi(r)\big)^2 v(a) \, da \, dr$$

$$\qquad | \int_{\mathbb{R}^d} \int_{0<|a|\leq 1} a^2 |\varphi_n(r) - \varphi(r)| \cdot |\varphi(r)| v(a) \, da \, dr$$

$$\leq \left(\int_{0<|a|\leq 1} a^2 v(a) \, da \right) \left(\|\varphi_n - \varphi\|_{L_2}^2 + \|\varphi_n - \varphi\|_{L_2} \|\varphi\|_{L_2} \right).$$

Since $\|\varphi_n - \varphi\|_{L_p} \to 0$ for all $p > 0$ as φ_n converges to φ in $\mathscr{S}(\mathbb{R}^d)$, we conclude that $\lim_{n \to \infty} |G(\varphi_n) - G(\varphi)| = 0$, which proves the continuity of $\widehat{\mathscr{P}_w}(\varphi)$. $\qquad \square$

Note that (4.14), which will be referred to as Lévy–Schwartz admissibility, is a very slight restriction on the classical condition ($\epsilon = 0$) for id laws (see (4.1) in Definition 4.3). The fact that ϵ can be chosen arbitrarily small reflects the property that the functions in $\mathscr{S}$ have a faster-than-algebraic decay. Another equivalent formulation of Lévy–Schwartz admissibility is

$$\mathbb{E}\{|\langle \varphi, w \rangle|^{\epsilon}\} < \infty \quad \text{for some } \epsilon > 0, \tag{4.15}$$

which follows from (9.10) and Proposition 9.10 in Chapter 9. Along the same lines, it can be shown that the finiteness of the ϵth-order moment in (4.15) for any non-trivial φ_0 implies that the same holds true for all $\varphi \in \mathscr{S}(\mathbb{R}^d)$.

From now on, we implicitly assume that the Lévy–Schwartz admissibility condition is met.

To exemplify the procedure, we select a quadratic exponent which is trivially admissible (since $v(a) = 0$). This results in

$$\widehat{\mathscr{P}}_{w\text{Gauss}}(\varphi) = \exp\left(-\frac{b_2}{2} \|\varphi\|_{L_2}^2 \right),$$

which is the functional that completely characterizes the white Gaussian noise of the classical theory of stationary processes.

4.4.2 Impulsive Poisson noise

We have already alluded to the fact that the continuous-domain white-noise processes w are highly singular and generally too rough to admit an interpretation as conventional functions of the index variable $r \in \mathbb{R}^d$. The realizations (or sample paths) are generalized functions that can only be probed indirectly through their scalar products $\langle \varphi, w \rangle$ with test functions or observation windows, as illustrated in Section 4.1. While the use of such an indirect approach is unavoidable in the mathematical formulation, it is possible to provide an explicit pointwise description of a noise realization in the special case where the Lévy exponent f is associated with a compound-Poisson distribution [UT11]. The corresponding impulsive Poisson noise model is

$$w(r) = \sum_{k \in \mathbb{Z}} A_k \delta(r - r_k), \tag{4.16}$$

where r_k are random point locations in $\mathbb{R}^d$ and where the A_k are i.i.d. random variables with pdf p_A. The random events are indexed by k (using some arbitrary ordering); they are mutually independent and follow a spatial Poisson distribution. Specifically, let Π be any finite-measure subset of $\mathbb{R}^d$; then the probability of observing $N(\Pi) = n$ events in Π is

$$\text{Prob}\,(N(\Pi) = n) = \frac{e^{-\lambda \text{Vol}(\Pi)} (\lambda \text{Vol}(\Pi))^n}{n!},$$

where $\text{Vol}(\Pi)$ is the measure (or spatial volume) of Π. This is to say that the Poisson parameter λ represents the average number of random impulses per unit hyper-volume. The link with the formal specification of Lévy noise in Definition 4.5 is as follows.

THEOREM 4.9 *The characteristic functional of the impulsive Poisson noise specified by (4.16) is*

$$\widehat{\mathscr{P}}_{w_{\text{Poisson}}}(\varphi) = \mathbb{E}\{e^{j\langle \varphi, w \rangle}\} = \exp\left(\int_{\mathbb{R}^d} f_{\text{Poisson}}(\varphi(r))\, dr \right) \tag{4.17}$$

with

$$f_{\text{Poisson}}(\omega) = \lambda \int_{\mathbb{R}} (e^{ja\omega} - 1) p_A(a)\, da = \lambda(\widehat{p}_A(\omega) - 1), \tag{4.18}$$

where λ is the Poisson density parameter, p_A the amplitude pdf of the Dirac impulses, and $\widehat{p}_A$ the corresponding characteristic function.

Proof We select an arbitrary test function $\varphi \in \mathscr{D}(\mathbb{R}^d)$ of compact support, with its support included in the centered cube $\Pi_\varphi = [-c_\varphi, c_\varphi]^d$. We denote by $N_{w,\varphi}$ the number of Poisson points of w in Π_φ; by definition, it is a Poisson random variable with parameter $\lambda \text{Vol}(\Pi_\varphi)$. The restriction of w to Π_φ corresponds to the random sum

$$\sum_{n=1}^{N_{w,\varphi}} A'_n \delta(r - r'_n),$$

where we have used an appropriate relabeling of the variables $\{(A_k, r_k) \,|\, r_k \in \Pi_\varphi\}$ in (4.16). Correspondingly, we have $\langle \varphi, w \rangle = \sum_{n=1}^{N_{w,\varphi}} A'_n \varphi(r'_n)$.

By the order-statistics property of Poisson processes, the r'_n are independent and all equivalent in distribution to a random variable r' that is uniform on Π_φ.

Using the law of total expectation, we expand the characteristic functional of w, $\widehat{\mathscr{P}}_w(\varphi) = \mathbb{E}\left\{ e^{j\langle \varphi, w\rangle} \right\}$, as

$$\widehat{\mathscr{P}}_w(\varphi) = \mathbb{E}\left\{ \mathbb{E}\left\{ e^{j\langle \varphi, w\rangle} \middle| N_{w,\varphi} \right\} \right\}$$

$$= \mathbb{E}\left\{ \mathbb{E}\left\{ \prod_{n=1}^{N_{w,\varphi}} e^{jA'_n \varphi(r'_n)} \middle| N_{w,\varphi} \right\} \right\}$$

$$= \mathbb{E}\left\{ \prod_{n=1}^{N_{w,\varphi}} \mathbb{E}\left\{ e^{jA' \varphi(r')} \right\} \right\} \qquad \text{(by independence)}$$

$$= \mathbb{E}\left\{ \prod_{n=1}^{N_{w,\varphi}} \mathbb{E}\left\{ \mathbb{E}\left\{ e^{jA \varphi(r')} \middle| A \right\} \right\} \right\} \qquad \text{(total expectation)}$$

$$= \mathbb{E}\left\{ \prod_{n=1}^{N_{w,\varphi}} \mathbb{E}\left\{ \frac{\int_{\Pi_\varphi} e^{jA \varphi(r')} \, dr'}{\mathrm{Vol}(\Pi_\varphi)} \right\} \right\} \qquad \text{(as } r' \text{ is uniform in } \Pi_\varphi\text{)}$$

$$= \mathbb{F}\left\{ \prod_{n=1}^{N_{w,\varphi}} \frac{\int_{\mathbb{R}} \int_{\Pi_\varphi} e^{ja \varphi(r)} \, dr \, p_A(a) \, da}{\mathrm{Vol}(\Pi_\varphi)} \right\}. \qquad (4.19)$$

The last expression has the inner expectation expanded in terms of the pdf p_A of the random variable A. Defining the auxiliary functional

$$M(\varphi) = \int_{\Pi_\phi} \int_{\mathbb{R}} e^{ja \varphi(r)} p_A(a) \, da \, dr,$$

we rewrite (4.19) as

$$\mathbb{E}\left\{ \prod_{n=1}^{N_{w,\varphi}} \frac{M(\varphi)}{\mathrm{Vol}(\Pi_\varphi)} \right\} = \mathbb{E}\left\{ \left(\frac{M(\varphi)}{\mathrm{Vol}(\Pi_\varphi)} \right)^{N_{w,\varphi}} \right\}.$$

Next, we use the fact that $N_{w,\varphi}$ is a Poisson random variable to compute the above expectation directly:

$$\mathbb{E}\left\{ \left(\frac{M(\varphi)}{\mathrm{Vol}(\Pi_\varphi)} \right)^{N_{w,\varphi}} \right\} = \sum_{n \geq 0} \left(\frac{M(\varphi)}{\mathrm{Vol}(\Pi_\varphi)} \right)^n \frac{e^{-\lambda \mathrm{Vol}(\Pi_\varphi)} \left(\lambda \mathrm{Vol}(\Pi_\varphi) \right)^n}{n!}$$

$$= e^{-\lambda \mathrm{Vol}(\Pi_\varphi)} \sum_{n \geq 0} \frac{(\lambda M(\varphi))^n}{n!}$$

$$= e^{-\lambda \mathrm{Vol}(\Pi_\varphi)} e^{\lambda M(\varphi)} \qquad \text{(Taylor)}$$

$$= \exp\left(\lambda \left(M(\varphi) - \mathrm{Vol}(\Pi_\varphi) \right) \right).$$

We now replace $M(\varphi)$ by its integral equivalent, noting also that $\text{Vol}(\Pi_\varphi) = \int_{\Pi_\varphi} \int_{\mathbb{R}} 1 \times p_A(a)\, da\, d\mathbf{r}$, whereupon we obtain the expression

$$\widehat{\mathscr{P}_w}(\varphi) = \exp\left(\lambda \int_{\Pi_\varphi} \int_{\mathbb{R}} (e^{ja\varphi(r)} - 1)p_A(a)\, da\, d\mathbf{r}\right).$$

As $(e^{ja\varphi(r)} - 1)$ vanishes outside the support of φ (and, therefore, outside Π_φ), we may enlarge the domain of the inner integral to all of $\mathbb{R}^d$, which yields (4.17). Finally, we use the fact that the derived Poisson functional is part of the Lévy family and invoke Theorem 4.8 to extend the domain of $\widehat{\mathscr{P}_w}$ from $\mathscr{D}(\mathbb{R}^d)$ to $\mathscr{S}(\mathbb{R}^d)$.   □

The interest of this result is twofold. First, it gives a concrete meaning to the compound-Poisson scenario in Figure 4.1c, allowing for a description in terms of conventional point processes. In the same vein, we can propose a physical analogy for the elementary Poisson term $f_3(\omega) = \lambda(e^{ja_0\omega} - 1)$ in Section 4.2.2 with $p_A(a) = \delta(a - a_0)$: the counting of photons impinging on the detectors of a CCD camera with the photon density being constant over $\mathbb{R}^d$ and the integration time proportional to λ. The corresponding process is usually termed "photon noise" in optical imaging. Second, the explicit noise model (4.16) suggests a practical mechanism for generating generalized Poisson processes as a weighted sum of shifted Green's functions of L, each Dirac impulse being replaced by the response of the inverse operator in the innovation model in Figure 2.1.

Note that the above description of generalized compound-Poisson processes is compatible with the usual definition of finite-rate-of-innovation signals. Yet, this is by far not the whole story since the impulsive Poisson noise is the only member of the Lévy family whose "innovation rate," as measured by λ, is finite.

4.4.3   Properties of white noise

To emphasize the parallel with the scalar formulation in Section 4.2, we start by introducing the functional counterpart of Definition 4.1.

DEFINITION 4.6 (Generalized Lévy exponent)   A continuous complex-valued functional F on the nuclear space $\mathscr{S}(\mathbb{R}^d)$ such that $F(0) = 0$, $F(\varphi) = \overline{F(-\varphi)}$ is called a *generalized Lévy exponent* if it is conditionally positive definite of order one; i.e.,

$$\sum_{m=1}^{N}\sum_{n=1}^{N} F(\varphi_m - \varphi_n)\xi_m\overline{\xi}_n \geq 0$$

under the condition $\sum_{n=1}^{N} \xi_n = 0$ for every possible choice $\varphi_1, \ldots, \varphi_N \in \mathscr{S}(\mathbb{R}^d)$, $\xi_1, \ldots, \xi_N \in \mathbb{C}$, and $N \in \mathbb{Z}^+$.

This definition is motivated by the infinite-dimensional counterpart of Schoenberg's correspondence theorem (Theorem 4.7) [PR70].

THEOREM 4.10 (Prakasa Rao)   *Let F be a complex-valued functional on the nuclear space $\mathscr{S}(\mathbb{R}^d)$ such that $F(0) = 0$, $F(\varphi) = \overline{F(-\varphi)}$. Then, the following conditions are equivalent.*

(1) The functional F is conditionally positive definite of order one.

(2) For every choice $\varphi_1, \ldots, \varphi_N \in \mathcal{S}(\mathbb{R}^d)$, $\xi_1, \ldots, \xi_N \in \mathbb{C}$, $\tau > 0$ and $N \in \mathbb{Z}^+$,

$$\sum_{m=1}^{N} \sum_{n=1}^{N} \exp\left(\tau F(\varphi_m - \varphi_n)\right) \xi_m \overline{\xi}_n \geq 0.$$

(3) For every choice $\varphi_1, \ldots, \varphi_N \in \mathcal{S}(\mathbb{R}^d)$, $\xi_1, \ldots, \xi_N \in \mathbb{C}$, and $N \in \mathbb{Z}^+$,

$$\sum_{m=1}^{N} \sum_{n=1}^{N} \left(F(\varphi_m - \varphi_n) - F(\varphi_m) - F(-\varphi_n)\right) \xi_m \overline{\xi}_n \geq 0.$$

In the white-Lévy-noise scenario of Definition 4.5, we have that

$$F(\varphi) = \int_{\mathbb{R}^d} f\bigl(\varphi(\boldsymbol{r})\bigr) \, \mathrm{d}\boldsymbol{r}. \tag{4.20}$$

It then comes as no surprise that the generalized Lévy exponent $F(\varphi)$ inherits the relevant properties of f, including conditional positive definiteness, with

$$\sum_{m=1}^{N} \sum_{n=1}^{N} F(\varphi_m - \varphi_n) \xi_m \overline{\xi}_n = \int_{\mathbb{R}^d} \underbrace{\sum_{m=1}^{N} \sum_{n=1}^{N} f\bigl(\varphi_m(\boldsymbol{r}) - \varphi_n(\boldsymbol{r})\bigr) \xi_m \overline{\xi}_n}_{\geq 0} \, \mathrm{d}\boldsymbol{r} \geq 0$$

subject to the constraint $\sum_{n=1}^{N} \xi_n = 0$.

The simple additive nature of the mapping (4.20) between generalized Lévy exponents and the classical ones translates into the following white-noise properties which are central to our formulation.

(1) Independent atoms and stationarity

PROPOSITION 4.11   *A white Lévy noise is stationary and independent at every point.*

Proof The stationarity property is expressed by $\widehat{\mathscr{P}}_w(\varphi) = \widehat{\mathscr{P}}_w(\varphi(\cdot - \boldsymbol{r}_0))$ for all $\boldsymbol{r}_0 \in \mathbb{R}^d$. It is established by simple change of variable in the defining integral. To investigate the independence at every point, we determine the joint characteristic function of the random variables $X_1 = \langle \varphi_1, w \rangle$ and $X_2 = \langle \varphi_2, w \rangle$, which is given by $\widehat{p}_{X_1, X_2}(\omega_1, \omega_2) = \exp\left(F(\omega_1 \varphi_1 + \omega_2 \varphi_2)\right)$, where F is defined by (4.20). When φ_1 and φ_2 have non-overlapping support, we use the fact that $f(0) = 0$ and decompose the exponent as

$$F(\omega_1 \varphi_1 + \omega_2 \varphi_2) = F(\omega_1 \varphi_1) + F(\omega_2 \varphi_2),$$

which implies that

$$\widehat{p}_{X_1, X_2}(\omega_1, \omega_2) = \exp\left(F(\omega_1 \varphi_1)\right) \times \exp\left(F(\omega_2 \varphi_2)\right)$$

$$= \widehat{p}_{X_1}(\omega_1) \times \widehat{p}_{X_2}(\omega_2),$$

where $\widehat{p}_X(\omega) = \exp\left(F(\omega \varphi)\right)$, which proves that X_1 and X_2 are independent. The independence at every point follows from the fact that one can consider contracting, Dirac-like sequences of functions φ_1 and φ_2 that are non-overlapping and whose support gets arbitrarily small. $\qquad\square$

(2) Infinite divisibility

PROPOSITION 4.12   *A white Lévy noise is uniquely specified by a canonical id distribution $p_{\text{id}}(x) = \int_{\mathbb{R}} e^{f(\omega) - j\omega x} \frac{d\omega}{2\pi}$, where f is the defining Lévy exponent in (4.20). The latter corresponds to the pdf of the observation $X = \langle \text{rect}(\cdot - r_0), w \rangle$ through a rectangular window at some arbitrary location r_0.*

The first part is just a restatement of the functional equivalence between Lévy noises and id distributions on the one hand, and Lévy exponents on the other. As for the second part, we recall that the characteristic function of the variable $X = \langle \varphi, w \rangle$ is given by

$$\mathbb{E}\{e^{j\omega X}\} = \mathbb{E}\{e^{j\omega\langle\varphi, w\rangle}\} = \mathbb{E}\{e^{j\langle\omega\varphi, w\rangle}\} = \widehat{\mathscr{P}}_w(\omega\varphi) = \exp\left(F(\omega\varphi)\right).$$

By choosing $\varphi(r) = \text{rect}(r - r_0)$ with $\text{rect}(r) = 1$ for $r \in (-\frac{1}{2}, \frac{1}{2}]^d$ and zero otherwise, we formally resolve the integral $\int_{\mathbb{R}^d} f(\omega\text{rect}(r)) \, dr = f(\omega)$; this implies that $\widehat{p}_{\text{id}}(\omega) = \exp\left(f(\omega)\right)$, which is the desired result.

Along the same line, we can show that the use of an arbitrary, non-rectangular analysis window does not fundamentally change the situation, in the sense that the pdf remains infinitely divisible.

PROPOSITION 4.13   *The observation $X = \langle \varphi, w \rangle$ of a white Lévy noise with Lévy exponent f through an arbitrary observation window φ – not necessarily in $\mathscr{S}(\mathbb{R}^d)$ – yields an infinitely divisible random variable X whose characteristic function is $\widehat{p}_X(\omega) = e^{f_\varphi(\omega)}$, where $f_\varphi(\omega) = \int_{\mathbb{R}^d} f(\omega\varphi(r)) \, dr$. The validity of f_φ requires some (mild) technical condition on f when $\varphi \in L_p(\mathbb{R}^d)$ is not rapidly decaying.*

This property is investigated in depth in Chapter 9 and exploited for deriving transform-domain statistics. The precise statement of this id result for $\varphi \in L_p(\mathbb{R}^d)$ is given in Theorem 9.1.

Another manifestation of the id property is that a continuous-domain Lévy noise can always be broken down into an arbitrary number of i.i.d components.

PROPOSITION 4.14   *A white Lévy noise w is infinitely divisible in the sense that it can be decomposed as $w = w_1 + \cdots + w_N$ for any $N \in \mathbb{Z}^+$, where the w_n are i.i.d. white-noise processes.*

This simply follows from the property that the characteristic functional of the sum of two independent processes is the product of their individual characteristic functionals. Specifically, we can write

$$\widehat{\mathscr{P}}_w(\varphi) = \left(\widehat{\mathscr{P}}_{w_N}(\varphi)\right)^N,$$

where $\widehat{\mathscr{P}}_{w_N}(\varphi) = \exp\left(\int_{\mathbb{R}^d} f(\varphi(r))/N \, dr\right)$ is the characteristic functional of a Lévy noise. The justification is that $f(\omega)/N$ is a valid Lévy exponent for any $N \geq 1$. In the impulsive Poisson case, this simply translates into the Poisson density parameter λ being divided by N.

Interestingly, there is also a converse to the statement in Proposition 4.14 [PR70]: a generalized process s in $\mathscr{S}'(\mathbb{R}^d)$ is infinitely divisible if and only if its characteristic functional can be written as $\widehat{\mathscr{P}}_s(\varphi) = \exp\left(F(\varphi)\right)$, where $F(\varphi)$ is a continuous, conditionally positive definite functional over $\mathscr{S}(\mathbb{R}^d)$ (or generalized Lévy exponent) as specified in Definition 4.6 [PR70, main theorem]. While this general characterization is nice conceptually, it is hardly discriminative since the underlying notion of infinite divisibility applies to all concrete families of generalized stochastic processes that are known to us. In particular, it does not require the "whiteness" property that is fundamental for defining proper innovations.

(3) Flat power spectrum

Strictly speaking, the properties of stationarity and independence at every point are not sufficient for specifying "white" noise. There is also some implicit idea of enforcing a flat Fourier spectrum. A simple example that satisfies the two first properties but fails to meet the latter is the weak derivative of a Lévy noise whose generalized power spectrum (when defined) is not flat but proportional to $|\omega|^2$.

The notion of power spectrum is based on second-order moments and only makes sense when the stochastic process is stationary with a well-defined autocorrelation. In Gelfand's theory, the second-order dependencies are captured by the correlation functional $\mathscr{B}_w(\varphi_1, \varphi_2) = \mathbb{E}\{\langle \varphi_1, w \rangle \cdot \langle \varphi_2, w \rangle\}$, where it is assumed that the generalized noise process w is real-valued and that its second-order moments are well defined. The latter second-order requirement is equivalent to imposing that the Lévy exponent f should be twice-differentiable at the origin or, equivalently, that the canonical id distribution p_{id} of the process has a finite second-order moment.

PROPOSITION 4.15   *Let w be a (second-order) white Lévy noise with Lévy exponent f such that $f'(0) = 0$ (zero-mean assumption) and $\sigma_w^2 = -f''(0) < +\infty$ (finite-variance assumption). Then,*

$$\mathscr{B}_w(\varphi_1, \varphi_2) = \sigma_w^2 \langle \varphi_1, \varphi_2 \rangle. \tag{4.21}$$

Formally, this corresponds to the statement that the autocorrelation of a second-order [5] Lévy noise is a Dirac impulse; i.e.,

$$R_w(r) = \mathbb{E}\{w(r_0)w(r_0 - r)\} = \mathscr{B}_w\big(\delta(\cdot - r_0), \delta(\cdot - r_0 + r)\big)$$
$$= \sigma_w^2 \delta(r),$$

where $r_0 \in \mathbb{R}^d$ can be arbitrary, as a consequence of the stationarity property. This also means that the generalized power spectrum of a second-order white Lévy noise is flat:

$$\Phi_w(\omega) = \mathscr{F}\{R_w(r)\}(\omega) = \sigma_w^2.$$

We recall that the term "white" is used in reference to white light, whose electromagnetic spectrum is distributed over the visible band in a way that stimulates all color receptors of the eye equally. This is in opposition to "colored" noise, whose spectral content is not equally distributed.

---

[5] In the statistical literature, a second-order process usually designates a stochastic process whose second-order moments are all well defined. In the case of generalized processes, the property refers to the existence of the correlation functional.

Proof We have that $\mathscr{B}_w(\varphi_1, \varphi_2) = \mathbb{E}\{X_1 X_2\}$, where $X_1 = \langle \varphi_1, w \rangle$ and $X_2 = \langle \varphi_2, w \rangle$ are real-valued random variables with joint characteristic function $\widehat{p}_{X_1, X_2}(\boldsymbol{\omega}) = \exp\left(F(\omega_1 \varphi_1 + \omega_2 \varphi_2)\right)$ with $\boldsymbol{\omega} = (\omega_1, \omega_2)$. We then invoke the moment-generating property of the Fourier transform, which translates into

$$\mathscr{B}_w(\varphi_1, \varphi_2) = \mathbb{E}\{X_1 X_2\} = (-j)^2 \left. \frac{\partial^2 \widehat{p}_{X_1, X_2}(\boldsymbol{\omega})}{\partial \omega_1 \partial \omega_2} \right|_{\omega_1 = 0, \omega_2 = 0}.$$

By applying the chain rule twice, we obtain

$$\frac{\partial^2 \widehat{p}_{X_1, X_2}(\boldsymbol{\omega})}{\partial \omega_1 \partial \omega_2} = e^{f_{X_1, X_2}(\boldsymbol{\omega})} \left(\frac{\partial f_{X_1, X_2}(\boldsymbol{\omega})}{\partial \omega_1} \frac{\partial f_{X_1, X_2}(\boldsymbol{\omega})}{\partial \omega_2} + \frac{\partial^2 f_{X_1, X_2}(\boldsymbol{\omega})}{\partial \omega_1 \partial \omega_2} \right),$$

where

$$f_{X_1, X_2}(\boldsymbol{\omega}) = \log \widehat{p}_{X_1, X_2}(\omega_1, \omega_2) = \int_{\mathbb{R}^d} f\left(\omega_1 \varphi_1(\boldsymbol{r}) + \omega_2 \varphi_2(\boldsymbol{r})\right) d\boldsymbol{r}$$

is the cumulant generating function of p_{X_1, X_2}. The required first derivative with respect to ω_i, $i = 1, 2$ is given by

$$\frac{\partial f_{X_1, X_2}(\boldsymbol{\omega})}{\partial \omega_i} = \int_{\mathbb{R}^d} f'\left(\omega_1 \varphi_1(\boldsymbol{r}) + \omega_2 \varphi_2(\boldsymbol{r})\right) \varphi_i(\boldsymbol{r}) \, d\boldsymbol{r},$$

which, when evaluated at the origin, simplifies to

$$\frac{\partial f_{X_1, X_2}(\mathbf{0})}{\partial \omega_i} = f'(0) \int_{\mathbb{R}^d} \varphi_i(\boldsymbol{r}) \, d\boldsymbol{r} = -j \, \mathbb{E}\{X_i\}. \tag{4.22}$$

Similarly, we get

$$\frac{\partial^2 f_{X_1, X_2}(\mathbf{0})}{\partial \omega_1 \partial \omega_2} = f''(0) \int_{\mathbb{R}^d} \varphi_1(\boldsymbol{r}) \varphi_2(\boldsymbol{r}) \, d\boldsymbol{r}.$$

By combining those results and using the property that $f_{X_1, X_2}(\mathbf{0}) = 0$, we conclude that

$$\mathbb{E}\{X_1 X_2\} = -f''(0) \langle \varphi_1, \varphi_2 \rangle - \left(f'(0)\right)^2 \langle \varphi_1, 1 \rangle \langle \varphi_2, 1 \rangle,$$

which is equivalent to (4.21) under the hypothesis that $f'(0) = 0$. It is also clear from (4.22) that this latter condition is equivalent to the zero-mean property of the noise; that is, $\mathbb{E}\{\langle \varphi, w \rangle\} = 0$ for all $\varphi \in \mathscr{S}(\mathbb{R}^d)$. Finally, we note that (4.21) is compatible with the more general cumulant formula (9.22) if we set $\boldsymbol{n} = (1, 1)$, $n = 2$, and $\kappa_2 = (-j)^2 f''(0)$. $\qquad\square$

Since $f(0) = 0$ by definition, another way of writing the hypotheses in Proposition 4.15 is $f(\omega) = -\frac{\sigma_w^2}{2}\omega^2 + O(|\omega|^3)$, which expresses an asymptotic equivalence with the symmetric Gaussian scenario (purely quadratic Lévy exponent). This *second-order assumption* ensures that the noise has zero mean and a finite variance σ_w^2 so that its correlation functional (4.21) is well defined. In what follows, it is made implicit whenever we are talking of correlations or power spectra.

(4) Stochastic counterpart of the Dirac impulse

From an engineering perspective, white noise is often viewed as the stochastic analog of the Dirac distribution δ, whose spectrum is flat in the literal sense (i.e., $\mathcal{F}\{\delta\} = 1$). The fundamental difference, of course, is that the generalized function δ is a deterministic entity. The simplest way of introducing randomness is by considering a shifted and weighted impulse $A\delta(\cdot - r_0)$ whose location r_0 is uniformly distributed over some compact subset of $\mathbb{R}^d$ and whose amplitude A is a random variable with pdf p_A. A richer form of excitation is obtained through the summation of such i.i.d. elementary contributions, which results in the construction of impulsive Poisson noise, as specified by (4.16). Theorem 4.9 ensures that this explicit way of representing noise is legitimate in the case where the Lévy density $v = \lambda p_A$ is integrable and the Gaussian part absent. We shall now see that this constructive approach can be pushed to the limit for the non-Poisson brands of innovations, including the Gaussian ones.

PROPOSITION 4.16 *A white Lévy noise is the limit of a sequence of impulsive Poisson-noise processes in the sense of the weak convergence of the underlying infinite-dimensional measures.*

Proof The technical part of the proof uses an infinite-dimensional generalization of Lévy's continuity theorem and will be reported elsewhere. The key idea is to consider the following sequence of Lévy exponents:

$$f_n(\omega) = n\left(e^{\frac{1}{n}f(\omega)} - 1\right) = f(\omega) + O\left(\frac{f^2(\omega)}{n}\right),$$

which are of the compound-Poisson type with $\lambda_n = n$ and $\widehat{p}_{A_n}(\omega) = e^{\frac{1}{n}f(\omega)}$ and which converge to $f(\omega)$ as n goes to infinity. This suggests forming the corresponding sequence of characteristic functionals

$$\widehat{\mathscr{P}}_{w_n}(\varphi) = \exp\left(\int_{\mathbb{R}^d} n\left(e^{\frac{1}{n}f(\varphi(r))} - 1\right) dr\right),$$

which are expected to converge to $\widehat{\mathscr{P}}_w(\varphi) = \exp\left(\int_{\mathbb{R}^d} f(\varphi(r)) \, dr\right)$ as $n \to +\infty$. The main point for the argument is that these are all of the impulsive Poisson type for n fixed (see Theorem 4.9). The crux of the proof is to control the convergence by specifying some appropriate bound and to verify that some basic equicontinuity conditions are met. □

The result is interesting because it gives us some insight into the nature of continuous domain noise. The limit process involves random Dirac impulses that get denser as n increases. When the variance of the noise $\sigma_w^2 = -f''(0)$ is finite, the increase of the average number of impulses per unit volume $\lambda_n = O(n)$ is compensated by a decrease of the variance of the amplitude distribution in inverse proportion: $\mathrm{Var}\{A_n\} = \sigma_w^2/n$. While any of the generalized noise processes in the sequence is as rough as a Dirac impulse, this picture suggests that the degree of singularity of the limit object in the non-Poisson scenario can be potentially reduced due to the accumulation of impulses and the fact that the variance of their amplitude distribution converges to zero. The particular example that we have in mind is Gaussian white noise, which can be obtained as a limit of compound-Poisson processes with contracting Gaussian amplitude distributions.

4.5   Generalized stochastic processes and linear models

As already mentioned, the class of generalized stochastic processes that are of interest to us are those defined through the generic innovation model $Ls = w$ (linear stochastic differential equation) where the differential operator L is shift-invariant and where the driving term w is a continuous-domain white Lévy noise. Having made sense of the latter, we can now proceed with the specification of the class of admissible whitening operators L. The key requirement here is that the model be invertible (in the sense of generalized functions) which, by duality, translates into some continuity constraint on the adjoint operator L^{-1*}. For the time being, we shall limit ourselves to making some general statements about L and its inverse that ensure existence, while deferring to Chapter 5 for concrete examples of admissible operators.

4.5.1   Innovation models

The interpretation of the above continuous-domain linear model in the sense of generalized functions is

$$\forall \varphi \in \mathscr{S}(\mathbb{R}^d), \quad \langle \varphi, Ls \rangle = \langle \varphi, w \rangle. \tag{4.23}$$

The generalized stochastic process $s = L^{-1}w$ is generated by solving this equation, which amounts to a linear transformation of the Lévy innovation w. Formally, this translates into

$$\forall \varphi \in \mathscr{S}(\mathbb{R}^d), \quad \langle \varphi, s \rangle = \langle \varphi, L^{-1}w \rangle = \langle L^{-1*}\varphi, w \rangle, \tag{4.24}$$

where L^{-1} is an appropriate right inverse of L. The above manipulation obviously only makes sense if the action of the adjoint operator L^{-1*} is well defined over Schwartz' class $\mathscr{S}(\mathbb{R}^d)$ of test functions – ideally, a continuous mapping from $\mathscr{S}(\mathbb{R}^d)$ into itself or, possibly, $L_p(\mathbb{R}^d)$ (or some variant) if one imposes suitable restrictions on the Lévy exponent f to maintain continuity.

We like to refer to (4.23) as the *analysis* statement of the model, and to (4.24) – or its shorthand, $s = L^{-1}w$ – as the *synthesis* description. Of course, this will only work properly if we have an exact equivalence, meaning that there is a proper and unique definition of L^{-1}. The latter will need to be made explicit on a case-by-case basis with the possible help of boundary conditions.

4.5.2   Existence and characterization of the solution

We shall now see that, under suitable conditions on L^{-1*} (see Theorem 4.17 below), one can completely specify such processes via their characteristic functional and ensure their existence as solutions of (4.23). We recall that the characteristic functional of a generalized stochastic process s is defined as

$$\widehat{\mathscr{P}_s}(\varphi) = \mathbb{E}\{e^{j\langle\varphi,s\rangle}\}$$
$$= \int_{\mathscr{S}'(\mathbb{R}^d)} e^{j\langle\varphi,s\rangle} \mathscr{P}_s(ds),$$

where the latter expression involves an abstract infinite-dimensional integral over the space of tempered distributions and provides the connection with the defining measure $\mathscr{P}_s$ on $\mathscr{S}'(\mathbb{R}^d)$. $\widehat{\mathscr{P}_s}$ is a functional $\mathscr{S}(\mathbb{R}^d) \to \mathbb{C}$ that associates the complex number $\widehat{\mathscr{P}_s}(\varphi)$ with each test function $\varphi \in \mathscr{S}(\mathbb{R}^d)$ and which is endowed with three fundamental properties: positive definiteness, continuity, and normalization (i.e., $\widehat{\mathscr{P}_s}(0) = 1$). It can also be specified using the more concrete formula

$$\widehat{\mathscr{P}_s}(\varphi) = \int_{\mathbb{R}} e^{jy}\, d\mathscr{P}_{\langle\varphi,s\rangle}(y), \qquad (4.25)$$

which involves a classical Stieltjes integral with respect to the probability law $\mathscr{P}_{Y=\langle\varphi,s\rangle} =$ Prob$(Y < y)$, where $Y = \langle\varphi,s\rangle$ is a conventional scalar random variable, once φ is fixed.

For completeness, we also recall the meaning of the underlying terminology in the context of a generic (normed or nuclear) space $\mathscr{X}$ of test functions.

DEFINITION 4.7 (positive definite functional) A complex-valued functional $G : \mathscr{X} \to \mathbb{C}$ defined over the function space $\mathscr{X}$ is said to be *positive definite* if

$$\sum_{m=1}^{N}\sum_{n=1}^{N} G(\varphi_m - \varphi_n)\xi_m\bar{\xi}_n \geq 0$$

for every possible choice of $\varphi_1, \ldots, \varphi_N \in \mathscr{X}$, $\xi_1, \ldots, \xi_N \in \mathbb{C}$, and $N \in \mathbb{N}^+$.

DEFINITION 4.8 (Continuous functional) A functional $G : \mathscr{X} \to \mathbb{R}$ (or $\mathbb{C}$) is said to be *continuous* (with respect to the topology of the function space $\mathscr{X}$) if, for any convergent sequence (φ_i) in $\mathscr{X}$ with limit $\varphi \in \mathscr{X}$, the sequence $G(\varphi_i)$ converges to $G(\varphi)$; that is,

$$\lim_{i} G(\varphi_i) = G(\lim_{i} \varphi_i).$$

An essential element of our formulation is that Schwartz' space of test functions $\mathscr{S}(\mathbb{R}^d)$ is nuclear (see Section 3.1.3), as required by the Minlos–Bochner theorem (Theorem 3.9). The latter expresses the one-to-one correspondence (in the form of an infinite-dimensional Fourier pair) between the characteristic functional $\widehat{\mathscr{P}_s} : \mathscr{S}(\mathbb{R}^d) \to \mathbb{C}$ and the measure $\mathscr{P}_s$ on $\mathscr{S}'(\mathbb{R}^d)$ that uniquely characterizes the generalized process s. The truly powerful aspect of the theorem is that it suffices to check that $\widehat{\mathscr{P}_s}$ satisfies the three defining conditions – positive definiteness, continuity over $\mathscr{S}(\mathbb{R}^d)$, and normalization – to prove that it is a valid characteristic functional, which then automatically ensures the existence of the process since the corresponding measure over $\mathscr{S}'(\mathbb{R}^d)$ is well defined.

The formulation of our generative model (4.24) in that context is

$$\widehat{\mathscr{P}}_s(\varphi) = \widehat{\mathscr{P}}_{L^{-1}w}(\varphi) = \widehat{\mathscr{P}}_w(L^{-1*}\varphi), \tag{4.26}$$

where $\widehat{\mathscr{P}}_w(\varphi)$ is the characteristic functional of the innovation process w.

THEOREM 4.17 (Generalized innovation model)   *Let* $U = L^{-1*}$ *be a linear operator that satisfies the two conditions*

(1) Left-inverse property: $UL^*\varphi = \varphi$ *for all* $\varphi \in \mathscr{S}(\mathbb{R}^d)$ *where* L^* *is the adjoint of some given (whitening) operator* L*;*
(2) Stability: U *is a continuous linear map from* $\mathscr{S}(\mathbb{R}^d)$ *into itself, or, by extension,* $\mathscr{S}(\mathbb{R}^d) \to \mathscr{R}(\mathbb{R}^d)$*.*

Then, the generalized stochastic process s that is characterized by $\mathbb{E}\{e^{j\langle \varphi,s \rangle}\} = \widehat{\mathscr{P}}_s(\varphi) = \exp\left(\int_{\mathbb{R}^d} f\left(L^{-1*}\varphi(r)\right) dr\right)$ *is well defined in* $\mathscr{S}'(\mathbb{R}^d)$ *and satisfies the innovation model* $Ls = w$*, where w is a Lévy innovation with exponent f.*
   When f is p-admissible (see Definition 4.4) with $p \geq 1$, the second condition can be replaced by the weaker requirement that U *is a continuous linear map from* $\mathscr{S}(\mathbb{R}^d)$ *into* $L_p(\mathbb{R}^d)$*.*

Proof   First, we prove that s is a bona fide generalized stochastic process in $\mathscr{S}'(\mathbb{R}^d)$ by showing that $\widehat{\mathscr{P}}_s(\varphi)$ is a continuous, positive definite functional on $\mathscr{S}(\mathbb{R}^d)$ such that $\widehat{\mathscr{P}}_s(0) = 1$ (by the Minlos–Bochner theorem).
   The Lévy noise functional $\widehat{\mathscr{P}}_w(\varphi) = \exp\left(\int_{\mathbb{R}^d} f\left(\varphi(r)\right) dr\right)$ is continuous over $\mathscr{S}(\mathbb{R}^d)$ by construction (see Theorem 4.8). This, together with the assumption that L^{-1*} is a continuous operator on $\mathscr{S}(\mathbb{R}^d)$, implies that the composed functional $\widehat{\mathscr{P}}_s(\varphi) = \widehat{\mathscr{P}}_w(L^{-1*}\varphi)$ is continuous on $\mathscr{S}(\mathbb{R}^d)$. The reasoning is also applicable when L^{-1*} is a continuous operator $\mathscr{S}(\mathbb{R}^d) \to L_p(\mathbb{R}^d)$ and $\widehat{\mathscr{P}}_w(\varphi)$ is continuous over $L_p(\mathbb{R}^d)$ – see the triangular diagram in Figure 3.1 with $\mathscr{X} = \mathscr{S}(\mathbb{R}^d)$ and $\mathscr{Y} = L_p(\mathbb{R}^d)$. This latter scenario is covered by Theorem 8.2, which establishes the positive definiteness and continuity of $\widehat{\mathscr{P}}_w$ over $L_p(\mathbb{R}^d)$ when f is p-admissible (see Section 8.2). The case where L^{-1*} is a continuous operator $\mathscr{S}(\mathbb{R}^d) \to \mathscr{R}(\mathbb{R}^d)$ is handled in the same fashion by invoking Proposition 8.1.
   Next, for any given set of functions $\varphi_1, \ldots, \varphi_N \in \mathscr{S}(\mathbb{R}^d)$ and coefficients $\xi_1, \ldots, \xi_N \in \mathbb{C}$, we have

$$\sum_{m=1}^{N}\sum_{n=1}^{N} \widehat{\mathscr{P}}_s(\varphi_m - \varphi_n)\xi_m\bar{\xi}_n$$

$$= \sum_{m=1}^{N}\sum_{n=1}^{N} \widehat{\mathscr{P}}_w\left(L^{-1*}(\varphi_m - \varphi_n)\right)\xi_m\bar{\xi}_n$$

$$= \sum_{m=1}^{N}\sum_{n=1}^{N} \widehat{\mathscr{P}}_w(L^{-1*}\varphi_m - L^{-1*}\varphi_n)\xi_m\bar{\xi}_n \qquad \text{(by linearity)}$$

$$\geq 0, \qquad \text{(by the positivity of } \widehat{\mathscr{P}}_w)$$

which shows that $\widehat{\mathscr{P}_s}$ is positive definite on $\mathscr{S}(\mathbb{R}^d)$. Finally, $\widehat{\mathscr{P}_s}(0) = \widehat{\mathscr{P}_w}(L^{-1*}0) = \widehat{\mathscr{P}_w}(0) = 1$, which completes the first part of the proof.

The above result and stability conditions ensure that the action of the inverse operator L^{-1} is well defined over the relevant subset of tempered distributions, which justifies the formal manipulation made in (4.24). Now, if L^{-1*} is a proper left inverse of L^*, we have that

$$\langle \varphi, w \rangle = \langle \underbrace{L^{-1*}L^*}_{\text{Id}} \varphi, w \rangle = \langle L^*\varphi, \underbrace{L^{-1}w}_{s} \rangle = \langle \varphi, Ls \rangle,$$

which proves that the generalized process $s = L^{-1}w$ satisfies (4.23). $\qquad\square$

The next chapters are devoted to the investigation of specific instances of this model and to making sure that the conditions for existence in Theorem 4.17 are met. We shall also discuss the conceptual connection with splines and wavelets. This connection is fundamental to our purpose. In Chapter 9, we shall then use the generalized innovation model to show that the primary statistical features of the input noise are essentially transferred to the signal as well as to the transform domain. The main point is that the marginal distributions are all part of the same infinitely divisible family as long as the composition of the mixing procedure (L^{-1}) and the signal analysis remains linear. On the other hand, the amount of coupling and level of interdependence will strongly depend on the nature of the transformation.

4.6   Bibliographical notes

Sections 4.2 and 4.3

Infinitely divisible distributions were introduced by de Finetti in 1929, and their primary properties established by Kolmogorov, Lévy, Khintchine, and Feller in the 1930s [BDR02, SVH03, MR06]. They constitute a classical topic in probability theory in tight connection with the central-limit theorem [GK68, Fel71]. The general expression (4.3) for the exponent of the characteristic function of an infinitely divisible random variable was given by Lévy [Lév34]. Shortly after, Khintchine provided a purely analytical derivation [Khi37b, Khi37a]. The sketch of proof in Section 4.2.4 is adapted from [Khi37a], whose translation is given in [MR06].

We have chosen to name Theorem 4.1 after Lévy and Schoenberg because it essentially results from the combination of two fundamental theorems in harmonic analysis named after these authors [Lév34, Sch38]. While their groundwork dates back to the 1930s, it took until the late 1960s to reformulate the Lévy–Khintchine characterization in terms of the (conditional) positive definiteness of the exponent [Joh66].

The p-admissibility condition was introduced in [UTS14] in order to simplify the derivation of bounds and continuity properties related to Lévy exponents. The argument concerning the compatibility of infinite divisibility with the notion of sparsity is also adapted from this paper.

For additional information on id laws, we refer the reader to [SVH03, Sat94, CT04]; these works also contain the ground material for the specification of the symmetric id distributions in Table 4.1. Further distributional properties relating to decay and the effect of repeated convolution are exposed in Chapter 9.

Section 4.4

The specification of white Lévy noise by means of its characteristic functional (see Definition 4.5) is based on a series of theorems by Gelfand and Vilenkin [GV64]. Interestingly, the generic form (4.13) is not only sufficient for defining a (stationary) innovation, as proven by these authors, but also necessary if one adds the observability constraint that $X_{\mathrm{id}} = \langle \mathrm{rect}, w \rangle$ is a well-defined random variable [AU14]. The restriction of the family to the space of tempered distributions was investigated by Fageot *et al.* [Fag14]. Theorem 4.9 is adapted from [UT11].

The abstract characterization of infinite divisibility and the full generalization of the Lévy–Khintchine formula for measures over topological vector spaces is covered in the works of Fernique and Prakasa Rao [Fer67, PR70].

Section 4.5

The innovation or filtered-white-noise model has a long tradition in communication and statistical signal processing in relation to time-series analysis [BS50, WM57, Kai70]. The classical assumption is that the excitation noise (innovation) is Gaussian and that the shaping filter is causal with a causal inverse. The innovation, as defined by Wiener and Masani, then corresponds to the unpredictable part of the signal; that is, the difference between the value of the signal at given time t and the optimal linear forecast of that value based on the information available prior to t. Thanks to the Gaussian hypothesis, one can then formulate a coherent correlation theory of such processes, using standard Fourier and Hilbert-space techniques, in which continuous-domain white noise only intervenes as a formal entity; that is, a Gaussian process whose power spectrum is a constant. This purely spectral description of white noise is consistent with the Wiener–Khintchine theorems,[6] which explains its popularity among engineers [Pap91, Yag86].

The non-Gaussian extension of the innovation model that is presented in this chapter is conceptually similar, but relies on the more elaborate definition of continuous-domain white noise and the functional tools that were developed by Gelfand to formulate his theory of generalized stochastic processes [Gel55, GV64]. A slightly more restrictive version of the model with Gaussian and/or impulsive Poisson excitation was presented in [UT11]. The original statement of Theorem 4.17 for $d = 1$ can be found in [UTS14]. While the level of generality of this result is sufficient for our purpose, we must warn the reader that the framework cannot directly handle non-linear transformations because the underlying objects are generalized functions, which are intrinsically linear. For completeness, we mention the existence of an extended theory of white noise, due to Hida, which is aimed at overcoming this limitation [HKPS93, HS04, HS08]. This theory gives a meaning to certain classes of non-linear white-noise functionals – in analogy with Itô's calculus – but it is mathematically quite involved and beyond the scope of this book.

---

[6] The Wiener–Khintchine theorem states that the autocorrelation function of a second-order stationary process is the inverse Fourier transform of its power spectrum.

5 Operators and their inverses

In this chapter we review three classes of linear shift-invariant (LSI) operators: convolution operators with stable LSI inverses, operators that are linked with ordinary differential equations, and fractional operators.

The first class, considered in Section 5.2, is composed of the broad family of multidimensional operators whose inverses are stable convolution operators – or filters. Convolution operators play a central role in signal processing. They are easy to characterize mathematically via their impulse response. The corresponding generative model for stochastic processes amounts to LSI filtering of a white noise, which automatically yields stationary processes.

Our second class is the 1-D family of ordinary differential operators with constant coefficients, which is relevant to a wide range of modeling applications. In the "stable" scenario, reviewed in Section 5.3, these operators admit stable LSI inverses on $\mathscr{S}'$ and are therefore included in the previous category. On the other hand, when the differential operators have one or more zeros on the imaginary axis (the marginally stable/unstable case), they find a non-trivial null space in $\mathscr{S}'$, which consists of (exponential) polynomials. This implies that they are no longer unconditionally invertible on $\mathscr{S}'$, and that we can at best identify left- or right-side inverses, which should additionally fulfill appropriate "boundedness" requirements in order to be usable in the definition of stochastic processes. However, as we shall see in Section 5.4, obtaining an inverse with the required boundedness properties is feasible but requires giving up shift-invariance. As a consequence, stochastic processes defined by these operators are generally non-stationary.

The third class of LSI operators, investigated in Section 5.5, consists of fractional derivatives and/or Laplacians in one and several dimensions. Our focus is on the family of linear operators on $\mathscr{S}$ that are simultaneously homogeneous (scale-invariant up to a scalar coefficient) and invariant under shifts and rotations. These operators are intimately linked to self-similar processes and fractals [BU07, TVDVU09]. Once again, finding a stable inverse operator to be used in the definition of self-similar processes poses a mathematical challenge since the underlying system is inherently unstable. The difficulty is evidenced by the fact that statistical self-similarity is generally not compatible with stationarity, which means that a non-shift-invariant inverse operator needs to be constructed. Here again, a solution may be found by extending the approach used for the previous class of operators. From our first example in Section 5.1, we shall actually see that one is able to reconcile the classical theory of stationary processes with that of self-similar ones by viewing the latter as a limit case of the former.

Before we begin our discussion of operators, let us formalize some notions of invariance.

DEFINITION 5.1 (Translation-invariance)   An operator T is *shift-* (or *translation-*) *invariant* if and only if, for any function φ in its domain and any $r_0 \in \mathbb{R}^d$,

$$T\{\varphi(\cdot - r_0)\}(r) = T\{\varphi\}(r - r_0).$$

DEFINITION 5.2 (Scale-invariance)   An operator T is *scale-invariant* (homogeneous) of order γ if and only if, for any function φ in its domain,

$$T\{\varphi\}(r/a) = |a|^\gamma T\{\varphi(\cdot/a)\}(r),$$

where $a \in \mathbb{R}^+$ is the dilation factor.

An alternative version of the scale-invariance condition is

$$T\{\varphi(a\cdot)\}(r) = |a|^\gamma T\{\varphi\}(ar), \tag{5.1}$$

where a now represents a contraction factor.

DEFINITION 5.3 (Rotation-invariance)   An operator T is scalarly *rotation-invariant* if and only if, for any function φ in its domain,

$$T\{\varphi\}(\mathbf{R}^T r) = T\{\varphi(\mathbf{R}^T \cdot)\}(r),$$

where $\mathbf{R}$ is any orthogonal matrix in $\mathbb{R}^{d \times d}$ (by using orthogonal matrices in the definition, we take into account both proper and improper rotations, with respective determinants 1 and -1).

5.1   Introductory example: first-order differential equation

To fix ideas, let us consider the generic first-order differential operator $L = D - \alpha Id$ with $\alpha \in \mathbb{C}$, where $D = \frac{d}{dr}$ and Id are the derivative and identity operators, respectively. Clearly, L is LSI, but generally not scale-invariant unless $\alpha = 0$. The corresponding linear system with (deterministic or stochastic) output s and input w is defined by the differential equation $\frac{d}{dr}s(r) - \alpha s(r) = w(r)$. Under the classical stability assumption $\mathrm{Re}(\alpha) < 0$ (pole in the left half of the complex plane), its impulse response is given by

$$\rho_\alpha(r) = \mathscr{F}^{-1}\left\{\frac{1}{j\omega - \alpha}\right\}(r) = \mathbb{1}_+(r)e^{\alpha r} \tag{5.2}$$

and is rapidly decaying. This provides us with an explicit characterization of the inverse operator, which reduces to a simple convolution with a decreasing causal exponential:

$$(D - \alpha Id)^{-1}\varphi = \rho_\alpha * \varphi.$$

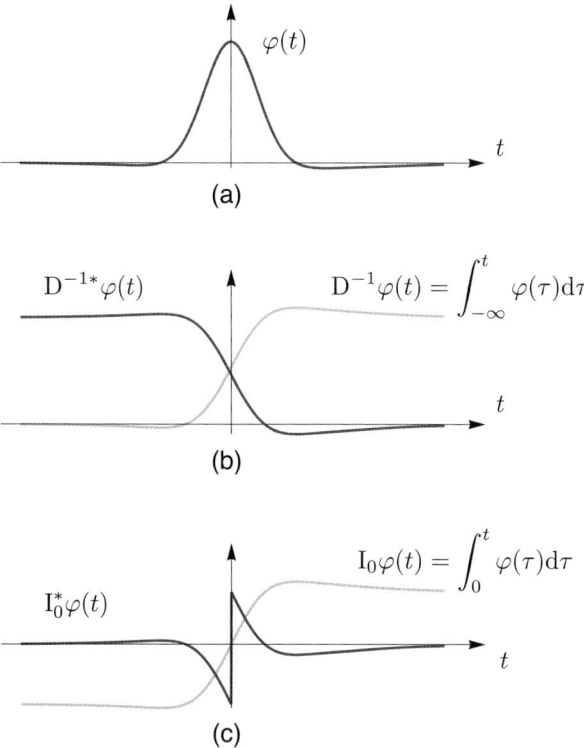

Figure 5.1 Comparison of antiderivative operators. (a) Input signal. (b) Result of shift-invariant integrator and its adjoint. (c) Result of scale-invariant integrator I_0 and its L_p-stable adjoint I_0^*; the former yields a signal that vanishes at the origin, while the latter enforces the decay of the output as $t \to -\infty$ at the cost of a jump discontinuity at the origin.

Likewise, it is easy to see that the corresponding adjoint inverse operator L^{-1*} is specified by

$$(D - \alpha Id)^{-1*}\varphi = \rho_\alpha^\vee * \varphi,$$

where $\rho_\alpha^\vee(r) = \rho_\alpha(-r)$ is the time-reversed version of ρ_α. Thanks to its rapid decay, $\rho_\alpha^\vee$ defines a continuous linear translation-invariant map from $\mathscr{S}(\mathbb{R})$ into itself.

This allows us to express the solution of the first-order SDE as a filtered version of the input noise $s_\alpha = \rho_\alpha * w$. It follows that s_α is a stationary process that is completely specified by its characteristic form $\mathbb{E}\{e^{j\langle\varphi, s_\alpha\rangle}\} = \exp\left(\int_{\mathbb{R}} f\left((\rho_\alpha^\vee * \varphi)(r)\right) dr\right)$, where f is the Lévy exponent of the innovation w (see Section 4.5).

Let us now focus our attention on the limit case $\alpha = 0$, which yields an operator $L = D$ that is scale-invariant. Here too, it is possible to specify the LSI inverse (integrator)

$$D^{-1}\varphi(r) = \int_{-\infty}^{r} \varphi(\tau)\, d\tau = (\mathbb{1}_+ * \varphi)(r),$$

whose output is well defined pointwise when $\varphi \in \mathscr{S}(\mathbb{R})$. The less favorable aspect is that the classical LSI integrator does not fulfill the usual stability requirement due to the non-integrability of its impulse response $\mathbb{1}_+ \notin L_1(\mathbb{R})$. This implies that $D^{-1}*\varphi = \mathbb{1}_+^\vee * \varphi$ is generally not in $L_p(\mathbb{R})$. Thus, we are no longer fulfilling the admissibility condition in Theorem 4.17. The source of the problem is the lack of decay of $D^{-1}*\varphi(r)$ as $r \to -\infty$ when $\int_\mathbb{R} \varphi(\tau)\,d\tau = \widehat{\varphi}(0) \neq 0$ (see Figure 5.1b). Fortunately, this can be compensated by defining the modified antiderivative operator

$$I_0^*\varphi(r) = \int_r^\infty \varphi(\tau)\,d\tau - \mathbb{1}_+(-r)\widehat{\varphi}(0) = (\mathbb{1}_+^\vee * \varphi)(r) - \int_\mathbb{R} \varphi(\tau)\,d\tau\,\mathbb{1}_+^\vee(r)$$

$$= D^{-1}*\varphi(r) - (D^{-1}*\varphi)(-\infty)\mathbb{1}_+^\vee(r), \tag{5.3}$$

which happens to be the *only* left inverse of $D^* = -D$ that is both scale-invariant and L_p-stable for any $p > 0$. The adjoint of I_0^* specifies the adjusted [1] integrator

$$I_0\varphi(r) = \int_0^r \varphi(\tau)\,d\tau = D^{-1}\varphi(r) - (D^{-1}\varphi)(0), \tag{5.4}$$

which is the correct scale-invariant inverse of D for our formulation, to be applied to elements of $\mathscr{S}'(\mathbb{R})$. It follows that the solution of the corresponding (unstable) SDE can be expressed as $s = I_0 w$, which is a well-defined stochastic process as long as the input noise is Lévy p-admissible with $p \geq 1$ (by Theorem 4.17). We note that the price to pay for the stabilization of the solution is to give up on shift-invariance. Indeed, the adjusted integrator is such that it imposes the boundary condition $s(0) = (I_0 w)(0) = 0$ (see Figure 5.1c), which is incompatible with stationarity, but a necessary condition for self-similarity. The so-constructed processes are fully specified by their characteristic form $\exp(\int_\mathbb{R} f(I_0\varphi(r))\,dt)$, where f is a Lévy exponent. Based on this representation, we can show that these are equivalent to the Lévy processes that are usually defined for $r \geq 0$ only [Sat94]. While this connection with the classical theory of Lévy processes is already remarkable, it turns out that the underlying principle is quite general and applicable to a much broader family of operators, provided that we can ensure L_p-stability.

5.2   Shift-invariant inverse operators

The association of LSI operators with convolution integrals will be familiar to most readers. In effect, we saw in Section 3.3.5 that, as a consequence of Schwartz' kernel theorem, every continuous LSI operator $L : \mathscr{S}(\mathbb{R}^d) \to \mathscr{S}'(\mathbb{R}^d)$ corresponds to a convolution

$$L : \varphi \mapsto L\{\delta\} * \varphi$$

with a kernel (*impulse response*) $L\{\delta\} \in \mathscr{S}'(\mathbb{R}^d)$. Moreover, by the convolution-product rule, we may also characterize L by a multiplication in the Fourier domain:

$$L : \varphi \mapsto \mathscr{F}^{-1}\{\widehat{L}(\omega)\widehat{\varphi}(\omega)\}.$$

---

[1] Any two valid right inverses can only differ by a component (constant) that is in the null space of the operator. The scale-invariant solution is the one that forces the output to vanish at the origin.

We call $\widehat{L}$ the *Fourier multiplier* or *symbol* associated with the operator L.

From the Fourier-domain characterization of L, we see that, if $\widehat{L}$ is smooth and does not grow too fast, then L maps $\mathscr{S}(\mathbb{R}^d)$ back into $\mathscr{S}(\mathbb{R}^d)$. This is in particular true if L{δ} (the impulse response) is an ordinary locally integrable function with rapid decay. It is also true if L is a 1-D linear differential operator with constant coefficients, in which case L{δ} is a finite sum of derivatives of the Dirac distribution and the corresponding Fourier multiplier $\widehat{L}(\omega)$ is a polynomial in $j\omega$.

For operators with smooth Fourier multipliers that are nowhere zero in $\mathbb{R}^d$ and not decaying (or decaying slowly) at ∞, we can define the inverse $L^{-1} : \mathscr{S}(\mathbb{R}^d) \to \mathscr{S}(\mathbb{R}^d)$ of L by

$$L^{-1} : \varphi \mapsto \mathscr{F}^{-1}\left\{\frac{\widehat{\varphi}(\omega)}{\widehat{L}(\omega)}\right\}.$$

This inverse operator is also linear and shift-invariant, and has the convolution kernel

$$\rho_L = \mathscr{F}^{-1}\left\{\frac{1}{\widehat{L}(\omega)}\right\}, \tag{5.5}$$

which is in effect the *Green's function* of the operator L. Thus, we may write

$$L^{-1} : \varphi \mapsto \rho_L * \varphi.$$

For the cases in which $\widehat{L}(\omega)$ vanishes at some points, its inverse $1/\widehat{L}(\omega)$ is not in general a locally integrable function, but even in the singular case we may still be able to regularize the singularities at the zeros of $\widehat{L}(\omega)$ and obtain a singular "generalized function" whose inverse Fourier transform, per (5.5), once again yields a convolution kernel ρ_L that is a Green's function of L. The difference is that in this case, for an arbitrary $\varphi \in \mathscr{S}(\mathbb{R}^d)$, $\rho_L * \varphi$ may no longer belong to $\mathscr{S}(\mathbb{R}^d)$.

As in our introductory example, the simplest scenario occurs when the inverse operator L^{-1} is shift-invariant with an impulse response ρ_L that has sufficient decay for the system to be BIBO-stable (bounded input, bounded output).

PROPOSITION 5.1   *Let* $L^{-1}\varphi(r) = (\rho_L * \varphi)(r) = \int_{\mathbb{R}^d} \rho_L(r')\varphi(r - r')\, dr'$ *with* $\rho_L \in L_1(\mathbb{R}^d)$ *(or, more generally, where* ρ_L *is a complex-valued Borel measure of bounded variation). Then,* L^{-1} *and its adjoint specified by* $L^{-1*}\varphi(r) = (\rho_L^{\vee} * \varphi)(r) = \int_{\mathbb{R}^d} \rho_L(-r') \varphi(r - r')\, dr'$ *are both* L_p-*stable in the sense that*

$$\|L^{-1}\varphi\|_{L_p} \le \|\rho_L\|_{L_1} \|\varphi\|_{L_p}$$
$$\|L^{-1*}\varphi\|_{L_p} \le \|\rho_L\|_{L_1} \|\varphi\|_{L_p}$$

for all $p \ge 1$. *In particular, this ensures that* L^{-1*} *continuously maps* $\mathscr{S}(\mathbb{R}^d) \to L_p(\mathbb{R}^d)$.

The result follows from Theorem 3.5. For the sake of completeness, we shall establish the bound based on the two extreme cases $p = 1$ and $p = +\infty$.

Proof   To obtain the L_1 bound, we manipulate the norm of the convolution integral as

$$\|\rho_L * \varphi\|_{L_1} = \int_{\mathbb{R}^d} \left| \int_{\mathbb{R}^d} \rho_L(r)\varphi(r' - r) \, dr \right| dr'$$

$$\leq \int_{\mathbb{R}^d} \int_{\mathbb{R}^d} |\rho_L(r)\varphi(s)| \, dr \, ds \qquad \text{(change of variable } s = r' - r)$$

$$= \int_{\mathbb{R}^d} |\rho_L(r)| \, dr \int_{\mathbb{R}^d} |\varphi(s)| \, ds = \|\rho_L\|_{L_1} \|\varphi\|_{L_1},$$

where the exchange of integrals is justified by Fubini's theorem. The corresponding L_∞ bound follows from the simple pointwise estimate

$$|(\rho_L * \varphi)(r)| \leq \|\varphi\|_{L_\infty} \int_{\mathbb{R}^d} |\rho_L(r')| \, dr' = \|\rho_L\|_{L_1} \|\varphi\|_{L_\infty}$$

for any $r \in \mathbb{R}^d$. The final step is to invoke the Riesz–Thorin theorem (Theorem 3.4) which yields Young's inequality for L_p functions (see (3.11)), and hence proves the desired result for $1 \leq p \leq \infty$. The inequality also applies to the adjoint operator since the latter amounts to a convolution with the reversed impulse response $\rho_L^\vee(r) = \rho(-r) \in L_1(\mathbb{R}^d)$. The final statement simply follows from the fact that the convergence of a sequence of functions in the (strong) topology of $\mathscr{S}(\mathbb{R}^d)$ implies convergence in all L_p norms for $p > 0$. $\square$

Note that the L_1 condition in Proposition 5.1 is the standard hypothesis that is made in the theory of linear systems to ensure the BIBO stability of an analog filter. It is slightly stronger than the total variation (TV) condition in Theorem 3.5, which is *necessary and sufficient* for both BIBO ($p = \infty$) and L_1 stabilities.

If, in addition, $\rho_L(r)$ decays faster than any polynomial (e.g., is compactly supported or decays exponentially), then we can actually ensure $\mathscr{S}$-continuity so that there is no restriction on the class of corresponding stochastic processes.

PROPOSITION 5.2   *Let* $L^{-1}\varphi(r) = (\rho_L * \varphi)(r)$ *with* $|\rho_L(r)| \leq \frac{C_n}{1 + \|r\|^n}$ *for all* $n \in \mathbb{Z}^+$ *and* $r \in \mathbb{R}^d$. *Then,* L^{-1} *and* L^{-1*} *are* $\mathscr{S}$-*continuous in the sense that* $\varphi \in \mathscr{S} \Rightarrow L^{-1}\varphi, L^{-1*}\varphi \in \mathscr{S}$ *with both operators being bounded in an appropriate sequence of seminorms.*

The key here is that the convolution with ρ_L preserves the rapid decay of the test function φ. The degree of smoothness of the output is not an issue because, for non-constant functions, the convolution operation commutes with differentiation.

The good news is that the entire class of stable 1-D differential systems with rational transfer functions and poles in the left half of the complex plane falls into the category of Proposition 5.2. The application of such operators provides us with a convenient mechanism for solving ordinary differential equations, as detailed in Section 5.3.

The $\mathscr{S}$-continuity property is important for our formulation. It also holds for all shift-invariant differential operators whose impulse response is a point distribution; e.g., $(D - \mathrm{Id})\{\delta\} = \delta' - \delta$. It is preserved under convolution, which justifies the factorization of operators into simpler constituents.

5.3 Stable differential systems in 1-D

The generic form of a linear shift-invariant differential equation in 1-D with (deterministic or random) output s and driving term w is

$$\sum_{n=0}^{N} a_n D^n s = \sum_{m=0}^{M} b_m D^m w, \tag{5.6}$$

where the a_n and b_m are arbitrary complex coefficients with the normalization constraint $a_N = 1$. Equation (5.6) thus covers the general 1-D case of $Ls = w$, where L is a shift-invariant operator with the rational transfer function

$$\widehat{L}(\omega) = \frac{(j\omega)^N + a_{N-1}(j\omega)^{N-1} + \cdots + a_1(j\omega) + a_0}{b_M(j\omega)^M + \cdots + b_1(j\omega) + b_0} = \frac{p_N(j\omega)}{q_M(j\omega)}. \tag{5.7}$$

The poles of the system, which are the roots of the characteristic polynomial $p_N(\zeta) = \zeta^N + a_{N-1}\zeta^{N-1} + \cdots + a_0$ with Laplace variable $\zeta \in \mathbb{C}$, are denoted by $\{\alpha_n\}_{n=1}^{N}$. In the standard causal-stable scenario where $\mathrm{Re}(\alpha_n) < 0$ for $n = 1, \ldots, N$, the solution is obtained as

$$s(r) = L^{-1} w(r) = (\rho_L * w)(r),$$

where ρ_L is the causal Green's function of L specified by (5.5).

In practice, the determination of ρ_L is based on the factorization of the transfer function of the system as

$$\frac{1}{\widehat{L}(\omega)} = q_M(j\omega) \prod_{n=1}^{N} \frac{1}{j\omega - \alpha_n} \tag{5.8}$$

$$= b_M \frac{\prod_{m=1}^{M}(j\omega - \gamma_m)}{\prod_{n=1}^{N}(j\omega - \alpha_n)}, \tag{5.9}$$

which is then broken into simple constituents, either by serial composition of first-order factors or by decomposition into simple partial fractions. We are providing the fully factorized form (5.9) of the transfer function to recall the property that a stable Nth-order system is completely characterized by its poles $\{\alpha_n\}_{n=1}^{N}$ and zeros $\{\gamma_m\}_{m=1}^{M}$, up to the proportionality factor b_M.

Since the $\mathscr{S}$-continuity property is preserved through the composition of convolution operators, we shall rely on (5.9) to factorize L^{-1} into elementary operators. To that end, we shall study the effect of simple constituents (first-order differential operators with stable inverses) before considering their composition into higher-order operators. We shall also treat the leading polynomial factor $q_M(j\omega)$ in (5.8) separately because it corresponds to a convolution operator whose impulse response is the point distribution $\sum_{m=0}^{M} b_m \delta^{(m)}$. The latter is $\mathscr{S}$-continuous, irrespective of the choice of coefficients b_m (or, equivalently, the zeros γ_m in (5.9)).

5.3.1   First-order differential operators with stable inverses

The first-order differential operator

$$P_\alpha = D - \alpha \mathrm{Id}$$

corresponds to the convolution kernel (*impulse response*) $(\delta' - \alpha\delta)$ and Fourier multiplier $(j\omega - \alpha)$. For $\mathrm{Re}(\alpha) \neq 0$ (the *stable* case in signal-processing parlance), the Fourier multiplier, $(j\omega - \alpha)^{-1}$, is non-singular, with the well-defined inverse Fourier transform

$$\rho_\alpha(r) = \mathscr{F}^{-1}\left\{\frac{1}{j\omega - \alpha}\right\}(r) = \begin{cases} e^{\alpha r}\mathbb{1}_{[0,\infty)}(r) & \text{if } \mathrm{Re}(\alpha) < 0 \\ -e^{\alpha r}\mathbb{1}_{(-\infty,0]}(r) & \text{if } \mathrm{Re}(\alpha) > 0. \end{cases}$$

Clearly, in either the causal $(\mathrm{Re}(\alpha) < 0)$ or the anti-causal $(\mathrm{Re}(\alpha) > 0)$ case, these functions decay rapidly at infinity, and convolutions with them map $\mathscr{S}(\mathbb{R})$ back into itself. Moreover, a convolution with ρ_α inverts $P_\alpha : \mathscr{S}(\mathbb{R}) \to \mathscr{S}(\mathbb{R})$ from both the left and the right, for which we can write

$$P_\alpha^{-1}\varphi = \rho_\alpha * \varphi \tag{5.10}$$

for $\mathrm{Re}(\alpha) \neq 0$.

Since P_α and P_α^{-1}, $\mathrm{Re}(\alpha) \neq 0$, are both $\mathscr{S}$-continuous (continuous from $\mathscr{S}(\mathbb{R})$ into $\mathscr{S}(\mathbb{R})$), their action can be transferred to the space $\mathscr{S}'(\mathbb{R})$ of Schwartz distributions by identifying the adjoint operator P_α^* with $(-P_{-\alpha})$, in keeping with the identity

$$\langle P_\alpha\varphi, \phi \rangle = -\langle \varphi, P_{-\alpha}\phi \rangle$$

on $\mathscr{S}(\mathbb{R})$. We recall that, in this context, $\langle \varphi, \phi \rangle$ denotes the bilinear form $\int_{\mathbb{R}} \varphi(r)\phi(r)\,dr$, not the Hermitian product.

5.3.2   Higher-order differential operators with stable inverses

As noted earlier, the preservation of continuity by composition facilitates the study of differential systems by permitting us to decompose higher-order operators into first-order factors. Specifically, let us consider the equivalent factorized representation of the Nth-order differential equation (5.6) given by

$$P_{\alpha_1} \cdots P_{\alpha_N}\{s\}(r) = q_M(D)\{w\}(r), \tag{5.11}$$

where $\{\alpha_n\}_{n=1}^N$ are the poles of the system and where $q_M(D) = \sum_{m=0}^M b_m D^m$ is the Mth-order differential operator acting on the right-hand side of (5.6). Under the assumption that $\mathrm{Re}(\alpha_n) \neq 0$ (causal or anti-causal stability), we can invert the operators acting on the left-hand side of (5.11), which allows us to express the solution of the differential equation as

$$s(r) = \underbrace{P_{\alpha_N}^{-1} \cdots P_{\alpha_1}^{-1}}_{L^{-1}} q_M(D)\{w\}(r). \tag{5.12}$$

This translates into the following formulas for the corresponding inverse operator L^{-1}:

$$L^{-1} = P_{\alpha_N}^{-1} \cdots P_{\alpha_1}^{-1} \, q_M(D)$$
$$= b_M P_{\alpha_N}^{-1} \cdots P_{\alpha_1}^{-1} P_{\gamma_1} \cdots P_{\gamma_N},$$

which are consistent with (5.8) and (5.9), respectively. These operator-based manipulations are legitimate since all the elementary constituents in (5.11) and (5.12) are $\mathscr{S}$-continuous convolution operators. We also recall that all L_p-stable and, a fortiori, $\mathscr{S}$-continuous, convolution operators satisfy the properties of commutativity, associativity, and distributivity, so that the ordering of the factors in (5.11) and (5.12) is immaterial. Interestingly, this divide-and-conquer approach to the problem of inverting a differential operator is also extendable to the unstable scenarios (with $\mathrm{Re}(\alpha_n) = 0$ for some values of n), the main difference being that the ordering of operators becomes important (partial loss of commutativity).

5.4 Unstable Nth-order differential systems

Classically, a differential system is categorized as being unstable when some of its poles arc in the right complex half-plane, including the imaginary axis. Mathematically, there is no fundamental reason for excluding the cases $\mathrm{Re}(\alpha_n) > 0$ because one can simply switch to an anti-causal configuration which preserves the exponential decay of the response, as we did in defining these inverses in Section 5.3.1.

The only tricky situation occurs for purely imaginary poles of the form $\alpha_m = j\omega_0$ with $\omega_0 \in \mathbb{R}$, to which we now turn our attention.

5.4.1 First-order differential operators with unstable shift-invariant inverses

Once again, we begin with first-order operators. Unlike the case of $\mathrm{Re}(\alpha) \neq 0$, for purely imaginary $\alpha = j\omega_0$, $P_{j\omega_0}$ is not a surjective operator from $\mathscr{S}(\mathbb{R})$ to $\mathscr{S}(\mathbb{R})$ (meaning it maps $\mathscr{S}(\mathbb{R})$ into a proper subset of itself, not into the entire space), because it introduces a frequency-domain zero at ω_0. This implies that we cannot find an operator $U : \mathscr{S}(\mathbb{R}) \to \mathscr{S}(\mathbb{R})$ that is a right inverse of $P_{j\omega_0}$. In other words, we cannot fulfill $P_{j\omega_0} U\varphi = \varphi$ for all $\varphi \in \mathscr{S}(\mathbb{R})$ subject to the constraint that $U\varphi \in \mathscr{S}(\mathbb{R})$.

If we now consider $P_{j\omega_0}$ as an operator on $\mathscr{S}'(\mathbb{R})$ $\big($as the adjoint of $(-P_{-j\omega_0})\big)$, then this operator is not one-to-one. More precisely, it has a non-trivial null space consisting of multiples of the complex sinusoid $e^{j\omega_0 r}$. Consequently, on $\mathscr{S}'(\mathbb{R})$, $P_{j\omega_0}$ does not have a left inverse $U' : \mathscr{S}'(\mathbb{R}) \to \mathscr{S}'(\mathbb{R})$ with $U'P_{j\omega_0}f = f$ for all $f \in \mathscr{S}'(\mathbb{R})$.

The main conclusion of concern to us is that $P_{j\omega_0} : \mathscr{S}(\mathbb{R}) \to \mathscr{S}(\mathbb{R})$ and its adjoint $P_{j\omega_0}^* : \mathscr{S}'(\mathbb{R}) \to \mathscr{S}'(\mathbb{R})$ are not invertible in the usual sense of the word (i.e., from both sides). However, we are able to properly invert $P_{j\omega_0}$ on its image (range), as we discuss now.

Let $\mathscr{S}_{j\omega_0}$ denote the image of $\mathscr{S}(\mathbb{R})$ under $P_{j\omega_0}$. This is the same as the subspace of $\mathscr{S}(\mathbb{R})$ consisting of functions φ for which [2]

$$\int_{-\infty}^{+\infty} e^{-j\omega_0 r}\varphi(r)\, dr = 0.$$

In particular, for $j\omega_0 = 0$, we obtain $\mathscr{S}_0$, the space of Schwartz test functions with vanishing zeroth-order moment. We may then view $P_{j\omega_0}$ as an operator $\mathscr{S}(\mathbb{R}) \to \mathscr{S}_{j\omega_0}$, and this operator now has an inverse $P_{j\omega_0}^{-1}$ from $\mathscr{S}_{j\omega_0} \to \mathscr{S}(\mathbb{R})$ defined by

$$P_{j\omega_0}^{-1}\varphi(r) = (\rho_{j\omega_0} * \varphi)(r), \tag{5.13}$$

where

$$\rho_{j\omega_0}(r) = \mathscr{F}^{-1}\left\{ \frac{1}{j\omega - j\omega_0} \right\}(r) = \frac{1}{2}\operatorname{sign}(r)e^{j\omega_0 r}. \tag{5.14}$$

Specifically, this LSI operator satisfies the right- and left-inverse relations

$$P_{j\omega_0}P_{j\omega_0}^{-1}\varphi = \varphi \text{ for all } \varphi \in \mathscr{S}_{j\omega_0}$$
$$P_{j\omega_0}^{-1}P_{j\omega_0}\varphi = \varphi \text{ for all } \varphi \in \mathscr{S}(\mathbb{R}).$$

In order to be able to use such inverse operators for defining stochastic processes, we need to extend $P_{j\omega_0}^{-1}$ to all of $\mathscr{S}(\mathbb{R})$.

Note that, unlike the case of P_α^{-1} with $\operatorname{Re}(\alpha) \neq 0$, here the extension of $P_{j\omega_0}^{-1}$ to an operator $\mathscr{S}(\mathbb{R}) \to \mathscr{S}'(\mathbb{R})$ is in general not unique. For instance, $P_{j\omega_0}^{-1}$ may also be specified as

$$P_{j\omega_0,+}^{-1}\varphi(r) = (\rho_{j\omega_0,+} * \varphi)(r), \tag{5.15}$$

with causal impulse response

$$\rho_{j\omega_0,+}(r) = e^{j\omega_0 r}\mathbb{1}_+(r), \tag{5.16}$$

which defines the same operator on $\mathscr{S}_{j\omega_0}$ but not on $\mathscr{S}(\mathbb{R})$. In fact, we could as well have taken any impulse response of the form $\rho_{j\omega_0}(r) + p_0(r)$, where $p_0(r) = c_0 e^{j\omega_0 r}$ is an oscillatory component that is in the null space of $P_{j\omega_0}$. By contrast, the L_p-continuous inverses that we define below remain the same for all of these extensions. To convey the idea, we shall first consider the extension based on the causal operator $P_{j\omega_0,+}^{-1*}$ defined by (5.15). Its adjoint is denoted by $P_{j\omega_0,+}^{-1*}$ and amounts to an (anti-causal) convolution with $\overset{\vee}{\rho}_{j\omega_0,+}$.

To solve the stochastic differential equation $P_{j\omega_0}s = w$, we need to find a left inverse of the adjoint $P_{j\omega_0}^*$ acting on the space of test functions, which maps $\mathscr{S}(\mathbb{R})$ into the required L_p space (Theorem 4.17). The problem with the "shift-invariant" [3] extensions of the inverse defined above is that their image inside $\mathscr{S}'(\mathbb{R})$ is not contained in arbitrary L_p spaces. For this reason, we now introduce a different, "corrected" extension of the

---

[2] To see this, note that $(D - j\omega_0\operatorname{Id})\varphi(r) = e^{j\omega_0 r}D\{e^{-j\omega_0 r}\varphi(r)\}$.
[3] These operators are shift-invariant because they are defined by means of convolutions.

inverse of $P^*_{j\omega_0}$ that maps $\mathscr{S}(\mathbb{R})$ to $\mathscr{R}(\mathbb{R})$ – therefore, a fortiori, also into all L_p spaces. This corrected *left inverse*, which we shall denote as $I^*_{\omega_0}$, is constructed as

$$I^*_{\omega_0}\varphi(r) = P^{-1*}_{j\omega_0,+}\varphi(r) - \left(\lim_{y \to -\infty} P^{-1*}_{j\omega_0,+}\varphi(y)\right)\overset{\vee}{\rho}_{j\omega_0,+}(r)$$

$$= (\overset{\vee}{\rho}_{j\omega_0,+} * \varphi)(r) - \widehat{\varphi}(-\omega_0)\overset{\vee}{\rho}_{j\omega_0,+}(r), \tag{5.17}$$

in direct analogy with (5.3). As in our introductory example, the role of the second term is to remove the tail of $P^{-1}_{j\omega_0,+}\varphi(r)$. This ensures that the output decays fast enough to belong to $\mathscr{R}(\mathbb{R})$. It is not difficult to show that the convolutional definition of $I^*_{\omega_0}$ given by (5.17) does not depend on the specific choice of the impulse response within the class of admissible LSI inverses of $P_{j\omega_0}$ on $\mathscr{S}_{j\omega}$. Then, we may simplify the notation by writing

$$I^*_{\omega_0}\varphi(r) = (\overset{\vee}{\rho}_{j\omega_0} * \varphi)(r) - \widehat{\varphi}(-\omega_0)\overset{\vee}{\rho}_{j\omega_0}(r), \tag{5.18}$$

for any Green's function $\rho_{j\omega_0}$ of the operator $P_{j\omega}$. While the left-inverse operator $I^*_{\omega_0}$ fixes the decay, it fails to be a right inverse of $P^*_{j\omega_0}$ unless $\varphi \in \mathscr{S}_{j\omega_0}$ or, equivalently, $\widehat{\varphi}(-\omega_0) = 0$.

The corresponding *right inverse* of $P_{j\omega_0}$ is provided by the adjoint of $I^*_{\omega_0}$. It is identified via the scalar product manipulation

$$\langle \varphi, I^*_{\omega_0}\phi \rangle = \langle \varphi, P^{-1*}_{j\omega_0}\phi \rangle - \widehat{\phi}(-\omega_0)\langle \varphi, \overset{\vee}{\rho}_{j\omega_0}\rangle \qquad \text{(by linearity)}$$

$$= \langle P^{-1}_{j\omega_0}\varphi, \phi \rangle - \langle e^{j\omega_0 r}, \phi \rangle \ (P^{-1}_{j\omega_0}\varphi)(0) \qquad \text{(using (5.13))}$$

$$= \langle P^{-1}_{j\omega_0}\varphi, \phi \rangle - \langle e^{j\omega_0 r}(P^{-1}_{j\omega_0}\varphi)(0), \phi \rangle.$$

Since the above is equal to $\langle I_{\omega_0}\varphi, \phi \rangle$ by definition, we find that

$$I_{\omega_0}\varphi(r) = P^{-1}_{j\omega_0}\varphi(r) - e^{j\omega_0 r} \ (P^{-1}_{j\omega_0}\varphi)(0)$$

$$= (\rho_{j\omega_0} * \varphi)(r) - e^{j\omega_0 r} \ (\rho_{j\omega_0} * \varphi)(0), \tag{5.19}$$

where $\rho_{j\omega_0}$ is defined by (5.14). The specificity of I_{ω_0} is to impose the boundary condition $s(0) = 0$ on the output $s = I_{\omega_0}\varphi$, irrespective of the input function φ. This is achieved by the addition of a component that is in the null space of $P_{j\omega}$. This also explains why we may replace $\rho_{j\omega_0}$ in (5.19) by any other Green's function of $P_{j\omega}$, including the causal one given by (5.16).

In particular, for $j\omega_0 = 0$ (that is, for $P_{j\omega_0} = D$), we have

$$I_0\varphi(r) = \int_{-\infty}^{r} \varphi(t) \, dt - \int_{-\infty}^{0} \varphi(t) \, dt = \begin{cases} \int_0^r \varphi(t) \, dt & r \geq 0 \\ -\int_r^0 \varphi(t) \, dt & r < 0, \end{cases}$$

while the adjoint is given by

$$I_0^*\varphi(r) = \int_r^{\infty} \varphi(t) \, dt - \mathbb{1}_{(-\infty,0]}(r)\int_{-\infty}^{\infty} \varphi(t) \, dt = \begin{cases} \int_r^{\infty} \varphi(t) \, dt & r \geq 0 \\ -\int_{-\infty}^{r} \varphi(t) \, dt & r < 0. \end{cases}$$

These are equivalent to the solution described in Section 5.1 (see Figure 5.1).

The Fourier-domain counterparts of (5.19) and (5.18) are

$$I_{\omega_0}\varphi(r) = \int_{\mathbb{R}} \widehat{\varphi}(\omega) \frac{e^{j\omega r} - e^{j\omega_0 r}}{j\omega - j\omega_0} \frac{d\omega}{2\pi} \tag{5.20}$$

$$I_{\omega_0}^*\varphi(r) = \int_{\mathbb{R}} \frac{\widehat{\varphi}(\omega) - \widehat{\varphi}(-\omega_0)}{-j\omega - j\omega_0} e^{j\omega r} \frac{d\omega}{2\pi}, \tag{5.21}$$

respectively. One can observe that the form of the numerator in both integrals is such that it tempers the singularity of the denominator at $\omega = \omega_0$. The relevance of these corrected inverse operators for the construction of stochastic processes is due to the following theorem.

THEOREM 5.3   *The operator* $I_{\omega_0}^*$ *defined by (5.18) is continuous from* $\mathscr{S}(\mathbb{R})$ *to* $\mathscr{R}(\mathbb{R})$ *and extends continuously to a linear operator* $\mathscr{R}(\mathbb{R}) \to \mathscr{R}(\mathbb{R})$. *It is the dual of the operator* I_{ω_0} *defined by (5.19) in the sense that* $\langle I_{\omega_0}\phi, \varphi \rangle = \langle \phi, I_{\omega_0}^*\varphi \rangle$. *Moreover,*

$$I_{\omega_0}\varphi(0) = 0 \qquad\qquad \text{(zero boundary condition)}$$
$$I_{\omega_0}^*(\mathrm{D} - j\omega_0\mathrm{Id})^*\varphi = \varphi \qquad\qquad \text{(left-inverse property)}$$
$$(\mathrm{D} - j\omega_0\mathrm{Id})I_{\omega_0}\phi = \phi \qquad\qquad \text{(right-inverse property)}$$

for all $\varphi, \phi \in \mathscr{S}(\mathbb{R})$.

The first part of the theorem follows from Proposition 5.4 below, which indicates that $I_{\omega_0}^*$ preserves rapid decay. The statements in the second part have already been discussed. The left-inverse property, for instance, follows from the fact that $(\mathrm{D} - j\omega_0\mathrm{Id})^*\varphi \in \mathscr{S}_{-j\omega_0}$, which is the subspace of $\mathscr{S}(\mathbb{R})$ for which all the inverses of $\mathrm{P}_{j\omega_0}^* = -\mathrm{P}_{-j\omega_0}$ are equivalent. The right-inverse property of I_{ω_0} is easily verified by applying $\mathrm{P}_{j\omega_0}$ to (5.19).

To qualify the rate of decay of functions, we rely on the $L_{\infty,\alpha}$ norm, defined as

$$\|\varphi\|_{L_{\infty,\alpha}} = \operatorname*{ess\,sup}_{r \in \mathbb{R}} \left|\varphi(r)(1 + |r|)^\alpha\right|. \tag{5.22}$$

Hence, the inclusion $\varphi \in L_{\infty,\alpha}(\mathbb{R})$ is equivalent to

$$|\varphi(r)| \leq \frac{\|\varphi\|_{L_{\infty,\alpha}}}{(1 + |r|)^\alpha} \quad \text{a.e.,}$$

which is to say that φ has an algebraic decay of order α. We also recall that $\mathscr{R}(\mathbb{R})$ is the space of rapidly decaying functions, which is the intersection of all $L_{\infty,\alpha}(\mathbb{R})$ spaces with $\alpha \geq 0$. The relevant embedding relations are $\mathscr{S}(\mathbb{R}) \subset \mathscr{R}(\mathbb{R}) \subset L_{\infty,\alpha}(\mathbb{R}) \subset L_p(\mathbb{R})$ for any $\alpha > 1$ and $p \geq 1$. Moreover, since $\mathscr{S}(\mathbb{R})$ has the strictest topology in the chain, a sequence that converges in $\mathscr{S}(\mathbb{R})$ is also convergent in $\mathscr{R}(\mathbb{R})$, $L_{\infty,\alpha}(\mathbb{R})$, or $L_p(\mathbb{R})$.

PROPOSITION 5.4   *Let* $I_{\omega_0}^*$ *be the linear operator defined by (5.17). Then, for all* $\varphi \in L_{\infty,\alpha}(\mathbb{R})$ *with* $\alpha > 1$, *there exists a constant C such that*

$$\|I_{\omega_0}^*\varphi\|_{L_{\infty,\alpha-1}} \leq C \|\varphi\|_{L_{\infty,\alpha}}.$$

Hence, $I_{\omega_0}^*$ *is a continuous operator from* $L_{\infty,\alpha}(\mathbb{R})$ *into* $L_{\infty,\alpha-1}(\mathbb{R})$ *and, by restriction of its domain, from* $\mathscr{R}(\mathbb{R}) \to \mathscr{R}(\mathbb{R})$ *or* $\mathscr{S}(\mathbb{R}) \to \mathscr{R}(\mathbb{R})$.

Proof For $r < 0$, we rewrite (5.17) as

$$I_{\omega_0}^* \varphi(r) = P_{j\omega_0}^{-1*} \varphi(r) - e^{-j\omega_0 r} \widehat{\varphi}(-\omega_0)$$

$$= \int_r^{+\infty} e^{-j\omega_0(r-\tau)} \varphi(\tau) \, d\tau - e^{-j\omega_0 r} \int_{-\infty}^{\infty} e^{j\omega_0 \tau} \varphi(\tau) \, d\tau$$

$$= -e^{-j\omega_0 r} \int_{-\infty}^{r} e^{j\omega_0 \tau} \varphi(\tau) \, d\tau.$$

This implies that

$$|I_{\omega_0}^* \varphi(r)| = \left| \int_{-\infty}^{r} e^{j\omega_0 \tau} \varphi(\tau) \, d\tau \right|$$

$$\leq \int_{-\infty}^{r} |\varphi(\tau)| \, d\tau$$

$$\leq \int_{-\infty}^{r} \frac{\|\varphi\|_{L_{\infty,\alpha}}}{(1 + |\tau|)^\alpha} \, d\tau \leq C \frac{\|\varphi\|_{L_{\infty,\alpha}}}{(1 + |r|)^{\alpha-1}}$$

for all $r < 0$. For $r > 0$, $I_{\omega_0}^* \varphi(r) = \int_r^\infty e^{-j\omega_0(r-\tau)} \varphi(\tau) \, d\tau$ so that the above upper bound remains valid. $\square$

While $I_{\omega_0}^*$ is continuous over $\mathscr{R}(\mathbb{R})$, it is not shift-invariant. Moreover, it will generally spoil the global smoothness of the functions in $\mathscr{S}(\mathbb{R})$ to which it is applied due to the discontinuity at the origin that is introduced by the correction. By contrast, its adjoint I_{ω_0} preserves the smoothness of the input but fails to return functions that are rapidly decaying at infinity. This lack of shift-invariance and the slow growth of the output at infinity is the price to pay for being able to solve unstable differential systems.

5.4.2 Higher-order differential operators with unstable shift-invariant inverses

Given that the operators $I_{\omega_0}^*$, $\omega_0 \in \mathbb{R}$, defined in Section 5.4.1 are continuous $\mathscr{R}(\mathbb{R}) \to \mathscr{R}(\mathbb{R})$, they may be composed to obtain higher-order continuous operators $\mathscr{R}(\mathbb{R}) \to \mathscr{R}(\mathbb{R})$ that serve as left inverses of the corresponding higher-order differential operators in the sense of Section 5.4.1. More precisely, given $(\omega_1, \ldots, \omega_K) \in \mathbb{R}^K$, we define the composite integration operator

$$I_{(\omega_1 : \omega_K)} = I_{\omega_1} \circ \cdots \circ I_{\omega_K}, \tag{5.23}$$

whose adjoint is given by

$$I_{(\omega_1 : \omega_K)}^* = \left(I_{\omega_1} \circ \cdots \circ I_{\omega_K} \right)^* = I_{\omega_K}^* \circ \cdots \circ I_{\omega_1}^*. \tag{5.24}$$

$I_{(\omega_1 : \omega_K)}^*$, which maps $\mathscr{R}(\mathbb{R})$ $\left(\text{and therefore } \mathscr{S}(\mathbb{R}) \subset \mathscr{R}(\mathbb{R})\right)$ continuously into $\mathscr{R}(\mathbb{R})$, is then a left inverse of

$$P_{(j\omega_K : j\omega_1)}^* = P_{j\omega_1}^* \circ \cdots \circ P_{j\omega_K}^*,$$

with $P_\alpha = D - \alpha \mathrm{Id}$ and $P_\alpha^* = -P_{-\alpha}$. Conversely, $I_{(\omega_K : \omega_1)}$ is a right inverse of

$$P_{(j\omega_1 : j\omega_K)} = P_{j\omega_1} \circ \cdots \circ P_{j\omega_K}.$$

Putting everything together, with the definitions of Section 5.3.2, we arrive at the following corollary of Theorem 5.3:

COROLLARY 5.5 *For $\alpha = (\alpha_1, \ldots, \alpha_M) \in \mathbb{C}^M$ with $\mathrm{Re}(\alpha_n) \neq 0$ and $(\omega_1, \ldots, \omega_K) \in \mathbb{R}^K$, the $(M+K)$th-order operator $\mathrm{L}^{-1*} = \mathrm{P}_{(\alpha_M:\alpha_1)}^{-1*}\mathrm{I}_{(\omega_K:\omega_1)}^*$ maps $\mathscr{S}(\mathbb{R})$ continuously into $\mathscr{R}(\mathbb{R}) \subseteq L_p(\mathbb{R})$ for any $p > 0$. It is a left inverse of $\mathrm{L}^* = \mathrm{P}_{(j\omega_1:j\omega_K)}^*\mathrm{P}_{(\alpha_1:\alpha_M)}^*$ in the sense that*

$$\mathrm{P}_{(\alpha_M:\alpha_1)}^{-1*}\mathrm{I}_{(\omega_K:\omega_1)}^*\mathrm{P}_{(j\omega_1:j\omega_K)}^*\mathrm{P}_{(\alpha_1:\alpha_M)}^*\varphi = \varphi$$

for all $\varphi \in \mathscr{S}(\mathbb{R})$.

We shall now use this result to solve the differential equation (5.11) in the non-stable scenario. To that end, we order the poles in such a way that the unstable ones come last with $\alpha_{N-K+m} = j\omega_m$, $1 \le m \le K$, where K is the number of purely imaginary poles. We thus specify the right-inverse operator

$$\mathrm{L}^{-1} = \mathrm{I}_{(\omega_K:\omega_1)}\mathrm{P}_{(\alpha_{N-K}:\alpha_1)}^{-1}q_M(\mathrm{D}),$$

which we then apply to w to obtain the solution $s = \mathrm{L}^{-1}w$. In effect, by applying $\mathrm{I}_{(\omega_K:\omega_1)}$ last, we are also enforcing the K linear boundary conditions

$$\begin{cases} s(0) &= 0 \\ \mathrm{P}_{j\omega_K}\{s\}(0) &= 0 \\ &\vdots \\ \mathrm{P}_{j\omega_2}\cdots\mathrm{P}_{j\omega_K}\{s\}(0) &= 0. \end{cases} \tag{5.25}$$

To show that $s = \mathrm{L}^{-1}w$ is a consistent solution, we proceed by duality and write

$$\langle \varphi, \mathrm{P}_{(\alpha_1:\alpha_{N-K})}\mathrm{P}_{(j\omega_1:j\omega_K)}s \rangle = \langle \varphi, \ \mathrm{P}_{(\alpha_1:\alpha_{N-K})}\mathrm{P}_{(j\omega_1:j\omega_K)}\mathrm{L}^{-1}w \rangle$$

$$= \langle \mathrm{P}_{(j\omega_1:j\omega_K)}^*\mathrm{P}_{(\alpha_1:\alpha_{N-K})}^*\varphi, \ \mathrm{I}_{(\omega_K:\omega_1)}\mathrm{P}_{(\alpha_{N-K}:\alpha_1)}^{-1}q_M(\mathrm{D})w \rangle$$

$$= \langle \underbrace{\mathrm{P}_{(\alpha_{N-K}:\alpha_1)}^{-1*}\mathrm{I}_{(\omega_K:\omega_1)}^*\mathrm{P}_{(j\omega_1:j\omega_K)}^*\mathrm{P}_{(\alpha_1:\alpha_{N-K})}^*}_{\mathrm{Id}} \varphi, \ q_M(\mathrm{D})w \rangle$$

$$= \langle \varphi, \ q_M(\mathrm{D})w \rangle,$$

where we have made use of Corollary 5.5. This proves that s satisfies the differential equation (5.11) with driving term w, subject to the boundary conditions (5.25).

5.4.3   Generalized boundary conditions

In the resolution method presented so far, the inverse operator I_{ω_0} was designed to impose zero boundary conditions at the origin. In more generality, one may consider inverse operators $\mathrm{I}_{\omega_0, \varphi_0}$ that incorporate conditions of the form

$$\langle \varphi_0, \mathrm{I}_{\omega_0, \varphi_0}w \rangle = 0 \tag{5.26}$$

on the solution $s = I_{\omega_0,\varphi_0} w$. This leads to the definition of the right-inverse operator

$$I_{\omega_0,\varphi_0}\varphi(r) = (\rho_{j\omega_0} * \varphi)(r) - e^{j\omega_0 r} \frac{\langle \rho_{j\omega_0} * \varphi, \varphi_0 \rangle}{\widehat{\varphi}_0(-\omega_0)}, \tag{5.27}$$

where $\rho_{j\omega_0}$ is a Green's function of $P_{j\omega_0}$ and φ_0 is some given rapidly decaying function such that $\widehat{\varphi}_0(-\omega_0) \neq 0$. In particular, if we set $\omega_0 = 0$ and $\varphi_0 = \delta$, we recover the scale-invariant integrator $I_0 = I_{0,\delta}$ that was used in our introductory example (Section 5.1) to provide the connection with the classical theory of Lévy processes. The Fourier-domain counterpart of (5.27) is

$$I_{\omega_0,\varphi_0}\varphi(r) = \int_{\mathbb{R}} \widehat{\varphi}(\omega) \left(\frac{e^{j\omega r} - e^{j\omega_0 r} \frac{\widehat{\varphi}_0(-\omega)}{\widehat{\varphi}_0(-\omega_0)}}{j(\omega - \omega_0)} \right) \frac{d\omega}{2\pi}. \tag{5.28}$$

The above operator is well defined pointwise for any $\varphi \in L_1(\mathbb{R})$. Moreover, it is a right inverse of $(D - j\omega_0 \mathrm{Id})$ on $\mathscr{S}(\mathbb{R})$ because the regularization in the numerator amounts to a sinusoidal correction that is in the null space of the operator. The adjoint of I_{ω_0,φ_0} is specified by the Fourier-domain integral

$$I^*_{\omega_0,\varphi_0}\varphi(r) = \int_{\mathbb{R}} \left(\frac{\widehat{\varphi}(\omega) - \frac{\widehat{\varphi}(\omega_0)}{\widehat{\varphi}_0(-\omega_0)} \widehat{\varphi}_0(\omega)}{-j(\omega + \omega_0)} \right) e^{j\omega r} \frac{d\omega}{2\pi}, \tag{5.29}$$

which is non-singular too, thanks to the regularization in the numerator. The beneficial effect of this adjustment is that $I^*_{\omega_0,\varphi_0}$ is $\mathscr{R}$-continuous and L_p-stable, unlike its more conventional shift-invariant counterpart $P^{-1*}_{j\omega_0}$. The time-domain counterpart of (5.29) is

$$I^*_{\omega_0,\varphi_0}\varphi(r) = (\rho^{\vee}_{j\omega_0} * \varphi)(r) - \frac{\widehat{\varphi}(-\omega_0)}{\widehat{\varphi}_0(-\omega_0)} (\varphi_0 * \rho^{\vee}_{j\omega_0})(r), \tag{5.30}$$

where $\rho_{j\omega_0}$ is a Green's function of $P_{j\omega_0}$. This relation is very similar to (5.18), with the notable difference that the second term is convolved by φ_0. This suggests that we can restore the smoothness of the output by picking a kernel φ_0 with a sufficient degree of differentiability. In fact, by considering a sequence of such kernels in $\mathscr{S}(\mathbb{R})$ that converge to the Dirac distribution (in the weak sense), we can specify a left-inverse operator that is arbitrarily close to $I^*_{\omega_0}$ and yet $\mathscr{S}$-continuous.

While the imposition of generalized boundary conditions of the form (5.26) has some significant implications for the statistical properties of the signal (non-stationary behavior), it is less of an issue for signal processing because of the use of analysis tools (wavelets, finite-difference operators) that stationarize these processes – in effect, filtering out the null-space components – so that the traditional tools of the trade remain applicable. Therefore, to simplify the presentation, we shall only consider boundary conditions at zero in what follows, and work with the operators $I^*_{\omega_0}$ and I_{ω_0}.

5.5   Fractional-order operators

5.5.1   Fractional derivatives in one dimension

In one dimension, we consider the general class of all LSI operators that are also scale-invariant. To motivate their definition, let us recall that the nth-order derivative D^n corresponds to the Fourier multiplier $(j\omega)^n$. This suggests the following fractional extension, going back to Liouville, whereby the exponent n is replaced by a non-negative real number γ:

$$D^\gamma \varphi(r) = \int_\mathbb{R} (j\omega)^\gamma \widehat{\varphi}(\omega) e^{j\omega r} \frac{d\omega}{2\pi}. \tag{5.31}$$

This definition is further generalized in the next proposition, which gives a complete characterization of scale-invariant convolution operators in 1-D.

PROPOSITION 5.6 (see [UB07, Proposition 2])   *The complete family of 1-D scale-invariant convolution operators of order $\gamma \in \mathbb{R}$ reduces to the fractional derivative ∂_τ^γ whose Fourier-based definition is*

$$\partial_\tau^\gamma \varphi(r) = \int_\mathbb{R} (j\omega)^{\frac{\gamma}{2}+\tau} (-j\omega)^{\frac{\gamma}{2}-\tau} \widehat{\varphi}(\omega) e^{j\omega r} \frac{d\omega}{2\pi},$$

where $\widehat{\varphi}$ is the 1-D Fourier transform of the input function φ under the implicit assumption that the inverse Fourier integral on the right-hand side is convergent.

While the above representation is appealing, it needs to be treated with caution since the underlying Fourier multipliers are unbounded (at infinity or at zero), which is incompatible with L_p-stability (see Theorem 3.5). The next theorem shows that the fractional-derivative operators ∂_τ^γ are well defined over $\mathscr{S}(\mathbb{R})$ for $\gamma > -1$ and $\tau \in \mathbb{R}$, but that they have a tendency to spoil the decay of the functions to which they are applied.

THEOREM 5.7   *The differential operator ∂_τ^γ is continuous from $\mathscr{S}(\mathbb{R})$ to $L_p(\mathbb{R})$ for $\gamma > \frac{1}{p} - 1$. Moreover, the fractional derivative $\partial_\tau^\gamma \varphi$ of a test function $\varphi \in \mathscr{S}(\mathbb{R})$ remains indefinitely differentiable, but its decay is constrained by*

$$\left|\partial_\tau^\gamma \varphi(r)\right| \le \frac{\text{Const}}{1 + |r|^{\gamma+1}}.$$

This is to be contrasted with the effect of D^n for $n \in \mathbb{N}$, which maps $\mathscr{S}(\mathbb{R})$ into itself and hence preserves rapid decay. To explain the effect, we observe that a fractional differentiation is equivalent to a convolution. Since φ decreases rapidly, the decay of the output is imposed by the tail of the impulse response. For instance, in the case of the operator D^γ (with γ non-integer), we have that

$$D^\gamma \{\delta\}(r) = \frac{r_+^{-\gamma-1}}{\Gamma(-\gamma)}, \tag{5.32}$$

which, for $\gamma > -1$, is a generalized function that decays like $1/|r|^{\gamma+1}$. Note that the (apparent) singularity at $r = 0$ is not damaging since it is tempered by the finite-part integral of the definition (see Table A.1 and (5.33) below).

The Fourier-domain characterization in Proposition 5.6 implies that these operators are endowed with a semigroup property: they satisfy the composition rule $\partial_\tau^\gamma \partial_{\tau'}^{\gamma'} = \partial_{\tau+\tau'}^{\gamma+\gamma'}$ for $\gamma', \gamma + \gamma' \in (-1, +\infty)$ and $\tau, \tau' \in \mathbb{R}$. The parameter τ is a phase factor that yields a progressive transition between the purely causal derivative $D^\gamma = \partial_{\gamma/2}^\gamma$ and its anti-causal counterpart $D^{\gamma*} = \partial_{-\gamma/2}^\gamma$, which happens to be the adjoint of the former. We also note that ∂_τ^0 is equivalent to the fractional Hilbert transform operator $\mathcal{H}_\tau$ investigated in [CU10]. A special case of the semigroup property is $\partial_\tau^\gamma = \partial_{\gamma/2}^\gamma \partial_{\tau-\gamma/2}^0 = D^\gamma \mathcal{H}_{\tau-\gamma/2}$, which indicates that the fractional derivatives of order γ are all related to D^γ via a fractional Hilbert transform. The latter is a unitary operator (all-pass filter) that essentially acts as a shift operator on the oscillatory part of a wavelet.

The property that ∂_τ^γ can be factorized as $\partial_\tau^\gamma = D^n \partial_{\tau'}^\alpha$, with $n \in \mathbb{N}$, $\alpha = \gamma - n$, $\tau' = \tau - n/2$, and $D^n : \mathscr{S}(\mathbb{R}) \to \mathscr{S}(\mathbb{R})$, has important consequences for the theory. In particular, it suggests several equivalent descriptions of the generalized function $\frac{r_+^\lambda}{\Gamma(\lambda+1)}$, as in

$$\langle \varphi, \frac{r_+^\lambda}{\Gamma(\lambda+1)} \rangle = \langle \varphi, D^n \{ \frac{r_+^{\lambda+n}}{\Gamma(\lambda+n+1)} \} \rangle$$

$$= \langle D^{n*}\varphi, \frac{r_+^{\lambda+n}}{\Gamma(\lambda+n+1)} \rangle$$

$$= \frac{(-1)^n}{\Gamma(\lambda+n+1)} \int_0^\infty r^{\lambda+n} \varphi^{(n)}(r)\, dr. \qquad (5.33)$$

The last equality of (5.33) with $n = \min(0, \lfloor -\lambda \rfloor)$ provides an operational definition that reduces to a conventional integral.

In principle, we can obtain the shift-invariant inverse of the derivative operator ∂_τ^γ with $\gamma \geq 0$ by taking the order to be negative (fractional integrator) and reversing the sign of τ. Yet, based on (5.32), which is valid for $\gamma \in \mathbb{R}\backslash\mathbb{N}$, we see that this is problematic because the impulse response becomes more delocalized as γ decreases. As in the case of the ordinary derivative D^n, this calls for a stabilized version of the inverse.

THEOREM 5.8   *The fractional integration operator*

$$\partial_{-\tau,p}^{-\gamma*}\varphi(r) = \frac{1}{2\pi} \int_\mathbb{R} \frac{\hat{\varphi}(\omega) - \sum_{k=0}^{\lfloor \gamma + \frac{1}{p} \rfloor - 1} \frac{\hat{\varphi}^{(k)}(0)\omega^k}{k!}}{(-j\omega)^{\frac{\gamma}{2}-\tau}(j\omega)^{\frac{\gamma}{2}+\tau}} e^{j\omega r}\, d\omega \qquad (5.34)$$

continuously maps $\mathscr{S}(\mathbb{R})$ into $L_p(\mathbb{R})$ for $p > 0$, $\tau \in \mathbb{R}$, and $\gamma \in \mathbb{R}^+$, subject to the restriction $\gamma + \frac{1}{p} \neq 1, 2, \ldots$. It is a linear operator that is scale-invariant and is a left inverse of $\partial_{-\tau}^\gamma = \partial_\tau^{\gamma*}$. The adjoint operator $\partial_{-\tau,p}^{-\gamma}$ is given by

$$\partial_{-\tau,p}^{-\gamma}\varphi(r) = \frac{1}{2\pi} \int_\mathbb{R} \frac{e^{j\omega r} - \sum_{k=0}^{\lfloor \gamma + \frac{1}{p} \rfloor - 1} \frac{(j\omega r)^k}{k!}}{(-j\omega)^{\frac{\gamma}{2}-\tau}(j\omega)^{\frac{\gamma}{2}+\tau}} \hat{\varphi}(\omega)\, d\omega$$

and is the proper scale-invariant right inverse of ∂_τ^γ to be applied to generalized functions.

The stabilization in (5.34) amounts to an adjustment of the Fourier transform of the input (subtraction of an adequate number of terms of its Taylor series at the origin) to counteract the frequency-domain division by zero. This correction has the desirable features of being linear with respect to the input and of preserving scale-invariance, which is essential for our purpose.

Proof We only give a sketch of the proof for $p \geq 1$, leaving out the derivation of Proposition 5.9, which is somewhat technical. The first observation is that the operator can be factorized as $\partial_{-\tau,p}^{-\gamma*} = \partial_{\tau'}^{\alpha}(\mathrm{I}_0^*)^{n_p}$ with $\alpha = n_p - \gamma$ and $\tau' = \tau - n_p/2$, where I_0^* is the corrected (adjoint) integrator defined by (5.21) with $\omega_0 = 0$. The integer order of pre-integration is $n_p = \lfloor \gamma + \frac{1}{p} \rfloor - 1 + 1 = \lfloor \gamma + \frac{1}{p} \rfloor$, which implies that the residual degree of differentiation α is constrained according to

$$\frac{1}{p} - 1 < \alpha = n_p - \gamma < \frac{1}{p}.$$

This allows us to write $\partial_{-\tau,p}^{-\gamma*}\varphi = \partial_{\tau'}^{\alpha}\phi$, where $\phi = (\mathrm{I}_0^*)^{n_p}\varphi$ is rapidly decaying by Corollary 5.5. The required ingredient to complete the proof is a result analogous to Theorem 5.7 which would ensure that $\partial_{\tau}^{\alpha}\phi \in L_p(\mathbb{R})$ for $\alpha > \frac{1}{p} - 1$. The easy scenario is when $\varphi \in \mathscr{S}(\mathbb{R})$ has n_p vanishing moments, in which case $\phi = (\mathrm{I}_0^*)^{n_p}\varphi \in \mathscr{S}(\mathbb{R})$ so that Theorem 5.7 is directly applicable. In general, however, ϕ is (only) rapidly decreasing; this is addressed by the following extension.

PROPOSITION 5.9   *The fractional operator $\partial_{\tau}^{\alpha}\mathrm{I}_0^*$ is continuous from $\mathscr{S}(\mathbb{R})$ to $L_p(\mathbb{R})$ for $p > 0$, $\tau \in \mathbb{R}$, and $\frac{1}{p} - 1 < \alpha < \frac{1}{p}$. It also admits a continuous extension $\mathscr{R}(\mathbb{R}) \to L_p(\mathbb{R})$ for $p \geq 1$.*

The reason for including the operator I_0^* in the statement is to avoid making explicit hypotheses about the derivative of ϕ, which is rapidly decaying but also exhibits a Dirac impulse at the origin with a weight $(-\hat{\varphi}(0))$. Since $\mathrm{I}_0^*:\mathscr{R}(\mathbb{R}) \to \mathscr{R}(\mathbb{R})$ (by Proposition 5.4), the global continuity result for $p \geq 1$ then follows from the chaining of these elementary operators.

Similarly, we establish the left-inverse property by considering the factorization

$$\partial_{-\tau}^{\gamma} = (\mathrm{D}^*)^{n_p}\partial_{-\tau'}^{-\alpha}$$

and by recalling that I_0^* is a left inverse of $\mathrm{D}^* = -\mathrm{D}$. The result then follows from the identity $\partial_{\tau}^{\alpha}\partial_{-\tau}^{-\alpha} = \mathrm{Id}$, which is a special case of the semigroup property of scale-invariant LSI operators, under the implicit assumption that the underlying operations are well defined in the L_p sense.   □

A final observation that gives insight into the design of L_p-stable inverse operators is that the form of (5.34) for $p = 1$ coincides with the finite-part definition (see Appendix A) of the generalized function

$$\hat{g}_r(\omega) = \frac{e^{j\omega r}}{(-j\omega)^{\frac{\gamma}{2}-\tau}(j\omega)^{\frac{\gamma}{2}+\tau}} \in \mathscr{S}'(\mathbb{R}).$$

Specifically, by using the property that $\widehat{\varphi} \in \mathscr{S}(\mathbb{R})$, we have that

$$\partial_{-\tau,1}^{-\gamma *}\varphi(r) = \frac{1}{2\pi}\langle \widehat{\varphi}, \widehat{g}_r \rangle$$

$$= \text{p.f.}\,\frac{1}{2\pi}\int_{\mathbb{R}} \widehat{\varphi}(\omega)\frac{e^{j\omega r}}{(-j\omega)^{\frac{\gamma}{2}-\tau}(j\omega)^{\frac{\gamma}{2}+\tau}}\,d\omega$$

$$= \frac{1}{2\pi}\int_{\mathbb{R}}\left(\widehat{\varphi}(\omega) - \sum_{k=0}^{\lfloor \gamma \rfloor}\frac{\widehat{\varphi}^{(k)}(0)\omega^k}{k!}\right)\frac{e^{j\omega r}}{(-j\omega)^{\frac{\gamma}{2}-\tau}(j\omega)^{\frac{\gamma}{2}+\tau}}\,d\omega,$$

where the finite-part regularization in the latter integral is the same as in the definition in (A.1) of the generalized function x_+^λ with $-\operatorname{Re}(\lambda) - 1 = \gamma$. The catch with (5.34) is that the number of regularization terms $n_p = \lfloor \gamma + \frac{1}{p} \rfloor$ is not solely dependent upon γ, but also on $1/p$.

5.5.2   Fractional Laplacians

The fractional Laplacian of order $\gamma \geq 0$ is defined by the inverse Fourier integral

$$(-\Delta)^{\frac{\gamma}{2}}\varphi(r) = \int_{\mathbb{R}^d}\|\omega\|^\gamma\widehat{\varphi}(\omega)e^{j\langle \omega, r \rangle}\frac{d\omega}{(2\pi)^d},$$

where $\widehat{\varphi}(\omega)$ is the d-dimensional Fourier transform of $\varphi(r)$. For $\gamma = 2$, this characterization coincides with the classical definition of the negative Laplacian: $-\Delta = -\sum \partial_{r_i}^2$.

In slightly more generality, we obtain a complete characterization of homogeneous shift- and rotation-invariant operators and their inverses in terms of convolutions with homogeneous rotation-invariant distributions, as given in Theorem 5.10. The idea and definitions may be traced back to [Duc77, GS68, Hör80].

THEOREM 5.10 ([Taf11, Corollary 2.ac])   *Any continuous linear operator $\mathscr{S}(\mathbb{R}^d) \to \mathscr{S}'(\mathbb{R}^d)$ that is simultaneously shift- and rotation-invariant, and homogeneous or scale-invariant of order γ in the sense of Definition 5.2, is a multiple of the operator*

$$L^\gamma : \varphi \mapsto \rho^{-\gamma-d} * \varphi = (2\pi)^{\frac{d}{2}}\mathscr{F}^{-1}\left\{\rho^\gamma\widehat{\varphi}\right\},$$

where ρ^γ, $\gamma \in \mathbb{C}$, is the distribution defined by

$$\rho^\gamma(\omega) = \frac{\|\omega\|^\gamma}{2^{\frac{\gamma}{2}}\Gamma\left(\frac{\gamma+d}{2}\right)} \tag{5.35}$$

with the property that

$$\mathscr{F}\{\rho^\gamma\} = (2\pi)^{\frac{d}{2}}\rho^{-\gamma-d}$$

(Γ denotes the gamma function.)

Note that L^γ is self-adjoint, with $L^{\gamma *} = L^\gamma$.

As is clear from the definition of ρ^γ, for $\gamma \neq -d, -d-2, -d-4, \ldots$, L^γ is simply a renormalized version of the fractional Laplacian introduced earlier. For $\gamma = -d - 2m$, $m = 0, 1, 2, 3, \ldots$, where the gamma function in the denominator of

ρ^γ has a pole, ρ^γ is proportional to $(-\Delta)^m \delta$, while the previous definition of the fractional Laplacian without normalization does not define a scale-invariant operator.

Also note that when $\text{Re}(\gamma) > -d$ $(\text{Re}(\gamma) \leq -d$, respectively), ρ^γ (its Fourier transform, respectively) is singular at the origin. This singularity is resolved in the manner described in Appendix A, which is equivalent to the analytical continuation of the formula

$$\Delta \rho^\gamma = \gamma \rho^{\gamma-2},$$

initially valid for $\text{Re}(\gamma) - 2 > -d$, in the variable γ. For the details of the previous two observations we refer the reader to Appendix A and Tafti [Taf11, Section 2.2]. [4]

Finally, it is important to remark that, unlike integer-order operators, unless γ is a positive even integer the image of $\mathscr{S}(\mathbb{R}^d)$ under L^γ is not contained in $\mathscr{S}(\mathbb{R}^d)$. Specifically, while for any $\varphi \in \mathscr{S}$, $\text{L}^\gamma \varphi$ is always an infinitely differentiable regular function, in the case of $\gamma \neq 2m$, $m = 1, 2, \ldots$, it may have slow (polynomial) decay or growth, in direct analogy with the general 1-D scenario characterized by Theorem 5.7.

Put more simply, the fractional Laplacian of a Schwartz function is not in general a Schwartz function.

5.5.3 L_p-stable inverses

From the identity

$$\rho^\gamma(\omega)\rho^{-\gamma}(\omega) = \frac{1}{\Gamma\left(\frac{d+\gamma}{2}\right)\Gamma\left(\frac{d-\gamma}{2}\right)} \qquad \text{for } \omega \neq 0,$$

we conclude that up to normalization, $\text{L}^{-\gamma}$ is the inverse of L^γ on the space of Schwartz' test functions with vanishing moments, in particular for $\text{Re}(\gamma) > -d$. [5] This inverse for $\text{Re}(\gamma) > -d$ can be further extended to a shift-invariant *left* inverse of L^γ acting on Schwartz functions. However, as was the case in Section 5.4.1, this shift-invariant inverse generally does not map $\mathscr{S}(\mathbb{R}^d)$ into a given $L_p(\mathbb{R}^d)$ space, and is therefore not suitable for defining generalized random fields in $\mathscr{S}'(\mathbb{R}^d)$.

The problem exposed in the previous paragraph is once again overcome by defining a "corrected" left inverse. Here again, unlike the simpler scenario of ordinary differential operators, it is not possible to have a single left-inverse operator that maps $\mathscr{S}(\mathbb{R}^d)$ into the intersection of all $L_p(\mathbb{R}^d)$ spaces, $p > 0$. Instead, we shall need to define a separate left-inverse operator for each $L_p(\mathbb{R}^d)$ space we are interested in. Under the constraints of scale- and rotation-invariance, such "non-shift-invariant" left inverses are identified in the following theorem.

---

[4] The difference in factors of $(2\pi)^{\frac{d}{2}}$ and $(2\pi)^d$ between the formulas given here and in Tafti [Taf11] is due to different normalizations used in the definition of the Fourier transform.

[5] Here we exclude the cases where the gamma functions in the denominator have poles, namely where their argument is a negative integer. For details see [Taf11, Section 2.2.2.]

THEOREM 5.11 ([Taf11], Theorem 2.aq and Corollary 2.am)   *The operator*

$$L_p^{-\gamma*}: \varphi \mapsto \rho^{\gamma-d} * \varphi - \sum_{|k| \leq \lfloor \mathrm{Re}(\gamma) + \frac{d}{p} \rfloor - d} \frac{\partial_k \rho^{\gamma-d}}{k!} \int_{\mathbb{R}^d} y^k \varphi(y) \, dy \qquad (5.36)$$

with $\mathrm{Re}(\gamma) + \frac{d}{p} \neq 1, 2, 3, \ldots$ *is rotation-invariant and homogeneous of order* $(-\gamma)$ *in the sense of Definition 5.2. It maps* $\mathscr{S}(\mathbb{R}^d)$ *continuously into* $L_p(\mathbb{R}^d)$ *for* $p \geq 1$.

The adjoint of $L_p^{-\gamma*}$ is given by

$$L_p^{-\gamma}: \varphi \mapsto \rho^{\gamma-d} * \varphi - \sum_{|k| \leq \lfloor \mathrm{Re}(\gamma) + \frac{d}{p} \rfloor - d} r^k \frac{\partial_k L^{-\gamma} \varphi(0)}{k!}.$$

If we exclude the cases where the denominator of (5.35) has a pole, we may normalize the above operators to find left and right inverses corresponding to the fractional Laplacian $(-\Delta)^{\frac{\gamma}{2}}$. The next theorem gives an equivalent Fourier-domain characterization of these operators.

THEOREM 5.12 (see [SU12, Theorem 3.7])   *Let* $I_p^{\gamma*}$ *with* $p \geq 1$ *and* $\gamma > 0$ *be the isotropic fractional integral operator defined by*

$$I_p^{\gamma*} \varphi(r) = \int_{\mathbb{R}^d} \frac{\widehat{\varphi}(\omega) - \sum_{|k| \leq \lfloor \gamma + \frac{d}{p} \rfloor - d} \frac{\partial^k \widehat{\varphi}(0) \omega^k}{k!}}{\|\omega\|^\gamma} e^{j\langle \omega, r \rangle} \frac{d\omega}{(2\pi)^d}. \qquad (5.37)$$

Then, under the condition that $\gamma \neq 2, 4, \ldots$ *and* $\gamma + \frac{d}{p} \neq 1, 2, 3, \ldots$, $I_p^{\gamma*}$ *is the unique left inverse of* $(-\Delta)^{\frac{\gamma}{2}}$ *that continuously maps* $\mathscr{S}(\mathbb{R}^d)$ *into* $L_p(\mathbb{R}^d)$ *for* $p \geq 1$ *and is scale-invariant. The adjoint operator* I_p^γ, *which is the proper scale-invariant right inverse of* $(-\Delta)^{\frac{\gamma}{2}}$, *is given by*

$$I_p^\gamma \varphi(r) = \int_{\mathbb{R}^d} \frac{e^{j\langle \omega, r \rangle} - \sum_{|k| \leq \lfloor \gamma + \frac{d}{p} \rfloor - d} \frac{j^{|k|} r^k \omega^k}{k!}}{\|\omega\|^\gamma} \widehat{\varphi}(\omega) \frac{d\omega}{(2\pi)^d}. \qquad (5.38)$$

The fractional integral operators I_p^γ and $I_p^{\gamma*}$ are both scale-invariant of order $(-\gamma)$, but they are not shift-invariant.

5.6 Discrete convolution operators

We conclude this chapter by providing a few basic results on discrete convolution operators and their inverses. These will turn out to be helpful for establishing the existence of certain spline interpolators which are required for the construction of operator-like wavelet bases in Chapter 6 and for the representation of autocorrelation functions in Chapter 7.

The convention in this book is to use square brackets to index sequences. This allows one to distinguish them from functions of a continuous variable. In other words, $h(\cdot)$ or $h(r)$ stands for a function defined on a continuum, while $h[\cdot]$ or $h[k]$ denotes some discrete sequence. The notation $h[\cdot]$ is often simplified to h when the context is clear.

The discrete convolution between two sequences $h[\cdot]$ and $a[\cdot]$ over $\mathbb{Z}^d$ is defined as

$$(h * a)[n] = \sum_{m \in \mathbb{Z}^d} h[m]a[n - m]. \tag{5.39}$$

This convolution describes how a digital filter with discrete impulse response h acts on some input sequence a. If $h \in \ell_1(\mathbb{Z}^d)$, then the map $a[\cdot] \mapsto (h * a)[\cdot]$ is a continuous operator $\ell_p(\mathbb{Z}^d) \to \ell_p(\mathbb{Z}^d)$. This follows from Young's inequality for sequences,

$$\|h * a\|_{\ell_p} \leq \|h\|_{\ell_1}\|a\|_{\ell_p}, \tag{5.40}$$

where

$$\|a\|_{\ell_p} = \begin{cases} \left(\sum_{n \in \mathbb{Z}^d} |a[n]|^p\right)^{\frac{1}{p}} & \text{for } 1 \leq p < \infty \\ \sup_{n \in \mathbb{Z}^d} |a[n]| & \text{for } p = \infty. \end{cases}$$

Note that the condition $h \in \ell_1(\mathbb{Z}^d)$ is the discrete counterpart of the (more involved) TV condition in Theorem 3.5. As in the continuous formulation, it is *necessary and sufficient* for stability in the extreme cases $p = 1, +\infty$. A simplifying aspect of the discrete setting is that the boundedness of the operator for $p = \infty$ implies all the other forms of ℓ_p-stability because of the embedding $\ell_p(\mathbb{Z}^d) \subset \ell_q(\mathbb{Z}^d)$ for any $1 \leq p < q \leq \infty$. The latter property is a consequence of the basic norm inequality

$$\|a\|_{\ell_p} \geq \|a\|_{\ell_q} \geq \|a\|_{\ell_\infty}$$

for all $a \in \ell_p(\mathbb{Z}^d) \subset \ell_q(\mathbb{Z}^d) \subset \ell_\infty(\mathbb{Z}^d)$.

In the Fourier domain, (5.39) maps into the multiplication of the discrete Fourier transforms of h and a as

$$b[n] = (h * a)[n] \quad \Leftrightarrow \quad B(\omega) = H(\omega)A(\omega),$$

where we are using capital letters to denote the discrete-time Fourier transforms of the underlying sequences. Specifically, we have that

$$B(\omega) = \mathcal{F}_d\{b\}(\omega) = \sum_{k \in \mathbb{Z}^d} b[k]e^{-j\langle \omega, k\rangle},$$

which is 2π-periodic, while the corresponding inversion formula is

$$b[n] = \mathcal{F}_d^{-1}\{B\}[n] = \int_{[-\pi,\pi]^d} B(\omega)e^{j\langle \omega, n\rangle}\frac{d\omega}{(2\pi)^d}.$$

The stability condition $h \in \ell_1(\mathbb{Z}^d)$ ensures that the frequency response $H(\omega)$ of the digital filter h is bounded and continuous.

The task of specifying discrete inverse filters is greatly facilitated by a theorem, known as Wiener's lemma, which ensures that the inverse convolution operator is ℓ_p-stable whenever the frequency response of the original filter is non-vanishing.

THEOREM 5.13 (Wiener's lemma)   *Let $H(\omega) = \sum_{k \in \mathbb{Z}^d} h[k]e^{-j\langle \omega, k \rangle}$, with $h \in \ell_1(\mathbb{Z}^d)$, be a stable discrete Fourier multiplier such that $H(\omega) \neq 0$ for $\omega \in [-\pi, \pi]^d$. Then, $G(\omega) = 1/H(\omega)$ has the same property in the sense that it can be written as $1/H(\omega) = \sum_{k \in \mathbb{Z}^d} g[k]e^{-j\langle \omega, k \rangle}$ with $g \in \ell_1(\mathbb{Z}^d)$.*

The so-defined filter g identifies a stable convolution inverse of h with the property that

$$(g * h)[n] = (h * g)[n] = \delta[n],$$

where

$$\delta[n] = \begin{cases} 1 & \text{for } n = 0 \\ 0 & \text{for } n \in \mathbb{Z}^d \setminus \{0\} \end{cases}$$

is the Kronecker unit impulse.

5.7 Bibliographical notes

Section 5.2
The specification of L_p-stable convolution operators is a central topic in harmonic analysis [SW71, Gra08]. The basic result in Proposition 5.1 relies on Young's inequality with $q = r = 1$ [Fou77]. The complete class of functions that result in $\mathscr{S}$-continuous convolution kernels is provided by the inverse Fourier transform of the space of smooth slowly increasing Fourier multipliers, which play a crucial role in the theory of generalized functions [Sch66]. They extend $\mathscr{R}(\mathbb{R}^d)$ in the sense that they also contain point distributions such as δ and its derivatives.

Section 5.3
The operational calculus that is used for solving ordinary differential equations (ODEs) can be traced to Heaviside, who also introduced the symbol D for the derivative operator. It was initially met with skepticism because Heaviside's exposition lacked rigor. Nowadays, the preferred method for solving ODEs is based on the Laplace transform or the Fourier transform. The operator-based formalism that is presented in Section 5.3 is a standard application of distribution theory and Green's functions [Kap62, Zem10].

Section 5.4
The extension of the operational calculus for solving unstable ODE/SDEs is a more recent development. It was initiated in [BU07] in an attempt to link splines and fractals. The material presented in this section is based on [UTS14], for the most part. The generalized boundary conditions of Section 5.4.3 were introduced in [UTAK14].

Section 5.5
Fractional derivatives and Laplacians play a central role in the theory of splines, the primary reason being that these operators are scale-invariant [Duc77, UB00]. The proof

of Proposition 5.6 can be found in [UB07, Proposition 2], while Theorem 5.7 is a slight extension of [UB07, Theorem 3]. The derivation of the corresponding left and right inverses for $p = 2$ is carried out in the companion paper [BU07]. These results were extended to the multivariate setting for both the Gaussian [TVDVU09] and the generalized Poisson setting [UT11]. A detailed investigation of the L_p-stable left inverses of the fractional Laplacian, together with some unicity results, is provided in [SU12]. Further results and proofs can be found in [Taf11].

Section 5.6

Discrete convolution is a central topic in digital signal processing [OSB99]. The discrete version of Young's inequality can be established by using the same technique as for the proof of Proposition 5.1. In that respect, the condition $h \in \ell_1(\mathbb{Z}^d)$ is the standard criterion for stability in the theory of discrete-time linear systems [OSB99, Lat98]. It is necessary and sufficient for BIBO stability ($p = \infty$) and for the preservation of absolute summability ($p = 1$). Wiener stated his famous lemma in 1932 [Wie32, Lemma IIe]. Other relevant references are [New75, Sun07].

6 Splines and wavelets

A fundamental aspect of our formulation is that the whitening operator L is naturally tied to some underlying B-spline function, which will play a crucial role in what follows. The spline connection also provides a strong link with wavelets [UB03].

In this chapter, we review the foundations of spline theory and show how one can construct B-spline basis functions and wavelets that are tied to some specific operator L. The chapter starts with a gentle introduction to wavelets that exploits the analogy with Lego blocks. This naturally leads to the formulation of a multiresolution analysis of $L_2(\mathbb{R})$ using piecewise-constant functions and a *de visu* identification of Haar wavelets. We then proceed in Section 6.2 with a formal definition of our generalized brand of splines – the cardinal L-splines – followed by a detailed discussion of the fundamental notion of the Riesz basis. In Section 6.3, we systematically cover the first-order operators with the construction of exponential B-splines and wavelets, which have the convenient property of being orthogonal. We then address the theory in its full generality and present two generic methods for constructing B-spline basis functions (Section 6.4) and semi-orthogonal wavelets (Section 6.5). The pleasing aspect is that these results apply to the whole class of shift-invariant differential operators L whose null space is finite-dimensional (possibly trivial), which are precisely those that can be safely inverted to specify sparse stochastic processes.

6.1   From Legos to wavelets

It is instructive to get back to our introductory example of piecewise-contant splines in Chapter 1 (§1.3) and to show how these are naturally connected to wavelets. The fundamental idea in wavelet theory is to construct a series of fine-to-coarse approximations of a function $s(r)$ and to exploit the structure of the multiresolution approximation errors, which are orthogonal across scale. Here, we shall consider a series of approximating signals $\{s_i\}_{i \in \mathbb{Z}}$, where s_i is a piecewise-constant spline with knots positioned on $2^i \mathbb{Z}$. These multiresolution splines are represented by their B-spline expansion

$$s_i(r) = \sum_{k \in \mathbb{Z}} c_i[k]\phi_{i,k}(r), \tag{6.1}$$

where the B-spline basis functions (rectangles) are dilated versions of the cardinal ones by a factor of 2^i:

$$\phi_{i,k}(r) = \beta_+^0\left(\frac{r - 2^i k}{2^i}\right) = \begin{cases} 1, & \text{for } r \in [2^i k, 2^i(k+1)) \\ 0, & \text{otherwise.} \end{cases} \tag{6.2}$$

The variable i is the scale index that specifies the resolution (or knot spacing) $a = 2^i$, while the integer k encodes for the spatial location. The B-spline of degree zero, $\phi = \phi_{0,0} = \beta_+^0$, is the *scaling function* of the representation. Interestingly, it is the identification of a proper scaling function that constitutes the most fundamental step in the construction of a wavelet basis of $L_2(\mathbb{R})$.

DEFINITION 6.1 (Scaling function) $\phi \in L_2(\mathbb{R})$ is a valid scaling function if and only if it satisfies the following three properties:

• Two-scale relation:

$$\phi(r/2) = \sum_{k \in \mathbb{Z}} h[k]\phi(r - k), \tag{6.3}$$

where the sequence $h \in \ell_1(\mathbb{Z})$ is the so-called *refinement mask*
• Partition of unity:

$$\sum_{k \in \mathbb{Z}} \phi(r - k) = 1 \tag{6.4}$$

• The set of functions $\{\phi(\cdot - k)\}_{k \in \mathbb{Z}}$ forms a Riesz basis.

In practice, a given brand of (orthogonal) wavelets (e.g., Daubechies or splines) is often summarized by its refinement filter h since the latter uniquely specifies ϕ, subject to the admissibility constraints (6.4) and $\phi \in L_2(\mathbb{R})$. In the case of the B-spline of degree zero, we have that $h[k] = \delta[k] + \delta[k - 1]$, where

$$\delta[k] = \begin{cases} 1, & \text{for } k = 0 \\ 0, & \text{otherwise} \end{cases}$$

is the discrete Kronecker impulse. This translates into what we jokingly refer to as the *Lego–Duplo* relation [1]

$$\beta_+^0(r/2) = \beta_+^0(r) + \beta_+^0(r - 1). \tag{6.5}$$

The fact that β_+^0 satisfies the partition of unity is obvious. Likewise, we already observed in Chapter 1 that β_+^0 generates an orthogonal system which is a special case of a Riesz basis.

By considering the rescaled version of such a basis, we specify the subspace of splines at scale i as

$$V_i = \left\{ s_i(r) = \sum_{k \in \mathbb{Z}} c_i[k]\phi_{i,k}(r) : c_i \in \ell_2(\mathbb{Z}) \right\} \subset L_2(\mathbb{R}),$$

---

[1] Duplos are the larger-scale versions of Lego building blocks and are more suitable for smaller children to play with. The main point of the analogy with wavelets is that Legos and Duplos are compatible; they can be combined to build more complex shapes. The enabling property is that a Duplo is equivalent to two smaller Legos placed next to each other, as expressed by (6.5).

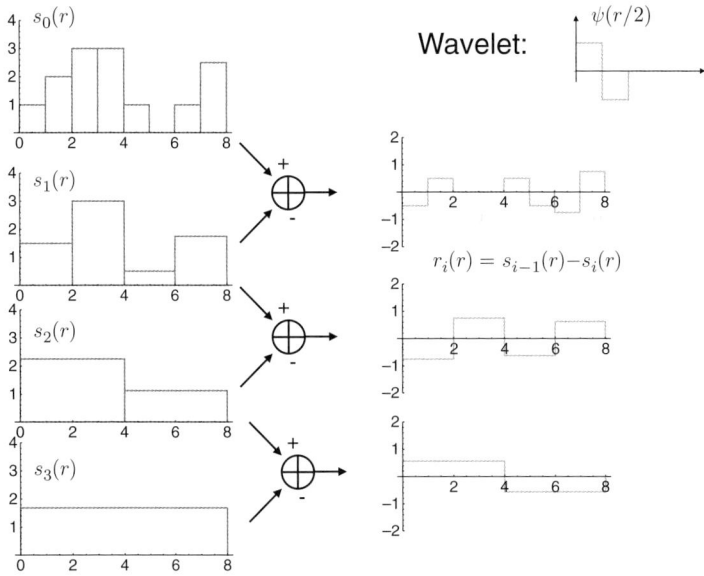

Figure 6.1 Multiresolution signal analysis using piecewise-constant splines with a dyadic scale progression. Left: multiresolution pyramid. Right: error signals between two successive levels of the pyramid.

which, in our example, contains all the finite-energy functions that are piecewise-constant on the intervals $[2^i k, 2^i(k+1))$ with $k \in \mathbb{Z}$. The two-scale relation (6.3) implies that the basis functions at scale $i = 1$ are contained in V_0 (the original space of cardinal splines) and, by extension, in V_i for $i \leq 0$. This translates into the general inclusion property $V_{i'} \subset V_i$ for any $i' > i$, which is fundamental to the theory. A subtler point is that the closure of $\bigcup_{i \in \mathbb{Z}} V_i$ is equal to $L_2(\mathbb{R})$, which follows from the fact that any finite-energy function can be approximated arbitrarily well by a piecewise-constant spline when the sampling step 2^i tends to zero ($i \rightarrow -\infty$). The necessary and sufficient condition for this asymptotic convergence is the partition of unity (6.4), which ensures that the representation is complete.

Having set the notation and specified the underlying hierarchy of function spaces, we now proceed with the multiresolution approximation procedure starting from the fine-scale signal $s_0(x)$, as illustrated in Figure 6.1. Given the sequence $c_0[\cdot]$ of fine-scale coefficients, our task is to construct the best spline approximation at scale 1 which is specified by its B-spline coefficients $c_1[\cdot]$ in (6.1) with $i = 1$. It is easy to see that the minimum-error solution (orthogonal projection of s_0 onto V_1) is obtained by taking the mean of two consecutive samples. The procedure is then repeated at the next-coarser scale and so forth until one reaches the bottom of the pyramid, as shown on the left-hand side of Figure 6.1. The description of this coarsening algorithm is

$$c_i[k] = \frac{1}{2}c_{i-1}[2k] + \frac{1}{2}c_{i-1}[2k+1] = (c_{i-1} * \tilde{h})[2k]. \tag{6.6}$$

It is run recursively for $i = 1, \ldots, i_{max}$ where i_{max} denotes the bottom level of the pyramid. The outcome is a multiresolution analysis of our input signal s_0.

In order to uncover the wavelets, it is enlightening to look at the residual signals $r_i(r) = s_{i-1}(r) - s_i(r) \in V_{i-1}$ on the right side of Figure 6.1. While these are splines that live at the same resolution as s_{i-1}, they actually have half the apparent degrees of freedom. These error signals exhibit a striking sign-alternation pattern due to the fact that two consecutive samples $(c_{i-1}[2k], c_{i-1}[2k+1])$ are at an equal distance from their mean value $(c_i[k])$. This suggests rewriting the residuals more concisely in terms of oscillating basis functions (wavelets) at scale i, like

$$r_i(r) = s_{i-1}(r) - s_i(r) = \sum_{k \in \mathbb{Z}} d_i[k]\psi_{i,k}(r), \tag{6.7}$$

where the (non-normalized) Haar wavelets are given by

$$\psi_{i,k}(r) = \psi_{\text{Haar}}\left(\frac{r - 2^i k}{2^i}\right),$$

with the Haar wavelet being defined by (1.19). The wavelet coefficients $d_i[\cdot]$ are given by the consecutive half-differences

$$d_i[k] = \frac{1}{2}c_{i-1}[2k] - \frac{1}{2}c_{i-1}[2k+1] = (c_{i-1} * \tilde{g})[2k]. \tag{6.8}$$

More generally, since the wavelet template at scale $i = 1$, $\psi_{1,0} \in V_0$, we can write

$$\psi(r/2) = \sum_{k \in \mathbb{Z}} g[k]\phi(r - k), \tag{6.9}$$

which is the wavelet counterpart of the two-scale relation (6.3). In the present example, we have $g[k] = (-1)^k h[k]$, which is a general relation that is characteristic of an orthogonal design. Likewise, in order to gain in generality, we have chosen to express the decomposition algorithms (6.6) and (6.8) in terms of discrete convolution (filtering) and downsampling operations where the corresponding Haar analysis filters are $\tilde{h}[k] = \frac{1}{2}h[-k]$ and $\tilde{g}[k] = \frac{1}{2}(-1)^k h[-k]$. The Hilbert-space interpretation of this approximation process is that $r_i \in W_i$, where W_i is the orthogonal complement of V_i in V_{i-1}; that is, $V_{i-1} = V_i + W_i$ with $V_i \perp W_i$ (as a consequence of the orthogonal-projection theorem).

Finally, we can close the loop by observing that

$$s_0(r) = s_{i_{max}}(r) + \sum_{i=1}^{i_{max}} \underbrace{\left(s_{i-1}(r) - s_i(r)\right)}_{r_i(r)}$$

$$= \sum_{k \in \mathbb{Z}} c_{i_{max}}[k]\phi_{i_{max},k}(r) + \sum_{i=1}^{i_{max}} \sum_{k \in \mathbb{Z}} d_i[k]\psi_{i,k}(r), \tag{6.10}$$

which provides an equivalent, one-to-one representation of the signal in an orthogonal wavelet basis, as illustrated in Figure 6.2.

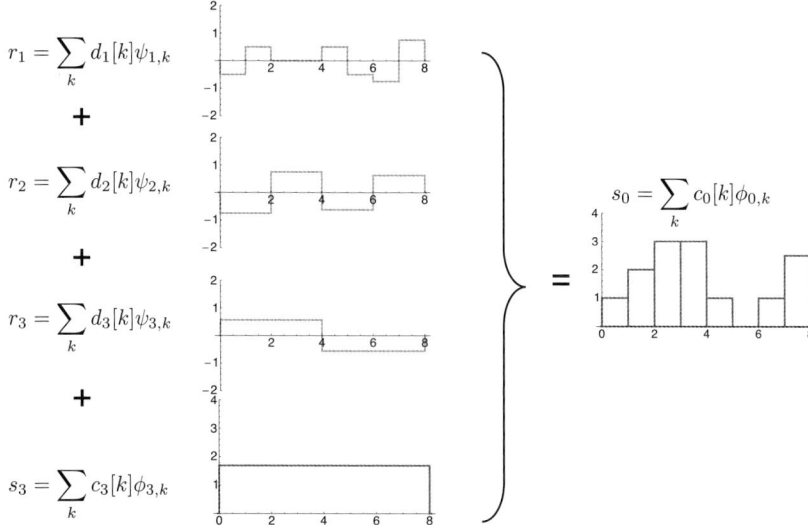

Figure 6.2 Decomposition of a signal into orthogonal scale components. The error signals $r_i = s_{i-1} - s_i$ between two successive signal approximations are expanded using a series of properly scaled wavelets.

More generally, we can push the argument to the limit and apply the decomposition to any finite-energy function

$$\forall s \in L_2(\mathbb{R}), \quad s = \sum_{i \in \mathbb{Z}} \sum_{k \in \mathbb{Z}} d_i[k]\psi_{i,k}, \tag{6.11}$$

where $d_i[k] = \langle s, \tilde{\psi}_{i,k}\rangle_{L_2}$ and $\{\tilde{\psi}_{i,k}\}_{(i,k)\in\mathbb{Z}^2}$ is a suitable (bi-)orthogonal wavelet basis with the property that $\langle \psi_{i,k}, \psi_{i',k'}\rangle_{L_2} = \delta_{k-k',i-i'}$.

Remarkably, the whole process described above – except the central expressions in (6.6) and (6.8), and the equations explicitly involving β_+^0 – is completely generic and applicable to any other wavelet basis of $L_2(\mathbb{R})$ that is specified in terms of a wavelet filter g and a scaling function ϕ (or, equivalently, an admissible refinement filter h). The bottom line is that the wavelet decomposition and reconstruction algorithm is fully described by the four digital filters $(h, g, \tilde{h}, \tilde{g})$ that form a perfect reconstruction filterbank. The Haar transform is associated with the shortest possible filters. Its less favorable aspects are that the basis functions are discontinuous and that the scale-truncated error decays only like the first power of the sampling step $a = 2^i$ (first order of approximation).

The fundamental point of our formulation is that the Haar wavelet is matched to the pure derivative operator $D = \frac{d}{dt}$, which goes hand-in-hand with Lévy processes (see Chapter 1). In that respect, the critical observations relating to spline and wavelet theory are as follows:

- All piecewise-constant functions can be interpreted as D-splines.
- The Haar wavelet acts as a smoothed version of the derivative in the sense that $\Psi_{\text{Haar}} = D\phi$, where ϕ is an appropriate kernel (triangle function).

- The B-spline of degree zero can be expressed as $\beta_+^0 = \beta_{\mathrm{D}} = \mathrm{D_d D^{-1}}\delta$, where the finite-difference operator $\mathrm{D_d}$ is the discrete counterpart of D.

We shall now show how these ideas are extendable to a much broader class of differential operators L.

6.2 Basic concepts and definitions

6.2.1 Spline-admissible operators

Let $\mathrm{L} : \mathscr{S}(\mathbb{R}^d) \to \mathscr{S}'(\mathbb{R}^d)$ be a generic Fourier-multiplier operator with frequency response $\widehat{L}(\omega)$. We further assume that L has a continuous extension $\mathrm{L} : \mathscr{X}(\mathbb{R}^d) \to \mathscr{S}'(\mathbb{R}^d)$ to some larger space of functions $\mathscr{X}(\mathbb{R}^d)$ with $\mathscr{S}(\mathbb{R}^d) \subset \mathscr{X}(\mathbb{R}^d)$.
The null space of L is denoted by $\mathscr{N}_{\mathrm{L}}$ and defined as

$$\mathscr{N}_{\mathrm{L}} = \{p_0(r) : Lp_0(r) = 0\}.$$

The immediate consequence of L being LSI is that $\mathscr{N}_{\mathrm{L}}$ is shift-invariant as well, in the sense that $p_0(r) \in \mathscr{N}_{\mathrm{L}} \Leftrightarrow p_0(r - r_0) \in \mathscr{N}_{\mathrm{L}}$ for any $r_0 \in \mathbb{R}^d$. We shall use this property to argue that $\mathscr{N}_{\mathrm{L}}$ generally consists of generalized functions whose Fourier transforms are point distributions. In the space domain, they correspond to modulated polynomials, which are linear combinations of exponential monomials of the form $e^{j\langle\omega_0,r\rangle}r^n$ with $\omega_0 \in \mathbb{R}^d$ and multi-index $n = (n_1, \ldots, n_d) \in \mathbb{N}^d$. It actually turns out that the existence of a single such element in $\mathscr{N}_{\mathrm{L}}$ has direct implications for the structure and dimensionality of the underlying function space.

PROPOSITION 6.1 (Characterization of null space)   *If L is LSI and $p_n(r) = e^{\langle z_0,r\rangle}r^n \in \mathscr{N}_{\mathrm{L}}$ with $z_0 \in \mathbb{C}^d$, then $\mathscr{N}_{\mathrm{L}}$ does necessarily include all exponential monomials of the form $p_m(r) = e^{\langle z_0,r\rangle}r^m$ with $m \leq n$. In addition, if $\mathscr{N}_{\mathrm{L}}$ is finite-dimensional, it can only consist of atoms of that particular form.*

Proof   The LSI property implies that $p_n(r - r_0) \in \mathscr{N}_{\mathrm{L}}$ for any $r_0 \in \mathbb{R}^d$. To make our point about the inclusion of the lower-order exponential polynomials in $\mathscr{N}_{\mathrm{L}}$, we start by expanding the scalar term $(r_i - r_{0,i})^{n_i}$ as

$$(r_i - r_{0,i})^{n_i} = \sum_{m=0}^{n_i} \binom{n_i}{m} r_i^m (-1)^{n_i-m} r_{0,i}^{n_i-m} = \sum_{m+k=n_i} \frac{n_i!}{m!\,k!}(-1)^k\, r_i^m\, r_{0,i}^k.$$

By proceeding in a similar manner with the other monomials and combining the results, we find that

$$(r - r_0)^n = \sum_{m+k=n} \frac{n!}{m!\,k!}(-1)^{|k|}\, r^m\, r_0^k$$

$$= \sum_{m \leq n} b_m(r_0)r^m,$$

with polynomial coefficients $b_m(r_0)$ that depend upon the multi-index m and the shift r_0. Finally, we note that the exponential factor $e^{\langle z_0,r\rangle}$ can be shifted by r_0 by simple multiplication with a constant (see (6.12) below). These facts taken together establish the structure of the underlying vector space. As for the last statement, we rely on the theory of Lie groups that tells us that the only finite-dimensional collection of functions that are translation-invariant is made of exponential polynomials. The pure exponentials $e^{\langle z_0, r \rangle}$ (with $n = 0$) are special in that respect: they are the eigenfunctions of the shift operator in the sense that

$$e^{\langle z_0, r - r_0 \rangle} = \lambda(r_0) \, e^{\langle z_0, r \rangle} \tag{6.12}$$

with the (complex) eigenvalue $\lambda(r_0) = e^{\langle z_0, r_0 \rangle}$, and hence the only elements that specify shift-invariant subspaces of dimension 1. □

Since our formulation relies on the theory of generalized functions, we shall focus on the restriction of $\mathcal{N}_L$ to $\mathcal{S}'(\mathbb{R}^d)$. This rules out the exponential factors $z_0 = \alpha_0 + j\omega_0$ in Proposition 6.1 with $\alpha_0 \in \mathbb{R}^d \setminus \{0\}$, for which the Fourier-multiplier operator is not necessarily well defined. We are then left with null-space atoms of the form $e^{j\langle \omega_0, r \rangle} r^n$ with $\omega_0 \in \mathbb{R}^d$, which are functions of slow growth.

The next important ingredient is the Green's function ρ_L of the operator L. Its defining property is $L\rho_L = \delta$, where δ is the d-dimensional Dirac distribution. Since there are many equivalent Green's functions of the form $\rho_L + p_0$ where $p_0 \in \mathcal{N}_L$ is an arbitrary component of the null space, we resolve the ambiguity by defining the (primary) Green's function of L as

$$\rho_L(r) = \mathscr{F}^{-1} \left\{ \frac{1}{\widehat{L}(\omega)} \right\} (r), \tag{6.13}$$

with the requirement that $\rho_L \in \mathcal{S}'(\mathbb{R}^d)$ is an *ordinary function* of *slow growth*. In other words, we want $\rho_L(r)$ to be defined pointwise for any $r \in \mathbb{R}^d$ and to grow no faster than a polynomial. The existence of the generalized inverse Fourier transform (6.13) imposes some minimal continuity and decay conditions on $1/\widehat{L}(\omega)$ and also puts some restrictions on the number and nature of its singularities $\left(\text{e.g., the zeros of } \widehat{L}(\omega)\right)$.

DEFINITION 6.2 (Spline admissibility) The Fourier-multiplier operator L : $\mathcal{X}(\mathbb{R}^d) \to \mathcal{S}'(\mathbb{R}^d)$ with frequency response $\widehat{L}(\omega)$ is called *spline admissible* if (6.13) is well defined and $\rho_L(r)$ is an ordinary function of slow growth.

An important characteristic of spline-admissible operators is the rate of growth of their frequency response at infinity.

DEFINITION 6.3 (Order of a Fourier multiplier) The Fourier multiplier $\widehat{L}(\omega)$ is of (asymptotic) order $\gamma \in \mathbb{R}^+$ if there exists a radius $R \in \mathbb{R}^+$ and a constant C such that

$$C|\omega|^\gamma \leq |\widehat{L}(\omega)| \tag{6.14}$$

for all $|\omega| \geq R$, where γ is critical in the sense that the condition fails for any larger value.

The order is in direct relation with the degree of smoothness of the Green's function ρ_L. In the case of a scale-invariant operator, it also coincides with the scaling order (or degree of homogeneity) of $\widehat{L}(\omega)$. For instance, the fractional-derivative operator D^γ, which is defined via the Fourier multiplier $(j\omega)^\gamma$, is of order γ. Its Green's function is given by (see Table A.1)

$$\rho_{D^\gamma}(r) = \mathcal{F}^{-1}\left\{\frac{1}{(j\omega)^\gamma}\right\}(r) = \frac{r_+^{\gamma-1}}{\Gamma(\gamma)}, \qquad (6.15)$$

where Γ is Euler's gamma function (see Appendix C) and $r_+^{\gamma-1} = \max(0, r)^{\gamma-1}$. Clearly, the latter is a function of slow growth. It has a single singularity at the origin whose Hölder exponent is $(\gamma - 1)$, and is infinitely differentiable everywhere else. It follows that ρ_{D^γ} is uniformly Hölder-continuous of degree $(\gamma-1)$. This is one less than the order of the operator. On the other hand, the null space of D^γ consists of the polynomials of degree $N = \lceil \gamma - 1 \rceil$ since $\frac{d^n (j\omega)^\gamma}{d\omega^n} \propto (j\omega)^{\gamma-n}$ is vanishing at the origin up to order N with $(\gamma - 1) \leq N < \gamma$ (see argument in Section 6.4.1).

A fundamental result is that all partial differential operators with constant coefficients are spline-admissible. This follows from the Malgrange–Ehrenpreis theorem, which guarantees the existence of their Green's function [Mal56, Wag09]. The generic form of such operators is

$$L_N = \sum_{|n|<N} a_n \partial^n$$

with $a_n \in \mathbb{R}^d$, where ∂^n is the multi-index notation for $\frac{\partial^{n_1 + \cdots + n_d}}{\partial r_1^{n_1} \cdots \partial r_d^{n_d}}$. The corresponding Fourier multiplier is $\widehat{L}_N(\omega) = \sum_{|n|<N} a_n j^{|n|} \omega^n$, which is a polynomial of degree $N = |n|$. The operator is *elliptic* if $\widehat{L}_N(\omega)$ vanishes at the origin and nowhere else. More generally, it is called *quasi-elliptic of order* γ if $\widehat{L}_N(\omega)$ fulfills the growth condition in Definition 6.3. For $d = 1$, it is fairly easy to determine ρ_L using standard Fourier-inversion techniques (see Chapter 5). Moreover, the condition for quasi-ellipticity of order N is automatically satisfied. When moving to higher dimensions, the study of partial differential operators and the determination of their Green's functions becomes more challenging because of the absence of a general multidimensional factorization mechanism. Yet, it is possible to treat special cases in full generality, such as the scale-invariant operators (with homogeneous, but not necessarily, rotation-invariant, Fourier multipliers) and the class of rotation-invariant operators that are polynomials of the Laplacian $(-\Delta)$.

6.2.2   Splines and operators

The foundation of our formulation is the direct correspondence between a spline-admissible operator L and a particular brand of splines.

DEFINITION 6.4 (Cardinal L-spline)   A function $s(r)$ (possibly of slow growth) is called a *cardinal* L-*spline* if and only if

$$Ls(r) = \sum_{k\in\mathbb{Z}^d} a[k]\delta(r - k).$$

The location of the Dirac impulses specifies the spline discontinuities (or knots). The term "cardinal" refers to the particular setting where these are located on the Cartesian grid $\mathbb{Z}^d$.

The remarkable aspect of this definition is that the operator L has the role of a mathematical A-to-D converter since it maps a continuously defined signal s into a discrete sequence $a = (a[k])$. Also note that the weighted sum of Dirac impulses on the right-hand side of the above equation can be interpreted as the continuous-domain representation of the discrete signal a – it is a hybrid-type representation that is commonly used in the theory of linear systems to model ideal sampling (multiplication with a train of Dirac impulses).

The underlying concept of a spline is fairly general and it naturally extends to non-uniform grids.

DEFINITION 6.5 (Non-uniform spline)   Let $\{r_k\}_{k \in S}$ be a set of points (not necessarily finite) that specifies a (non-uniform) grid in $\mathbb{R}^d$. Then, a function $s(r)$ (possibly of slow growth) is a *non-uniform* L-*spline* with knots $\{r_k\}_{k \in S}$ if and only if

$$\mathrm{L}s(r) = \sum_{k \in S} a_k \delta(r - r_k).$$

The direct implication of this definition is that a (non-uniform) L-spline with knots $\{r_k\}$ can generally be expressed as

$$s(r) = p_0(r) + \sum_{k \in S} a_k \rho_{\mathrm{L}}(r - r_k), \tag{6.16}$$

where $\rho_{\mathrm{L}} = \mathrm{L}^{-1}\delta$ is the Green's function of L and $p_0 \in \mathcal{N}_{\mathrm{L}}$ is an appropriate null-space component that is typically selected to fulfill some boundary conditions.

In the case where the grid is uniform, it is usually more convenient to express splines in terms of localized B-spline functions which are shifted replicates of a simple template β_{L}, or some other equivalent generator. An important requirement is that the set of B-spline functions constitutes a Riesz basis.

6.2.3   Riesz bases

To quote Ingrid Daubechies [Dau92], a Riesz basis is the next-best thing after an orthogonal basis. The reason for not enforcing orthogonality is to leave more room for other desirable features such as simplicity of construction, maximum localization of the basis function (e.g., compact support), and, last but not least, fast computational solutions.

DEFINITION 6.6 (Riesz basis)   A sequence of functions $\{\phi_k(r)\}_{k \in \mathbb{Z}}$ in $L_2(\mathbb{R}^d)$ forms a *Riesz basis* if and only if there exist two constants A and B such that

$$A \|c\|_{\ell_2} \leq \| \sum_{k \in \mathbb{Z}} c_k \phi_k(r) \|_{L_2(\mathbb{R}^d)} \leq B \|c\|_{\ell_2}$$

for any sequence $c = (c_k) \in \ell_2$. More generally, the basis is L_p-stable if there exist two constants A_p and B_p such that

$$A_p \|c\|_{\ell_p} \le \Big\| \sum_{k \in \mathbb{Z}} c_k \phi_k(r) \Big\|_{L_p(\mathbb{R}^d)} \le B_p \|c\|_{\ell_p}.$$

This definition imposes an equivalence between the L_2 (L_p, resp.) norm of the continuously defined function $s(r) = \sum_{k \in \mathbb{Z}} c_k \phi_k(r)$ and the ℓ_2 (ℓ_p, resp.) norm of its expansion coefficients (c_k). This ensures that the representation is stable in the sense that a small perturbation of the expansion coefficients results in a perturbation of comparable magnitude on $s(r)$ and vice versa. Also note that the lower inequality implies that the functions $\{\phi_k\}$ are linearly independent (by setting $s(r) = 0$), which is the defining property of a basis in finite dimensions – but which, on its own, does not ensure stability in infinite dimensions. When $A = B = 1$, we have a perfect norm equivalence, which translates into the basis being orthonormal (Parseval's relation). Finally, we point out that the existence of the bounds A and B ensures that the (infinite) Gram matrix is positive definite so that it can be readily diagonalized to yield an equivalent orthogonal basis.

In the (multi-)integer shift-invariant case where the basis functions are given by $\phi_k(r) = \phi(r - k), k \in \mathbb{Z}^d$, there is a simpler equivalent reformulation of the Riesz basis requirement of Definition 6.6.

THEOREM 6.2   *Let $\phi(r) \in L_2(\mathbb{R}^d)$ be a B-spline-like generator whose Fourier transform is denoted by $\hat{\phi}(\omega)$. Then, $\{\phi(r - k)\}_{k \in \mathbb{Z}^d}$ forms a Riesz basis with Riesz bounds A and B if and only if*

$$0 < A^2 \le \sum_{n \in \mathbb{Z}^d} |\hat{\phi}(\omega + 2\pi n)|^2 \le B^2 < \infty \tag{6.17}$$

for almost every ω. Moreover, the basis is L_p-stable for all $1 \le p \le +\infty$ if, in addition,

$$\sup_{r \in [0,1]^d} \sum_{k \in \mathbb{Z}^d} |\phi(r - k)| = A_{2,\infty} < +\infty. \tag{6.18}$$

Under such condition(s), the induced function space

$$V_\phi = \Big\{ s(r) = \sum_{k \in \mathbb{Z}^d} c[k] \phi(r - k) : c \in \ell_p(\mathbb{Z}^d) \Big\}$$

is a closed subspace of $L_p(\mathbb{R}^d)$, including the standard case $p = 2$.

Observe that the central quantity in (6.17) corresponds to the discrete-domain Fourier transform of the Gram sequence $a_\phi[k] = \langle \phi(\cdot - k), \phi \rangle_{L_2}$. Indeed, we have that

$$A_\phi(e^{j\omega}) = \sum_{k \in \mathbb{Z}^d} a_\phi[k] e^{-j\langle \omega, k \rangle} = \sum_{n \in \mathbb{Z}^d} |\hat{\phi}(\omega + 2\pi n)|^2, \tag{6.19}$$

where the right-hand side follows from Poisson's summation formula applied to the sampling at the integers of the autocorrelation function $(\overline{\phi}^\vee * \phi)(r)$. Equation (6.19) is especially advantageous in the case of compactly supported B-splines, for which the autocorrelation is often known explicitly (as a B-spline of twice the order), since it reduces the calculation to a finite sum over the support of the Gram sequence (discrete-domain Fourier transform).

Theorem 6.2 is a fundamental result in sampling and approximation theory [Uns00]. It is instructive here to briefly run through the L_2 part of the proof, which also serves as a refresher on some of the standard properties of the Fourier transform. In particular, we emphasize the interaction between the continuous and discrete aspects of the problem.

Proof We start by computing the Fourier transform of $s(r) = \sum_{k \in \mathbb{Z}^d} c[k] \phi(r - k)$, which gives

$$\mathcal{F}\{s\}(\omega) = \sum_{k \in \mathbb{Z}^d} c[k]\, e^{-j\langle \omega, k \rangle} \widehat{\phi}(\omega) \qquad \text{(by linearity and shift property)}$$

$$= C(e^{j\omega}) \cdot \widehat{\phi}(\omega),$$

where $C(e^{j\omega})$ is recognized as the discrete-domain Fourier transform of $c[\cdot]$. Next, we invoke Parseval's identity and manipulate the Fourier-domain integral as follows:

$$\|s\|_{L_2}^2 = \int_{\mathbb{R}^d} \left|C(e^{j\omega})\right|^2 \left|\widehat{\phi}(\omega)\right|^2 \frac{d\omega}{(2\pi)^d}$$

$$= \sum_{n \in \mathbb{Z}^d} \int_{[0,2\pi]^d} \left|C(e^{j(\omega+2\pi n)})\right|^2 \left|\widehat{\phi}(\omega + 2\pi n)\right|^2 \frac{d\omega}{(2\pi)^d}$$

$$= \int_{[0,2\pi]^d} \left|C(e^{j\omega})\right|^2 \sum_{n \in \mathbb{Z}^d} \left|\widehat{\phi}(\omega + 2\pi n)\right|^2 \frac{d\omega}{(2\pi)^d}$$

$$= \int_{[0,2\pi]^d} \left|C(e^{j\omega})\right|^2 A_\phi(e^{j\omega}) \frac{d\omega}{(2\pi)^d}.$$

Here, we have used the fact that $C(e^{j\omega})$ is 2π-periodic and the non-negativity of the integrand to interchange the summation and the integral (Fubini). This naturally leads to the inequality

$$\inf_{\omega \in [0,2\pi]^d} A_\phi(e^{j\omega}) \cdot \|c\|_{\ell_2}^2 \leq \|s\|_{L_2}^2 \leq \sup_{\omega \in [0,2\pi]^d} A_\phi(e^{j\omega}) \cdot \|c\|_{\ell_2}^2,$$

where we are now making use of Parseval's identity for sequences, so that

$$\|c\|_{\ell_2}^2 = \int_{[0,2\pi]^d} \left|C(e^{j(\omega)})\right|^2 \frac{d\omega}{(2\pi)^d}.$$

The final step is to show that these bounds are sharp. This can be accomplished through the choice of some particular (bandlimited) sequence $c[\cdot]$. $\square$

Note that the "almost everywhere" part of (6.17) can be dropped when $\phi \in L_1(\mathbb{R}^d)$ because the Fourier transform of such a function is continuous (Riemann–Lebesgue lemma).

While the result of Theorem 6.2 is restricted to the classical L_p spaces, there is no fundamental difficulty in extending it to wider classes of weighted (with negative powers) L_p spaces by imposing some stricter condition than (6.18) on the decay of ϕ. For instance, if ϕ has exponential decay, then the definition of the function space V_ϕ can be extended for all sequences c that are growing no faster than a polynomial. This happens to be the appropriate framework for sampling generalized stochastic processes which do not live in the L_p spaces since they are not decaying at infinity.

6.2.4   Admissible wavelets

The other important tool for analyzing stochastic processes is the wavelet transform, whose basis functions must be "tuned" to the object under investigation.

DEFINITION 6.7   A wavelet function ψ is called L-*admissible* if it can be expressed as $\psi = L^H \phi$ with $\phi \in L_1(\mathbb{R}^d)$.

Observe that we are now considering the Hermitian transpose operator $L^H = \overline{L}^*$, which is distinct from the adjoint operator L^* when the impulse response has some imaginary component. The reason for this is that the wavelet-analysis step involves a Hermitian inner product $\langle \cdot, \cdot \rangle_{L_2}$ whose definition differs by a complex conjugation from that of the distributional scalar product $\langle \cdot, \cdot \rangle$ used in our formulation of stochastic processes when the second argument is complex-valued; specifically, $\langle f, g \rangle_{L_2} = \langle f, \overline{g} \rangle = \int_{\mathbb{R}^d} f(r)\overline{g(r)}\, dr$.

The best-matched wavelet is the one for which the wavelet kernel ϕ is the most localized – ideally, the shortest possible support assuming that it is at all possible to construct a compactly supported wavelet basis. The very least is that ϕ should be concentrated around the origin and exhibit a sufficient rate of decay; for instance, $|\phi(r)| \le \frac{C}{1+\|r\|^\alpha}$ for some $\alpha > d$.

A direct implication of Definition 6.7 is that the wavelet ψ will annihilate all the components (e.g., polynomials) that are in the null space of L because $\langle \psi(\cdot - r_0), p_0 \rangle = \langle \phi(\cdot - r_0), Lp_0 \rangle = 0$, for all $p_0 \in \mathcal{N}_L$ and $r_0 \in \mathbb{R}^d$. In conventional wavelet theory, this behavior is achieved by designing "Nth-derivative-like" wavelets with vanishing moments up to polynomial degree $N - 1$.

6.3   First-order exponential B-splines and wavelets

Rather than aiming for the highest level of generality right away, we propose to first examine the 1-D first-order scenario in some detail. First-order differential models are important theoretically because they go hand-in-hand with the Markov property. In that respect, they constitute the next level of generalization beyond the Lévy processes. Mathematically, the situation is still quite comparable to that of the derivative operator in the sense that it leads to a nice, self-contained construction of (exponential) B-splines and wavelets. The interesting aspect, though, is that the underlying basis functions are no longer conventional wavelets that are dilated versions of a single prototype: they now fall into the lesser-known category of *non-stationary*[2] wavelets.

---

[2] In the terminology of wavelets, the term "non-stationary" refers to the property that the shape of the wavelet changes with scale, but not with respect to the location, as the more usual statistical meaning of the term would suggest.

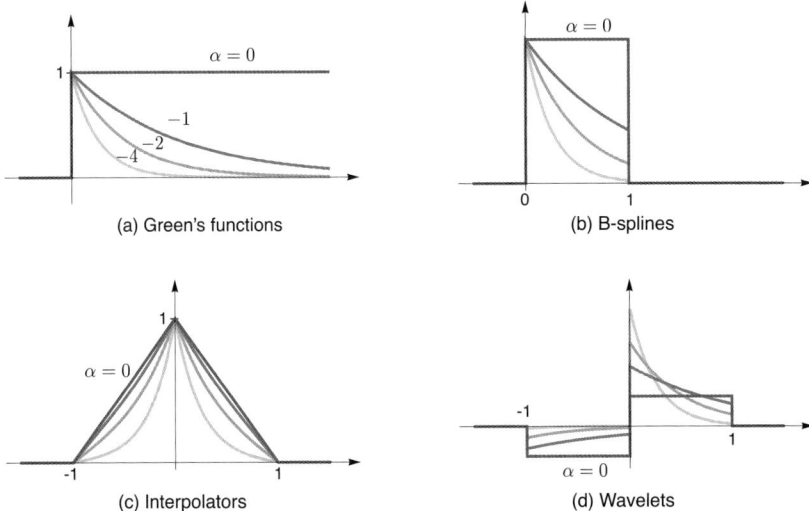

(a) Green's functions　　　(b) B-splines

(c) Interpolators　　　(d) Wavelets

Figure 6.3 Comparison of basis functions related to the first-order differential operator $P_\alpha = D - \alpha I$ for $\alpha = 0, -1, -2, -4$ (dark to light). (a) Green's functions $\rho_\alpha(r)$. (b) Exponential B-splines $\beta_\alpha(r)$. (c) Augmented spline interpolators $\varphi_{\text{int}}(r)$. (d) Orthonormalized versions of the exponential spline wavelets $\psi_\alpha(r) = P_\alpha^* \varphi_{\text{int}}(r)$.

The (causal) Green's function of our canonical first-order operator $P_\alpha = (D - \alpha \text{Id})$ is identical to the impulse response ρ_α of the corresponding differential system, while the (1-D) null space of the operator is given by $\mathcal{N}_\alpha = \{a_0 e^{\alpha r} : a_0 \in \mathbb{R}\}$. Some examples of such Green's functions are shown in Figure 6.3. The case $\alpha = 0$ (dark curve) is the classical one already treated in Section 6.1.

6.3.1 B-spline construction

The natural discrete approximation of the differential operator $P_\alpha = (D - \alpha \text{Id})$ is the first-order weighted difference operator

$$\Delta_\alpha s(r) = s(r) - e^\alpha s(r - 1). \tag{6.20}$$

Observe that Δ_α annihilates the exponentials $a_0 e^{\alpha r}$ so that its null space includes $\mathcal{N}_\alpha$. The corresponding B-spline is obtained by applying Δ_α to ρ_α, which yields

$$\beta_\alpha(r) = \mathscr{F}^{-1}\left\{\frac{1 - e^\alpha e^{-j\omega}}{j\omega - \alpha}\right\}(r) = \begin{cases} e^{\alpha r}, & \text{for } 0 \le r < 1 \\ 0, & \text{otherwise.} \end{cases} \tag{6.21}$$

In effect, the localization by Δ_α results in a "chopped-off" version of the causal Green's function that is restricted to the interval $[0, 1)$ (see Figure 6.3b). Importantly, the scheme remains applicable in the unstable scenario $\text{Re}(\alpha) \ge 0$. It always results in a well-defined Fourier transform due to the convenient pole–zero cancellation in the central expression of (6.21). The marginally unstable case $\alpha = 0$ results in the rectangular function shown in Figure 6.3b, which is the standard basis function for representing piecewise-constant signals. Likewise, β_α generates an orthogonal basis for the space of cardinal P_α-splines in accordance with Definition 6.4. This allows us to specify our prototypical exponential spline space as $V_0 = \text{span}\{\beta_\alpha(\cdot - k)\}_{k\in\mathbb{Z}}$ with knot spacing $2^0 = 1$.

6.3.2 Interpolator in augmented-order spline space

The second important ingredient is the interpolator for the "augmented-order" spline space generated by the autocorrelation $(\overline{\beta}_\alpha^\vee * \beta_\alpha)(r)$ of the B-spline. Constructing it is especially easy in the first-order case because it involves the simple normalization

$$\varphi_{\text{int},\alpha}(r) = \frac{1}{(\overline{\beta}_\alpha^\vee * \beta_\alpha)(0)} (\overline{\beta}_\alpha^\vee * \beta_\alpha)(r). \tag{6.22}$$

Specifically, $\varphi_{\text{int},\alpha}$ is the unique cardinal $P_\alpha^H P_\alpha$-spline function that vanishes at all the integers, except at the origin where it takes the value one (see Figure 6.3c). Its classical use is to provide a sinc-like kernel for the representation of the corresponding family of splines, and also for the reconstruction of spline-related signals, including special brands of stochastic processes, from their integer samples [UB05b]. Another remarkable and lesser-known property is that this function provides the proper smoothing kernel for defining an operator-like wavelet basis.

6.3.3 Differential wavelets

In the generalized spline framework, instead of specifying a hierarchy of multiresolution subspaces of $L_2(\mathbb{R})$ (the space of finite-energy functions) via the dilation of a scaling function, one considers the fine-to-coarse sequence of L-spline spaces

$$V_i = \{s(r) \in L_2(\mathbb{R}) : \mathrm{L}s(r) = \sum_{k\in\mathbb{Z}} a_i[k]\delta(r - 2^i k)\},$$

where the embedding $V_i \supseteq V_j$ for $i \leq j$ is obvious from the (dyadic) hierarchy of spline knots, so that $s_j \in V_j$ implies that $s_j \in V_i$ with an appropriate subset of its coefficients $a_i[k]$ being zero.

We now detail the construction of a wavelet basis at resolution 1 such that $W_1 = \text{span}\{\psi_{1,k}\}_{k\in\mathbb{Z}}$ with $W_1 \perp V_1$ and $V_1 + W_1 = V_0 = \text{span}\{\beta_\alpha(\cdot - k)\}_{k\in\mathbb{Z}}$. The recipe is to take $\psi_{1,k}(r) = \psi_\alpha(r - 1 - 2k)/\|\psi_\alpha\|_{L_2}$, where ψ_α is the mother wavelet given by

$$\psi_\alpha(r) = P_\alpha^H \varphi_{\text{int},\alpha}(r) \propto \Delta_\alpha^H \beta_\alpha(r).$$

Here, Δ_α^H is the Hermitian adjoint of the finite-difference operator Δ_α. Examples of such exponential-spline wavelets are shown in Figure 6.3d, including the classical Haar wavelet (up to a sign change) which is obtained for $\alpha = 0$. The basis functions $\psi_{1,k}$ are shifted versions of ψ_α that are centered at the odd integers and normalized to have a unit norm. Since these wavelets are non-overlapping, they form an orthonormal basis. Moreover, the basis is orthogonal to the coarser spline space V_1 as a direct consequence of the interpolating property of $\varphi_{\text{int},\alpha}$ (Proposition 6.6 in Section 6.5). Finally, based on the fact that $\psi_{1,k} \in V_0$ for all $k \in \mathbb{Z}$, one can show that these wavelets span W_1, which translates into

$$W_1 = \left\{ v(r) = \sum_{k \in \mathbb{Z}} v_1[k] \psi_{1,k}(r) : v_1 \in \ell_2(\mathbb{Z}) \right\}.$$

This method of construction extends to the other wavelet subspaces W_i provided that the interpolating kernel $\varphi_{\text{int},a}$ is replaced by its proper counterpart at resolution $a = 2^{i-1}$ and the sampling grid adjusted accordingly. Ultimately, this results in a wavelet basis of $L_2(\mathbb{R})$ whose members are all P_α-splines – that is, piecewise-exponential with parameter α – but not dilates of the same prototype unless $\alpha = 0$. Otherwise, the corresponding decomposition is not fundamentally different from a conventional wavelet expansion. The basis functions are equally well localized and the scheme admits the same type of fast reversible filterbank algorithm, albeit with scale-dependent filters [KU06].

6.4   Generalized B-spline basis

The procedure of Section 6.3.1 remains applicable for the broad class of spline-admissible operators (see Definition 6.2) in one or multiple dimensions. The two ingredients for constructing a generalized B-spline basis are: (1) the knowledge of the Green's function ρ_L of the operator L, and (2) the availability of a discrete approximation (finite-difference-like) of the operator of the form

$$L_d s(r) = \sum_{k \in \mathbb{Z}^d} d_L[k] s(r - k) \qquad (6.23)$$

with $d_L \in \ell_1(\mathbb{Z}^d)$ that fulfills the null-space matching constraint [3]

$$L_d p_0(r) = L p_0(r) = 0 \quad \text{for all } p_0 \in \mathcal{N}_L. \qquad (6.24)$$

The generalized B-spline associated with the operator L is then given by

$$\beta_L(r) = L_d \rho_L(r) = \mathcal{F}^{-1} \left\{ \frac{\sum_{k \in \mathbb{Z}^d} d_L[k] e^{-j\langle k, \omega \rangle}}{\widehat{L}(\omega)} \right\}(r), \qquad (6.25)$$

where the numerator and denominator in the right-hand expression correspond to the frequency responses of L_d and L, respectively. The null-space matching constraint is especially helpful for the unstable cases where $\rho_L \notin L_1(\mathbb{R}^d)$: it ensures that the zeros of $\widehat{L}(\omega)$ (singularities) are cancelled by some corresponding zeros of $\widehat{L}_d(\omega)$ so that the Fourier transform of β_L remains bounded.

DEFINITION 6.8   The function β_L specified by (6.25) is an admissible B-spline for L if and only if (1) $\beta_L \in L_1(\mathbb{R}^d) \cap L_2(\mathbb{R}^d)$ and (2) it generates a Riesz basis of the space of cardinal L-splines.

---

[3] We want the null space of L_d to include $\mathcal{N}_L$ and to remain as small as possible. In that respect, it is worth noting that the null space of a discrete operator will always be much larger than that of its continuous-domain counterpart. For instance, the derivative operator D suppresses constant signals, while its finite-difference counterpart annihilates all 1-periodic functions, including the constants.

In light of Theorem 6.2, the latter property requires the existence of the two Riesz bounds A and B such that

$$0 < A^2 \le \sum_{n \in \mathbb{Z}^d} |\hat{\beta}_L(\omega + 2\pi n)|^2 = \frac{\left|\sum_{k \in \mathbb{Z}^d} d_L[k] e^{-j\langle k,\omega \rangle}\right|^2}{\sum_{n \in \mathbb{Z}^d} |\hat{L}(\omega + 2\pi n)|^2} \le B^2. \tag{6.26}$$

A direct consequence of (6.25) is that

$$L\beta_L(r) = \sum_{k \in \mathbb{Z}^d} d_L[k]\delta(r - k), \tag{6.27}$$

so that β_L is itself a cardinal L-spline in accordance with Definition 6.4. The bottom line in Definition 6.8 is that any cardinal L-spline admits a unique representation in the B-spline basis $\{\beta_L(\cdot - k)\}_{k \in \mathbb{Z}^d}$ as

$$s(r) = \sum_{k \in \mathbb{Z}^d} c[k]\beta_L(r - k), \tag{6.28}$$

where the $c[k]$ are the *B-spline coefficients* of s.

While (6.25) provides us with a nice recipe for constructing B-splines, it does not guarantee that the Riesz-basis condition (6.26) is satisfied. This needs to be established on a case-by-case basis. The good news for the present theory of stochastic processes is that B-splines are available for virtually all the operators that have been discussed so far.

6.4.1   B-spline properties

To motivate the use of B-splines, we shall first restrict our attention to the space V_L of cardinal L-splines with finite energy, which is formally defined as

$$V_L = \left\{ s(r) \in L_2(\mathbb{R}^d) : Ls(r) = \sum_{k \in \mathbb{Z}^d} a[k]\delta(r - k) \right\}. \tag{6.29}$$

The foundation of spline theory is that there are two complementary ways of representing splines using different types of basis functions: Green's functions vs. B-splines. The first representation follows directly from Definition 6.4 (see also (6.16)) and is given by

$$s(r) = p_0(r) + \sum_{k \in \mathbb{Z}^d} a[k]\rho_L(r - k), \tag{6.30}$$

where $p_0 \in \mathcal{N}_L$ is a suitable element of the null space of L and where $\rho_L = L^{-1}\delta$ is the Green's function of the operator. The functions $\rho_L(\cdot - k)$ are non-local and very far from being orthogonal. In many cases, they are not even part of V_L, which raises fundamental issues concerning the L_2 convergence of the infinite sum [4] in (6.30) and the conditions that must be imposed upon the expansion coefficients $a[\cdot]$. The second type of B-spline expansion (6.28) does not have such stability problems. This is the primary reason why it is favored by practitioners.

---

[4] Without further assumptions on ρ_L and a, (6.30) is only valid in the weak sense of distributions.

Stable representation of cardinal L-splines

The equivalent B-spline specification of the space V_L of cardinal splines is

$$V_L = \left\{ s(r) = \sum_{k \in \mathbb{Z}^d} c[k] \beta_L(r - k) : c[\cdot] \in \ell_2(\mathbb{Z}^d) \right\},$$

where the generalized B-spline β_L satisfies the conditions in Definition 6.8. The Riesz-basis property ensures that the representation is stable in the sense that, for all $s \in V_L$, we have that

$$A\|c\|_{\ell_2} \leq \|s\|_{L_2} \leq B\|c\|_{\ell_2}. \tag{6.31}$$

Here, $\|c\|_{\ell_2} = \left(\sum_{k \in \mathbb{Z}^d} |c[k]|^2 \right)^{\frac{1}{2}}$ is the ℓ_2-norm of the B-spline coefficients c. The fact that the underlying functions are cardinal L-splines is a simple consequence of the atoms being splines themselves. Moreover, we can easily make the link with (6.30) by using (6.27), which yields

$$Ls(r) = \sum_{k \in \mathbb{Z}^d} c[k] L\beta_L(r - k) = \sum_{k \in \mathbb{Z}^d} \underbrace{(c * d_L)[k]}_{a[k]} \delta(r - k).$$

The less obvious aspect, which is implicit in the definition of the B-spline, is the *completeness* of the representation in the sense that the B-spline basis spans the space V_L defined by (6.29). We shall establish this by showing that the B-splines are capable of reproducing ρ_L as well as any component $p_0 \in \mathcal{N}_L$ in the null space of L. The implication is that any function of the form (6.30) admits a unique expansion in a B-spline basis. This is also true when the function is not in $L_2(\mathbb{R}^d)$, in which case the B-spline coefficients c are no longer in $\ell_2(\mathbb{Z}^d)$ due to the discrete–continuous norm equivalence (6.31).

Reproduction of Green's functions

The reproduction of Green's functions follows from the special form of (6.25). To reveal it, we consider the inverse L_d^{-1} of the discrete localization operator L_d specified by (6.23), whose continuous-domain impulse response is written as

$$L_d^{-1}\delta(r) = \sum_{k \in \mathbb{Z}^d} p[k]\delta(r - k) = \mathscr{F}^{-1} \left\{ \frac{1}{\sum_{k \in \mathbb{Z}^d} d_L[k] e^{-j\langle k, \omega \rangle}} \right\}.$$

The sequence p, which can be determined by generalized inverse Fourier transform, is of slow growth with the property that $(p * d)[k] = \delta[k]$. The Green's function reproduction formula is then obtained by applying L_d^{-1} to the B-spline β_L and making use of the left-inverse property of L_d^{-1}. Thus,

$$L_d^{-1}\beta_L(r) = L_d^{-1} L_d \rho_L(r) = \rho_L(r)$$

results in

$$\rho_L(r) = \sum_{k \in \mathbb{Z}^d} p[k] \beta_L(r - k). \tag{6.32}$$

To illustrate the concept, let us get back to our introductory example in Section 6.3.1, with $L = P_\alpha = (D - \alpha Id)$ where $\mathrm{Re}(\alpha) < 0$. The frequency response of this first-order operator is

$$\widehat{P}_\alpha(\omega) = j\omega - \alpha,$$

while its Green's function is given by

$$\rho_\alpha(r) = \mathbb{1}_+(r)e^{\alpha r} = \mathscr{F}^{-1}\left\{\frac{1}{j\omega - \alpha}\right\}(r).$$

On the discrete side of the picture, we have the finite-difference operator Δ_α with

$$\widehat{\Delta}_\alpha(\omega) = 1 - e^{\alpha - j\omega},$$

and its inverse Δ_α^{-1} whose expansion coefficients are

$$p_\alpha[k] = \mathbb{1}_+[k]e^{\alpha k} = \mathscr{F}_d^{-1}\left\{\frac{1}{1 - e^\alpha e^{-j\omega}}\right\}[k],$$

where $\mathscr{F}_d^{-1}$ denotes the discrete-domain inverse Fourier transform. [5] The application of (6.32) then yields the exponential-reproduction formula

$$\mathbb{1}_+(r)e^{\alpha r} = \sum_{k=0}^{\infty} e^{\alpha k}\beta_\alpha(r - k), \qquad (6.33)$$

where β_α is the exponential B-spline defined by (6.21). Note that the range of applicability of (6.33) extends to $\mathrm{Re}(\alpha) \leq 0$.

Reproduction of null-space components
A fundamental property of B-splines is their ability to reproduce the components that are in the null space of their defining operator. In the case of our working example, we can simply extrapolate (6.33) for negative indices, which yields

$$e^{\alpha r} = \sum_{k\in\mathbb{Z}} e^{\alpha k}\beta_\alpha(r - k).$$

It turns out that this reproduction property is induced by the matching null-space constraint (6.24) that is imposed upon the localization filter. While the reproduction of exponentials is interesting in its own right, we shall focus here on the important case of polynomials and provide a detailed Fourier-based analysis. We start by recalling that the general form of a multidimensional polynomial of total degree N is

$$q_N(r) = \sum_{|n|\leq N} a_n r^n,$$

---

[5] Our definition of the inverse discrete Fourier transform in 1-D is $\mathscr{F}_d^{-1}\left\{H(e^{j\omega})\right\}[k] = \frac{1}{2\pi}\int_{-\pi}^{\pi} H(e^{j\omega})e^{j\omega k}\, d\omega$ with $k \in \mathbb{Z}$.

using the multi-index notation with $\boldsymbol{n} = (n_1, \ldots, n_d) \in \mathbb{N}^d$, $\boldsymbol{r^n} = r_1^{n_1} \cdots r_d^{n_d}$, and $|\boldsymbol{n}| = n_1 + \cdots + n_d$. The generalized Fourier transform of $q_N \in \mathscr{S}'(\mathbb{R}^d)$ (see Table 3.3 and entry $\boldsymbol{r^n} f(\boldsymbol{r})$ with $f(\boldsymbol{r}) = 1$) is given by

$$\widehat{q}_N(\boldsymbol{\omega}) = \sum_{|\boldsymbol{n}| \leq N} (2\pi)^d a_{\boldsymbol{n}} \, \mathrm{j}^{|\boldsymbol{n}|} \partial^{\boldsymbol{n}} \delta(\boldsymbol{\omega}),$$

where $\partial^{\boldsymbol{n}} \delta$ denotes the $\boldsymbol{n}$th partial derivative of the multidimensional Dirac impulse δ. Hence, the Fourier multiplier $\widehat{L}$ will annihilate the polynomials of order N if and only if $\widehat{L}(\boldsymbol{\omega}) \partial^{\boldsymbol{n}} \delta(\boldsymbol{\omega}) = 0$ for all $|\boldsymbol{n}| \leq N$. To understand when this condition is met, we expand $\widehat{L}(\boldsymbol{\omega}) \partial^{\boldsymbol{n}} \delta(\boldsymbol{\omega})$ in terms of $\partial^{\boldsymbol{k}} \widehat{L}(0)$, $|\boldsymbol{k}| \leq |\boldsymbol{n}|$, by using the general product rule for the manipulation of Dirac impulses and their derivatives, given by

$$f(\boldsymbol{r}) \, \partial^{\boldsymbol{n}} \delta(\boldsymbol{r} - \boldsymbol{r}_0) = \sum_{\boldsymbol{k}+\boldsymbol{l}=\boldsymbol{n}} \frac{\boldsymbol{n}!}{\boldsymbol{k}! \, \boldsymbol{l}!} (-1)^{|\boldsymbol{n}|+|\boldsymbol{l}|} \partial^{\boldsymbol{k}} f(\boldsymbol{r}_0) \, \partial^{\boldsymbol{l}} \delta(\boldsymbol{r} - \boldsymbol{r}_0).$$

The latter follows from Leibniz' rule for partially differentiating a product of functions,

$$\partial^{\boldsymbol{n}} (f\varphi) = \sum_{\boldsymbol{k}+\boldsymbol{l}=\boldsymbol{n}} \frac{\boldsymbol{n}!}{\boldsymbol{k}! \, \boldsymbol{l}!} \partial^{\boldsymbol{k}} f \, \partial^{\boldsymbol{l}} \varphi,$$

and the adjoint relation $\langle \varphi, f \, \partial^{\boldsymbol{n}} \delta(\cdot - \boldsymbol{r}_0) \rangle = \langle \partial^{\boldsymbol{n}*}(f\varphi), \delta(\cdot - \boldsymbol{r}_0) \rangle$ with $\partial^{\boldsymbol{n}*} = (-1)^{|\boldsymbol{n}|} \partial^{\boldsymbol{n}}$. This allows us to conclude that the necessary and sufficient condition for the inclusion of the polynomials of order N in the null space of L is

$$\partial^{\boldsymbol{n}} \widehat{L}(0) = 0, \quad \text{for all } \boldsymbol{n} \in \mathbb{N}^d \text{ with } |\boldsymbol{n}| \leq N, \tag{6.34}$$

which is equivalent to $\widehat{L}(\boldsymbol{\omega}) = O(\|\boldsymbol{\omega}\|^{N+1})$ around the origin. Note that this behavior is prototypical of scale-invariant operators such as fractional derivatives and Laplacians. The same condition has obviously to be imposed upon the localization filter $\widehat{L}_{\mathrm{d}}$ in order for the Fourier transform of the B-spline in (6.25) to be non-singular at the origin. Since $\widehat{L}_{\mathrm{d}}(\boldsymbol{\omega})$ is 2π-periodic, we have that

$$\partial^{\boldsymbol{n}} \widehat{L}_{\mathrm{d}}(2\pi \boldsymbol{k}) = 0, \quad \boldsymbol{k} \in \mathbb{Z}^d, \boldsymbol{n} \in \mathbb{N}^d \text{ with } |\boldsymbol{n}| \leq N. \tag{6.35}$$

For practical convenience, we shall assume that the B-spline β_{L} is normalized to have a unit integral [6] so that $\widehat{\beta}_{\mathrm{L}}(0) = 1$. Based on (6.35) and $\widehat{\beta}_{\mathrm{L}}(\boldsymbol{\omega}) = \widehat{L}_{\mathrm{d}}(\boldsymbol{\omega})/\widehat{L}(\boldsymbol{\omega})$, we find that

$$\begin{cases} \widehat{\beta}_{\mathrm{L}}(0) = 1 \\ \partial^{\boldsymbol{n}} \widehat{\beta}_{\mathrm{L}}(2\pi \boldsymbol{k}) = 0, \quad \boldsymbol{k} \in \mathbb{Z}^d \backslash \{0\}, \boldsymbol{n} \in \mathbb{N}^d \text{ with } |\boldsymbol{n}| \leq N, \end{cases} \tag{6.36}$$

which are the so-called *Strang–Fix conditions of order N*. Recalling that $\mathrm{j}^{|\boldsymbol{n}|} \partial^{\boldsymbol{n}} \widehat{\beta}_{\mathrm{L}}(\boldsymbol{\omega})$ is the Fourier transform of $\boldsymbol{r^n} \beta_{\mathrm{L}}(\boldsymbol{r})$ and that periodization in the signal domain corresponds to a sampling in the Fourier domain, we finally deduce that

$$\sum_{\boldsymbol{k} \in \mathbb{Z}^d} (\boldsymbol{r} - \boldsymbol{k})^{\boldsymbol{n}} \beta_{\mathrm{L}}(\boldsymbol{r} - \boldsymbol{k}) = \mathrm{j}^{|\boldsymbol{n}|} \partial^{\boldsymbol{n}} \widehat{\beta}_{\mathrm{L}}(0) = C_{\boldsymbol{n}}, \quad \boldsymbol{n} \in \mathbb{N}^d \text{ with } 0 < |\boldsymbol{n}| \leq N, \tag{6.37}$$

---

[6] This is always possible thanks to (6.24), which ensures that $\widehat{\beta}_{\mathrm{L}}(0) \neq 0$ due to a proper cancellation of poles and zeros in the right-hand side of (6.25).

with the implicit assumption that β_L has a sufficient order of algebraic decay for the above sums to be convergent. The special case of (6.37) with $n = 0$ reads

$$\sum_{k \in \mathbb{Z}^d} \beta_L(r - k) = 1 \tag{6.38}$$

and is called the *partition of unity*. It reflects the fact that β_L reproduces the constants. More generally, (6.37) (or (6.36)) is equivalent to the existence of sequences p_n such that

$$r^n = \sum_{k \in \mathbb{Z}^d} p_n[k] \beta_L(r - k) \quad \text{for all } |n| \leq N, \tag{6.39}$$

which is a more direct statement of the polynomial-reproduction property. For instance, (6.37) with $n = (1, \ldots, 1)$ implies that

$$r \underbrace{\left(\sum_{k \in \mathbb{Z}^d} \beta_L(r - k) \right)}_{p_{(0,\ldots,0)}(k) = 1} - \sum_{k \in \mathbb{Z}^d} k \beta_L(r - k) = C_{(1,\ldots,1)},$$

from which one deduces that $p_{(1,\ldots,1)}[k] = k + C_{(1,\ldots,1)}$. The other sequences p_n, which are polynomials in k, may be determined in a similar fashion by proceeding recursively. Another equivalent way of stating the Strang–Fix conditions of order N is that the sums

$$\sum_{k \in \mathbb{Z}^d} k^n \beta_L(r - k) = \sum_{l \in \mathbb{Z}^d} j^{|n|} \partial_\omega^n \left(e^{-j\langle \omega, r \rangle} \widehat{\beta}_L(-\omega) \right) \Big|_{\omega = 2\pi l}$$

are polynomials with leading term r^n for all $|n| \leq N$. The left-hand-side expression follows from Poisson's summation formula[7] applied to the function $f(x) = x^n \beta_L(r - x)$ with r being considered as a constant shift.

Localization
The guiding principle for designing B-splines is to produce basis functions that are maximally localized on $\mathbb{R}^d$. Ideally, B-splines should have the smallest possible support, which is the property that makes them so useful in applications. When it is not possible to construct compactly supported basis functions, the B-spline should at least be concentrated around the origin and satisfy some decay bound with the tightest possible constants. The primary types of spatial localization, by order of preference, are:

(1) Compact support: $\beta_L(r) = 0$ for all $r \notin \Omega$ where $\Omega \subset \mathbb{R}^d$ is a bounded set with the smallest possible Lebesgue measure
(2) Exponential decay: $|\beta_L(r)| \leq C \exp(-\alpha |r|)$ with the largest possible $\alpha \in \mathbb{R}^+$
(3) Algebraic decay: $|\beta_L(r)| \leq C \frac{1}{1 + \|r\|^\alpha}$ with the largest possible $\alpha \in \mathbb{R}^+$.

By relying on the classical relations that link spatial decay to the smoothness of the Fourier transform, one can get a good estimate of spatial decay based on the knowledge

---

[7] The standard form of Poisson's summation formula is $\sum_{k \in \mathbb{Z}^d} f(k) = \sum_{l \in \mathbb{Z}^d} \widehat{f}(2\pi l)$. It is valid for any Fourier pair $f, \widehat{f} = \mathcal{F}\{f\} \in L_1(\mathbb{R}^d)$ with sufficient decay for the two sums to be convergent.

of the Fourier transform $\widehat{\beta}_{\mathrm{L}}(\omega) = \widehat{L_{\mathrm{d}}}(\omega)/\widehat{L}(\omega)$ of the B-spline. Since the localization filter $\widehat{L_{\mathrm{d}}}(\omega)$ acts by compensating the (potential) singularities of $\widehat{L}(\omega)$, the guiding principle is that the rate of decay is essentially determined by the degree of differentiability of $\widehat{L}(\omega)$.

Specifically, if $\widehat{\beta}_{\mathrm{L}}(\omega)$ is differentiable up to order N, then the B-spline β_{L} is guaranteed to have an algebraic decay of order N. To show this, we consider the Fourier transform pair $r^{n}\beta_{\mathrm{L}}(r) \leftrightarrow j^{|n|}\partial^{n}\widehat{\beta}_{\mathrm{L}}$ subject to the constraint that $\partial^{n}\widehat{\beta}_{\mathrm{L}} \in L_{1}(\mathbb{R}^{d})$ for all $|n| < N$. From the definition of the inverse Fourier integral, it immediately follows that

$$\left| r^{n}\beta_{\mathrm{L}}(r) \right| \le \frac{1}{(2\pi)^{d}} \| \partial^{n}\widehat{\beta}_{\mathrm{L}} \|_{L_{1}},$$

which, when properly combined over all multi-indices $|n| < N$, yields an algebraic decay estimate with $\alpha = N$. By pushing the argument to the limit, we see that exponential decay (which is faster than any order of algebraic decay) requires that $\widehat{\beta}_{\mathrm{L}} \in C^{\infty}(\mathbb{R}^{d})$ (infinite order of differentiability), which is only possible if $\widehat{L}(\omega) \in C^{\infty}(\mathbb{R}^{d})$ as well.

The ultimate limit in Fourier-domain regularity is when $\widehat{\beta}_{\mathrm{L}}$ has an analytic extension that is an entire [8] function. In fact, by the Paley–Wiener theorem (Theorem 6.3 below), one achieves compact support of β_{L} if and only if $\widehat{\beta}_{\mathrm{L}}(\zeta)$ is an entire function of exponential type. To explain this concept, we focus on the 1-D case where the B-spline β_{L} is supported in the finite interval $[-A, +A]$. We then consider the holomorphic Fourier (or Fourier–Laplace) transform of the B-spline, given by

$$\widehat{\beta}_{\mathrm{L}}(\zeta) = \int_{-A}^{+A} \beta_{\mathrm{L}}(r) e^{-\zeta r} \, dr \tag{6.40}$$

with $\zeta = \sigma + j\omega \in \mathbb{C}$, which formally amounts to replacing $j\omega$ by ζ in the expression for the Fourier transform of β_{L}. In order to obtain a proper analytic extension, we need to verify that $\widehat{\beta}_{\mathrm{L}}(\zeta)$ satisfies the Cauchy–Riemann equation. We shall do so by applying a dominated-convergence argument. To that end, we construct the exponential bound

$$\left| \widehat{\beta}_{\mathrm{L}}(\zeta) \right| \le e^{A|\zeta|} \int_{-A}^{+A} |\beta_{\mathrm{L}}(r)| \, dr$$

$$\le e^{A|\zeta|} \sqrt{\int_{-A}^{+A} 1 \, dr} \sqrt{\int_{-A}^{+A} |\beta_{\mathrm{L}}(r)|^{2} \, dr}$$

$$= e^{A|\zeta|} \sqrt{2A} \, \|\beta_{\mathrm{L}}\|_{L_{2}},$$

where we have applied the Cauchy–Schwarz inequality to derive the lower inequality. Since $e^{-\zeta r}$ for r fixed is itself an entire function and (6.40) is convergent over the whole complex plane, the conclusion is that $\widehat{\beta}_{\mathrm{L}}(\zeta)$ is entire as well, in addition to being a function of exponential type A as indicated by the bound. The whole strength of the Paley–Wiener theorem is that the implication also works the other way around.

---

[8] An entire function is a function that is analytic over the whole complex plane $\mathbb{C}$.

THEOREM 6.3 (Paley–Wiener)   *Let $f \in L_2(\mathbb{R})$. Then, f is compactly supported in $[-A, A]$ if and only if its Laplace transform*

$$F(\zeta) = \int_{\mathbb{R}} f(r) e^{-\zeta r} \, dr$$

is an entire function of exponential type A, meaning that there exists a constant C such that

$$|F(\zeta)| \leq C e^{A|\zeta|}$$

for all $\zeta \in \mathbb{C}$.

The result implies that one can deduce the support of f from its Laplace transform. We can also easily extend the result to the case where the support is not centered around the origin by applying the Paley–Wiener theorem to the autocorrelation function $(f * f^\vee)(r)$. The latter is supported in the interval $[-2A, 2A]$, which is twice the size of the support of f, irrespective of its center. This suggests the following expression for the determination of the support of a B-spline:

$$\text{support}(\beta_L) = \limsup_{R \to \infty} \frac{\log \left(\sup_{|\zeta| \leq R} \left| \widehat{\beta}_L(\zeta) \widehat{\beta}_L(-\zeta) \right| \right)}{R}. \tag{6.41}$$

It returns twice the exponential type of the recentered B-spline, which gives $\text{support}(\beta_L) = 2A$. While this formula is only strictly valid when $\widehat{\beta}_L(\zeta)$ is an entire function, it can be used otherwise as an operational measure of localization when the underlying B-spline is not compactly supported. Interestingly, (6.41) provides a measure that is additive with respect to convolution and proportional to the order γ. For instance, the support of an (exponential) B-spline associated with an ordinary differential operator of order N is precisely N, as a consequence of the factorization property of such B-splines (see Sections 6.4.2 and 6.4.4).

To get some insight into (6.41), let us consider the case of the polynomial B-spline of order one (or degree zero) with $\beta_D(r) = \mathbb{1}_{[0,1)}(r)$ and Laplace transform

$$\widehat{\beta}_D(\zeta) = \left(\frac{1 - e^{-\zeta}}{\zeta} \right).$$

The required product in (6.41) is

$$\widehat{\beta}_D(\zeta) \widehat{\beta}_D(-\zeta) = \frac{-e^\zeta + 2 - e^{-\zeta}}{\zeta^2},$$

which is analytic over the whole complex plane because of the pole–zero cancellation at $\zeta = 0$. For R sufficiently large, we clearly have that

$$\max_{|\zeta| \leq R} \left| \widehat{\beta}_D(\zeta) \widehat{\beta}_D(-\zeta) \right| = \frac{e^R + 2 + e^{-R}}{R^2} \to \frac{e^R}{R^2}.$$

By plugging the above expression into (6.41), we finally get

$$\text{support}(\beta_D) = \limsup_{R \to \infty} \frac{R - 2 \log R}{R} = 1,$$

which is the desired result. While the above calculation may look like overkill for the determination of the already-known support of β_D, it becomes quite handy for making predictions for higher-order operators. To illustrate the point, we now consider the B-spline of order γ associated with the (possibly fractional) derivative operator D^γ whose Fourier–Laplace transform is

$$\widehat{\beta}_{D^\gamma}(\zeta) = \left(\frac{1 - e^{-\zeta}}{\zeta}\right)^\gamma.$$

We can then essentially replicate the previous manipulation while moving the order out of the logarithm to deduce that

$$\text{support}(\beta_{D^\gamma}) = \limsup_{R \to \infty} \frac{\gamma R - 2\gamma \log R}{R} = \gamma.$$

This shows that the "support" of the B-spline is equal to its order, with the caveat that the underlying Fourier–Laplace transform $\widehat{\beta}_{D^\gamma}(\zeta)$ is only analytic (and entire) when the order γ is a positive integer. This points to the fundamental limitation that a B-spline associated with a fractional operator – that is, when $\widehat{L}(\zeta)$ is not an entire function – cannot be compactly supported.

Smoothness

The smoothness of a B-spline refers to its degree of continuity and/or differentiability. Since a B-spline is a linear combination of shifted Green's functions, its smoothness is the same as that of ρ_L.

Smoothness descriptors come in two flavors – Hölder continuity vs. Sobolev differentiability – depending on whether the analysis is done in the signal or Fourier domain. Due to the duality between Fourier decay and order of differentiation, the smoothness of β_L may be predicted from the growth of $\widehat{L}(\omega)$ at infinity, without need for the explicit calculation of ρ_L. To that end, one considers the Sobolev spaces W_2^α, which are defined as

$$W_2^\alpha(\mathbb{R}^d) = \left\{ f : \int_{\mathbb{R}^d} (1 + \|\omega\|^2)^\alpha |\widehat{f}(\omega)|^2 \, d\omega < \infty \right\}.$$

Since the partial differentiation operator ∂^n corresponds to a Fourier-domain multiplication by $(j\omega)^n$, the inclusion of f in $W_2^\alpha(\mathbb{R}^d)$ requires that its (partial) derivatives be well defined in the L_2 sense up to order α. The same is also true for the "Bessel potential" operators $(\text{Id} - \Delta)^{\alpha/2}$ of order α, or, alternatively, the fractional Laplacians $(-\Delta)^{\alpha/2}$ with Fourier multiplier $\|\omega\|^\alpha$.

PROPOSITION 6.4 *Let β_L be an admissible B-spline that is associated with a Fourier multiplier $\widehat{L}(\omega)$ of order γ. Then, $\beta_L \in W_2^\alpha(\mathbb{R}^d)$ for any $\alpha < \gamma - d/2$.*

Proof Because of Parseval's identity, the statement $\beta_L \in W_2^\alpha(\mathbb{R}^d)$ is equivalent [9] to $\beta_L \in L_2(\mathbb{R}^d)$ and $(-\Delta)^{\alpha/2}\beta_L \in L_2(\mathbb{R}^d)$. Since the first inclusion is part of the definition,

---

[9] The underlying Fourier multipliers are of comparable size in the sense that there exist two constants c_1 and c_2 such that $c_1(1 + \|\omega\|^2)^\alpha \le 1 + \|\omega\|^{2\alpha} \le c_2(1 + \|\omega\|^2)^\alpha$.

it is sufficient to check for the second. To that end, we recall the stability conditions $\beta_L \in L_1(\mathbb{R}^d)$ and $d_L \in \ell_1(\mathbb{R}^d)$, which are implicit to the B-spline construction (6.25). These, together with the order condition (6.14), imply that

$$|\widehat{L}_d(\omega)| \leq \|d_L\|_{\ell_1}$$

$$|\widehat{\beta}_L(\omega)| = \left|\frac{\widehat{L}_d(\omega)}{\widehat{L}(\omega)}\right| \leq \min\left(\|\beta_L\|_{L_1}, C\frac{\|d_L\|_{\ell_1}}{\|\omega\|^\gamma}\right).$$

This latter bound allows us to control the L_2 norm of $(-\Delta)^{\alpha/2}\beta_L$ by splitting the spectral range of integration as

$$\|(-\Delta)^{\alpha/2}\beta_L\|_{L_2}^2 = \int_{\mathbb{R}^d} \|\omega\|^{2\alpha}|\widehat{\beta}_L(\omega)|^2 \frac{d\omega}{(2\pi)^d}$$

$$= \int_{\|\omega\|<R} \|\omega\|^{2\alpha}|\widehat{\beta}_L(\omega)|^2 \frac{d\omega}{(2\pi)^d} + \int_{\|\omega\|>R} \|\omega\|^{2\alpha}|\widehat{\beta}_L(\omega)|^2 \frac{d\omega}{(2\pi)^d}$$

$$\leq \|\beta_L\|_{L_1}^2 \underbrace{\int_{\|\omega\|<R} \|\omega\|^{2\alpha} \frac{d\omega}{(2\pi)^d}}_{I_1} + C^2\|d\|_{\ell_1}^2 \underbrace{\int_{\|\omega\|>R} \|\omega\|^{2\alpha-2\gamma} \frac{d\omega}{(2\pi)^d}}_{I_2}.$$

The first integral I_2 is finite due to the boundedness of the domain. As for I_2, it is convergent provided that the rate of decay of the argument is faster than d, which corresponds to the critical Sobolev exponent $\alpha = \gamma - d/2$. $\square$

As the final step of the analysis, we invoke the Sobolev embedding theorems to infer that β_L is Hölder-continuous of order r with $r < \alpha - \frac{d}{2} = (\gamma - d)$, which essentially means that β_L is differentiable up to order r with bounded derivatives. One should keep in mind, however, that the latter estimate is a lower bound on Hölder continuity, unlike the Sobolev exponent in Proposition 6.4, which is sharp. For instance, in the case of the 1-D Fourier multiplier $(j\omega)^\gamma$, we find that the corresponding (fractional) B-spline – if it exists – should have a Sobolev smoothness $(\gamma - \frac{1}{2})$, and a Hölder regularity $r < (\gamma - 1)$. Note that the latter is arbitrarily close (but not equal) to the true estimate $r_0 = (\gamma - 1)$ that is readily deduced from the Green's function (6.15).

6.4.2   B-spline factorization

A powerful aspect of spline theory is that it is often possible to exploit the factorization properties of differential operators to recursively generate whole families of B-splines. Specifically, if the operator can be decomposed as $L = L_1 L_2$, where the B-splines associated with L_1 and L_2 are already known, then $\beta_L = \beta_{L_1} * \beta_{L_2}$ is the natural choice of B-spline for L.

PROPOSITION 6.5   *Let* β_{L_1}, β_{L_2} *be admissible B-splines for the operators* L_1 *and* L_2, *respectively. Then,* $\beta_L(r) = (\beta_{L_1} * \beta_{L_2})(r)$ *is an admissible B-spline for* $L = L_1 L_2$ *if and only if there exists a constant* $A > 0$ *such that*

$$\sum_{n \in \mathbb{Z}^d} \left|\widehat{\beta}_{L_1}(\omega + 2\pi n)\widehat{\beta}_{L_2}(\omega + 2\pi n)\right|^2 \geq A > 0.$$

for all $\boldsymbol{\omega} \in [0, 2\pi]^d$. *When* $L_1 = L_2^\gamma$ *for* $\gamma \geq 0$, *then the auxiliary condition is automatically satisfied.*

Proof Since $\beta_{L_1}, \beta_{L_2} \in L_1(\mathbb{R}^d)$, the same holds true for β_L (by Young's inequality). From the Fourier-domain definition (6.25) of the B-splines, we have

$$\widehat{\beta}_{L_i}(\boldsymbol{\omega}) = \frac{\sum_{k \in \mathbb{Z}^d} d_{L_i}[k] e^{-j\langle k, \boldsymbol{\omega} \rangle}}{\widehat{L}_i(\boldsymbol{\omega})} = \frac{\widehat{L}_{d,i}(\boldsymbol{\omega})}{\widehat{L}_i(\boldsymbol{\omega})},$$

which implies that

$$\beta_L = \mathscr{F}^{-1} \left\{ \frac{\widehat{L}_{d,1}(\boldsymbol{\omega}) \widehat{L}_{d,2}(\boldsymbol{\omega})}{\widehat{L}_1(\boldsymbol{\omega}) \widehat{L}_2(\boldsymbol{\omega})} \right\} = \mathscr{F}^{-1} \left\{ \frac{\widehat{L}_d(\boldsymbol{\omega})}{\widehat{L}(\boldsymbol{\omega})} \right\} = L_d L^{-1} \delta,$$

with $L^{-1} = L_2^{-1} L_1^{-1}$ and $L_d = L_{d,1} L_{d,2}$. This translates into the combined localization operator $L_d s(\boldsymbol{r}) = \sum_{k \in \mathbb{Z}^d} d_L[k] s(\boldsymbol{r} - \boldsymbol{k})$ with $d_L[k] = (d_1 * d_2)[k]$, which is factorizable by construction. To establish the existence of the upper Riesz bound for β_L, we perform the manipulation

$$A_{\beta_L}(\boldsymbol{\omega}) = \sum_{\boldsymbol{n} \in \mathbb{Z}^d} |\widehat{\beta}_L(\boldsymbol{\omega} + 2\pi \boldsymbol{n})|^2$$

$$= \sum_{\boldsymbol{n} \in \mathbb{Z}^d} |\widehat{\beta}_{L_1}(\boldsymbol{\omega} + 2\pi \boldsymbol{n})|^2 |\widehat{\beta}_{L_2}(\boldsymbol{\omega} + 2\pi \boldsymbol{n})|^2$$

$$\leq \left(\sum_{\boldsymbol{n} \in \mathbb{Z}^d} |\widehat{\beta}_{L_1}(\boldsymbol{\omega} + 2\pi \boldsymbol{n})| \, |\widehat{\beta}_{L_2}(\boldsymbol{\omega} + 2\pi \boldsymbol{n})| \right)^2$$

$$\leq \sum_{\boldsymbol{n} \in \mathbb{Z}^d} |\widehat{\beta}_{L_1}(\boldsymbol{\omega} + 2\pi \boldsymbol{n})|^2 \sum_{\boldsymbol{n} \in \mathbb{Z}^d} |\widehat{\beta}_{L_2}(\boldsymbol{\omega} + 2\pi \boldsymbol{n})|^2$$

$$\leq B_1^2 B_2^2 < +\infty,$$

where the third line follows from the norm inequality $\|a\|_{\ell_2} \leq \|a\|_{\ell_1}$ and the fourth from Cauchy–Schwarz; B_1 and B_2 are the upper Riesz bounds of β_{L_1} and β_{L_2}, respectively. The additional condition in the proposition takes care of the lower Riesz bound.   □

6.4.3   Polynomial B-splines

The factorization property is directly applicable to the construction of the polynomial B-splines (we use the equivalent notation $\beta_+^n = \beta_{D^{n+1}}$ in Section 1.3.2) via the iterated convolution of a B-spline of degree zero. Specifically,

$$\beta_{D^{n+1}}(\boldsymbol{r}) = (\beta_D * \beta_{D^n})(\boldsymbol{r}) = \underbrace{(\beta_D * \cdots * \beta_D)}_{n+1}(\boldsymbol{r})$$

with $\beta_D = \beta_+^0 = \mathbb{1}_{[0,1)}$ and the convention that $\beta_{D^0} = \beta_{\text{Id}} = \delta$.

6.4.4   Exponential B-splines

More generally, one can consider a generic Nth-order differential operator of the form $P_\alpha = P_{\alpha_1} \cdots P_{\alpha_N}$ with parameter vector $\alpha = (\alpha_1, \ldots, \alpha_N) \in \mathbb{C}^N$ and $P_{\alpha_n} = D - \alpha_n \mathrm{Id}$. The corresponding basis function is an exponential B-spline of order N with parameter vector α, which can be decomposed as

$$\beta_\alpha(r) = \left(\beta_{\alpha_1} * \beta_{\alpha_2} * \cdots * \beta_{\alpha_N} \right)(r), \tag{6.42}$$

where $\beta_\alpha = \beta_{P_\alpha}$ is the first-order exponential spline defined by (6.21). The Fourier-domain counterpart of (6.42) is

$$\hat{\beta}_\alpha(\omega) = \prod_{n=1}^{N} \frac{1 - e^{\alpha_n} e^{-j\omega}}{j\omega - \alpha_n}, \tag{6.43}$$

which also yields

$$\beta_\alpha(r) = \Delta_\alpha \rho_\alpha(r), \tag{6.44}$$

where $\Delta_\alpha = \Delta_{\alpha_1} \cdots \Delta_{\alpha_N}$ (with Δ_α defined by (6.20)) is the corresponding Nth-order localization operator (weighted differences) and ρ_α the causal Green's function of P_α. Note that the complex parameters α_n, which are the roots of the characteristic polynomial of P_α, are the poles of the exponential B-spline, as seen in (6.43). The actual recipe for localization is that each pole is cancelled by a corresponding (2π-periodic) zero in the numerator.

Based on the above equations, one can infer the following properties of the exponential B-splines (see [UB05a] for a complete treatment of the topic):

- They are causal, bounded, and compactly supported in $[0, N]$, simply because all elementary constituents in (6.42) are bounded and supported in $[0, 1)$.
- They are piecewise-exponential with joining points at the integers and a maximal degree of smoothness (spline property). The first part follows from (6.44) using the well-known property that the causal Green's function of an Nth-order ordinary differential operator is an exponential polynomial restricted to the positive axis. As for the statement about smoothness, the B-splines are Hölder-continuous of order $(N - 1)$. In other words, they are differentiable up to order $(N - 1)$ with bounded derivatives. This follows from the fact that $D\beta_{\alpha_n}(r) = \delta(r) - e^{\alpha_n}\delta(r - 1)$, which implies that every additional elementary convolution factor in (6.42) improves the differentiability of the resulting B-spline by one.
- They are the shortest elementary constituents of exponential splines (maximally localized kernels) and they each generate a valid Riesz basis (by integer shifting) of the spaces of cardinal P_α-splines if and only if $\alpha_n - \alpha_m \neq j2\pi k, k \in \mathbb{Z}$, for all distinct, purely imaginary poles.
- They reproduce the exponential polynomials that are in the null space of the operator P_α, as well as any of its Green's functions ρ_α, which all happen to be special types of P_α-splines (with a minimum number of singularities).

- For $\boldsymbol{\alpha} = (0,\ldots,0)$, one recovers Schoenberg's classical polynomial B-splines of degree $(N-1)$ [Sch46, Sch73a], as expressed by the notational equivalence

$$\beta_+^n(r) = \beta_{\mathrm{D}^{n+1}}(r) = \beta_{\underbrace{(0,\ldots,0)}_{n+1}}(r).$$

The system-theoretic interpretation is that the classical polynomial spline of degree n has a pole of multiplicity $(n+1)$ at the origin: It corresponds to an (unstable) linear system that is an $(n+1)$-fold integrator.

There is also a corresponding B-spline calculus whose main operations are

- Convolution by concatenation of parameter vectors:

$$(\beta_{\boldsymbol{\alpha}_1} * \beta_{\boldsymbol{\alpha}_2})(r) = \beta_{(\boldsymbol{\alpha}_1 : \boldsymbol{\alpha}_2)}(r)$$

- Mirroring by sign change:

$$\beta_{\boldsymbol{\alpha}}(-r) = \left(\prod_{n=1}^{N} e^{\alpha_n}\right) \beta_{-\boldsymbol{\alpha}}(r+N)$$

- Complex conjugation:

$$\overline{\beta_{\boldsymbol{\alpha}}(r)} = \beta_{\overline{\boldsymbol{\alpha}}}(r)$$

- Modulation by parameter shifting:

$$e^{j\omega_0 r}\beta_{\boldsymbol{\alpha}}(r) = \beta_{\boldsymbol{\alpha} + \mathbf{j}\omega_0}(r),$$

with the convention that $\mathbf{j} = (\mathrm{j},\ldots,\mathrm{j})$.

Finally, we point out that exponential B-splines can be computed explicitly on a case-by-case basis using the mathematical software described in [Uns05, Appendix A].

6.4.5   Fractional B-splines

The fractional splines are an extension of the polynomial splines for all non-integer degrees $\alpha > -1$. The most notable members of this family are the causal fractional B-splines β_+^α whose basic constituents are piecewise-power functions of degree α [UB00]. These functions are associated with the causal fractional-derivative operator $\mathrm{D}^{\alpha+1}$, whose Fourier-based definition is

$$\mathrm{D}^\gamma \varphi(r) = \int_{\mathbb{R}} (\mathrm{j}\omega)^\gamma \widehat{\varphi}(\omega) e^{\mathrm{j}\omega r}\, \frac{\mathrm{d}\omega}{2\pi}$$

in the sense of generalized functions. The causal Green's function of D^γ is the one-sided power function of degree $(\gamma - 1)$ specified by (6.15). One constructs the corresponding B-splines through a localization process similar to the classical one, replacing finite differences by the fractional differences defined as

$$\Delta_+^\gamma \varphi(r) = \int_{\mathbb{R}} (1 - e^{-\mathrm{j}\omega})^\gamma \widehat{\varphi}(\omega) e^{\mathrm{j}\omega r}\, \frac{\mathrm{d}\omega}{2\pi}. \tag{6.45}$$

In this respect, it is important to note that $(1 - e^{-j\omega})^\gamma = (j\omega)^\gamma + O(|\omega|^{2\gamma})$, which justifies this particular choice. By applying (6.25), we readily obtain the Fourier-domain representation of the fractional B–splines,

$$\widehat{\beta_+^\alpha}(\omega) = \frac{(1 - e^{-j\omega})^{\alpha+1}}{(j\omega)^{\alpha+1}}, \qquad (6.46)$$

which can then be inverted to provide the explicit time-domain formula

$$\beta_+^\alpha(r) = \frac{\Delta_+^{\alpha+1} r_+^\alpha}{\Gamma(\alpha + 1)} \qquad (6.47)$$

$$= \sum_{m=0}^{\infty} (-1)^m \binom{\alpha + 1}{m} \frac{(r - m)_+^\alpha}{\Gamma(\alpha + 1)},$$

where the generalized fractional binomial coefficients are given by

$$\binom{\alpha + 1}{m} = \frac{\Gamma(\alpha + 2)}{\Gamma(m + 1)\,\Gamma(\alpha + 2 - m)} = \frac{(\alpha + 1)!}{m!\,(\alpha + 1 - m)!}.$$

What is remarkable about this construction is the way in which the classical B-spline formulas of Section 1.3.2 carry over to the fractional case almost literally, by merely replacing n by α. This is especially striking when we compare (6.47) to (1.11), as well as the expanded versions of these formulas given below, which follow from the (generalized) binomial expansion of $(1 - e^{-j\omega})^{\alpha+1}$.

Likewise, it is possible to construct the (α, τ) extension of these B-splines. They are associated with the operators $L = \partial_\tau^{\alpha+1} \longleftrightarrow (j\omega)^{\frac{\alpha+1}{2}+\tau}(-j\omega)^{\frac{\alpha+1}{2}-\tau}$ and $\tau \in \mathbb{R}$ [BU03]. This family covers the entire class of translation- and scale-invariant operators in 1-D (see Proposition 5.6).

The fractional B-splines share virtually all the properties of the classical B-splines, including the two-scale relation, and can also be used to define fractional wavelet bases with an order $\gamma = \alpha + 1$ that varies continuously. They only lack positivity and compact support. Their most notable properties are summarized below.

- *Generalization.* For α integer, they are equivalent to the classical polynomial splines. The fractional B-splines interpolate the polynomial ones in very much the same way as the gamma function interpolates the factorials.
- *Stability.* All brands of fractional B-splines satisfy the Riesz-basis condition in Theorem 6.2.
- *Regularity.* The fractional splines are α-Hölder continuous; their critical Sobolev exponent (degree of differentiability in the L_2 sense) is $\alpha + 1/2$ (see Proposition 6.4).
- *Polynomial reproduction.* The fractional B-splines reproduce the polynomials of degree $N = \lceil \alpha \rceil$ that are in the null space of the operator $D^{\alpha+1}$ (see Section 6.2.1).
- *Decay.* The fractional B-splines decay at least like $|r|^{-\alpha-2}$; the causal ones are compactly supported for α integer.
- *Order of approximation.* The fractional splines have the non-integer order of approximation $\alpha + 1$, a property that is rather unusual in approximation theory.

- *Fractional derivatives.* Simple formulas are available for obtaining the fractional derivatives of B-splines. In addition, the corresponding fractional spline wavelets behave essentially like fractional-derivative operators.

6.4.6   Additional brands of univariate B-splines

To be complete, we briefly mention some additional types of univariate B-splines that have been investigated systematically in the literature:

- The generalized exponential B-splines of order N that cover the whole class of differential operators with rational transfer functions [Uns05]. These are parameterized by their poles and zeros. Their properties are very similar to those of the exponential B-splines of Section 6.4.4, which are included as a special case.
- The Matérn splines of (fractional) order γ and parameter $\alpha \in \mathbb{R}^+$ with $L = (D + \alpha \mathrm{Id})^\gamma \longleftrightarrow (j\omega + \alpha)^\gamma$ [RU06]. These constitute the fractionalization of the exponential B-spline with a single pole of multiplicity N.

In principle, it is possible to construct even broader families via the convolution of existing components. The difficulty is that it may not always be possible to obtain explicit signal-domain formulas, especially when some of the constituents are fractional.

6.4.7   Multidimensional B-splines

While the construction of B-splines is well understood and covered systematically in 1-D, the task becomes more challenging in multiple dimensions because of the inherent difficulty of imposing compact support. Apart from the easy cases where the operator L can be decomposed in a succession of 1-D operators (tensor-product B-splines and box splines), the available collection of multidimensional B-splines is much more restricted than in the univariate case. The construction of B-splines is still considered an art where the ultimate goal is to produce the most localized basis functions. The primary families of multidimensional B-splines that have been investigated so far are:

- the polyharmonic B-splines of (fractional) order γ with $L = (-\Delta)^{\frac{\gamma}{2}} \longleftrightarrow \|\omega\|^\gamma$ [MN90b, Rab92a, Rab92b, VDVBU05]
- the box splines of multiplicity $N \geq d$ with $L = D_{u_1} \cdots D_{u_N} \longleftrightarrow \prod_{n=1}^{N} \langle j\omega, u_n \rangle$ with $\|u_n\| = 1$, where $D_{u_n} = \langle \nabla, u_n \rangle$ is the directional derivative along u_n [dBHR93]. The box splines are compactly supported functions in $L_1(\mathbb{R}^d)$ if and only if the set of orientation vectors $\{u_n\}_{n=1}^{N}$ forms a frame of $\mathbb{R}^d$.

We encourage the reader who finds the present list incomplete to work on expanding it. The good news for the present study is that the polyharmonic B-splines are particularly relevant for image-processing applications because they are associated with the class of operators that are scale- and rotation-invariant. They naturally come into play when considering isotropic fractal-type random fields.

The principal message of this section is that B-splines – no matter the type – are localized functions with an equivalent width that increases in proportion to the order. In general, the fractional brands and the non-separable multidimensional ones are not compactly supported. The important issue of localization and decay is not yet fully resolved in higher dimensions. Also, since $L_d s = \beta_L * Ls$, it is clear that the search for a "good" B-spline β_L is intrinsically related to the problem of finding an accurate numerical approximation L_d of the differential operator L. Looking at the discretization issue from the B-spline perspective leads to new insights and sometimes to non-conventional solutions. For instance, in the case of the Laplacian $L = \Delta$, the continuous-domain localization requirement points to the choice of the 2-D discrete operator Δ_d described by the 3×3 filter mask

$$\text{Isotropic discrete Laplacian:} \quad \frac{1}{6} \begin{pmatrix} -1 & -4 & -1 \\ -4 & 20 & -4 \\ -1 & -4 & -1 \end{pmatrix},$$

which is not the standard version used in numerical analysis. This particular set of weights produces a much nicer, bell-shaped polyharmonic B-spline than the conventional finite-difference mask, which induces significant directional artifacts, especially when one starts iterating the operator [VDVBU05].

6.5   Generalized operator-like wavelets

In direct analogy with the first-order scenario in Section 6.3.3, we shall now take advantage of the general B-spline formalism to construct a wavelet basis that is matched to some generic operator L.

6.5.1   Multiresolution analysis of $L_2(\mathbb{R}^d)$

The first step is to lay out a fine-to-coarse sequence of (multidimensional) L-spline spaces in essentially the same way as in our first-order example. Specifically,

$$V_i = \{s(r) \in L_2(\mathbb{R}^d) : Ls(r) = \sum_{k \in \mathbb{Z}^d} a_i[k]\delta(r - \mathbf{D}^i k)\},$$

where $\mathbf{D}$ is a proper dilation matrix with integer entries (e.g., $\mathbf{D} = 2\mathbf{I}$ in the standard dyadic configuration). These spline spaces satisfy the general embedding relation $V_i \supseteq V_j$ for $i \leq j$.

The reference space ($i = 0$) is the space of cardinal L splines which admits the standard B-spline representation

$$V_0 = \{s(r) = \sum_{k \in \mathbb{Z}^d} c[k]\beta_L(r - k) : c \in \ell_2(\mathbb{Z}^d)\},$$

where β_L is given by (6.25). Our implicit assumption is that each V_i admits a similar B-spline representation

$$V_i = \{s(r) = \sum_{k \in \mathbb{Z}^d} c_i[k]\beta_{L,i}(r - \mathbf{D}^i k) : c_i \in \ell_2(\mathbb{Z}^d)\},$$

which involves the multiresolution generators $\beta_{L,i}$ described in Section 6.5.2.

6.5.2   Multiresolution B-splines and the two-scale relation

In direct analogy with (6.25), the multiresolution B-splines $\beta_{L,i}$ are localized versions of the Green's function ρ_L with respect to the grid $\mathbf{D}^i\mathbb{Z}^d$. Specifically, we have that

$$\beta_{L,i}(\mathbf{r}) = \sum_{k\in\mathbb{Z}^d} d_i[k]\rho_L(\mathbf{r} - \mathbf{D}^i k) = L_{d,i}L^{-1}\delta(\mathbf{r}),$$

where $L_{d,i}$ is the discretized version of L on the grid $\mathbf{D}^i\mathbb{Z}^d$. The Fourier-domain counterpart of this equation is

$$\widehat{\beta}_{L,i}(\boldsymbol{\omega}) = \frac{\sum_{k\in\mathbb{Z}^d} d_i[k]e^{-j\langle\boldsymbol{\omega},\mathbf{D}^i k\rangle}}{\widehat{L}(\boldsymbol{\omega})}. \tag{6.48}$$

The implicit requirement for the multiresolution decomposition scheme to work is that $\beta_{L,i}$ generates a Riesz basis. This needs to be asserted on a case-by-case basis.

A particularly favorable situation occurs when the operator L is scale-invariant with $\widehat{L}(a\boldsymbol{\omega}) = |a|^{\gamma}\widehat{L}(\boldsymbol{\omega})$. Let $i' > i$ be two multiresolution levels of the pyramid such that $\mathbf{D}^{i'-i} = m\mathbf{I}$, where m is a proportionality constant. It is then possible to relate the B-spline at resolution i' to the one at the finer level i via the simple dilation relation

$$\beta_{L,i'}(\mathbf{r}) \propto \beta_{L,i}(\mathbf{r}/m).$$

This is shown by considering the Fourier transform of $\beta_{L,i}(\mathbf{r}/m)$, which is written as

$$|m|^d\widehat{\beta}_{L,i}(m\boldsymbol{\omega}) = |m|^d\frac{\sum_{k\in\mathbb{Z}^d} d_i[k]e^{-j\langle\boldsymbol{\omega},m\mathbf{D}^i k\rangle}}{\widehat{L}(m\boldsymbol{\omega})}$$

$$= |m|^{d-\gamma}\frac{\sum_{k\in\mathbb{Z}^d} d_i[k]e^{-j\langle\boldsymbol{\omega},\mathbf{D}^{i'-i}\mathbf{D}^i k\rangle}}{\widehat{L}(\boldsymbol{\omega})}$$

$$= |m|^{d-\gamma}\frac{\sum_{k\in\mathbb{Z}^d} d_i[k]e^{-j\langle\boldsymbol{\omega},\mathbf{D}^{i'} k\rangle}}{\widehat{L}(\boldsymbol{\omega})}$$

and found to be compatible with the form of $\widehat{\beta}_{L,i'}(\boldsymbol{\omega})$ given by (6.48) by taking $d_{i'}[k] \propto d_i[k]$. The prototypical scenario is the dyadic configuration $\mathbf{D} = 2\mathbf{I}$ for which the B-splines at level i are all constructed through the dilation of the single prototype $\beta_L = \beta_{L,0}$, subject to the scale-invariance constraint on L. This happens, for instance, for the classical polynomial splines which are associated with the Fourier multipliers $(j\omega)^N$.

A crucial ingredient for the fast wavelet-transform algorithm is the two-scale relation that links the B-spline basis functions at two successive levels of resolution. Specifically, we have that

$$\beta_{L,i+1}(\mathbf{r}) = \sum_{k\in\mathbb{Z}^d} h_i[k]\beta_{L,i}(\mathbf{r} - \mathbf{D}^i k),$$

where the sequence h_i specifies the scale-dependent refinement filter. The frequency response of h_i is obtained by taking the ratios of the Fourier transforms of the corresponding B-splines as

$$\widehat{h}_i(\omega) = \frac{\widehat{\beta}_{L,i+1}(\omega)}{\widehat{\beta}_{L,i}(\omega)} \tag{6.49}$$

$$= \frac{\sum_{k \in \mathbb{Z}^d} d_{i+1}[k]e^{-j\langle \omega, \mathbf{D}^{i+1}k\rangle}}{\sum_{k \in \mathbb{Z}^d} d_i[k]e^{-j\langle \omega, \mathbf{D}^i k\rangle}}, \tag{6.50}$$

which is $2\pi(\mathbf{D}^T)^{-i}$-periodic and hence defines a valid digital filter with respect to the spatial grid $\mathbf{D}^i \mathbb{Z}^d$.

To illustrate those relations, we return to our introductory example in Section 6.1: the Haar wavelet transform, which is associated with the Fourier multipliers $j\omega$ (derivative) and $(1 - e^{-j\omega})$ (finite-difference operator). The dilation matrix is $\mathbf{D} = 2$ and the localization filter is the same at all levels because the underlying derivative operator is scale-invariant. By plugging those entities into (6.48), we obtain the Fourier transform of the corresponding B-spline at resolution i as

$$\widehat{\beta}_{D,i}(\omega) = 2^{-i/2} \frac{1 - e^{j2^i \omega}}{j\omega},$$

where the normalization by $2^{-i/2}$ is included to standardize the norm of the B-splines. The application of (6.49) then yields

$$\widehat{h}_i(\omega) = \frac{1}{\sqrt{2}} \frac{1 - e^{j2^{i+1}\omega}}{1 - e^{j2^i \omega}}$$

$$= \frac{1}{\sqrt{2}}(1 + e^{j2^i \omega}),$$

which, up to the normalization by $\sqrt{2}$, is the expected refinement filter with coefficients proportional to $(1, 1)$ that are independent of the scale.

6.5.3   Construction of an operator-like wavelet basis

To keep the notation simple, we concentrate on the specification of the wavelet basis at the scale $i = 1$, with $W_1 = \text{span}\{\psi_{1,k}\}_{k \in \mathbb{Z}^d \setminus \mathbf{D}\mathbb{Z}^d}$ such that $W_1 \perp V_1$ and $V_0 = V_1 + W_1$, where $V_0 = \text{span}\{\beta_L(\cdot - k)\}_{k \in \mathbb{Z}^d}$ is the space of cardinal L-splines.

The relevant smoothing kernel is the interpolation function $\varphi_{\text{int}} = \varphi_{\text{int},0}$ for the space of cardinal $L^H L$-splines, which is generated by $(\overline{\beta}_L^\vee * \beta_L)(r)$ (autocorrelation of the generalized B-spline). This interpolator is best described in the Fourier domain using the formula

$$\varphi_{\text{int}}(r) = \mathscr{F}^{-1} \left\{ \frac{|\widehat{\beta}_L(\omega)|^2}{\sum_{n \in \mathbb{Z}^d} |\widehat{\beta}_L(\omega + 2\pi n)|^2} \right\}(r), \tag{6.51}$$

where $\widehat{\beta_L}$ (resp., $\overline{\widehat{\beta_L}}$) is the Fourier transform of the generalized B-spline β_L (resp., $\overline{\beta_L^\vee}$). It satisfies the fundamental interpolation property

$$\varphi_{\text{int}}(k) = \delta[k] = \begin{cases} 1, & \text{for } k = 0 \\ 0, & \text{for } k \in \mathbb{Z}^d \setminus \{0\}. \end{cases} \qquad (6.52)$$

The existence of such a function is guaranteed whenever $\beta_L = \beta_{L,0}$ is an admissible B-spline. In particular, the Riesz basis condition (6.26) implies that the denominator of $\widehat{\varphi}_{\text{int}}(\omega)$ in (6.51) is non-vanishing.

The sought-after wavelets are then constructed as $\psi_{1,k}(r) = \psi_L(r-k)/\|\psi_L\|_{L_2}$, where the operator-like mother wavelet ψ_L is given by

$$\psi_L(r) = L^H \varphi_{\text{int}}(r), \qquad (6.53)$$

where L^H is the adjoint of L with respect to the Hermitian-symmetric L_2 inner product. Also, note that we are removing the functions located on the next-coarser resolution grid $\mathbf{D}\mathbb{Z}^d$ associated with V_1 (critically sampled configuration).

The proof of the following result is illuminating because it relies heavily on the notion of duality, which is central to our whole argument.

PROPOSITION 6.6 *The operator-like wavelet $\psi_L = L^H \varphi_{\text{int}}$ satisfies the property $\langle s_1, \psi_L(\cdot - k)\rangle_{L_2} = \langle s_1, \overline{\psi_L(\cdot - k)}\rangle = 0, \forall k \in \mathbb{Z}^d \setminus \mathbf{D}\mathbb{Z}^d$ for any spline $s_1 \in V_1$. Moreover, it can be written as $\psi_L(r) = L_d^H \tilde{\beta}_L(r) = \sum_{k \in \mathbb{Z}^d} d_L[-k] \tilde{\beta}_L(r-k)$, where $\{\tilde{\beta}_L(\cdot - k)\}_{k \in \mathbb{Z}^d}$ is the dual basis of V_0 such that $\langle \tilde{\beta}_L(\cdot - k), \beta_L(\cdot - k')\rangle_{L_2} = \delta[k - k']$. This implies that $W_1 = \text{span}\{\psi_{1,k}\}_{k \in \mathbb{Z}^d \setminus \mathbf{D}\mathbb{Z}^d} \subset V_0$ and $W_1 \perp V_1$.*

Proof We pick an arbitrary spline $s_1 \in V_1$ and perform the inner-product manipulation

$$\langle s_1, \psi_L(\cdot - k_0)\rangle_{L_2} = \langle s_1, L^* \{\overline{\varphi_{\text{int}}(\cdot - k_0)}\}\rangle \qquad \text{(by shift-invariance of L)}$$

$$= \langle Ls_1, \overline{\varphi_{\text{int}}(\cdot - k_0)}\rangle \qquad \text{(by duality)}$$

$$= \left\langle \sum_{k \in \mathbb{Z}^d} a_1[k] \delta(\cdot - \mathbf{D}k), \overline{\varphi_{\text{int}}(\cdot - k_0)}\right\rangle \qquad \text{(by definition of } V_1)$$

$$= \sum_{k \in \mathbb{Z}^d} a_1[k] \overline{\varphi_{\text{int}}(\mathbf{D}k - k_0)}. \qquad \text{(by definition of } \delta)$$

Due to the interpolation property of φ_{int}, the kernel values in the sum are vanishing if $\mathbf{D}k - k_0 \in \mathbb{Z}^d \setminus \{0\}$ for all $k \in \mathbb{Z}^d$, which proves the first part of the statement.

As for the second claim, we consider the Fourier-domain expression for ψ_L:

$$\widehat{\psi}_L(\omega) = \overline{\widehat{L}(\omega)}\widehat{\varphi}_{\text{int}}(\omega) = \overline{\widehat{L}(\omega)}\ \widehat{\tilde{\beta}}_L(\omega)\ \widehat{\beta}_L(\omega),$$

where

$$\widehat{\tilde{\beta}}_L(\omega) = \frac{\widehat{\beta}_L(\omega)}{\sum_{n \in \mathbb{Z}^d} |\widehat{\beta}_L(\omega + 2\pi n)|^2}$$

is the Fourier transform of the dual B-spline $\tilde{\beta}_L$. The above factorization implies that $\varphi_{\text{int}}(r) = (\tilde{\beta}_L * \overline{\beta_L^\vee})(r)$, which ensures the biorthonormality $\langle \tilde{\beta}_L(\cdot - k), \beta_L(\cdot - k')\rangle_{L_2} = \delta[k] = \varphi_{\text{int}}(k - k')$ of the basis functions.

Finally, by replacing $\widehat{\beta}_{\mathrm{L}}(\boldsymbol{\omega})$ by its explicit expression (6.25), we show that $\widehat{L}(\boldsymbol{\omega})\widehat{\beta}_{\mathrm{L}}(\boldsymbol{\omega}) = \widehat{L}_{\mathrm{d}}(\boldsymbol{\omega})$, where $\widehat{L}_{\mathrm{d}}(\boldsymbol{\omega}) = \sum_{k\in\mathbb{Z}^d} d_{\mathrm{L}}[k]e^{-\langle k,\boldsymbol{\omega}\rangle}$ is the frequency response of the (discrete) operator L_{d}. This implies that $\widehat{\psi}_{\mathrm{L}} = \widehat{L_{\mathrm{d}}\tilde{\beta}_{\mathrm{L}}}$, which is the Fourier equivalent of $\psi_{\mathrm{L}} = L_{\mathrm{d}}^H \tilde{\beta}_{\mathrm{L}}$. □

This interpolation-based method of construction is applicable to all the wavelet subspaces W_i and leads to the specification of operator-like Riesz bases of $L_2(\mathbb{R}^d)$ under relatively mild assumptions on L [KUW13]. Specifically, we have that $W_i = \mathrm{span}\{\psi_{i,k}\}_{k\in\mathbb{Z}^d\setminus\mathbf{D}\mathbb{Z}^d}$ with

$$\psi_{i,k}(\boldsymbol{r}) \propto L^H \varphi_{\mathrm{int},i-1}(\boldsymbol{r} - \mathbf{D}^{i-1}\boldsymbol{k}), \tag{6.54}$$

where $\varphi_{\mathrm{int},i-1}$ is the $L^H L$-spline interpolator on the grid $\mathbf{D}^{i-1}\mathbb{Z}^d$. The fact that the interpolator is specified with respect to the grid of the next-finer spline space $V_{i-1} = \mathrm{span}\{\beta_{\mathrm{L},i-1}(\cdot - \mathbf{D}^{i-1}\boldsymbol{k})\}_{k\in\mathbb{Z}^d}$ is essential to ensure that $W_i \subset V_{i-1}$. This kernel satisfies the fundamental interpolation property

$$\varphi_{\mathrm{int},i-1}(\mathbf{D}^{i-1}\boldsymbol{k}) = \delta[\boldsymbol{k}], \tag{6.55}$$

which results in W_i being orthogonal to $V_i = \mathrm{span}\{\beta_{\mathrm{L},i}(\cdot - \mathbf{D}^i\boldsymbol{k})\}_{k\in\mathbb{Z}^d}$ (the reasoning is the same as in the proof of Proposition 6.6, which covers the case $i = 1$). For completeness, we also provide the general expression for the Fourier transform of $\varphi_{\mathrm{int},i}$,

$$
\begin{aligned}
\widehat{\varphi}_{\mathrm{int},i}(\boldsymbol{\omega}) &= |\det(\mathbf{D})|^i \frac{|\widehat{\beta}_{\mathrm{L},i}(\boldsymbol{\omega})|^2}{\sum_{n\in\mathbb{Z}^d}|\widehat{\beta}_{\mathrm{L},i}(\boldsymbol{\omega}+2\pi(\mathbf{D}^T)^{-i}\boldsymbol{n})|^2} \\
&= |\det(\mathbf{D})|^i \frac{|\widehat{L}_{\mathrm{d},i}(\boldsymbol{\omega})|^2 / |\widehat{L}(\boldsymbol{\omega})|^2}{|\widehat{L}_{\mathrm{d},i}(\boldsymbol{\omega})|^2 / \sum_{n\in\mathbb{Z}^d}|\widehat{L}(\boldsymbol{\omega}+2\pi(\mathbf{D}^T)^{-i}\boldsymbol{n})|^2} \\
&= |\det(\mathbf{D})|^i \frac{|\widehat{L}(\boldsymbol{\omega})|^{-2}}{\sum_{n\in\mathbb{Z}^d}|\widehat{L}(\boldsymbol{\omega}+2\pi(\mathbf{D}^T)^{-i}\boldsymbol{n})|^{-2}}, \tag{6.56}
\end{aligned}
$$

which can be used to show that $L^H \varphi_{\mathrm{int},i}(\cdot - \mathbf{D}^i\boldsymbol{k}) \propto L_{\mathrm{d},i}^H \tilde{\beta}_{\mathrm{L},i}(\cdot - \mathbf{D}^i\boldsymbol{k}) \in V_i$ for any $\boldsymbol{k} \in \mathbb{Z}^d$.

While we have seen that this scheme produces an orthonormal basis for the first-order operator P_α in Section 6.3.3, the general procedure does only guarantee semi-orthogonality. More precisely, it ensures the orthogonality between the wavelet subspaces W_i. If necessary, one can always fix the intra-scale orthogonality a posteriori by forming appropriate linear combinations of wavelets at a given resolution. The resulting orthogonal wavelets will still be L-admissible in the sense of Definition 6.7. However, for $d > 1$, intra-scale orthogonalization is likely to spoil the simple, convenient structure of the above construction, which uses a single generator per scale irrespective of the number of dimensions. Indeed, the examples of multidimensional orthogonal wavelet transforms that can be found in the literature – either separable or non-separable – systematically involve $M = (\det(\mathbf{D}) - 1)$ distinct wavelet generators per scale. Moreover, unlike the present operator-like wavelets, they generally do not admit an explicit analytical description.

In summary, wavelets generally behave like differential operators and it is possible to match them to a given class of stochastic processes. The wavelet transforms that are currently most widely used in applications act as multiscale derivatives or Laplacians. They are therefore best suited for the representation of fractal-type stochastic processes that are defined by scale-invariant SDEs [TVDVU09].

The general theme that emerges is that a signal transform will behave appropriately if it has the ability to suppress the signal components (polynomial or sinusoidal trends) that are in the null space of the whitening operator L. This will result in a stationarizing effect that is well documented in the Gaussian context [Fla89, Fla92]. This is the fundamental reason why vanishing moments are so important.

6.6   Bibliographical notes

Section 6.1
Alfréd Haar constructed the orthogonal Haar system as part of his Ph.D. thesis, which he defended in 1909 under the supervision of David Hilbert [Haa10]. From then on, the Haar system remained relatively unnoticed until it was revitalized by the discovery of wavelets nearly one century later. Stéphane Mallat set the foundation of the multiresolution theory of wavelets in [Mal89] with the help of Yves Meyer, while Ingrid Daubechies constructed the first orthogonal family of compactly supported wavelets [Dau88]. Many of the early constructions of wavelets are based on splines [Mal89, CW91, UAE92, UAE93]. The connection with splines is actually quite fundamental in the sense that all multiresolution wavelet bases, including the non-spline brands such as Daubechies', necessarily include a B-spline as a convolution factor – the latter is responsible for their primary mathematical properties such as vanishing moments, differentiability, and order of approximation [UB03]. Further information on wavelets can be found in several textbooks [Dau92, Mey90, Mal09].

Section 6.2
Splines constitute a beautiful topic of investigation in their own right, with hundreds of papers specifically devoted to them. The founding father of the field is Schoenberg, who, during wartime, was asked to develop a computational solution for constructing an analytic function that fits a given set of equidistant noisy data points [Sch88]. He came up with the concept of spline interpolation and proved that polynomial spline functions have a unique expansion in terms of B-splines [Sch46]. While splines can also be specified for non-uniform grids and extended in a variety of ways [dB78, Sch81a], the cardinal setting is especially pleasing because it lends itself to systematic treatment with the aid of the Fourier transform [Sch73a]. The relation between splines and differential operators was recognized early on and led to the generalization known as L-splines [SV67].

The classical reference on partial differential operators and Fourier multipliers is [Hör80]. A central result of the theory is the Malgrange–Ehrenpreis theorem [Mal56,

Ehr54], and its extension stating that the convolution with a compactly supported generalized function is invertible [Hör05].

The concept of a Riesz basis is standard in functional analysis and approximation theory [Chr03]. The special case where the basis functions are integer translates of a single generator is treated in [AU94]. See also [Uns00] for a review of such representations in the context of sampling theory.

Section 6.3

The first-order illustrative example is borrowed from [UB05a, Figure 1] for the construction of the exponential B-spline, and from [KU06, Figure 1] for the wavelet part of the story.

Section 6.4

The 1-D theory of cardinal L-splines for ordinary differential operators with constant coefficients is due to Micchelli [Mic76]. In the present context, we are especially concerned with ordinary differential equations, which go hand-in-hand with the extended family of cardinal exponential splines [Uns05]. The properties of the relevant B-splines are investigated in full detail in [UB05a], which constitutes the ground material for Section 6.4. A key property of B-splines is their ability to reproduce polynomials. It is ensured by the Strang–Fix conditions (6.37) which play a central role in approximation theory [dB87, SF71]. While there is no fundamental difficulty in specifying cardinal-spline interpolators in multiple dimensions, it is much harder to construct compactly supported B-splines, except for the special cases of the box splines [dBH82, dBHR93] and exponential box splines [Ron88]. For elliptic operators such as the Laplacian, it is possible to specify exponentially decaying B-splines, with the caveat that the construction is not unique [MN90b, Rab92a, Rab92b]. This calls for some criterion to identify the most localized solution [VDVBU05]. B-splines, albeit non-compactly supported ones, can also be specified for fractional operators [UB07]. This line of research was initiated by Unser and Blu with the construction of the fractional B-splines [UB00]. As suggested by the name, the (Gaussian) stochastic counterparts of these B-splines are Mandelbrot's fractional Brownian motions [MVN68], as we shall see in Chapters 7 and 8. The association is essentially the same as the connection between the B-spline of degree zero (rect) and Brownian motion, or, by extension, the whole family of Lévy processes (see Section 1.3).

Section 6.5

de Boor *et al.* were among the first to extend the notion of multiresolution analysis beyond the idea of dilation and to propose a general framework for constructing "non-stationary" wavelets [dBDR93]. Khalidov and Unser proposed a systematic method for constructing wavelet-like basis functions based on exponential splines and proved that these wavelets behave like differential operators [KU06]. The material in Section 6.5 is an extension of those ideas to the case of a generic Fourier-mutiplier operator in multiple dimensions; the full technical details can be found in [KUW13]. Operator-like wavelets have also been specified within the framework of conventional multiresolution analysis; in particular, for the Laplacian and its iterates [VDVBU05, TVDVU09] and for the various brands of 1-D fractional derivatives [VDVFHUB10], which have the common property of being scale-invariant. Finally, we mention that each exponential-spline wavelet has a compactly supported Daubechies counterpart that is orthogonal and operator-like in the sense of having the same vanishing exponential moments [VBU07].

7 Sparse stochastic processes

Having dealt with the technicalities of defining acceptable inverse operators, we can now apply the framework to characterize – and also generate – relevant families of sparse processes. As in the previous chapters, we start with a simple example to expose the key ideas. Then, in Section 7.2, we develop the generalized version of the innovation model that covers the complete spectrum of Gaussian and sparse stochastic processes. We characterize the solution(s) in full generality, while pinpointing the conditions under which the so-defined processes are stationary or self-similar. In Section 7.3, we provide a complete description of the stationary processes, including the CARMA (continuous-time autoregressive moving average) family which constitutes the non-Gaussian extension of the classical ARMA processes. In Section 7.4, we turn our attention to non-stationary signals and characterize the important class of Lévy-type processes that are defined by unstable linear SDEs. Finally, in Section 7.5, we investigate fractal-type processes (not necessarily Gaussian) that are solutions of fractional, scale-invariant SDEs.

7.1 Introductory example: non-Gaussian AR(1) processes

A Lévy-driven AR(1) process with parameter $\alpha \in \mathbb{C}$ is defined by a first-order SDE with $L = P_\alpha = (D - \alpha\mathrm{Id})$, as given by

$$P_\alpha s_\alpha = w,$$

where w is a white Lévy noise excitation. We have already seen that the solution for $\mathrm{Re}(\alpha) < 0$ is given by $s_\alpha = P_\alpha^{-1} w = \rho_\alpha * w$, where ρ_α is the impulse response of the underlying system. We have also shown in Section 5.3.1 that the concept remains applicable for $\mathrm{Re}(\alpha) > 0$ using the extended definition (5.10) of P_α^{-1}. Since P_α^{-1} is a $\mathscr{S}$-continuous convolution operator, this results in a well-defined stationary process, the Gaussian version of which is often referred to as the Ornstein–Uhlenbeck process.

To make the connection with splines, we observe that the first-order impulse response can be written as a sum of exponential B-splines,

$$\rho_\alpha(r) = \sum_{k\geq 0} e^{\alpha k} \beta_\alpha(r - k), \tag{7.1}$$

as illustrated in Figure 7.1a. The B-spline generator $\beta_\alpha(r)$ is defined by (6.21) and is supported in the interval $[0, 1)$. A key observation is that the B-spline coefficients $e^{\alpha k}$

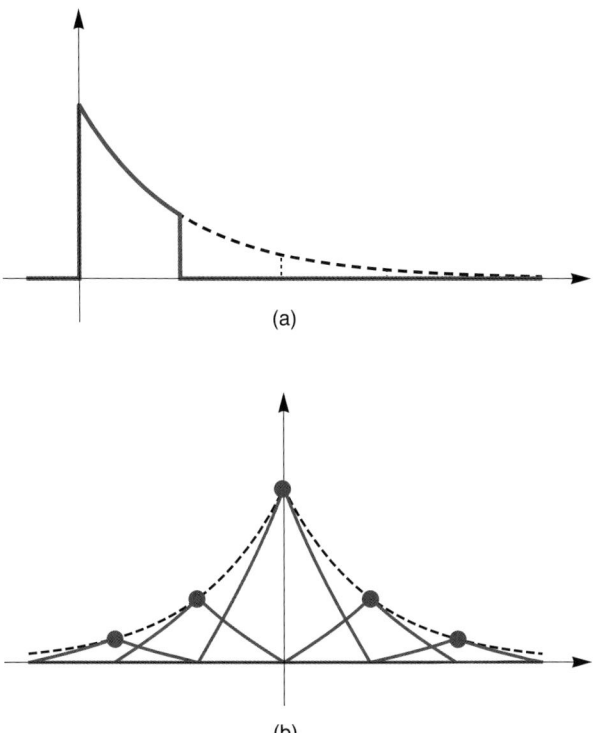

(a)

(b)

Figure 7.1 Spline-based representation of the impulse response and autocorrelation function of a differential system with a pole at $\alpha = -1$. (a) The impulse response (dashed line) is decomposed as a linear combination of the integer shifts of an exponential B-spline (solid line). (b) The autocorrelation is synthesized by interpolating its sample values at the integers (discrete autocorrelation); the corresponding (second-order) spline interpolation kernels are represented using solid lines.

in (7.1) correspond to the impulse response of the digital filter Δ_α^{-1} described by the transfer function $\frac{1}{1-e^\alpha z^{-1}}$, which is the natural discrete counterpart of P_α^{-1}.

The operator version of (7.1) therefore reads $\rho_\alpha = P_\alpha^{-1}\delta = \Delta_\alpha^{-1}\beta_\alpha$, which makes an interesting connection between the analog and discrete versions of a first-order operator. We have shown in prior work that this type of relation is fundamental to the theory of linear systems and that it carries over for higher-order systems [Uns05].

Since the driving term is white, the correlation structure of the process (second-order statistics) is fully characterized by the (Hermitian-symmetric) autocorrelation of the impulse response. In the case where α is real-valued and negative, we get

$$R_{\rho_\alpha}(r) = \langle \rho_\alpha(\cdot + r), \overline{\rho_\alpha} \rangle = (\overline{\rho}_\alpha^\vee * \rho_\alpha)(r) \propto e^{\alpha|r|} = \sum_{k\in\mathbb{Z}} e^{\alpha|k|} \varphi_{\text{int},\alpha}(r - k), \qquad (7.2)$$

which can also be expanded in terms of (augmented-order) B-splines. For simplicity, we have left out the proportionality factor so that the right-hand side can be identified as the normalized autocorrelation function of the AR(1) process

$$c_{s_\alpha}(r) = \frac{\mathbb{E}\{s_\alpha(\cdot + r)\,\overline{s_\alpha(\cdot)}\}}{\mathbb{E}\{|s_\alpha|^2\}} = e^{\alpha|r|}.$$

The rightmost sum in (7.2) expresses the fact that the continuous-domain correlation function can be reconstructed from the discrete-domain one, $c_{s_\alpha}[k] = c_{s_\alpha}(r)\big|_{r=k} = e^{\alpha|k|}$ (the sampled version of the former), by using a Shannon-type interpolation formula, which involves the "augmented-order" spline kernel $\varphi_{\text{int},\alpha}$ introduced in Section 6.3.2.

Let $\sigma_w^2 < +\infty$ denote the variance of a zero-mean white input noise w observed through a normalized observation window $\varphi/\|\varphi\|$. Then, it is well known (Wiener–Khintchine theorem) that the power spectrum of s_α is given by

$$\Phi_{s_\alpha}(\omega) = \sigma_w^2\, \mathscr{F}\{\overline{\rho_\alpha^\vee} * \rho_\alpha\}(\omega) = \sigma_w^2\, |\widehat{\rho}_\alpha(\omega)|^2,$$

with $\widehat{\rho}_\alpha(\omega) = \frac{1}{j\omega-\alpha}$. We note, however, that the power spectrum provides a complete characterization of the process only when the input noise is Gaussian.

7.2   General abstract characterization

The generic solution of the general innovation model (4.23) is given by $s = L^{-1}w$, where w is a particular brand of white noise with Lévy exponent f (see Definition 4.1) and L^{-1} a proper right inverse of L. The theoretical results of Section 4.5.2 guarantee the existence of this solution as a generalized stochastic process over $\mathscr{S}'(\mathbb{R}^d)$ provided that L^{-1*} (the adjoint of L^{-1}) and f satisfy some joint regularity conditions. The three configurations of interest that balance the range of acceptable innovations are listed below for further reference.

DEFINITION 7.1 (Conditions for existence)

- *Condition S:* L^{-1*} is a continuous operator $\mathscr{S}(\mathbb{R}^d) \to \mathscr{S}(\mathbb{R}^d)$ and f is Lévy–Schwartz admissible (see Theorem 4.8)
- *Condition R:* L^{-1*} is a continuous operator $\mathscr{S}(\mathbb{R}^d) \to \mathscr{R}(\mathbb{R}^d)$ and f is Lévy–Schwartz admissible
- *Condition L_p:* L^{-1*} is a continuous operator $\mathscr{S}(\mathbb{R}^d) \to L_p(\mathbb{R}^d)$ and f is p-admissible (see Definition 4.4) for some $p \in [1,2]$.

When any one of these conditions is satisfied, the couple (L^{-1*}, f) is said to be admissible.

Note that the conditions of Definition 7.1 are given in order of increasing level of complexity in the inversion of the whitening operator L when the problem is ill-posed over $\mathscr{S}(\mathbb{R}^d)$.

Now, if (L^{-1*}, f) is admissible, then the underlying generalized stochastic processes, or random fields when $d > 1$, are well defined and completely specified by their characteristic functional (see Theorem 4.17)

$$\widehat{\mathcal{P}_s}(\varphi) = \mathbb{E}\{e^{j\langle\varphi,s\rangle}\} = \exp\left(\int_{\mathbb{R}^d} f\left(L^{-1*}\varphi(r)\right) dr\right). \tag{7.3}$$

In order to represent second-order dependencies, Gelfand and Vilenkin define the correlation functional of a generalized (complex-valued) process s as

$$\mathcal{B}_s(\varphi_1, \varphi_2) = \mathbb{E}\{\langle\varphi_1, s\rangle \, \overline{\langle\varphi_2, s,\rangle}\}, \tag{7.4}$$

where $\overline{\langle\varphi_2, s\rangle}$ stands for the complex conjugate of $\langle\varphi_2, s\rangle$.

These two description modes are easy to relate when the process s is real-valued (see Section 3.5.2). To that end, we observe that $\mathcal{B}_s(\varphi_1, \varphi_2) = \mathbb{E}\{X_1 X_2\}$, where $X_1 = \langle\varphi_1, s\rangle$ and $X_2 = \langle\varphi_2, s\rangle = \overline{\langle\varphi_2, s\rangle}$ are (conventional) real-valued random variables. We then invoke the moment-generating property of the joint characteristic function of X_1 and X_2, $\mathbb{E}\{e^{j(\omega_1 X_1 + \omega_2 X_2)}\} = \widehat{\mathcal{P}_s}(\omega_1\varphi_1 + \omega_2\varphi_2)$, which yields

$$\mathcal{B}_s(\varphi_1, \varphi_2) = -\left.\frac{\partial^2 \widehat{\mathcal{P}_s}(\omega_1\varphi_1 + \omega_2\varphi_2)}{\partial\omega_1 \partial\omega_2}\right|_{\omega_1=0,\omega_2=0}. \tag{7.5}$$

We have already used this mechanism in Section 4.4.3, Proposition 4.15 to determine the covariance of a white Lévy noise under the standard zero-mean and finite-variance assumptions. Specifically, we showed that the correlation/covariance form of the innovation process w is given by

$$\mathcal{B}_w(\varphi_1, \varphi_2) = \sigma_w^2 \langle\varphi_1, \varphi_2\rangle, \tag{7.6}$$

with $\sigma_w^2 = -f''(0)$. Here, we rely on duality (i.e., $\langle\varphi, L^{-1}w\rangle = \langle L^{-1*}\varphi, w\rangle$) to transfer this result to the output of the general innovation model (4.23) as

$$\begin{aligned}
\mathcal{B}_s(\varphi_1, \varphi_2) &= \mathcal{B}_{L^{-1}w}(\varphi_1, \varphi_2) \\
&= \mathcal{B}_w(L^{-1*}\varphi_1, \overline{L^{-1*}}\varphi_2) \\
&= \sigma_w^2 \langle\overline{L^{-1}}L^{-1*}\varphi_1, \varphi_2\rangle,
\end{aligned} \tag{7.7}$$

which is consistent with (7.5) under the implicit assumptions that $\sigma_w^2 = -f''(0)$ and $f'(0) = 0$. Finally, we recover the autocorrelation function of s by making the substitution $\varphi_1 = \delta(\cdot - r_1)$ and $\varphi_2 = \delta(\cdot - r_2)$ in (7.7), which leads to

$$\begin{aligned}
R_s(r_1, r_2) &= \mathbb{E}\{s(r_1)\, \overline{s(r_2)}\} \\
&= \mathcal{B}_s(\delta(\cdot - r_1), \delta(\cdot - r_2)) \\
&= \sigma_w^2 \langle\overline{L^{-1}}L^{-1*}\delta(\cdot - r_1), \delta(\cdot - r_2)\rangle.
\end{aligned} \tag{7.8}$$

This is justified by the kernel theorem (see Section 3.3.4), which allows one to express the correlation functional as

$$\mathcal{B}_s(\varphi_1, \varphi_2) = \int_{\mathbb{R}^d}\int_{\mathbb{R}^d} \varphi_1(r_1) R_s(r_1, r_2)\overline{\varphi_2(r_2)}\, dr_1 dr_2,$$

where $R_s(r_1, r_2) \in \mathcal{S}'(\mathbb{R}^d \times \mathbb{R}^d)$ is the (generalized) correlation function of the generalized stochastic process s, by definition.

The bottom line is that the correlation structure of the process is entirely determined by the impulse response of the Hermitian symmetric operator $\mathbf{L}^{-1}\mathbf{L}^{-1*}$, which may or may not be shift-invariant.

For formalization purposes, it is useful to categorize stochastic processes based on whether or not they are invariant to the elementary coordinate transformations. Invariance here is not meant literally, but probabilistically, in the sense that the application of a given spatial transformation (translation, rotation, or scaling) leaves the probability laws of the process unchanged.

However, since the objects of interest are generalized functions, we need to properly define the underlying notions. The translation by $r_0 \in \mathbb{R}^d$ of a generalized function $\phi \in \mathcal{S}'(\mathbb{R}^d)$ is denoted by $\phi(\cdot - r_0)$, while its scaling by a is written as $\phi(\cdot/a)$. The definition of these operations (see Section 3.3.2) is

$$\langle \varphi, \phi(\cdot - r_0) \rangle = \langle \varphi(\cdot + r_0), \phi \rangle \qquad \text{(translation by } r_0 \in \mathbb{R}^d)$$
$$\langle \varphi, \phi(\mathbf{T}^{-1}\cdot) \rangle = |\det(\mathbf{T})| \, \langle \varphi(\mathbf{T}\cdot), \phi \rangle \qquad \text{(affine transformation)}$$

for any pair $(\varphi, \phi) \in \mathcal{S}(\mathbb{R}^d) \times \mathcal{S}'(\mathbb{R}^d)$ where it is implicitly assumed that the $(d \times d)$ coordinate transformation matrix $\mathbf{T}$ is invertible. The scaling by $a > 0$ is obtained by selecting $\mathbf{T} = a\mathbf{I}$ whose determinant is a^d.

DEFINITION 7.2 (Stationary process) A generalized stochastic process s is *stationary* if it has the same probability laws as any of its translated versions $s(\cdot - r_0)$ or, equivalently, if the characteristic functional $\widehat{\mathscr{P}}_s(\varphi) = \mathbb{E}\{e^{j\langle \varphi, s \rangle}\}$ satisfies

$$\widehat{\mathscr{P}}_s(\varphi) = \widehat{\mathscr{P}}_s\big(\varphi(\cdot + r_0)\big) \qquad (7.9)$$

for any $\varphi \in \mathcal{S}(\mathbb{R}^d)$ and any $r_0 \in \mathbb{R}^d$.

DEFINITION 7.3 (Isotropic process) A generalized stochastic process s is *isotropic* if it has the same probability laws as any of its rotated versions $s(\mathbf{R}^T\cdot)$ or, equivalently, if its characteristic functional $\widehat{\mathscr{P}}_s(\varphi) = \mathbb{E}\{e^{j\langle \varphi, s \rangle}\}$ satisfies

$$\widehat{\mathscr{P}}_s(\varphi) = \widehat{\mathscr{P}}_s\big(\varphi(\mathbf{R}\cdot)\big)$$

for any $\varphi \in \mathcal{S}(\mathbb{R}^d)$ and any $(d \times d)$ rotation matrix $\mathbf{R}$.

DEFINITION 7.4 (Self-similar process) A generalized stochastic process s is *self-similar* of scaling order H if it has the same probability laws as any of its scaled and renormalized versions $a^H s(\cdot/a)$ or, equivalently, if $\widehat{\mathscr{P}}_s(\varphi) = \mathbb{E}\{e^{j\langle \varphi, s \rangle}\}$ satisfies

$$\widehat{\mathscr{P}}_s(\varphi) = \widehat{\mathscr{P}}_s\big(a^{H+d}\varphi(a\cdot)\big)$$

for any $\varphi \in \mathcal{S}(\mathbb{R}^d)$ and any dilation factor $a > 0$.

The scaling order H is also called the *Hurst exponent* of the process. Here, the relevant adjoint relation is $\langle \varphi, a^H s(\cdot/a) \rangle = \langle a^H |a|^d \varphi(a\cdot), s \rangle$, which follows from the definition of the affine transformation and the linearity of the duality product.

One can readily check that all Lévy noises are stationary and isotropic. Self-similarity, by contrast, is a more restrictive property that is only shared by the stable members of the family for which the exponent f is a homogeneous function of degree α.

Similarly, one can also define weaker forms of invariance by considering the effect of a transformation on the first- and second-order moments only. This leads to the notions of wide-sense stationarity, isotropy, and self-similarity under the implicit assumption that the variances are finite (second-order process).

DEFINITION 7.5 (Wide-sense stationarity)   A generalized stochastic process s is *wide-sense stationary* (WSS) if

$$\mathbb{E}\{\langle \varphi, s \rangle\} = \mathbb{E}\{\langle \varphi(\cdot + r_0), s \rangle\}$$
$$\mathscr{B}_s(\varphi_1, \varphi_2) = \mathscr{B}_s\big(\varphi_1(\cdot + r_0), \varphi_2(\cdot + r_0)\big)$$

for any $\varphi, \varphi_1, \varphi_2 \in \mathscr{S}(\mathbb{R}^d)$ and any $r_0 \in \mathbb{R}^d$, or, equivalently, if its (generalized) mean $\mathbb{E}\{s(r)\}$ is constant and its (generalized) autocorrelation is a function of the relative displacement only; that is,

$$\mathbb{E}\{s(r_1)\overline{s(r_2)}\} = R_s(r_1 - r_2).$$

If, in addition, $R_s(r_1 - r_2) = R_s(|r_1 - r_2|)$, then the process is WSS isotropic.

For the innovation model $s = \mathrm{L}^{-1}w$, we have $\mathbb{E}\{s(r)\} = 0$ whenever $f'(0) = 0$, which is a property that is shared by all symmetric Lévy exponents (and all concrete examples considered in this book). This zero-mean assumption facilitates the treatment of second-order processes (with minimal loss in generality).

DEFINITION 7.6 (Wide-sense self-similarity)   A generalized stochastic process s (with zero mean) is *wide-sense self-similar* with scaling order H if it has the same second-order moments as its scaled and renormalized version $a^H s(\cdot/a)$ with $a > 0$. The condition is met if the correlation functional satisfies

$$\mathscr{B}_s(\varphi_1, \varphi_2) = a^{2H+2d} \mathscr{B}_s\big(\varphi_1(a\cdot), \varphi_2(a\cdot)\big) \tag{7.10}$$

for any $\varphi_1, \varphi_2 \in \mathscr{S}(\mathbb{R}^d)$ or, equivalently, if the (generalized) autocorrelation function is such that

$$a^{2H} R_s\Big(\frac{r_1}{a}, \frac{r_2}{a}\Big) = R_s(r_1, r_2).$$

Note that wide-sense self-similarity with $H \neq 0$ implies that $R_s(0, 0) = \mathbb{E}\{s^2(0)\}$ is either zero or infinite, so the property is incompatible with wide-sense stationarity (unless $s = 0$).

In the case of our generalized innovation model, the invariance properties of the stochastic process are primarily determined by the choice of the operator L. The precise statement is given in the theorem below, which also makes the link with the more conventional specification of a linear process in terms of a stochastic integral against some kernel h.

THEOREM 7.1   *Let* $s = L^{-1}w$ *be a generalized stochastic process whose character-istic functional is given by (7.3), where* (L^{-1*}, f) *is an admissible pair in reference to Definition 7.1. We also define the kernel*

$$h(\cdot, r') = L^{-1}\{\delta(\cdot - r')\} \in \mathscr{S}'(\mathbb{R}^d \times \mathbb{R}^d). \tag{7.11}$$

Then, depending on the characteristics of L^{-1} *(or, equivalently,* L^{-1*}*), the process* s *enjoys the following properties.*

(1) *If* L^{-1} *is linear shift-invariant, then* s *is stationary and*

$$h(r, r') = h(r - r', 0) = \rho_L(r - r'),$$

   where $\rho_L = L^{-1}\{\delta\}$ *is the Green's function of* L.
(2) *If* L^{-1} *is shift- and rotation-invariant, then* s *is stationary isotropic and* $h(r, r') = \rho_L(|r - r'|)$, *where* $\rho_L(|r|) = L^{-1}\{\delta\}(r)$ *is a purely radial function.*
(3) *If* L^{-1*} *is scale-invariant of order* $(-\gamma)$ *and* $\sigma_w^2 = -f''(0) < \infty$, *then* s *is wide-sense self-similar with Hurst exponent* $H = \gamma - d/2$.
(4) *If* L^{-1*} *is scale-invariant of order* $(-\gamma)$ *and* f *is homogeneous* [1] *of degree* $0 < \alpha \le 2$, *then* s *is self-similar with Hurst exponent* $H = \gamma - d + d/\alpha$.

Moreover, if h *is an ordinary function of* $\mathbb{R}^d \times \mathbb{R}^d$ *with* $h(r, \cdot) \in \mathscr{R}(\mathbb{R}^d)$ *(resp.,* $h(r, \cdot) \in L_p(\mathbb{R}^d)$*) for* $r \in \mathbb{R}^d$, *then* s *admits the pointwise representation*

$$s(r) = \langle h(r, \cdot), w \rangle. \tag{7.12}$$

Proof   The existence of the generalized stochastic process s is ensured by Theorem 4.17. Moreover, the kernel theorem (Theorem 3.1) allows one to express the measurement $X = \langle \varphi, L^{-1}w \rangle = \langle L^{-1*}\varphi, w \rangle$ as

$$X = \langle \varphi, s \rangle = \langle \int_{\mathbb{R}^d} h(r, \cdot)\varphi(r) \, dr, w \rangle,$$

where $h(\cdot, \cdot) \in \mathscr{S}'(\mathbb{R}^d \times \mathbb{R}^d)$ is the kernel defined by (7.11).

Statement (1): If L^{-1} is LSI, then the same holds true for L^{-1*}, whose impulse response is $L^{-1*}\{\delta\}(r) = h(0, r) = \rho_L(-r)$. By duality, this is equivalent to $L^{-1}\{\delta\}(r) = h(r, 0) = \rho_L(r)$ with the property that $LL^{-1}\{\delta\} = L\rho_L = \delta$, so that ρ_L is the Green's function of L. From the definition of translation-invariance, we have that $L^{-1*}\{\varphi(\cdot + r_0)\}(r) = L^{-1*}\{\varphi\}(r+r_0)$, which is then used to show that $\int_{\mathbb{R}^d} f(L^{-1*}\{\varphi(\cdot + r_0)\}(r)) \, dr = \int_{\mathbb{R}^d} f(L^{-1*}\{\varphi\}(x)) \, dx$ with the change of variable $x = r+r_0$. Hence the condition in Definition 7.2 is fulfilled.

Statement (2): This is a special case of Statement (1) where L^{-1} is also rotation-invariant, which translates into $\rho_L(r) = \rho_L(|r|)$. The condition in Definition 7.3 is established by simple change of variable in the exponent of the characteristic functional.

Statement (3): Our strategy there is to verify (7.10) starting from (7.7). The fact that L^{-1} is scale-invariant of order $(-\gamma)$ implies that both L^{-1} and L^{-1*} enjoy the same

---

[1] The class of such admissible Lévy exponents are the α-stable ones; the symmetric members of the family are $f_\alpha(\omega; s_0) = -|s_0\omega|^\alpha$.

property, as expressed by (5.1). Therefore,

$$\mathscr{B}_s\big(\varphi_1(a\cdot),\varphi_2(a\cdot)\big) = \sigma_w^2\langle L^{-1*}\{\varphi_1(a\cdot)\},\overline{L^{-1*}\{\varphi_2(a\cdot)\}}\rangle$$
$$= \sigma_w^2\langle a^{-\gamma}L^{-1*}\{\varphi_1\}(a\cdot),\overline{a^{-\gamma}L^{-1*}\{\varphi_2\}(a\cdot)}\rangle \quad \text{(by scale-invariance)}$$
$$= \sigma_w^2 a^{-2\gamma}\langle L^{-1*}\{\varphi_1\}(a\cdot),\overline{L^{-1*}\{\varphi_2\}(a\cdot)}\rangle \quad \text{(by bilinearity)}$$
$$= a^{-2\gamma}a^{-d}\sigma_w^2\langle L^{-1*}\varphi_1,\overline{L^{-1*}\varphi_2}\rangle \quad \text{(change of variable)}$$
$$= a^{-2\gamma-d}\mathscr{B}_s(\varphi_1,\varphi_2),$$

so that the condition in Definition 7.6 is satisfied if and only if $2H + 2d = 2\gamma + d$, which we rewrite as $H = \gamma - d/2$.

Statement (4): The result is obtained by evaluating the required characteristic exponent where $\widehat{\mathscr{P}}_s(\varphi)$ is specified by (7.3) with f such that $f(a\omega) = a^\alpha f(\omega)$ (α homogeneous). Specifically,

$$\log\widehat{\mathscr{P}}_s\big(a^{H+d}\varphi(a\cdot)\big) = \int_{\mathbb{R}^d} f\left(a^{H+d}L^{-1*}\{\varphi(a\cdot)\}(r)\right)dr \quad \text{(by linearity of } L^{-1*})$$
$$= \int_{\mathbb{R}^d} f\left(a^{H+d}a^{-\gamma}L^{-1*}\{\varphi\}(ar)\right)dr \quad \text{(by scale-invariance)}$$
$$= \int_{\mathbb{R}^d} a^{-d}f\left(a^{H+d-\gamma}L^{-1*}\{\varphi\}(x)\right)dx \quad \text{(change of variable)}$$
$$= a^{\alpha H+\alpha d-\alpha\gamma-d}\int_{\mathbb{R}^d} f\left(L^{-1*}\varphi(r)\right)dr. \quad (\alpha\text{-homogeneity of }f)$$

The latter expression is equal to $\log\widehat{\mathscr{P}}_s(\varphi)$ if and only if $\alpha H + \alpha d - \alpha\gamma - d = 0$, which yields $H = \gamma - d + d/\alpha$.

As for the final result, we consider the observation $X_0 = \langle\varphi_0, w\rangle$ where $\varphi_0 = h(r_0,\cdot)$ with $\varphi_0 \in \mathscr{R}(\mathbb{R}^d)$ $\big($resp., $\varphi_0 \in L_p(\mathbb{R}^d)\big)$. Since $\widehat{\mathscr{P}}_w$ admits a continuous extension over $\mathscr{R}(\mathbb{R}^d)$ (see Proposition 8.1 in Chapter 8), we can specify the characteristic function of X_0 as $\widehat{p}_{X_0}(\omega) = \mathbb{E}\{e^{j\omega X_0}\} = \widehat{\mathscr{P}}_w(\omega\varphi_0)$. Then, $\widehat{p}_{X_0}$ is a continuous, positive definite function of ω so that we can invoke Bochner's theorem (Theorem 3.7), which ensures that $X_0 = \langle L^{-1*}\{\delta(\cdot - r_0)\}, w\rangle = \langle\delta(\cdot - r_0), L^{-1}w\rangle = s(r_0)$ is a well-defined (conventional) random variable. The result also extends to $\varphi_0 \in L_p(\mathbb{R}^d)$ whenever f is p-admissible (see Theorem 8.2, which will be proven later on). □

Since the Lévy innovations w are all intrinsically stationary, there is no distinction in this model between stationarity and WSS, except for the fact that the latter requires the variance σ_w^2 to be finite. This is not so for self-similarity, which is a more-demanding property. In that respect, we note that there is no contradiction between Statements (3) and (4) in Theorem 7.1 because the second-order moments of α-stable processes (for which f is homogeneous of order α) are undefined for $\alpha < 2$ (due to the unbounded variance of the noise). The intersection occurs for $\alpha = 2$ (Gaussian scenario) while larger homogeneity indices ($\alpha > 2$) are excluded by the Lévy admissibility condition.

The last result in Theorem 7.1 is fundamental, for it tells us when $s(r)$ can be interpreted as a conventional stochastic process; that is, as a random function of the index variable r. With a slight abuse of notation, we may rewrite (7.12) as

$$s(r) = \int_{\mathbb{R}^d} h(r, r')w(r')\,dr' = \int_{\mathbb{R}^d} h(r, r')W(dr'), \tag{7.13}$$

which shows the connection with conventional stochastic integration (Itô calculus). Here, W is a random measure over $\mathbb{R}^d$ that is formally defined as $W(E) = \langle \mathbb{1}_E, w \rangle$ for any Borel set $E \subseteq \mathbb{R}^d$.

While we have already pointed out the incompatibility between stationarity and self-similarity, there is a formal way to bypass this limitation by enforcing stationarity selectively through the test functions whose moments are vanishing (up to a certain order). Specifically, we shall specify in Section 7.5 processes that fulfill this quasi-stationarity condition with the help of the L_p-stable, scale-invariant inverse of L^* defined in Section 5.5. This construction results in self-similar processes with stationary increments, the prototypical example being fractional Brownian motion. But before that we shall investigate other concrete examples of generalized stochastic processes, starting with the simpler stationary ones.

7.3   Non-Gaussian stationary processes

If the inverse operator L^{-1} is shift-invariant with generic impulse response $\rho_L \in L_1(\mathbb{R}^d)$, then (7.12) is equivalent to a convolutional system with $s(r) = (\rho_L * w)(r)$. We can then apply (7.3) in conjunction with Proposition 5.1 to obtain the characteristic functional of this process, which reads

$$\widehat{\mathscr{P}_s}(\varphi) = \exp\left(\int_{\mathbb{R}^d} f\left((\rho_L^\vee * \varphi)(r)\right) dr\right). \tag{7.14}$$

More generally, we may consider generalized processes that are obtained by LSI filtering of an innovation process w and are not necessarily solutions of a stochastic differential equation.

PROPOSITION 7.2 (Generalized stationary processes)   *Let $s = h * w$, where $\|\mu_h\|_{TV} < \infty$ (bounded variation) (resp., h is rapidly decaying) and w is a white-noise process over $\mathscr{S}'(\mathbb{R}^d)$ whose Lévy exponent f is p-admissible (resp., Lévy–Schwartz admissible). Then, s is a generalized stochastic process in $\mathscr{S}'(\mathbb{R}^d)$ that is stationary and completely specified by the characteristic functional $\widehat{\mathscr{P}_s}(\varphi) = \exp\left(\int_{\mathbb{R}^d} f\left((h^\vee * \varphi)(r)\right) dr\right)$. In general, the process is non-Gaussian unless $f(\omega) = -\frac{\sigma_w^2}{2}|\omega|^2$.*

The proof is the same as that of Statement (1) in Theorem 7.1. As for the existence of the process when f is p-admissible, we rely on the convolution inequality, $\|h^\vee * \varphi\|_{L_p} = \|h * \varphi^\vee\|_{L_p} \le \|\mu_h\|_{TV}\,\|\varphi\|_{L_p}$, which ensures that the L_p condition in Definition 7.1 is satisfied. In that respect, we note that the bounded variation hypothesis on h (which is less stringent than $h \in L_1$) is the minimal requirement for stability when $p = 1$, while it can be weakened to $\|\widehat{h}\|_{L_\infty} < \infty$ for $p = 2$ (see Statements (1) and (3) in Theorem 3.5).

In addition, when $h \in L_p(\mathbb{R}^d)$ $\left(\text{resp., } h \in \mathcal{R}(\mathbb{R}^d)\right)$, we can invoke the last part of Theorem 7.1 to show that the point values $s(\boldsymbol{r})$ of the process are well defined, so that s also admits a classical interpretation.

7.3.1 Autocorrelation function and power spectrum

Next, we determine the autocorrelation function of the stochastic process s and make a link with splines by following essentially the same path as for our introductory AR(1) example. Our basic requirement is that the whitening operator L is spline-admissible in the sense of Definition 6.8, which implies the existence of the generalized B-spline β_L and the interpolation function φ_{int} specified by (6.51).

Let $\widehat{L_d}(\boldsymbol{\omega}) = \sum_{k \in \mathbb{Z}^d} d_L[\boldsymbol{k}] e^{-j\langle \boldsymbol{\omega}, \boldsymbol{k} \rangle}$ with $d_L \in \ell_1(\mathbb{Z}^d)$ be the transfer function of L_d (the discrete version of the whitening operator L). Under the condition that $\widehat{L_d}(\boldsymbol{\omega}) \neq 0$ for $\boldsymbol{\omega} \in [-\pi, \pi]^d$, we invoke Wiener's lemma (Theorem 5.13), which ensures that the (discrete) inverse operator L_d^{-1} is L_p-stable. This naturally leads to the generalized B-spline reproduction formula for the impulse response of the system,

$$\rho_L(\boldsymbol{r}) = L_d^{-1}\beta_L(\boldsymbol{r}) = \sum_{k \in \mathbb{Z}^d} p[\boldsymbol{k}]\beta_L(\boldsymbol{r} - \boldsymbol{k}), \tag{7.15}$$

with $p[\boldsymbol{k}] = \int_{[-\pi,\pi]^d} \frac{e^{j\langle \boldsymbol{\omega}, \boldsymbol{k}\rangle}}{\widehat{L_d}(\boldsymbol{\omega})} \frac{d\boldsymbol{\omega}}{(2\pi)^d} \in \ell_1(\mathbb{Z}^d)$. The concept also generalizes for the specification of the second-order moments.

PROPOSITION 7.3 *The autocorrelation function of the stationary stochastic process* $s(\boldsymbol{r}) = (\rho_L * w)(\boldsymbol{r})$, *where w is a white Lévy noise with variance σ_w^2, is given by*

$$R_s(\boldsymbol{r}) = \mathbb{E}\{s(\cdot + \boldsymbol{r})\,\overline{s(\cdot)}\} = \sigma_w^2\,(\rho_L * \overline{\rho_L^\vee})(\boldsymbol{r}).$$

Moreover, it satisfies the Shannon-like interpolation formula

$$R_s(\boldsymbol{r}) = \sigma_w^2(\rho_L * \overline{\rho_L^\vee})(\boldsymbol{r}) = \sum_{k \in \mathbb{Z}^d} R_s[\boldsymbol{k}]\varphi_{\text{int}}(\boldsymbol{r} - \boldsymbol{k}),$$

where the interpolation function is defined by (6.51) and where the expansion coefficients $R_s[\boldsymbol{k}] = \mathbb{E}\{s(\cdot + \boldsymbol{k})\,\overline{s(\cdot)}\} = R_s(\boldsymbol{r})|_{\boldsymbol{r}=\boldsymbol{k}}$ *correspond to the discrete-domain version of the correlation.*

The power spectrum of the process is the Fourier transform of the autocorrelation function,

$$\Phi_s(\boldsymbol{\omega}) = \mathscr{F}\{R_s\}(\boldsymbol{\omega}) = \frac{\sigma_w^2}{|\widehat{L}(\boldsymbol{\omega})|^2}, \tag{7.16}$$

an expression that is consistent with the interpretation of the signal as a filtered white noise.

Proof   The first part of the statement is obtained from (7.8) as

$$\mathbb{E}\{s(r_1)\,\overline{s(r_2)}\} = \sigma_w^2 \langle L^{-1}L^{-1*}\delta(\cdot - r_1), \delta(\cdot - r_2)\rangle$$

$$= \sigma_w^2 \langle L^{-1}L^{-1*}\delta, \delta(\cdot + r_1 - r_2)\rangle \qquad \text{(by shift-invariance)}$$

$$= \sigma_w^2 (\overline{\rho}_L * \rho_L^{\vee})(-r_1 + r_2) \qquad \text{(by definition of } \delta)$$

$$= \sigma_w^2 (\rho_L * \overline{\rho}_L^{\vee})(r_1 - r_2). \qquad \text{(by Hermitian symmetry)}$$

As expected, this yields a correlation function that only depends on the relative differ-ence $r = (r_1 - r_2)$ of the index variables. To establish the validity of the interpolation formula, we consider (6.51) and manipulate the Fourier-domain expression for φ_{int} as

$$\widehat{\varphi}_{\text{int}}(\omega) = \frac{|\widehat{\beta}_L(\omega)|^2}{\sum_{n\in\mathbb{Z}^d} |\widehat{\beta}_L(\omega + 2\pi n)|^2}$$

$$= \frac{\frac{|\widehat{L}_d(\omega)|^2}{|\widehat{L}(\omega)|^2}}{\sum_{n\in\mathbb{Z}^d} \frac{|\widehat{L}_d(\omega+2\pi n)|^2}{|\widehat{L}(\omega+2\pi n)|^2}} \qquad \text{(from definition of generalized B-spline)}$$

$$= \frac{\frac{1}{|\widehat{L}(\omega)|^2}}{\sum_{n\in\mathbb{Z}^d} \frac{1}{|\widehat{L}(\omega+2\pi n)|^2}}$$

$$= \frac{\Phi_s(\omega)}{\sum_{n\in\mathbb{Z}^d} \Phi_s(\omega + 2\pi n)},$$

where the simplification in the third line results from the property that $\widehat{L}_d(\omega)$ (transfer function of a digital filter) is 2π-periodic. The final formula is the ratio of $\Phi_s(\omega)$ (the continuous-domain power spectrum of s given by (7.16)) and its discrete-domain coun-terpart $\sum_{k\in\mathbb{Z}^d} R_s[k]e^{-j\langle\omega,k\rangle} = \sum_{n\in\mathbb{Z}^d} \Phi_s(\omega + 2\pi n)$ (by Poisson's summation formula), which proves the desired result.   □

7.3.2   Generalized increment process

A standard side-effect of innovation models is the induction of long-range signal depen-dencies due to the non-compact nature (IIR) of the impulse response of the system L^{-1}. We have already pointed out that those could be partially suppressed by application of the discrete form of the whitening operator L_d. The good news is that the resulting out-put (generalized increment process) will not only be approximately decoupled but also stationary, irrespective of the properties of the input process.

PROPOSITION 7.4 (Generalized increment process)   *Let* $s = L^{-1}w$ *be a generalized stochastic process (possibly non-stationary) where the whitening operator* L *is spline-admissible with generalized B-spline* $\beta_L \in \mathcal{R}(\mathbb{R}^d)$ *such that* $\beta_L(\cdot - r_0) = L_d L^{-1}\delta(\cdot - r_0)$ *for all* $r_0 \in \mathbb{R}^d$. *Then,* $u = L_d s$ *is stationary with characteristic functional* $\widehat{\mathscr{P}}_u(\varphi) = \exp\left(\int_{\mathbb{R}^d} f((\beta_L^{\vee} * \varphi)(r))\,dr\right)$, *where f is the Lévy exponent of the innovation w. Its auto-correlation function is* $\mathbb{E}\{u(\cdot + r)\,\overline{u(\cdot)}\} = \sigma_w^2(\beta_L * \overline{\beta}_L^{\vee})(r)$.

Note that we can also write an extended version of this result for the class of spline-admissible operators L with $\beta_L \in L_1(\mathbb{R}^d)$ which requires that f be p-admissible for some $p \in [1, 2]$.

Proof The characterization of the B-spline in the statement of the proposition ensures that the composition of L_d and L^{-1} is LSI with impulse response $\beta_L = L_d L^{-1} \delta$. Since L_d is LSI, the condition is trivially satisfied when L^{-1} is shift-invariant as well. Generally, the condition will also be met in the non-stationary scenarios considered later in this chapter (see Proposition 7.6), the fundamental reason being that L_d must annihilate the components that are in the null space of L (see Section 6.4 on the construction of generalized B-splines). This property allows us to write $u = L_d s = L_d L^{-1} w = \beta_L * w$. The result then follows as a direct consequence of Propositions 7.2 and 7.3. $\square$

An important implication of Proposition 7.4 is that the quality of the decoupling is solely dependent upon the localization properties of β_L. This theme is further developed in Sections 7.4.4 and 8.3.

7.3.3 Generalized stationary Gaussian processes

In light of Proposition 7.2, we observe that the complete class of stationary Gaussian processes is specifiable via the basic convolutional model $s = h * w_{\text{Gauss}}$, where w_{Gauss} is a zero-mean Gaussian innovation process and h an L_2-stable convolution operator with Fourier multiplier $H(\omega) = \hat{h}(\omega) \in L_\infty(\mathbb{R}^d)$. This follows from the $p = 2$ admissibility of the Gaussian Lévy exponent $f_{\text{Gauss}}(\omega) = -\sigma_w^2 |\omega|^2/2$, the necessity and sufficiency of the condition $H(\omega) \in L_\infty(\mathbb{R}^d)$ for the convolution operator to be continuous over $L_2(\mathbb{R}^d)$ (see Theorem 3.5), and the last existence result in Theorem 4.17.

The resulting generalized Gaussian process is uniquely specified by its autocorrelation function

$$R_{s_{\text{Gauss}}}(r) = \sigma_w^2 (h * \overline{h}^\vee)(r),$$

or, equivalently, by its spectral density

$$\Phi_{s_{\text{Gauss}}}(\omega) = \mathcal{F}\{R_{s_{\text{Gauss}}}(r)\}(\omega) = \sigma_w^2 |H(\omega)|^2. \tag{7.17}$$

Its characteristic functional is given by

$$\widehat{\mathscr{P}}_{s_{\text{Gauss}}}(\varphi) = \exp\left(-\frac{\sigma_w^2}{2}\|h^\vee * \varphi\|^2\right), \tag{7.18}$$

which, by using Parseval's identity, can also be rewritten as

$$\widehat{\mathscr{P}}_{s_{\text{Gauss}}}(\varphi) = \exp\left(-\frac{1}{2}\int_{\mathbb{R}^d} \Phi_{s_{\text{Gauss}}}(\omega)|\widehat{\varphi}(\omega)|^2 \frac{d\omega}{(2\pi)^d}\right).$$

The necessary and sufficient condition for the existence of such generalized Gaussian processes is that $\Phi_{s_{\text{Gauss}}}(\omega)$ be bounded almost everywhere. This is less restrictive than the requirement $\Phi_{s_{\text{Gauss}}} \in L_2(\mathbb{R}^d)$ of the classical formulation, which ensures that the variance of the process $\mathbb{E}\{|s_{\text{Gauss}}(r_0)|^2\}$ is finite. The simplest example that fits the generic filtered-white-noise model but does not meet the latter condition is w_{Gauss} with $h = \delta$.

7.3.4   CARMA processes

The acronym ARMA traditionally refers to discrete Gaussian processes that are solutions of stable Nth-order difference equations driven by discrete white Gaussian noise. The corresponding discrete system is characterized by a rational transfer function whose denominator determines the Nth-order AR (autoregressive) part of the filter and the numerator the MA (moving average) part. The Gaussian CARMA(N, M) (continuous-ARMA) processes are the continuous-domain counterparts of these discrete processes. They are solutions of the generic Nth-order differential equation

$$p_N(\mathrm{D})s(t) = q_M(\mathrm{D})w(t), \tag{7.19}$$

with defining polynomials

$$p_N(\zeta) = \zeta^N + a_{N-1}\zeta^{N-1} + \cdots + a_0 = \prod_{n=1}^{N}(\zeta - \alpha_n)$$

$$q_M(\zeta) = b_M\zeta^M + b_{M-1}\zeta^{M-1} + \cdots + b_0 = b_M\prod_{m=1}^{M}(\zeta - \gamma_m),$$

where $a_n, b_m \in \mathbb{R}$, $M < N$, and D is the derivative operator. Traditionally, the driving term w is a Gaussian white noise with variance σ_w^2. The underlying linear system is characterized by its poles $\boldsymbol{\alpha} = (\alpha_1, \ldots, \alpha_N)$ and zeros $\boldsymbol{\gamma} = (\gamma_1, \ldots, \gamma_M)$ and is known to be causal-stable if and only if $\mathrm{Re}(\alpha_n) < 0$, for $n = 1, \ldots, N$. Under those conditions, the solution of (7.19) is given by

$$s_{\boldsymbol{\alpha}}(t) = (h_{\boldsymbol{\alpha},\boldsymbol{\gamma}} * w)(t)$$

with $h_{\boldsymbol{\alpha},\boldsymbol{\gamma}}(t) = \mathscr{F}^{-1}\{H_{\boldsymbol{\alpha},\boldsymbol{\gamma}}(\omega)\}(t)$, where

$$H_{\boldsymbol{\alpha},\boldsymbol{\gamma}}(\omega) = \frac{q_M(j\omega)}{p_N(j\omega)} = b_M\frac{\prod_{m=1}^{M}(j\omega - \gamma_m)}{\prod_{n=1}^{N}(j\omega - \alpha_n)}$$

is the frequency response of the underlying system. The link with the operator formalism of Section 5.3.2 is

$$h_{\boldsymbol{\alpha},\boldsymbol{\gamma}}(t) = \underbrace{b_M\mathrm{P}_{\gamma_1}\cdots\mathrm{P}_{\gamma_M}\mathrm{P}_{\alpha_N}^{-1}\cdots\mathrm{P}_{\alpha_1}^{-1}}_{\mathrm{L}^{-1}}\delta(t)$$

with $\mathrm{P}_\alpha = \mathrm{D} - \alpha\mathrm{Id}$ and $\mathrm{P}_\alpha^{-1}\delta(t) = \mathbb{1}_+(t)e^{\alpha t}$. Moreover, since $M < N$, we can decompose $H_{\boldsymbol{\alpha},\boldsymbol{\gamma}}(\omega)$ into simple partial fractions and obtain an expression for the impulse response of the system as a sum of elementary components that decay exponentially, which shows that $h_{\boldsymbol{\alpha},\boldsymbol{\gamma}}(t)$ is rapidly decreasing. We are therefore meeting the least constraining condition $h_{\boldsymbol{\alpha},\boldsymbol{\gamma}} \in \mathscr{R}(\mathbb{R})$ of Proposition 7.2. This ensures that we can apply the framework to specify not only Gaussian CARMA processes, but also a whole variety of non-Gaussian variants associated with more general Lévy innovations. These extended CARMA processes are stationary and completely characterized by the characteristic functional (7.14) with $\rho_L = h_{\alpha,\gamma}$ and $d = 1$ without any restriction on the Lévy exponent f. Moreover, since the underlying kernel $h(t_0, \tau) = h_{\alpha,\gamma}(t_0 - \tau)$ for t_0 fixed is bounded and exponentially decreasing, they admit an interpretation as an ordinary stochastic process (by Theorem 7.1).

The autocorrelation function of the process is defined under the additional second-order hypotheses $f''(0) = -\sigma_w^2$ and $f'(0) = 0$. It is given by

$$R_{s_\alpha}(t) = \sigma_w^2 (h_{\alpha,\gamma} * \overline{h}_{\alpha,\gamma}^\vee)(t),$$

which is a sum of symmetric exponentials with parameters α. The corresponding power spectrum is

$$\Phi_{s_\alpha}(\omega) = \sigma_w^2 \frac{b_M^2 \prod_{m=1}^{M} |j\omega - \gamma_m|^2}{\prod_{n=1}^{N} |j\omega - \alpha_n|^2} = \sigma_w^2 \left| \frac{q_M(j\omega)}{p_N(j\omega)} \right|^2,$$

which is consistent with (7.17).

7.4 Lévy processes and their higher-order extensions

Lévy processes and their extensions can also be defined in the introduced framework, but their specification is more delicate due to the fact that their underlying SDE is unstable. This requires the use of the "regularized" bounded inverse operators that we presented in Section 5.4. As this constitutes a significant departure from the traditional shift-invariant setting, we shall detail the construction in Section 7.4.1 and provide the connection with the classical theory.

7.4.1 Lévy processes

In the standard time-domain formulation of stochastic processes, the solution of a linear differential equation is usually expressed as the stochastic integral

$$s(t) = \int_0^{+\infty} h(t, \tau)\, \mathrm{d}W(\tau),$$

where W is a random measure which is a Brownian motion or, by extension, a Lévy process. In keeping with the above notation, we shall now show that a Lévy process $W(t)$ can be generated via the integration of white noise as $W(t) = \int_0^t w(\tau)\, \mathrm{d}\tau = \int_0^t \mathrm{d}W(\tau)$, which is consistent with the Lévy innovation $w = \dot{W}$ being the derivative of W (in the "weak" sense of generalized functions).

To establish this connection, we recall the classical definition of Lévy processes.

DEFINITION 7.7 (Lévy process)   The stochastic process $W = \{W(t) : t \in \mathbb{R}^+\}$ is a *Lévy process* if it fulfills the following requirements:

(1) $W(0) = 0$ almost surely.
(2) Given $0 \leq t_1 < t_2 < \ldots < t_n$, the increments $W(t_2) - W(t_1)$, $W(t_3) - W(t_2)$, ..., $W(t_n) - W(t_{n-1})$ are mutually independent.
(3) For any given step T, the increment process $\delta_T W(t)$, where δ_T is the operator that associates $W(t)$ with $\big(W(t) - W(t - T)\big)$, is stationary.

In addition, one typically requires the process to fulfill some form of probabilistic continuity.

In our framework, the equivalent of the above processes is obtained as a solution of the stochastic differential equation

$$DW = \dot{W} = w, \tag{7.20}$$

subject to the boundary condition $W(0) = 0$, where $D = \frac{d}{dt}$ is the derivative operator and W is to be defined as a random element of $\mathscr{S}'(\mathbb{R})$. The driving term w in (7.20) is a 1-D Lévy innovation, as defined in Section 4.4, with characteristic functional

$$\mathbb{E}\{e^{j\langle \varphi, w\rangle}\} = \widehat{\mathscr{P}}_w(\varphi) = \exp\left(\int_{\mathbb{R}} f(\varphi(r))\, dr\right). \tag{7.21}$$

We recall that w has the property of independence at every point, meaning that any pair of random variables $\langle \varphi, w\rangle$ and $\langle \psi, w\rangle$, for test functions φ and ψ of disjoint support, are independent. In terms of the characteristic functional, this property translates into having $\widehat{\mathscr{P}}_w(\omega_1 \varphi + \omega_2 \psi)$ factorize as

$$\widehat{\mathscr{P}}_w(\omega_1 \varphi + \omega_2 \psi) = \widehat{\mathscr{P}}_w(\omega_1 \varphi)\widehat{\mathscr{P}}_w(\omega_2 \psi) \quad \text{for disjointly supported } \varphi \text{ and } \psi. \tag{7.22}$$

To say that a generalized random process W fulfills (7.20) is, for us, to have

$$\langle D^*\varphi, W\rangle = \langle \varphi, w\rangle \quad \text{for all } \varphi \in \mathscr{S}(\mathbb{R}), \tag{7.23}$$

where $D^* = -D$ is the adjoint of D. For W to be fully characterized as a random element of $\mathscr{S}'(\mathbb{R})$, we need to give a consistent definition of $\langle \varphi, s\rangle$ for all $\varphi \in \mathscr{S}(\mathbb{R})$. We next show that we find a particular solution of (7.20) by defining

$$\langle \varphi, W\rangle = \langle I_0^*\varphi, w\rangle, \tag{7.24}$$

where I_0^* is the left inverse $\mathscr{S}(\mathbb{R}) \to \mathscr{R}(\mathbb{R})$ of D^* specified in Section 5.4.1. In view of (7.24), $\langle \varphi, W\rangle$ is probabilistically characterized by the functional

$$\widehat{\mathscr{P}}_W(\varphi) = \widehat{\mathscr{P}}_w(I_0^*\varphi). \tag{7.25}$$

To see that (7.24) implies (7.23), note, first, that for any $\varphi' \in \mathscr{S}(\mathbb{R})$ that can be written as $D^*\varphi$ for some $\varphi \in \mathscr{S}(\mathbb{R})$, we have

$$\langle \varphi', W\rangle = \langle D^*\varphi, W\rangle = \langle I_0^* D^*\varphi, w\rangle.$$

Now, since I_0^* is a left inverse $\mathscr{S}(\mathbb{R}) \to \mathscr{R}(\mathbb{R})$ of $D^* : \mathscr{S}(\mathbb{R}) \to \mathscr{S}(\mathbb{R})$, we find

$$\langle D^*\varphi, W\rangle = \langle \varphi, w\rangle$$

(where φ can be arbitrarily chosen), which is the same as (7.23).

We symbolically represent the particular solution W defined by (7.24) and (7.25) as

$$W = \mathrm{I}_0 w, \tag{7.26}$$

where I_0 is the adjoint of I_0^*. For completeness, we also determine the corresponding kernel $h(t, \tau)$ in (7.11), which takes the form

$$\mathrm{I}_0\{\delta(\cdot - \tau)\}(t) = \mathbb{1}_+(t - \tau) - \mathbb{1}_+(-\tau). \tag{7.27}$$

It is the "transpose" of the generalized impulse response of I_0^* given by

$$\mathrm{I}_0^*\{\delta(\cdot - \tau)\}(t) = \mathbb{1}_+(\tau - t) - \mathbb{1}_+(-t) = \begin{cases} \mathbb{1}_{(0,\tau]}(t), & \text{for } \tau \geq 0 \\ -\mathbb{1}_{(\tau,0]}(t), & \text{for } \tau < 0, \end{cases} \tag{7.28}$$

which follows from (5.17) with $\omega_0 = 0$ and φ being replaced by $\delta(\cdot - \tau)$. While (7.27) and (7.28) are equivalent identities with the role of the variables t and τ being interchanged, the main point is that the kernel on the right of (7.28) for τ fixed is compactly supported. This allows us to invoke Theorem 7.1 to show that the point values of the process, $W(t_n) = \langle \mathbb{1}_{(0,t_n]}, w \rangle$, are ordinary random variables.

Having defined a particular solution of (7.20) as $\mathrm{I}_0 w$, let us now show that it is consistent with the axiomatic definition of a Lévy process given by Definition 7.7. The zero boundary condition at the origin (Property (1) in Definition 7.7) is imposed by the operator I_0 (see Theorem 5.3 with $\omega_0 = 0$). As for the other two properties, we recall that, for $t \geq 0$,

$$\mathrm{I}_0^* \varphi(t) = \int_t^{+\infty} \varphi(\tau) \, d\tau,$$

from which we deduce

$$\mathrm{I}_0^* \delta_T^* \varphi(t) = \int_t^\infty \varphi(\tau) - \varphi(t + T) \, d\tau = \int_t^{t+T} \varphi(\tau) \, d\tau = \left(\mathbb{1}_{[-T,0)} * \varphi\right)(t).$$

From there, we see that, for the increment process $\delta_T W$,

$$\langle \varphi, \delta_T W \rangle = \langle \delta_T^* \varphi, W \rangle = \langle \mathrm{I}_0^* \delta_T^* \varphi, w \rangle = \langle \mathbb{1}_{[-T,0)} * \varphi, w \rangle,$$

which is equivalent to

$$\delta_T W = W(\cdot) - W(\cdot - T) = \mathbb{1}_{(0,T]} * w$$

because $\mathbb{1}_{(0,T]} = \mathbb{1}_{[-T,0)}^{\vee}$. Now, since w is stationary, we have that

$$X_t = \langle \varphi(\cdot - t), \delta_T W \rangle = \langle \mathbb{1}_{[-T,0)} * \varphi(\cdot - t), w \rangle \overset{w}{=} \langle \mathbb{1}_{[-T,0)} * \varphi, w \rangle = \langle \varphi, \delta_T W \rangle = X_0$$

for all $t \in \mathbb{R}$, where $\overset{w}{=}$ denotes equivalence in law. This proves that $\delta_T W$ is stationary. Finally, by writing $\left(W(t_m) - W(t_{m-1})\right) = \langle \mathbb{1}_{(t_{m-1},t_m]}, w \rangle$ and using Proposition 3.10 in combination with (7.22), we see that the joint characteristic function of the increments $U_1 = W(t_2) - W(t_1)$, $U_2 = W(t_3) - W(t_2)$, ..., $U_{n-1} = W(t_n) - W(t_{n-1})$ with $0 \leq t_1 < t_2 < \ldots < t_n$ separates as

$$\widehat{p}_{(U_1:U_{n-1})}(\omega_1, \ldots, \omega_{n-1}) = \widehat{\mathscr{P}}_w(\omega_1 \mathbb{1}_{(t_1,t_2]}) \, \widehat{\mathscr{P}}_w(\omega_2 \mathbb{1}_{(t_2,t_3]}) \, \cdots \, \widehat{\mathscr{P}}_w(\omega_{n-1} \mathbb{1}_{(t_{n-1},t_n]}),$$

which implies their independence (second property in Definition 7.7).

We close this section with a demonstration of the use of (7.25) for the determination of the statistical distribution of the sample values of a Lévy process. In our formalism, the sampling operation is expressed as

$$W(t_1) = \langle \delta(\cdot - t_1), W \rangle = \langle I_0^* \delta(\cdot - t_1), w \rangle = \langle \mathbb{1}_{(0,t_1]}, w \rangle.$$

This allows us to deduce the characteristic function of $W(t_1)$ by direct substitution [2] of $\varphi = \omega_1 \delta(\cdot - t_1)$ in (7.25) as

$$\widehat{p}_{W(t_1)}(\omega_1) = \mathbb{E}\{e^{j\omega_1 W(t_1)}\} = \widehat{\mathscr{P}}_w(\omega_1 \mathbb{1}_{(0,t_1]})$$

$$= \exp\left(\int_{\mathbb{R}} f(\omega_1 \mathbb{1}_{(0,t_1]}(\tau)) \, d\tau\right).$$

Recalling that $f(0) = 0$, we then use the fact that $\mathbb{1}_{(0,t_1]}(t)$ is equal to 1 on its support of size t_1 and zero otherwise to simplify the above integral, which yields

$$\widehat{p}_{W(t_1)}(\omega_1) = e^{t_1 f(\omega_1)}. \tag{7.29}$$

This final result is the celebrated Lévy–Khintchine characterization of a Lévy process, which shows that the pdf of $W(t_1)$ is infinitely divisible and that W is non-stationary because the underlying Lévy exponent $t_1 f(\omega_1)$ is dependent upon the sampling instant t_1.

7.4.2   Higher-order extensions of Lévy processes

We saw in (7.20) that Lévy processes can be characterized as solutions of first-order SDEs involving a simple derivative operator. Naturally, it is of interest to extend this definition to unstable higher-order SDEs of the same form as (7.19) with K poles on the imaginary axis. To that end, we adopt the operator notation of Section 5.4.2 and an ordering of the poles where the purely imaginary ones come last with $\alpha_{N-K-k} = j\omega_k$, $1 \leq k \leq K$. We then factorize $p_N(\mathrm{D})$ into first-order terms, separating the unstable terms of the left-hand side, which allows us to rewrite (7.19) as

$$(\mathrm{P}_{\alpha_1} \cdots \mathrm{P}_{\alpha_{N-K}}) (\mathrm{P}_{j\omega_1} \cdots \mathrm{P}_{j\omega_K}) s_\alpha = q_M(\mathrm{D}) \, w,$$

where $\mathrm{P}_{\alpha_n} = \mathrm{D} - \alpha_n \mathrm{Id}$ and $q_M(\mathrm{D})$ is a polynomial in D of degree M. The above SDE can be solved by inverting each of the first-order operators on the left individually, in the manner described in Section 5.4.2. The formal solution is thus given by

$$s_\alpha = \underbrace{(\mathrm{I}_{\omega K} \cdots \mathrm{I}_{\omega_1})}_{\mathrm{I}_{(\omega K : \omega_1)}} \underbrace{(\mathrm{P}_{\alpha_{N-K}}^{-1} \cdots \mathrm{P}_{\alpha_1}^{-1})}_{\mathrm{P}_{(\alpha_1 : \alpha_{N-K})}^{-1}} q_M(\mathrm{D}) w$$

$$= \mathrm{I}_{(\omega K : \omega_1)} \mathrm{T}_{\mathrm{LSI}} w, \tag{7.30}$$

where the elementary inverse operators $\mathrm{P}_{\alpha_n}^{-1}$ and $\mathrm{I}_{\omega k}$ are defined by (5.10) and (5.19), respectively. Note that the proposed method of resolution involves the two composite

---

[2] This is equivalent to plugging $\varphi = \omega_1 \mathbb{1}_{(0,t_1]}$ into $\widehat{\mathscr{P}}_w(\varphi)$, which is legitimate since the latter is a continuous, positive definite functional over $\mathscr{R}(\mathbb{R})$ (see Proposition 8.1).

operators $I_{(\omega_K:\omega_1)}$ and $T_{LSI} = P^{-1}_{(\alpha_1:\alpha_{N-K})} q_M(D)$, the last of which is linear shift-invariant and $\mathcal{S}$-continuous. While the ordering of the factors of T_{LSI} is immaterial (due to the commutativity of convolution), this is not so for the Kth-order integration operator $I_{(\omega_K:\omega_1)} = I_{\omega_K} \cdots I_{\omega_1}$, which is shift-variant and directly responsible for the boundary conditions. Specifically, the prescribed ordering of the imaginary poles imposes the K linear boundary conditions at the origin

$$
\begin{cases}
s_\alpha(0) &= 0 \\
P_{j\omega_K} s_\alpha(0) &= 0 \\
&\vdots \\
P_{j\omega_2} \cdots P_{j\omega_K} s_\alpha(0) &= 0,
\end{cases}
\tag{7.31}
$$

which are part of the definition of the underlying generalized Lévy process s_α.

Finally, we invoke Corollary 5.5 and Theorem 4.17 to get the full characterization of the Nth-order generalized Lévy process in terms of its characteristic functional

$$
\widehat{\mathscr{P}}_{s_\alpha}(\varphi) = \exp\left(\int_{\mathbb{R}} f\left(T^*_{LSI} I^*_{(\omega_1:\omega_K)} \varphi(t)\right) dt \right),
\tag{7.32}
$$

where f is the Lévy–Schwartz exponent of the innovation, $I^*_{(\omega_1:\omega_K)}$ is the composite stabilized integration operator defined by (5.24) and (5.18), and T_{LSI} is the convolution operator corresponding to the stable part of the system with Fourier multiplier

$$
\widehat{T}_{LSI}(\omega) = \frac{q_M(j\omega)\prod_{k=1}^K(j\omega - j\omega_k)}{p_N(j\omega)} = \frac{q_M(j\omega)}{\prod_{n=1}^{N-K}(j\omega - \alpha_n)},
$$

where $Re(\alpha_n) \neq 0$ for $1 \leq n \leq N-K$. Clearly, the extended filtered-white-noise model (7.30) is compatible with the definition of CARMA processes of Section 7.3.4 if we simply set $K = 0$; it also yields the classical Lévy processes when $N = K = 1$ and $\omega_1 = 0$. In that respect, we observe that the derived process $P_{j\omega_1} \cdots P_{j\omega_K} s_\alpha = T_{LSI} w$ is stationary (by Proposition 7.2 and the right-inverse property of $I_{(\omega_K:\omega_1)}$), so that we may interpret K as the order of stationary deficiency.

7.4.3 Non-stationary Lévy correlations

In Section 7.3.1, we showed that there is a simple relation between the autocorrelation function of a second-order stationary process and the Green's function ρ_L of the whitening operator of L. We also derived a spline interpolation formula that connects the continuous- and discrete-domain versions of the autocorrelation function (see Proposition 7.3). In principle, it is possible to obtain similar results for the higher-order extensions of the Lévy processes from Section 7.4.2, but the correlation formulas are more involved and somewhat harder to get to. We shall illustrate the concept by considering the case of an Nth-order process with a single order of non-stationarity ($K = 1$ and $\alpha_N = j\omega_0$) subject to the boundary condition $s_\alpha(0) = 0$.

PROPOSITION 7.5 (Correlation structure of processes with one order of non-stationarity) *Let $s_\alpha(t)$ be an Nth-order generalized Lévy process with whitening* *operator* L *that meets the boundary condition* $s_\alpha(0) = 0$ *imposed by the presence of the pole* $j\omega_0$ *on the imaginary axis. The non-stationary correlation of the continuous and discrete-domain versions of the process is fully specified by*

$$R_{s_\alpha}(t, t') = \mathbb{E}\{s_\alpha(t)\overline{s_\alpha(t')}\} = v_{s_\alpha}(t'-t) - e^{j\omega_0 t}v_{s_\alpha}(t') - e^{-j\omega_0 t'}v_{s_\alpha}(-t)$$

$$R_{s_\alpha}[k, k'] = \mathbb{E}\{s_\alpha[k]\overline{s_\alpha[k']}\} = v_{s_\alpha}[k'-k] - e^{j\omega_0 k}v_{s_\alpha}[k'] - e^{-j\omega_0 k'}v_{s_\alpha}[-k],$$

where $v_s(t) = \sigma_w^2 \rho_{\overline{L}L^*}(t)$ *and where* $v_{s_\alpha}[k] = v_{s_\alpha}(t)|_{t=k}$. *These two entities are linked through the exponential-spline interpolation formula*

$$v_{s_\alpha}(t) = \sum_{k\in\mathbb{Z}} v_{s_\alpha}[k]\varphi_{\text{int}}(t-k),$$

where $\varphi_{\text{int}}(t)$ *is specified by (6.51). Moreover,* $v_{s_\alpha}(t) = O(|t|)$ *is a function of slow growth whose generalized Fourier transform is given by*

$$\widehat{v}_{s_\alpha}(\omega) = \frac{\sigma_w^2}{|\widehat{L}(-\omega)|^2} = V_{s_\alpha}(e^{j\omega})\widehat{\varphi}_{\text{int}}(\omega),$$

where $\widehat{L}(\omega)$ *is the frequency response of* L *which exhibits a zero at* $\alpha_N = j\omega_0$.

Proof From (7.8), we have that $R_{s_\alpha}(t_1, t_2) = \sigma_w^2 \overline{L^{-1}}L^{-1*}\{\delta(\cdot - t_1)\}(t_2)$. Since the system has a single singularity on the imaginary axis at $j\omega_0$, (7.32) implies that $L^{-1*} = T_{\text{LSI}}^* I_{\omega_0}^*$, where T_{LSI} is a BIBO-stable LSI system whose transfer function is $\widehat{T}_{\text{LSI}}(\omega) = \frac{j(\omega+\omega_0)}{\widehat{L}(\omega)}$. Using the Fourier-domain formula (5.21) of $I_{\omega_0}^*$ and the fact that T_{LSI} is a standard convolution operator, we find that the Fourier transform of $T_{\text{LSI}}^* I_{\omega_0}^*\{\delta(\cdot - t_1)\}$ is given by

$$\widehat{T}_{\text{LSI}}(-\omega)\frac{e^{-j\omega t_1} - e^{j\omega_0 t_1}}{-j(\omega+\omega_0)}.$$

Likewise, we have that $\overline{L^{-1}} = \overline{I_{\omega_0}}\overline{T}_{\text{LSI}}$, which, by considering the complex-conjugate counterpart of (5.20) for I_{ω_0}, yields

$$\overline{L^{-1}}L^{-1*}\{\delta(\cdot - t_1)\}(t) = \int_{\mathbb{R}} \left(\frac{e^{-j\omega t_1} - e^{j\omega_0 t_1}}{-j(\omega+\omega_0)}\right)|\widehat{T}_{\text{LSI}}(-\omega)|^2 \left(\frac{e^{j\omega t} - e^{-j\omega_0 t}}{j(\omega+\omega_0)}\right)\frac{d\omega}{2\pi}$$

$$= \int_{\mathbb{R}} |\widehat{T}_{\text{LSI}}(-\omega)|^2 \left(\frac{e^{j\omega(t-t_1)} - e^{j\omega_0 t_1}e^{j\omega t} - e^{-j\omega_0 t}e^{-j\omega t_1} + e^{-j\omega_0(t-t_1)}}{|\omega+\omega_0|^2}\right)\frac{d\omega}{2\pi} \quad (7.33)$$

$$= \rho_{\overline{L}L^*}(t-t_1) - e^{j\omega_0 t_1}\rho_{\overline{L}L^*}(t) - e^{-j\omega_0 t}\rho_{\overline{L}L^*}(-t_1).$$

The critical step in this derivation is the evaluation of the integral in (7.33) which, contrary to appearances, is non-singular, due to the presence of the fourth term in the numerator. To make this explicit, we recall that the proper specification of the inverse

Fourier transform of $1/|\widehat{L}(-\omega)|^2$, which has a second-order singularity at $\omega = -\omega_0$, involves a finite-part integral (see Appendix A) that can be resolved as follows:

$$
\begin{aligned}
\rho_{\overline{\text{LL}}^*}(t) &= \mathscr{F}^{-1}\left\{ \frac{1}{|\widehat{L}(-\omega)|^2} \right\}(t) \\
&= \text{p.f.} \int_{\mathbb{R}} |\widehat{T}_{\text{LSI}}(-\omega)|^2 \frac{e^{j\omega t}}{|\omega + \omega_0|^2} \frac{d\omega}{2\pi} \\
&= \int_{\mathbb{R}} |\widehat{T}_{\text{LSI}}(-\omega)|^2 \frac{e^{j\omega t} - e^{-j\omega_0 t}\left(1 + j(\omega + \omega_0)t\right)}{|\omega + \omega_0|^2} \frac{d\omega}{2\pi},
\end{aligned}
$$

where the numerator is corrected by subtracting the first two terms of the Taylor series of $e^{j\omega t}$ around $\omega = -\omega_0$.

This means that, in order to neutralize the singularity in the denominator of (7.33), we need to correct the three leading exponentials in the numerator by subtracting their first-order development around $\omega = -\omega_0$. This is possible by rewriting the numerator as

$$
\begin{aligned}
\text{Numerator} &= e^{j\omega(t-t_1)} - e^{-j\omega_0(t-t_1)}\left[1 + j(\omega + \omega_0)(t - t_1)\right] \\
&\quad - e^{j\omega_0 t_1}\left(e^{j\omega t} - e^{-j\omega_0 t}\left[1 + j(\omega + \omega_0)t\right]\right) \\
&\quad - e^{-j\omega_0 t}\left(e^{-j\omega t_1} - e^{j\omega_0 t_1}\left[1 - j(\omega + \omega_0)t_1\right]\right) \\
&= e^{j\omega(t-t_1)} - e^{j\omega_0 t_1}e^{j\omega t} - e^{-j\omega_0 t}e^{-j\omega t_1} + e^{-j\omega_0(t-t_1)},
\end{aligned}
$$

which shows that the terms that are required to make the inverse-Fourier integral non-singular precisely add up to $e^{-j\omega_0(t-t_1)}$. $\qquad\square$

In particular, for $N = 1$ and $\omega_0 = 0$, we find that

$$
R_{\text{Wiener}}(t, t') = \frac{\sigma_w^2}{2}\left(|t' - t| - |t'| - |t|\right),
$$

which is the well-known autocorrelation function of Brownian motion (a.k.a. Wiener process).

7.4.4 Removal of long-range dependencies

The characteristic functional (7.32) provides a complete description of the Lévy processes and their extensions. While the formula is elegant conceptually, it is not quite as favorable for the derivation of the joint statistics of these processes. Even in the simplest case of correlations, the calculations can get quite involved, as we just saw in Section 7.4.3. The source of the difficulty is the non-stationary nature of the operator $\text{I}^*_{(\omega_1 : \omega_K)}$. Another confounding factor is the induction of long-range dependencies due to lack of decay of the Green's function ρ_{L} in the non-stable scenario. Fortunately, these dependencies can be suppressed, for the most part, by applying the decoupling procedure outlined in Proposition 7.4. The practical benefit is that this produces an equivalent signal – the generalized-increment process – that is stationary and maximally decoupled, which greatly simplifies the statistical analysis.

For an Nth-order generalized Lévy process, it is possible to characterize all relevant quantities explicitly by taking advantage of the functional link with exponential B-splines. Specifically, given the poles $\boldsymbol{\alpha} \in \mathbb{C}^N$ of the system, one defines the following spline-related entities:

(1) The (causal) exponential B-spline with parameter $\boldsymbol{\alpha}$ (poles)

$$\beta_{\boldsymbol{\alpha}}(t) = \int_{\mathbb{R}} e^{j\omega t} \prod_{n=1}^{N} \left(\frac{1 - e^{\alpha_n - j\omega}}{j\omega - \alpha_n} \right) \frac{d\omega}{2\pi} \tag{7.34}$$

(2) The Nth-order (causal) finite-difference (or localization) operator

$$\Delta_{\boldsymbol{\alpha}} \varphi(t) = \Delta_{\alpha_1} \cdots \Delta_{\alpha_N} \varphi(t), \tag{7.35}$$

which is obtained by cascading the first-order operators $\Delta_{\alpha_n} \varphi(t) = \varphi(t) - e^{\alpha_n} \varphi(t-1)$ with $n = 1, \ldots, N$.

The important point for our argument is that the exponential B-spline $\beta_{\boldsymbol{\alpha}}(t)$ is causal with a compact support of size N and differentiable up to order $(N - 1)$ (see Section 6.4.4).

We recall that the generalized Lévy process s_{α} satisfies the differential equation (5.6), which is associated with the whitening operator L whose frequency response in the sense of generalized functions is

$$\widehat{L}(\omega) = \frac{p_N(j\omega)}{q_M(j\omega)} = \frac{\prod_{n=1}^{N} (j\omega - \alpha_n)}{b_M \prod_{m=1}^{M} (j\omega - \gamma_m)}.$$

This suggests considering the generalized exponential B-spline

$$\beta_L(t) = \Delta_{\boldsymbol{\alpha}} \rho_L(t)$$
$$= q_M(\mathrm{D}) \beta_{\boldsymbol{\alpha}}(t), \tag{7.36}$$

where $\rho_L = \mathscr{F}^{-1} \{ 1/\widehat{L}(\omega) \}$ is the Green's function of L. The crucial property here is that $\beta_L \in \mathscr{R}(\mathbb{R})$ with the shortest possible support. Indeed, β_L has the same (minimal) support as $\beta_{\boldsymbol{\alpha}}$ and is bounded (because $q_M(\mathrm{D})$ is a differential operator of order $M < N$), independently of the location of the poles in the complex plane. Our next proposition ensures that the B-spline in (7.36) and the finite-difference operator $\Delta_{\boldsymbol{\alpha}}$ are the appropriate ingredients for decoupling generalized Lévy processes.

PROPOSITION 7.6 (Approximate inversion by finite differences)   *Let* L^{-1} *be the Nth-order inverse operator defined by*

$$\mathrm{L}^{-1} = \mathrm{I}_{\omega_K} \cdots \mathrm{I}_{\omega_1} \mathrm{P}_{\alpha_{N-K}}^{-1} \cdots \mathrm{P}_{\alpha_1}^{-1} q_M(\mathrm{D}),$$

with $\mathrm{Re}(\alpha_n) \neq 0$ *for* $1 \leq n \leq N - K$ *and* $\alpha_{K+k} = j\omega_k$ *for* $1 \leq k \leq K$, *where* I_{ω_k} *and* $\mathrm{P}_{\alpha_n}^{-1}$ *are specified by (5.19) and (5.10), respectively. Then, the generalized B-spline* β_L *defined by (7.36) and (7.34) has the following properties:*

$$\Delta_{\boldsymbol{\alpha}} \mathrm{L}^{-1} \varphi = \beta_L * \varphi$$
$$\mathrm{L}^{-1*} \Delta_{\boldsymbol{\alpha}}^* \varphi = \beta_L^{\vee} * \varphi$$

for all $\varphi \in \mathscr{S}(\mathbb{R})$.

Since β_L is maximally localized and close to an impulse, the interpretation is that Δ_α yields an approximate inverse of L^{-1}. In other words, Δ_α acts as a discrete proxy for the continuous-domain whitening operator L.

Proof The key is to rely on the factorization property (6.42) of the exponential B-spline β_α and to consider the elementary factors one at a time. To that end, we first establish that

$$\Delta_\alpha P_\alpha^{-1} \varphi = \beta_\alpha * \varphi \tag{7.37}$$

$$\Delta_{j\omega_k} I_{\omega_k} \varphi = \beta_{j\omega_k} * \varphi, \tag{7.38}$$

as well as the adjoint counterparts of these relations. The first identity is a direct consequence of the definition of the first-order exponential B-spline

$$\beta_\alpha(t) = \Delta_\alpha \rho_\alpha(t) = \mathscr{F}^{-1} \left\{ \frac{1 - e^{\alpha - j\omega}}{j\omega - \alpha} \right\}(t),$$

where $\rho_\alpha = \mathscr{F}^{-1}\{1/(j\omega - \alpha)\}$ is the Green's function of the operator P_α or, equivalently, the impulse response of the inverse operator P_α^{-1}. As for the second identity, we apply the time-domain definition (5.19) of I_{ω_k}, which yields

$$\Delta_{j\omega_k} I_{\omega_k} \varphi = \Delta_{j\omega_k}\{\rho_{j\omega_k} * \varphi\} - (\rho_{j\omega_k} * \varphi)(0) \underbrace{\Delta_{j\omega_k}\{e^{j\omega_k r}\}}_{=0}$$

$$= \Delta_{j\omega_k}\{\rho_{j\omega_k}\} * \varphi$$

$$= \beta_{j\omega_k} * \varphi,$$

where we have used the fact that $\Delta_{j\omega_n}$ annihilates the sinusoidal components that are in the null space of $P_{j\omega_k}$ and the associativity of convolution operators such as $\Delta_{j\omega_k}$. By applying (7.37) and (7.38) recursively and making use of the commutativity of $\mathscr{S}$-continuous LSI operators, we find that

$$\Delta_\alpha L^{-1} \varphi = \Delta_{(\alpha_1 : \alpha_{N-1})} \left(\Delta_{j\omega_K} I_{\omega_K} \right) I_{(\omega_{K-1} : \omega_1)} P_{\alpha_{N-K}}^{-1} \cdots P_{\alpha_1}^{-1} q_M(D)\{\varphi\}$$

$$= \beta_{j\omega_K} * \Delta_{(\alpha_1 : \alpha_{N-2})} \left(\Delta_{j\omega_{K-1}} I_{\omega_{K-1}} \right) I_{(\omega_{K-2} : \omega_1)} P_{\alpha_{N-K}}^{-1} \cdots P_{\alpha_1}^{-1} q_M(D)\{\varphi\}$$

$$\vdots$$

$$= \beta_{j\omega_K} * \cdots * \beta_{j\omega_1} * \beta_{\alpha_{N-K}} * \cdots * \beta_{\alpha_1} * q_M(D)\{\varphi\}$$

$$= q_M(D)\{\beta_{\alpha_1} * \cdots * \beta_{\alpha_N}\} * \varphi$$

$$= q_M(D)\{\beta_\alpha\} * \varphi$$

$$= \beta_L * \varphi.$$

The adjoint counterpart of this identity is obtained by applying the same divide-and-conquer strategy. $\square$

The idea is now to apply the localization operator Δ_α to s_α in order to partially decouple the process and, more importantly, to suppress its non-stationary components. This yields the *generalized-increment process* $u_\alpha = \Delta_\alpha s_\alpha$, which has a much simpler statistical structure than s_α. Indeed, by combining the result of Proposition 7.6 with (7.30), we show that

$$\langle \varphi, u_\alpha \rangle = \langle \varphi, \Delta_\alpha s_\alpha \rangle$$
$$= \langle \Delta_\alpha^* \varphi, L^{-1} w \rangle$$
$$= \langle L^{-1*} \Delta_\alpha^* \varphi, w \rangle = \langle \beta_L^\vee * \varphi, w \rangle$$

for all $\varphi \in \mathcal{S}(\mathbb{R})$. This is equivalent to

$$u_\alpha = \Delta_\alpha s_\alpha = \beta_L * w, \tag{7.39}$$

which is a form that is much more convenient than $s_\alpha = L^{-1} w$ because the convolution with β_L preserves stationarity. The other favorable aspect is that the generalized B-spline β_L (which has a compact support) is much better localized than the Green's function ρ_L, especially in the non-stable scenario where ρ_L exhibits polynomial growth. We are now in the position to invoke Proposition 7.4 to get the complete statistical characterization of u_α, including its correlation function which is proportional to $\beta_{\overline{L}L*}(t) = (\beta_L * \overline{\beta}_L^\vee)(t)$, where β_L is the generalized exponential B-spline defined by (7.36).

7.4.5 Examples of sparse processes

Examples of realizations of Gaussian vs. sparse stochastic processes are shown in Figures 7.2 to 7.5. These signals were generated using three types of driving

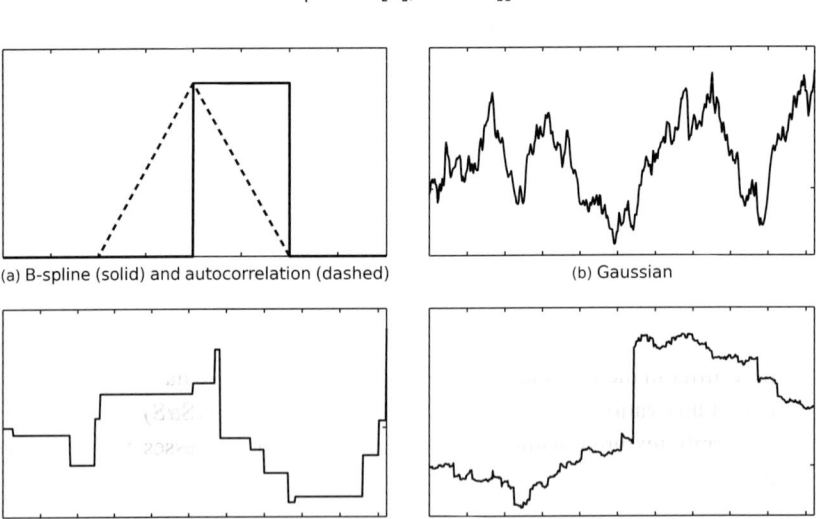

poles = [0], zeros = []

(a) B-spline (solid) and autocorrelation (dashed)

(b) Gaussian

(c) Poisson ($\lambda = 1/32$, Gaussian amplitudes)

(d) SαS ($\alpha = 1.2$)

Figure 7.2 Example 1: generation of generalized stochastic processes with whitening operator $L = D$ (pole vector $\alpha = (0)$). (a) B-spline functions $\beta_L(t) = \text{rect}(t - \frac{1}{2})$ and $\beta_{\overline{L}L*}(t) = \text{tri}(t)$. (b) Brownian motion. (c) Compound-Poisson process with $\lambda = 1/32$ and Gaussian amplitude distribution $p_A(a) = (2\pi)^{-1/2} e^{-a^2/2}$. (d) S$\alpha$S Lévy motion with $\alpha = 1.2$.

poles = [0, 0], zeros = []

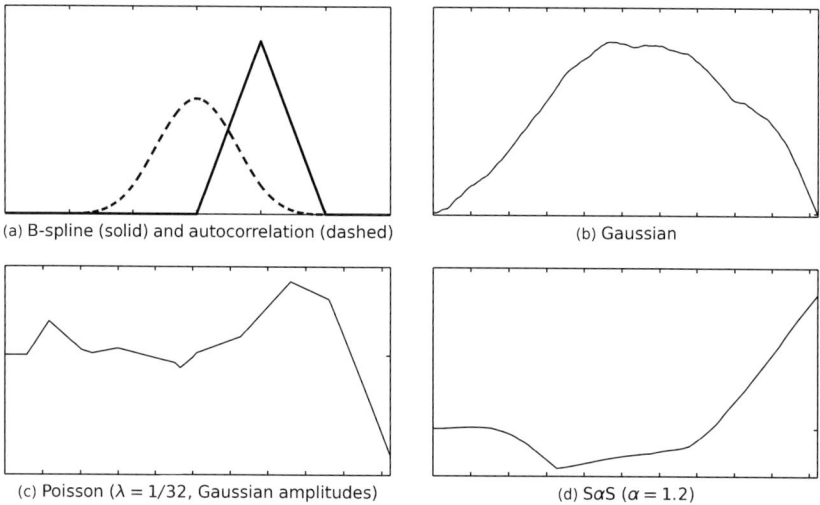

(a) B-spline (solid) and autocorrelation (dashed)   (b) Gaussian

(c) Poisson ($\lambda = 1/32$, Gaussian amplitudes)   (d) SαS ($\alpha = 1.2$)

Figure 7.3 Example 2: generation of generalized stochastic processes with whitening operator $L = D^2$ $\left(\text{pole vector } \boldsymbol{\alpha} = (0, 0)\right)$. (a) B-spline functions $\beta_L(t) = \text{tri}(t)$ and $\beta_{\overline{\text{L}}\text{L}*}(t)$ (cubic B-spline). (b) Gaussian process. (c) Generalized Poisson process with $\lambda = 1/32$ and Gaussian amplitude distribution $p_A(a) = (2\pi)^{-1/2}e^{-a^2/2}$. (d) Generalized S$\alpha$S process with $\alpha = 1.2$.

innovations: Gaussian (panel b), impulsive Poisson (panel c), and symmetric-alpha-stable (SαS) with $\alpha = 1.2$ (panel d).

The relevant operators are:

- Example 1: $L = D$ (Lévy process)
- Example 2: $L = D^2$ (second-order extension of Lévy process)
- Example 3: $L = (D - \alpha_1 \text{Id})(D - \alpha_2 \text{Id})$ and $\boldsymbol{\alpha} = (j3\pi/4, -j3\pi/4)$ (generalized Lévy process)
- Example 4: $L = (D - \alpha_1 \text{Id})(D - \alpha_2 \text{Id})$ and $\boldsymbol{\alpha} = (-0.05 + j\pi/2, -0.05 - j\pi/2)$ (CAR(2) process).

The corresponding B-splines (β_L and $\beta_{\overline{\text{L}}\text{L}*}$) are shown in the upper-left panel of each figure.

The signals that are displayed side-by-side share the same whitening operator, but they differ in their sparsity patterns, which come in three flavors: none (Gaussian), finite rate of innovation (Poisson), and heavy-tailed statistics (SαS). The Gaussian signals are uniformly textured, while the generalized Poisson processes are piecewise-smooth by construction.

Lowpass processes

The classical Lévy processes (Figure 7.2) are obtained by integration of white Lévy noise. They go hand-in-hand with the B-spline of degree zero (rect) and its autocorrelation (triangle function) which is a B-spline degree 1. The Gaussian version (Figure 7.2b) is a Brownian motion. It is quite rough and nowhere differentiable in the classical sense.

poles = [j3π/4, −j3π/4], zeros = []

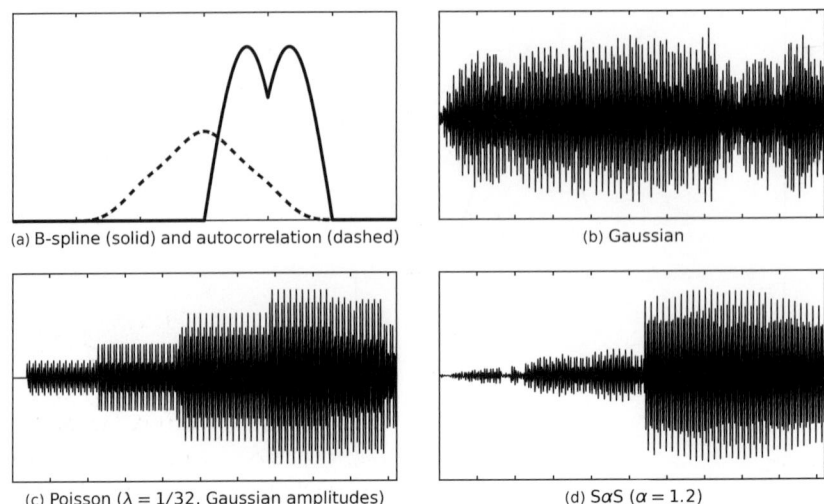

(a) B-spline (solid) and autocorrelation (dashed)

(b) Gaussian

(c) Poisson (λ = 1/32, Gaussian amplitudes)

(d) SαS (α = 1.2)

Figure 7.4 Example 3: generation of generalized stochastic processes with operator
$L = (D − α_1 \text{Id})(D − α_2 \text{Id})$ and $α = (j3π/4, −j3π/4)$. (a) B-spline functions $β_L$ and $β_{\overline{L}L*}$.
(b) Gaussian process. (c) Generalized Poisson process with $λ = 1/32$ and Gaussian amplitude
distribution. (d) Generalized SαS process with $α = 1.2$.

Yet, it is mean-square continuous due to the presence of the single pole at the origin.
The Poisson version (compound-Poisson process) is piecewise-constant, each jump corresponding to the occurrence of a Dirac impulse. The SαS Lévy motion exhibits local
fluctuations punctuated by large (but rare) jumps, as is characteristic for this type of
process [ST94, App09]. Overall, it is the jump behavior that dominates, making it even
sparser than its Poisson counterpart.

The example in Figure 7.3 (second-order extension of a Lévy process) corresponds to
one additional level of integration, which yields smoother signals (i.e., one-time differentiable in the classical sense). The corresponding Poisson process is piecewise-linear,
while the SαS version looks globally smoother than the Gaussian one, except for a few
sharp discontinuities in its slope. The basic B-spline here is a triangle, while $β_{\overline{L}L*}$ is
a cubic B-spline. The signals in Figures 7.2 and 7.3 are non-stationary; the underlying
processes have the remarkable property of being self-similar (fractals) due to the scale-
invariance of the pure-derivative operators. The Gaussian and SαS processes are strictly
self-similar in the sense that the statistics are preserved through rescaling. By contrast,
the scaling of the Poisson processes necessitates some corresponding adjustment of the
rate parameter $λ$ [UT11].

Bandpass processes

The second-order signals in Figure 7.4 are non-stationary as well, but no longer lowpass nor self-similar. They are real-valued, and C^1-continuous almost everywhere (pair
of complex-conjugate poles in the left complex plane). They constitute some kind

poles = $[-.05 + j\pi/2, -.05 - j\pi/2]$, zeros = $[\,]$

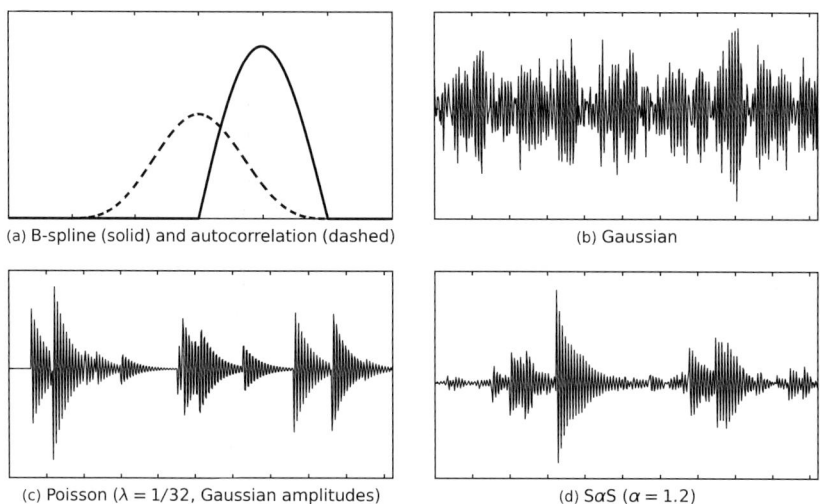

(a) B-spline (solid) and autocorrelation (dashed)

(b) Gaussian

(c) Poisson ($\lambda = 1/32$, Gaussian amplitudes)

(d) SαS ($\alpha = 1.2$)

Figure 7.5 Example 4: generation of generalized stochastic processes with whitening operator $L = (D - \alpha_1 \mathrm{Id})(D - \alpha_2 \mathrm{Id})$ and $\boldsymbol{\alpha} = (-0.05 + j\pi/2, -0.05 - j\pi/2)$. (a) B-spline functions β_L and $\beta_{\overline{L}*}$. (b) Gaussian AR(2) process. (c) Generalized Poisson process with $\lambda = 1/32$ and Gaussian amplitude distribution. (d) SαS AR(2) process with $\alpha = 1.2$.

of modulated (or bandpass) counterpart of the Lévy processes, which appears to be much better suited for the modeling of acoustic signals. As in the other examples, the Gaussian version is looking cluttered. The Poisson signal is somewhat stereotyped (stretches of pure oscillating regime) and not quite as realistic-looking as its SαS counterpart.

As soon as the poles are moved away from the imaginary axis, the processes become stationary. This is illustrated in Figure 7.5 with some CAR(2) (continuous autoregressive) examples, the non-Gaussian versions of which have a marked tendency to exhibit characteristic bursts associated with the impulse response of the system. These latter processes are part of the stationary CARMA family characterized in Section 7.3.4.

7.4.6 Mixed processes

One can also construct signals with a more intricate structure by simple addition of independent elementary processes. This results in the mixed process $s_{\mathrm{mix}} = s_1 + \cdots + s_M$ whose characteristic functional is the product of the characteristic functionals of the individual constituents. It is given by

$$\widehat{\mathscr{P}}_{s_{\mathrm{mix}}}(\varphi) = \prod_{m=1}^{M} \widehat{\mathscr{P}}_{s_m}(\varphi) = \exp\left(\int_{\mathbb{R}} \sum_{m=1}^{M} f_m\big(L_m^{-1*}\varphi(t)\big)\,dt\right),$$

where s_m is some elementary process with whitening operator L_m and Lévy function $f_m(\omega)$. As a demonstration of the concept, we have synthesized some acoustic samples by mixing random signals associated with elementary musical notes (pair of poles at the corresponding frequency). These can be downloaded from http://www.sparseprocesses.org/. The Gaussian versions are diffuse, cluttered, and boring to listen to. The generalized Poisson and SαS samples are more interesting perceptually – reminiscent of chimes – with the latter sounding less dry and more realistic. Note that mixing does not gain us anything in the Gaussian case because the resulting signal is still part of the traditional family of Gaussian ARMA processes (this follows from Parseval's relation and the fact that $\sum_{m=1}^{M} \frac{\sigma_w^2}{|\widehat{L}_m(-\omega)|^2}$ is expressible as an equivalent rational power spectrum). This is not so for the non-Gaussian members of the family, which are not decomposable in general, meaning that the mixing of sparse processes opens up new modeling perspectives. Interestingly, the Gaussian acoustic samples are almost impossible to compress using mp3/AAC, while the generalized Poisson and SαS ones can be faithfully reproduced at a much lower bit rate.

7.5   Self-similar processes

In order to construct self-similar stochastic processes in our framework, we seek operators L that are scale-invariant, and which have scale-invariant inverses, in the sense of Definition 5.2. This narrows down the class of suitable operators to the fractional derivatives ∂_τ^γ in 1-D and the fractional Laplacians $(-\Delta)^{\frac{\gamma}{2}}$ in multiple dimensions if rotation-invariance is required as well. This suggests that self-similar processes can be specified as solutions of fractional SDEs. Once again, the definition of these processes requires the use of regularized bounded inverses which, in this case, were introduced in Section 5.5. A difference with the Lévy processes is that, for fractional-order operators, the definition of the inverse depends not only on the forward operator but also on the innovation – more precisely, on the domain of the characteristic functional of the innovation.

An alternative non-isotropic multidimensional solution is a separable operator of the form $L = \partial_{r_1,\tau_1}^\gamma \cdots \partial_{r_d,\tau_d}^\gamma$, where $\partial_{r_n,\tau}^\gamma$ denotes the (γ, τ)-derivative along the coordinate axis r_n (see Section 5.5.1). The simplest example is the partial derivative operator $L = \partial_{r_1} \cdots \partial_{r_d}$, which results in the definition of the Mondrian process (see Section 7.5.3).

We now describe in more detail the first class of self-similar and isotropic processes and fields. The generation of these processes, as suggested above, is achieved by applying the L_p-continuous inverses defined in Section 5.5.3 to self-similar and rotation-invariant innovations.

For the processes that are defined using these operators to be self-similar in the strict sense (see Definition 7.4), the innovation to which the operator is applied needs to be self-similar as well, in the sense that $\langle \varphi, w \rangle$ and $c \langle \varphi, w(\frac{\cdot}{a}) \rangle$ have the same statistics for some appropriate normalization factor c. This narrows down the class of innovations to α-stable noises. More generally, however, we may apply a scale-invariant operator to a Lévy noise in order to obtain a process/field that is wide-sense self-similar (see Definition 7.6).

In Sections 7.5.1 and 7.5.2, we review genuine self-similar models[3] that result from the application of scale-invariant operators to SαS innovations, and then devote Section 7.5.3 to the case of Poisson innovations, which yield wide-sense self-similar models.

7.5.1 Stable fractal processes

The class of self-similar and rotation-invariant innovations used in our definition consists of symmetric-alpha-stable white noises defined (in one and several dimensions) by the characteristic functional

$$\widehat{\mathscr{P}}_{w_{\alpha,s_0}}(\varphi) = e^{-\|s_0\varphi\|_\alpha^\alpha} \tag{7.40}$$

with $\alpha \in (0,2]$, where $\|\cdot\|_\alpha^\alpha$ denotes the αth power of the L_α (quasi-)norm and s_0 is an arbitrary positive normalization constant. Note that the largest domain of definition of $\widehat{\mathscr{P}}_{\alpha,s_0}$ on which it remains finite is the Lebesgue space $L_\alpha(\mathbb{R}^d)$. The characteristic functional $\widehat{\mathscr{P}}_{w_{\alpha,s_0}}$ is also continuous with respect to the L_α topology (and, a fortiori, with respect to any finer topology such as those of $\mathscr{D}, \mathscr{S} \subset L_\alpha$).

The innovation w_{α,s_0} defined by (7.40) is stationary, isotropic, and self-similar with scaling index $\left(\frac{d}{\alpha} - d\right)$ in the following sense:

$$\langle \varphi, w_{\alpha,s_0}(\mathbf{R}^\mathsf{T}\cdot)\rangle = \langle \varphi(\mathbf{R}\cdot), w_{\alpha,s_0}\rangle$$

$$= \langle \varphi, w_{\alpha,s_0}\rangle \qquad \text{in probability law (rotation-invariance)}$$

$$\langle \varphi, w_{\alpha,s_0}(a^{-1}\cdot)\rangle = a^d \langle \varphi(a\cdot), w_{\alpha,s_0}\rangle$$

$$= a^{d-\frac{d}{\alpha}} \langle \varphi, w_{\alpha,s_0}\rangle. \qquad \text{in probability law (self-similarity).}$$

This, in combination with the rotation-invariance and scale-invariance of degree $(-\gamma)$ of the inverse Laplacian operator $\mathrm{L}_\alpha^{-\gamma*}$ introduced in Theorem 5.11, shows that the generalized random process/field

$$s_{\alpha,H} = \mathrm{L}_\alpha^{-H-d+\frac{d}{\alpha}} w_{\alpha,s_0}$$

defined by the characteristic functional

$$\widehat{\mathscr{P}}_{s_{\alpha,H}}(\varphi) = \widehat{\mathscr{P}}_{w_{\alpha,s_0}}(\mathrm{L}_\alpha^{-H-d+\frac{d}{\alpha}*}\varphi) \tag{7.41}$$

is isotropic and self-similar with Hurst exponent H for $H \neq 0,1,2,\dots$ (see Theorem 7.1, Statement (4) with $\gamma = H + d - d/\alpha$). Finally, we note that the existence of these processes is guaranteed since the operator $\mathrm{L}_\alpha^{-\gamma*}$ used above correctly maps $\mathscr{S}(\mathbb{R}^d)$ into $L_\alpha(\mathbb{R}^d)$, per Theorem 5.11.

The random processes/fields $s_{\alpha,H}$ are the so-called *fractional stable motions,* which generalize fractional *Brownian* motions corresponding to the Gaussian case ($\alpha = 2$). These processes have been widely applied in stochastic modeling, especially in their

---

[3] These models are studied in more detail in the thesis of P. D. Tafti [Taf11], which is entirely devoted to the study of self-similar fields, with emphasis on vectorial extensions (flow fields).

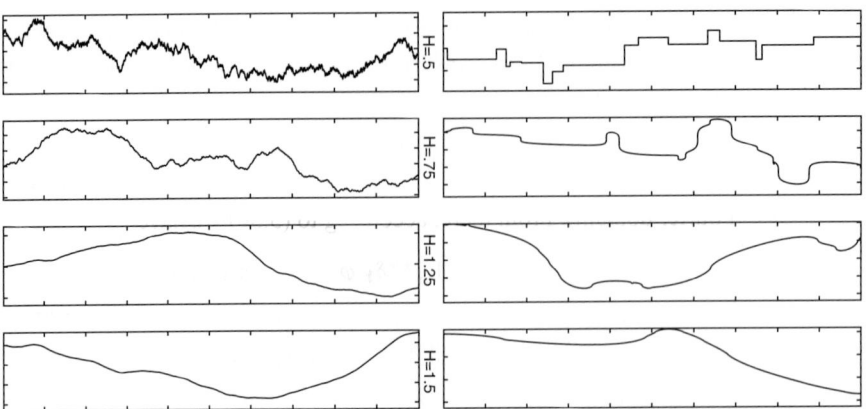

Figure 7.6 Gaussian vs. sparse fractal-like processes: comparison of fractional Brownian motion (left column) and generalized Poisson (right column) stochastic processes as the order increases. The processes that are side-by-side have the same order $\gamma = H + \frac{1}{2}$ and identical second-order statistics.

early Gaussian variety known as *fractional Brownian motion*, as discussed by Mandelbrot and Van Ness [MVN68] (some examples are shown in the left column of Figure 7.6). They are remarkable due to their fractal nature and long-range dependencies. Their most significant statistical properties are listed below.

(1) *Self-similarity.* The process $s_{\alpha,H}$ is equivalent in probability law to its scaled version $a^H s_{\alpha,H}(\frac{\cdot}{a})$. The property is established by showing that $s_{\alpha,H}$ and $a^H s_{\alpha,H}(\frac{\cdot}{a})$ have the same characteristic functional (see proof of Statement (4) in Theorem 7.1). It also justifies its parameterization by H, the *Hurst* or *self-similarity exponent* of the process.

(2) *Isotropy.* The process $s_{\alpha,H}$ is equivalent in probability law to $s_{\alpha,H}(\mathbf{R}^{\mathsf{T}}\cdot)$, where $\mathbf{R}$ is an arbitrary orthonormal matrix in $\mathbb{R}^{d\times d}$. The proof is similar to that of the previous property.

(3) *Non-stationarity and stationary* $(n + 1)th$ *increments.* The process $s_{\alpha,H}$ is non-stationary since it is the result of applying the non-shift-invariant operator $\mathrm{L}_\alpha^{-\gamma}$ with $\gamma = H + d - d/\alpha$ to a stationary white-noise process. However, despite its non-stationarity, any finite increment of $s_{\alpha,H}$ of order $\lfloor H \rfloor + 1$, as defined in the following theorem, is stationary.

THEOREM 7.7   *Let* $Y = \{y_0, \ldots, y_n\}$ *be a set of* $n + 1$ *vectors in* $\mathbb{R}^d$ *and denote by* δ_Y *the finite-difference operator defined as the composition of the operators*

$$\delta_{y_i} : f \mapsto f - f(\cdot - y_i).$$

Then, for $\lfloor H \rfloor \leq n$, *the random process/field*

$$\delta_Y^* s_{\alpha,H}$$

is stationary.

Proof We first observe that the $(n+1)$th-order finite-difference operator δ_Y is shift-invariant and annihilates the moments of a function φ up to degree n, so that

$$\int_{\mathbb{R}^d} y^k \delta_Y \varphi(y) \, dy = 0$$

for $|k| \le n$ (this is proved easily by induction on n or by differentiation in the Fourier domain). From there, according to (5.36) we have

$$L_\alpha^{-\gamma *} \delta_Y \varphi = \rho^{\gamma-d} * (\delta_Y \varphi)$$

because the correction term for $\delta_Y \varphi$ that normally distinguishes $L_\alpha^{-\gamma *}$ from the shift-invariant inverse is zero in this case. Consequently,

$$
\begin{aligned}
\langle \varphi(\cdot - h), \delta_Y^* s_{\alpha,H} \rangle &= \langle \delta_Y \varphi(\cdot - h), s_{\alpha,H} \rangle \\
&= \langle L_\alpha^{-H-d+\frac{d}{\alpha}*} \delta_Y \varphi(\cdot - h), w_\alpha \rangle \\
&= \langle L_\alpha^{-H-d+\frac{d}{\alpha}*} \delta_Y \varphi, w_\alpha(\cdot + h) \rangle \qquad \text{by shift-invariance} \\
&= \langle L_\alpha^{-H-d+\frac{d}{\alpha}*} \delta_Y \varphi, w_\alpha \rangle \qquad \text{by the stationarity of } w_\alpha \\
&= \langle \varphi, \delta_Y^* s_{\alpha,H} \rangle, \qquad \text{in probability law}
\end{aligned}
$$

which proves that $\delta_Y^* s_{\alpha,H}$ is stationary. $\qquad \square$

(4) *Variogram and covariance in the finite-variance case (with $0 < H < 1$).* For $\alpha < 2$, fractional stable motions have infinite variance. But for $\alpha = 2$, the covariance structure of fractional Brownian fields can be derived from the characteristic functional of its $(n + 1)$th-order increments. Here, we show this derivation for $0 < H < 1$.

PROPOSITION 7.8 (Self-similar variograms) *The variogram and the covariance function of a fractional Brownian field with $0 < H < 1$ are given by*

$$2\gamma_{s_{2,H}}(r, s) = \mathbb{E}\{|s_{2,H}(r) - s_{2,H}(s)|^2\} \propto 2\rho^{2H}(r - s)$$

and

$$R_{s_{2,H}}(r, s) = \mathbb{E}\{s_{2,H}(r)s_{s,H}(s)\} \propto 2\rho^{2H}(r) - 2\rho^{2H}(r - s) + 2\rho^{2H}(s),$$

respectively, where ρ^γ is the γ-homogeneous distribution defined in Theorem 5.10.

Proof Let us fix r and s and temporarily denote by u the increment process

$$u(h) = s_{2,H}(r + h) - s_{2,H}(s + h) = \delta_{r-s} s_{2,H}(r + h).$$

Then, the variogram of $s_{2,H}$ corresponds to the variance of u at $\mathbf{0}$. To compute it, we first observe that

$$
\begin{aligned}
\langle \varphi, \delta_{r-s} s_{2,H}(\cdot + r) \rangle &= \langle \varphi, \delta_{r-s} L_2^{-H-d/2} w(\cdot + r) \rangle \\
&= \langle L_2^{-H-d/2*} \{\delta_{r-s}^* \varphi(\cdot + r)\}, w \rangle \\
&= \langle \rho^{H-d/2} * \varphi(\cdot + r) - \rho^{H-d/2} * \varphi(\cdot + s), w \rangle \\
&= \langle (\rho^{H-d/2}(\cdot + r) - \rho^{H-d/2}(\cdot + s)) * \varphi, w \rangle.
\end{aligned}
$$

This shows that, for fixed r and s, u is a filtered Gaussian white noise with generalized covariance function

$$\left(\rho^{H-d/2}(\cdot+r)-\rho^{H-d/2}(\cdot+s)\right)^{\vee}*\left(\rho^{H-d/2}(\cdot+r)-\rho^{H-d/2}(\cdot+s)\right) \propto 2\rho^{2H}(r-s)-2\rho^{2H}(\cdot),$$

using the symmetry and convolution properties of $\rho^{\vee}$. In particular, for the variance at $\mathbf{0}$ of u (which is the same as the variance everywhere, due to stationarity), we have

$$2\gamma_{s_{2,H}}(r,s) = \mathbb{E}\{|s_{2,H}(r)-s_{2,H}(s)|^2\} \propto 2\rho^{2H}(r-s).$$

By developing the above result, we then find the generalized covariance

$$R_{s_{2,H}}(r,s) = \mathbb{E}\{s_{2,H}(r)s_{s,H}(s)\} \propto 2\rho^{2H}(r)-2\rho^{2H}(r-s)+2\rho^{2H}(s).$$

$\square$

The result of Proposition 7.8 generalizes to all wide-sense self-similar processes obtained by replacing the α-stable innovation in (7.41) with a general finite-variance stationary innovation.

7.5.2   Fractional Brownian motion through the looking-glass

To demonstrate the analytical power of the formalism, we now consider the special case $d=1$, $\alpha=2$, and $H \in (0,1)$. The corresponding self-similar Gaussian process is fractional Brownian motion (fBm), commonly denoted by B_H. The evaluation of (7.41) then yields

$$\mathbb{E}\{e^{j\langle\varphi,B_H\rangle}\} = \widehat{\mathscr{P}}_{B_H}(\varphi) = \exp\left(-\frac{1}{2}\int_{\mathbb{R}}\left|\frac{\widehat{\varphi}(\omega)-\widehat{\varphi}(0)}{(-j\omega)^\gamma}\right|^2\frac{d\omega}{2\pi}\right) \tag{7.42}$$

with $\gamma = H+\frac{1}{2}$, where we have applied (5.38) and used Parseval's relation to rewrite $\|L_2^{-\gamma*}\varphi\|_2^2$ in the Fourier domain. It is important to understand that (7.42) completely characterizes fBm. While there are several equivalent ways of writing the denominator in the Fourier-domain integral, we have chosen the form $|\omega|^{2\gamma} = |(-j\omega)^\gamma|^2$. In terms of operators, this translates into

$$\widehat{\mathscr{P}}_{B_H}(\varphi) = \exp\left(-\frac{1}{2}\|I_{0,2}^{\gamma*}\{\varphi\}\|_{L_2}\right),$$

where $I_{0,2}^{\gamma*}$ is the canonical left inverse of the fractional-derivative operator $D^{\gamma*}$ (see definitions in Table 7.1). This, in turn, implies that fBm is the solution of the fractional stochastic differential equation

$$D^\gamma B_H = w, \tag{7.43}$$

where w is a normalized Gaussian white noise.

By writing the explicit form of $I_{0,2}^{\gamma*}$ with $\gamma \in (0.5, 1.5)$, we obtain

$$I_{0,2}^{\gamma*}\{\varphi\}(\tau) = \int_{\mathbb{R}}\frac{\widehat{\varphi}(\omega)-\widehat{\varphi}(0)}{(-j\omega)^\gamma}e^{j\omega\tau}\frac{d\omega}{2\pi} \tag{7.44}$$

$$= (D^{-\gamma+1})^*I_0^*\{\varphi\}(\tau), \tag{7.45}$$

Table 7.1 Causal scale-invariant operators and their adjoint.

Definition	Properties
Fractional derivatives of order $\gamma \in (-1, +\infty)$	
$D^{\gamma}\{\varphi\}(\tau) = \int_{\mathbb{R}} \widehat{\varphi}(\omega)(j\omega)^{\gamma} e^{j\omega\tau} \dfrac{d\omega}{2\pi}$	Causal, LSI
$D^{\gamma*}\{\varphi\}(\tau) = \int_{\mathbb{R}} \widehat{\varphi}(\omega)(-j\omega)^{\gamma} e^{j\omega\tau} \dfrac{d\omega}{2\pi}$	Anti-causal, LSI
Fractional integrators of order $\gamma \in (1 - \frac{1}{p}, +\infty)$, $\gamma + \frac{1}{p} \neq 1, 2, 3, \ldots$	
$I^{\gamma}_{0,p}\{\varphi\}(\tau) = \int_{\mathbb{R}} \dfrac{e^{j\omega\tau} - \sum_{k=0}^{\lfloor \gamma - 1 + \frac{1}{p} \rfloor} \frac{(j\omega)^k}{k!}}{(j\omega)^{\gamma}} \widehat{\varphi}(\omega) \dfrac{d\omega}{2\pi}$	$I^{\gamma}_{0,p}\{\varphi\}(0) = 0$
$I^{\gamma*}_{0,p}\{\varphi\}(\tau) = \int_{\mathbb{R}} \dfrac{\widehat{\varphi}(\omega) - \sum_{k=0}^{\lfloor \gamma - 1 + \frac{1}{p} \rfloor} \frac{\widehat{\varphi}^{(k)}(0)\omega^k}{k!}}{(-j\omega)^{\gamma}} e^{j\omega\tau} \dfrac{d\omega}{2\pi}$	L_p-stable

where $I^*_0 = I^{1*}_{0,2}$ is the regularized integrator that we have already encountered during our investigation of Lévy processes. One can also verify that (7.44) coincides with the L_p-stable left inverse $\partial^{-\gamma*}_{\tau,2}$ of Theorem 5.8 with $\tau = -\gamma/2$ and $p = 2$. The interest of (7.45) is that it suggests a possible representation of fBm as

$$B_H = I_0 D^{-H+1/2} w,$$

where I_0 imposes the boundary condition $B_H(0) = 0$.

To determine the underlying kernel denoted by $h_{\gamma,2}(t, \tau)$, we recall that $h(t, \tau) = L^{-1}\{\delta(\cdot - \tau)\}(t) = L^{-1*}\{\delta(\cdot - t)\}(\tau)$. Specifically, by inserting the Fourier transform of $\delta(\cdot - t)$ into (7.44), we find that

$$
\begin{aligned}
h_{\gamma,2}(t, \tau) &= \int_{\mathbb{R}} \frac{e^{-j\omega t} - 1}{(-j\omega)^{\gamma}} e^{j\omega\tau} \frac{d\omega}{2\pi} \\
&= \frac{1}{\Gamma(\gamma)} \left(\left(-(\tau - t)\right)^{\gamma-1}_+ - (-\tau)^{\gamma-1}_+ \right) \\
&= \frac{1}{\Gamma(\gamma)} \left((t - \tau)^{\gamma-1}_+ - (-\tau)^{\gamma-1}_+ \right),
\end{aligned}
\tag{7.46}
$$

which is consistent with the relation

$$\mathscr{F}^{-1}\left\{ \frac{1}{(j\omega)^{\alpha+1}} \right\}(t) = \frac{t^{\alpha}_+}{\Gamma(\alpha + 1)}$$

for $\alpha \notin \mathbb{Z}$ (see Table A.1).

Examples of such kernel functions with $t = t_1 = 1$ (fixed) are shown in Figure 7.7 – the ones of interest with $\gamma \in (0.5, 1.5)$ have their area shaded. These are representatives of the two main regimes where the functions are either bounded $(\gamma \in [1, 3/2))$ or unbounded $(\gamma \in (1/2, 1))$ with two (square-integrable) singularities at $\tau = 0$ and $\tau = t_1$. While $h_{\gamma,2}(t, \tau)$ is made up of individual atoms (one-sided power functions) whose

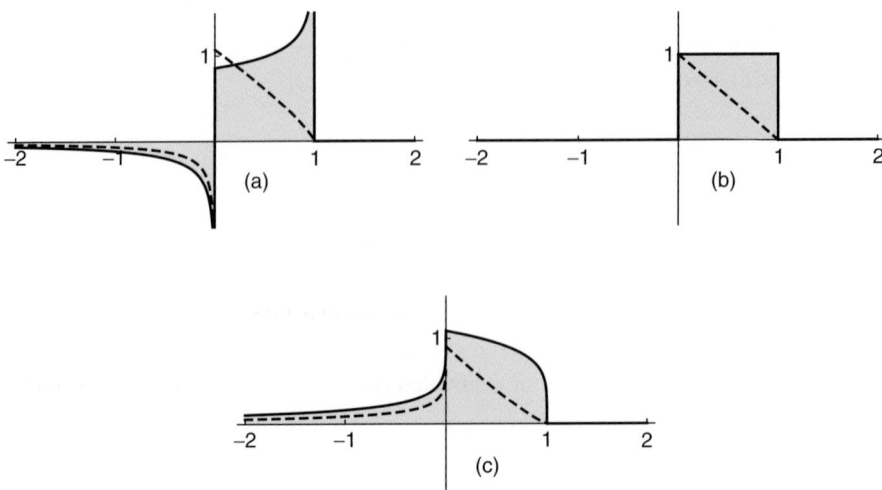

Figure 7.7 Comparison of the standard and higher-order fBm-related kernels $h_{\gamma_1,2}(1,\tau)$ (shaded) vs. $h_{\gamma_2,2}(1,\tau)$ (dashed line) with $\gamma_2 = \gamma_1 + 1$: (a) $(\gamma_1,\gamma_2) = (0.8, 1.8)$; (b) $(\gamma_1,\gamma_2) = (1,2)$; (c) $(\gamma_1,\gamma_2) = (1.2, 2.2)$.

energy is infinite, the remarkable twist is that the combination in (7.46) yields a function of τ that is square-integrable.

PROPOSITION 7.9 *The function $h_{\gamma,2}(t_1,\cdot)$, where $t_1 \in \mathbb{R}$ is a fixed parameter, belongs to $L_2(\mathbb{R})$ for $\gamma \in (0.5, 1.5)$. More generally, the right-hand side of (7.46) defines the kernel $h_{\gamma,p}(t_1,\cdot) \in L_p(\mathbb{R})$ with $p > 0$ if and only if $1 - \frac{1}{p} < \gamma < 2 - \frac{1}{p}$. On the other hand, none of these kernels is rapidly decaying, except for $h_{1,2}(t_1,\cdot) = \mathbb{1}_{(0,t_1]}$ (Brownian motion).*

Proof We consider the scenario $t_1 \geq 0$. First, we note that (7.47) with $\gamma = 1$ and $t = t_1$ simplifies to

$$h_{1,2}(t_1,\cdot) = \mathrm{I}_0^*\{\delta(\cdot - t_1)\}(\tau) = \mathbb{1}_{(0,t_1]}(\tau), \qquad (7.47)$$

which is compactly supported and hence rapidly decaying (see Figure 7.7b). For the other values of γ, we split the range of integration into two parts in order to handle the singularities separately from the decay of the tail.

(1) $\tau \in [-t_1, t_1]$ with $t_1 > 0$ finite.
   This part of the L_p-integral will be bounded if $\int_0^1 \tau^{p(\gamma-1)}\, d\tau < \infty$. This happens for $1 - \frac{1}{p} < \gamma$, which is always satisfied for $p = 2$ since $\gamma = H + \frac{1}{2} > \frac{1}{2}$.
(2) $\tau \in (-\infty, -t_1)$.
   Here, we base our derivation on the factorized representation $\mathrm{I}_{0,2}^\gamma = \mathrm{I}_0 \mathrm{D}^{-\gamma+1}$, which follows from (7.45). The operator $\mathrm{D}^{-\gamma+1}$ is a shift-invariant operator. Its impulse

response is given by

$$\rho_{D^{\gamma-1}}(t) = \mathscr{F}^{-1}\left\{\frac{1}{(j\omega)^{\gamma-1}}\right\}(t) = \frac{t_+^{\gamma-2}}{\Gamma(\gamma-1)},$$

which is also the Green's function of $D^{\gamma-1}$. Using (7.47), we then obtain

$$h_{\gamma,2}(t_1,\tau) = (D^{-\gamma+1})^*I_0^*\{\delta(\cdot - t_1)\}(\tau) = (\overset{\vee}{\rho}_{D^{\gamma-1}} * \mathbb{1}_{(0,t_1]})(\tau),$$

which implies that $h_{\gamma,2}(t_1,\tau)$ decays like $1/|\tau|^{2-\gamma}$ as $\tau \to -\infty$ (unless $\gamma = 1$, in which case $\rho_{\mathrm{Id}} = \delta$). It follows that the tail of $|h_{\gamma,2}(t_1,\tau)|^p$ is integrable provided that $(2-\gamma)p > 1$ or, equivalently, $\gamma < 2 - 1/p$, which corresponds to the whole range $H \in (0,1)$ for $p = 2$.

□

Taking advantage of Proposition 7.9, we invoke Theorem 7.1 to show that the generalized process $B_H(t) = I_0 D^{-H+1/2}w(t)$ admits a classical interpretation as a random function of t. The theorem also yields the stochastic integral representation

$$B_H(t) = \langle h_{H+1/2,2}(t,\cdot), w\rangle$$

$$= \int_{\mathbb{R}} \frac{1}{\Gamma(\gamma)}\left((t-\tau)_+^{\gamma-1} - (-\tau)_+^{\gamma-1}\right)\, dW(\tau)$$

$$= \frac{1}{\Gamma(\gamma)}\left(\int_{-\infty}^0 \left((t-\tau)^{\gamma-1} - (-\tau)^{\gamma-1}\right)\, dW(\tau) + \int_0^t (t-\tau)^{\gamma-1}\, dW(\tau)\right),$$

$$(7.48)$$

where the last equation with $\gamma = H + 1/2$ is the one that was originally used by Mandelbrot and Van Ness to define fBm. Here, $W(\tau) = B_{1/2}(\tau)$ is the standard Brownian motion whose derivative in the sense of generalized functions is w.

While this result is reassuring, the real power of the distributional approach is that it naturally lends itself to generalizations. For instance, we may extend the definition to larger values of H by applying an additional number of integrals. The relevant adjoint operator is $I_0^{n*} = (I_0^*)^n$, whose Fourier-domain expression is

$$(I_0^*)^n\{\varphi\}(\tau) = \int_{\mathbb{R}} \frac{\widehat{\varphi}(\omega) - \sum_{k=0}^n \frac{\widehat{\varphi}^{(k)}(0)\omega^k}{k!}}{(-j\omega)^n} e^{j\omega\tau}\, \frac{d\omega}{2\pi}. \qquad (7.49)$$

This formula, which is consistent with the form of $I_{2,0}^{\gamma*}$ in Table 7.1 with $\gamma = n$, follows from the definition of I_0^* and the property that the Fourier transform of $\varphi \in \mathscr{S}(\mathbb{R})$ admits the Taylor series representation $\widehat{\varphi}(\omega) = \sum_{k=0}^\infty \widehat{\varphi}^{(k)}(0)\frac{\omega^k}{k!}$, which yields the required limits at the origin. This allows us to consider the higher-order version of (7.43) for $\gamma = H + \frac{1}{2}$ with $H \in \mathbb{R}^+\setminus\mathbb{N}$, the solution of which is formally expressed as

$$B_H = I^n D^{-(H-n)+\frac{1}{2}}w,$$

where $n = \lceil H \rceil$. This construction translates into the extended version of fBm specified by the characteristic functional

$$\mathbb{E}\{e^{j\langle\varphi,B_H\rangle}\} = \widehat{\mathscr{P}_{B_H}}(\varphi) = \exp\left(-\frac{1}{2}\int_{\mathbb{R}}\left|\frac{\widehat{\varphi}(\omega) - \sum_{k=0}^{\lceil H \rceil}\frac{\widehat{\varphi}^{(k)}(0)\omega^k}{k!}}{(-j\omega)^{H+1/2}}\right|^2 \frac{d\omega}{2\pi}\right), \qquad (7.50)$$

where $H \in \mathbb{R}^+\backslash\mathbb{N}$ is the Hurst exponent of the process.

The relevant kernels are determined by using the same inversion technique as before and are given by

$$h_{\gamma,p}(t,\tau) = I_{0,p}^{\gamma}\{\delta(\cdot - t)\}(\tau)$$

$$= \frac{1}{\Gamma(\gamma)}(t-\tau)_+^{\gamma-1} - \sum_{k=0}^{\lfloor\gamma-1+1/p\rfloor}\frac{t^k}{k!}\frac{(-\tau)_+^{\gamma-1-k}}{\Gamma(\gamma-k)}, \qquad (7.51)$$

where the case of interest for fBm is $p = 2$ (see examples in Figure 7.7). For $\gamma = n \in \mathbb{N}$, the expression simplifies to

$$h_{n,p}(t,\tau) = I_0^{n*}\{\delta(\cdot - t)\}(\tau) = \begin{cases} \mathbb{1}_{(0,t]}(\tau)\frac{(t-\tau)^n}{n!}, & \text{for } 0 < t \\ -\mathbb{1}_{[t,0)}(\tau)\frac{(t-\tau)^n}{n!}, & \text{for } t < 0, \end{cases}$$

which, once again, is compactly supported.

The final ingredient for the theory is the extension of Proposition 7.9, which is proven in exactly the same fashion.

PROPOSITION 7.10   *The kernels defined by (7.51) can be decomposed as*

$$h_{\gamma,p}(t,\tau) = \left(\rho_{\mathrm{D}^{\gamma-n}}^{\vee} * \mathbb{1}_{(0,t]}\frac{(t-\cdot)^n}{n!}\right)(\tau),$$

where $n = \lfloor\gamma - 1 + 1/p\rfloor$. Moreover, $h_{\gamma,p}(t_1,\cdot) \in L_p(\mathbb{R})$ for any fixed value $t = t_1$ and $\gamma - 1 + \frac{1}{p} \in \mathbb{R}^+\backslash\mathbb{N}$.

For $p = 2$, this covers the whole range of non-integer Hurst exponents $H = \gamma - 1/2$ so that the classical interpretation of $B_H(t)$ remains applicable. Moreover, since I_0^n is a right inverse of D^n, these processes satisfy the extended boundary conditions

$$\mathrm{D}^k B_H(0) = B_{H-k}(0) = 0$$

for $k = 0, \ldots, \lceil H \rceil - 1$.

Similarly, we can fractionally integrate the SαS noise w_α to generate stable self-similar processes that are inherently sparse for $\alpha < 2$. The operation is acceptable as long as $I_{0,p}^{\gamma}$ meets the L_p-stability requirement of Theorem 5.8 with $p = \alpha$, which we restate as

$$H_\alpha = \gamma - 1 + \frac{1}{\alpha} \in \mathbb{R}^+\backslash\mathbb{N}.$$

Interestingly enough, this is the same condition as in Proposition 7.10. This ensures that the so-constructed fractional stable motions (fSm) are well defined in the generalized sense (by the Minlos-Bochner theorem), as well as in the ordinary sense (by Theorem 7.1). Statement (4) of Theorem 7.1 also indicates that the parameter H_α actually represents the Hurst exponent of these processes, which is consistent with the analysis of Section 7.5.1 (in particular, (7.41) with $d = 1$).

It is also possible to define stable fractal processes by considering any other variant ∂_τ^γ of fractional derivative within the family of scale-invariant operators (see Proposition 5.6). Unlike in the Gaussian case where the Fourier phase is irrelevant, the fractional SDE $\partial_\tau^\gamma s_{\alpha,H} = w_\alpha$ will specify a wider variety of self-similar processes with the same overall properties.

7.5.3   Scale-invariant Poisson processes

While not strictly self-similar, the realizations of random fields obtained by solving a stochastic PDE involving a scale-invariant (homogeneous) operator and a compound-Poisson innovation give rise to interesting patterns. After revisiting the classical compound-Poisson process, we review Poisson fields and give two examples of such processes and fields here, namely, those associated with the fractional Laplacian operator and with the partial derivative $\partial_{r_1} \cdots \partial_{r_d}$.

Compound-Poisson processes and fractional extensions
The compound-Poisson process is a Lévy process per (7.26) with a compound-Poisson innovation as the driving term of its SDE. Its restriction to the positive half of the real axis finds a simple representation as

$$\sum_{k\in\mathbb{Z}} A_k \mathbb{1}_{r\geq 0}(r - r_k), \tag{7.52}$$

where the sequence $\{r_k\}$ forms a Poisson point process in $\mathbb{R}^+$ and the amplitudes A_k are i.i.d. Clearly, realizations of the above process correspond to random polynomial splines of degree zero with random knots (see Figure 1.1). This characterization on the half-axis is consistent with the one given on the full axis in Equation (1.14).

More generally, we may define fractional processes with a compound-Poisson innovation and the fractional integrators identified in Section 5.5.1, which give rise to random fractional splines. The subclass of these generalized Poisson processes associated with causal fractional integrators finds the following representation as a piecewise fractional polynomial on the positive half-axis:

$$\sum_{k\in\mathbb{Z}} A_k (r - r_k)_+^{\gamma-1}.$$

They are whitened by the causal fractional-derivative operator D^γ whose Fourier symbol is $(j\omega)^\gamma$. Some examples of such processes are given in the right column of Figure 7.6.

Scale-invariant Poisson fields
As we saw in Section 4.4.2, a compound-Poisson innovation w can be modeled as an infinite collection of Dirac impulses with random amplitudes, with two important properties. First, the number of Diracs in any given compact neighborhood Π is a Poisson

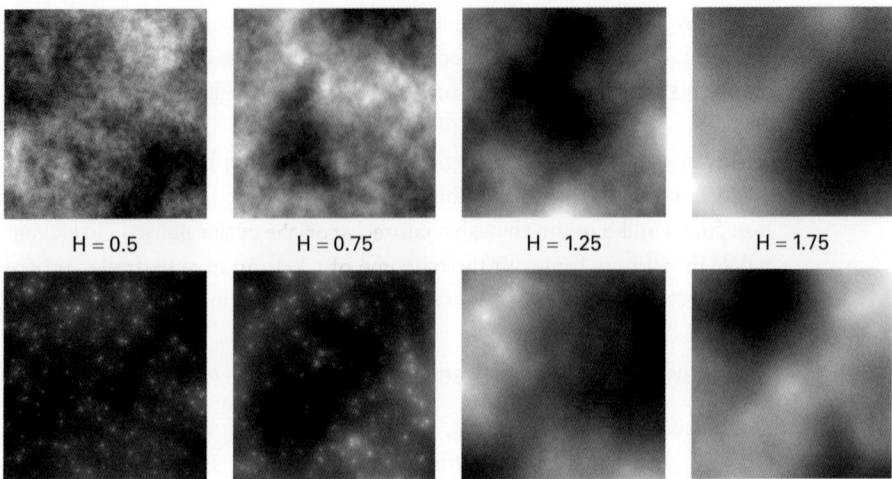

Figure 7.8 Gaussian fBm (top row) vs. sparse generalized Poisson images (bottom row) in 2-D. The fields in each column share the same operator and second-order statistics.

variable with parameter $\lambda \text{Vol}(\Pi)$, where λ is a fixed density parameter and $\text{Vol}(\Pi)$ is the volume (Lebesgue measure) of Π. Second, fixing Π and conditioning on the number of Diracs therein, the distribution of the position of each Dirac is uniform in Π and independent of all the other Diracs (their amplitudes are also independent). We can represent a realization of w symbolically as

$$w(\boldsymbol{r}) = \sum_{k \in \mathbb{Z}} A_k \delta(\boldsymbol{r} - \boldsymbol{r}_k),$$

where A_k are the i.i.d. amplitudes with some generalized probability density p_A, the $\boldsymbol{r}_k$ are the positions coming from a Poisson innovation (i.e., fulfilling the above require-ments) independent from the A_k, and the ordering (the index k) is insignificant. We recall that, by Theorem 4.9, the characteristic functional of the compound-Poisson process is related to the density of the said amplitudes by

$$\widehat{\mathscr{P}}_{w_{\text{Poisson}}}(\varphi) = \exp\left(\int_{\mathbb{R}^d} \lambda \int_{\mathbb{R}} \left(e^{ja\varphi(\boldsymbol{r})} - 1\right) p_A(a) \, \mathrm{d}a \, \mathrm{d}\boldsymbol{r}\right).$$

We shall limit ourselves here to amplitude distributions that have, at a minimum, a finite first moment.

The combination of $\widehat{\mathscr{P}}_{w_{\text{Poisson}}}$ with the operators noted previously involves no addi-tional subtleties compared to the case of α-stable innovations, at least for the processes with finite mean that we are considering here. It is, however, noteworthy to recall that the compound-Poisson Lévy exponents are p-admissible with $p = 1$ and, in the special case of symmetric and finite-variance amplitude distributions, for any $p \in [1, 2]$ (see Proposition 4.5). The inverse operator with which $\widehat{\mathscr{P}}_{w_{\text{Poisson}}}$ is composed therefore needs to be chosen/constructed such that it maps $\mathscr{S}(\mathbb{R}^d)$ continuously into $L_p(\mathbb{R}^d)$. Such op-erators were identified in Section 5.4.2 for integer-order SDEs and Sections 5.5.3, 5.5.1 for fractional-order derivatives and Laplacians.

The generic form of such Poisson processes is

$$s_{\text{Poisson}}(\boldsymbol{r}) = p_0(\boldsymbol{r}) + \sum_{k \in \mathbb{Z}} A_k \rho_{\text{L}}(\boldsymbol{r} - \boldsymbol{r}_k),$$

where ρ_{L} is the Green's function of L and $p_0(\boldsymbol{r}) \in \mathcal{N}_{\text{L}}$ is a component in the null space of the whitening operator. The reason for the presence of p_0 is to satisfy the boundary conditions at the origin that are imposed upon the process.

We provide examples of the outcome of applying inverse fractional Laplacians to a Poisson field in Figure 7.8. The fact that the Green's function of the underlying fractional Laplacian is proportional to $\|r\|^{H-1}$ explains the change of appearance of these processes when switching from the singular mode with $H < 1$ to the continuous one with $H > 1$. As H increases further, the generalized Poisson processes become smoother and more similar visually to their fractal Gaussian counterparts (fractional Brownian field), which are displayed on the upper row of Figure 7.8 for comparison. Next, we shall look at another interesting example, the *Mondrian processes.*

The Mondrian process

The Mondrian process is the name by which we refer to the process associated with the innovation model with operator $L = \partial_{r_1} \cdots \partial_{r_d}$ and a sparse innovation – typically, a compound-Poisson one [UT11]. The instance of primary interest to us is the 2-D variety with $L = \partial_{r_1} \partial_{r_2}$. In this case, the Green's function of L is the indicator function of the positive quarter-plane, given by

$$\rho_{\text{L}}(\boldsymbol{r}) = \mathbb{1}_{r_1, r_2 \geq 0}(\boldsymbol{r}),$$

which we may use to define an admissible inverse of L in the positive quarter-plane. Similar to the compound-Poisson processes from Section 7.5.3, the Mondrian process finds a simple description in the positive quarter-plane as the random sum

$$\sum_{k \in \mathbb{Z}} A_k \mathbb{1}_{r_1, r_2 \geq 0}(\boldsymbol{r} - \boldsymbol{r}_k),$$

in direct parallel to (7.52), where the $\boldsymbol{r}_k$ are distributed uniformly in any neighborhood in the positive quarter-plane (Poisson point distribution).

A sample realization of this process, which bears some resemblance to paintings by Piet Mondrian, is shown in Figure 7.9.

7.6   Bibliographical notes

Section 7.1

The first-order processes considered in Section 7.1 are often referred to as non-Gaussian Ornstein–Uhlenbeck processes; one of their privileged areas of application is financial modeling [BNS01]. The representation of the autocorrelation in terms of B-splines is discussed in [KMU11, Section II.B], albeit in the purely Gaussian context. The fact

Figure 7.9 Realization of the Mondrian process with $\lambda = 1/30$.

that exponential B-splines provide the formal connection between ordinary continuous-domain differential operators and their discrete-domain finite-difference counterpart was first made explicit in [Uns05].

Section 7.2

Our formulation relies heavily on Gelfand and Vilenkin's theory of generalized stochastic processes, in which stationarity plays a predominant role [GV64]. The main analytical tools are the characteristic and the correlation functionals which have been used with great effect by Russian probabilists in the 1970s and 1980s (e.g., [Yag86]), but which are lesser known in Western circles. The probable reason is that the dominant paradigm for the investigation of stochastic processes is measure-theoretic – as opposed to distributional – meaning that processes are defined through stochastic integrals (with the help of Itô's calculus and its non-Gaussian variants) [Øks07, Pro04]. Both approaches have their advantages and limitations. On one hand, the Itô calculus can handle certain non-linear operations on random processes that cannot be dealt with in the distributional approach. Gelfand's framework, on the other hand, is ideally suited for performing any kind of linear operations, including some, such as fractional derivatives, which are much more difficult to define in the other framework. Theorem 7.1 is fundamental in that respect because it provides the higher-level elements for performing the translation between the two modes of representation.

Section 7.3

The classical theory of stationary processes was developed for the most part in the 1930s. The concept of (second-order) stationarity is due to Khintchine, who established the correlation properties of such processes [Khi34]. Other key contributors include Wiener [Wie30], Doob [Doo37], Cramér [Cra40], and Kolmogorov [Kol41]. Classical textbooks on stochastic processes are [Bar61,Doo90,Wie64]. The manual by Papoulis is very popular among engineers [Pap91]. The processes described by Proposition 7.2 are often called linear processes (under the additional finite-variance hypothesis) [Bar61]. Their conventional representation as a stochastic integral is

$$s(t) = \int_{\mathbb{R}} h(t - \tau) \, dW(\tau) = \int_{-\infty}^{t} h(t - \tau) \, dW(\tau), \tag{7.53}$$

where $W(\tau) = \langle \mathbb{1}_{[0,\tau)}, w \rangle$ is a stationary finite-variance process with independent increments (that is, a second-order Lévy process) and where $h \in L_1(\mathbb{R})$ is such that $h(t) = 0$, for $t < 0$ (causality). Bartlett provides an expression for their characteristic functional that is formally equivalent to the result given in Proposition 7.2 [Bar61, Section 5.2, Equations (24)–(25)]. His result can be traced back to 1951 [BK51, Equation (34)] but does not address the issue of existence. It involves the auxiliary function

$$K_W(\omega) = \log \mathbb{E} \left\{ \exp \left(j\omega \int_0^1 dW(\tau) \right) \right\},$$

which coincides with the Lévy exponent f of our formulation (see Proposition 4.12). A special case of the model is the so-called shot noise, which is obtained by taking w to be a pure Poisson noise [Ric77, BM02].

Implicit in Property 7.3 is the fact that the autocorrelation of a second-order stationary process is a cardinal L*L-spline. This observation is the key to proving the equivalence between smoothing-spline techniques and linear mean-square estimation in the sense of Wiener [UB05b]. It also simplifies the identification of the whitening operator L from sampled data [KMU11].

The Gaussian CARMA processes are often referred to as Gaussian stationary processes with rational spectral density. They are the solutions of ordinary stochastic differential equations and are treated in most books on stochastic processes [Bar61, Doo90, Pap91]. They are usually defined in terms of a stochastic integral such as (7.53) where h is the impulse response of the underlying Nth-order differential system and $W(t)$ a standard Brownian motion. Alternatively, one can solve the SDE using state-space techniques; this is the approach taken by Brockwell to characterize the extended family of Lévy-driven CARMA processes [Bro01]. Brockwell makes the assumption that $\mathbb{E}\{|W(1)|^\epsilon\} < \infty$ for some $\epsilon > 0$, which is equivalent to our Lévy–Schwartz admissibility condition (see (9.10) and surrounding text) and makes the formulations perfectly compatible. The CARMA model can also be extended to $N \le M$, in which case it yields stationary processes that are no longer defined pointwise. The Gaussian theory of such generalized CARMA processes is given in [BH10]. In light of what has been said, such results are also transposable to the more general Lévy setting, since the underlying convolution operator remains $\mathscr{S}$-continuous (under the stability assumption that there is no pole on the imaginary axis).

Section 7.4

The founding paper that introduces Lévy processes – initially called additive processes – is the very same that uncovers the celebrated Lévy–Khintchine formula and specifies the family of α-stable laws [Lév34]. We can therefore only concur with Loève's statement about its historical importance (see Section 1.4). Besides Lévy's classical monograph [Lév65], other recommended references on Lévy processes are [Sat94, CT04, App09]. The book by Cont and Tankov is a good starting point, while Sato's treatise is remarkable for its precision and completeness.

The operator-based method of resolution of unstable stochastic differential equations that is deployed in this section was developed by the authors with the help of Qiyu Sun [UTS14]. The approach provides a rigorous backing for statements such as "a Lévy process is the integral of a non-Gaussian white noise," "a continuous-domain non-Gaussian white noise is the (weak) derivative of a Lévy process," or "a Lévy process is a non-Gaussian process with a $1/\omega$ spectral behavior," which are intuitively appealing, but less obviously related to the conventional definition. The main advantage of the framework is that it allows for the direct transposition of deterministic methods for solving linear differential equations to the stochastic setting, irrespective of any stability considerations. Most of the examples of sparse stochastic processes are taken from [UTAK14].

Section 7.5

Fractional Brownian motion (fBm) with Hurst exponent $0 < H < 1$ was introduced by Mandelbrot and Van Ness in 1968 [MVN68]. Interestingly, there is an early paper by Kolmogorov that briefly mentions the possibility of defining such stochastic objects [Kol40]. The multidimensional counterparts of these processes are fractional Brownian fields [Man01], which are also solution of the fractional SDE $(-\Delta)^{\gamma/2}s = w$ where w is a Gaussian white noise [TVDVU09]. The family of stable self-similar processes with $0 < H < 1$ is investigated in [ST94], while their higher-order extensions are briefly considered in [Taf11].

The formulation of fBms as generalized stochastic processes was initiated by Blu and Unser in order to establish an equivalence between MMSE estimation and spline interpolation [BU07]. A by-product of this study was the derivation of (7.42) and (7.50). The latter representation is compatible with the higher-order generalization of fBm for $H \in \mathbb{R}^+\backslash\mathbb{Z}^+$ proposed by Perrin *et al.* [PHBJ$^+$01], with the underlying kernel (7.51) being the same.

The scale-invariant Poisson processes of Section 7.5.3 were introduced by the authors [UT11]; they are the direct stochastic counterparts of spline functions where the knots and the strength of the singularities are assigned in a random fashion. The examples, including the Mondrian process, are taken from that paper on sparse stochastic processes.

8 Sparse representations

In order to obtain an uncoupled representation of a sparse process $s = \mathrm{L}^{-1}w$ of the type described in the previous chapters, it is essential that we somehow invert the integral operator L^{-1}. The ideal scenario would be to apply the differential operator $\mathrm{L} = (\mathrm{L}^{-1})^{-1}$ to uncover the innovation w that is independent at every point. Unfortunately, this is not feasible in practice because we do not have access to the signal $s(r)$ over the entire domain $r \in \mathbb{R}^d$, but only to its sampled values on a lattice or, more generally, to a series of coefficients in some appropriate basis. Our analysis options, as already alluded to in Chapter 2, are essentially twofold: the application of a discrete version of the operator, or an operator-like wavelet analysis.

This chapter is devoted to the in-depth investigation of these two modes of representation. As with the other chapters, we start with a concrete example (Lévy process) to lay down the key ideas in Section 8.1. Our primary tool for deriving the transform-domain pdfs is the characteristic functional, reviewed in Section 8.2 and properly extended so that it can handle arbitrary analysis functions $\varphi \in L_p(\mathbb{R}^d)$. In Section 8.3, we investigate the decoupling ability of finite-difference-type operators and determine the statistical distribution of the resulting *generalized increments*. In Section 8.4, we show how a sparse process can be expanded in a matched-wavelet basis and provide the complete multivariate description of the transform-domain statistics, including general formulas for the wavelet cumulants. Finally, in Section 8.5, we apply these methods to the representation of first-order processes. In particular, we focus on the sampled version of these processes and adopt a matrix-vector formalism to make the link with traditional transform-domain signal-processing techniques. Our main finding is that operator-like wavelets generally outperform the classical sinusoidal transforms (DCT and KLT) for the decoupling of sparse AR(1) and Lévy processes and that they essentially provide an *independent-component analysis* (ICA).

8.1 Decoupling of Lévy processes: finite differences vs. wavelets

To expose the reader to the concepts, it is instructive to return to our introductory example in Chapter 1: the Lévy process denoted by $W(t)$. As we already saw, the fundamental property of $W(t)$ is that its increments are independent (see Definition 7.7).

When the Lévy process is sampled uniformly on the integer grid, we have access to its equispaced increments

$$u[k] = W(t)|_{t=k} - W(t)|_{t=k-1},$$ (8.1)

which are i.i.d. Moreover, due to the property that $W(0) = 0$ (almost surely), the relation can be inverted as

$$W(k) = \sum_{n=1}^{k} u[n],$$ (8.2)

which provides a convenient recipe for generating Lévy processes.

To see how this fits the formalism of Chapter 7, we recall that the corresponding whitening operator is $D = \frac{d}{dt} = P_0$. Its discrete counterpart is the finite-difference operator $D_d = \Delta_0$. This notation is consistent with the pole vector of a (basic) Lévy process being $\alpha = (0)$.

According to the generic procedure outlined in Section 7.4.4, we then specify the *increment process* associated with $W(t), t \in \mathbb{R}$ as

$$u_0(t) = \Delta_0 W(t) = (\beta_0 * w)(t) = \langle \beta_0^\vee(\cdot - t), w \rangle,$$ (8.3)

where $\beta_0 = \beta_D = \beta_+^0$ is the rectangle B-spline defined by (1.4) (or (6.21) with $\alpha = 0$) and w is an innovation process with Lévy exponent f. Based on Proposition 7.4, we deduce that $u_0(t)$, which is defined for all $t \in \mathbb{R}$, is stationary with characteristic functional $\widehat{\mathscr{P}}_{u_0}(\varphi) = \widehat{\mathscr{P}}_w(\beta_0^\vee * \varphi)$, where $\widehat{\mathscr{P}}_w$ is defined by (7.21). Let us now consider the random variable $U = \langle \delta(\cdot - k), s \rangle = u[k]$. To obtain its characteristic function, we rely on the theoretical framework of Section 8.2 and simply substitute $\varphi = \omega\delta(\cdot - k)$ in $\widehat{\mathscr{P}}_{u_0}(\varphi)$, which yields

$$\begin{aligned}
\widehat{p}_U(\omega) &= \widehat{\mathscr{P}}_w(\omega\beta_0^\vee(\cdot - k)) = \widehat{\mathscr{P}}_w(\omega\beta_0^\vee) \qquad \text{(by stationarity)} \\
&= \exp\left(\int_{\mathbb{R}} f(\omega\beta_0^\vee(t)) \, dt\right) \\
&= e^{f(\omega)} = \widehat{p}_{id}(\omega).
\end{aligned}$$ (8.4)

The disappearance of the integral results from the binary nature of $\beta_0^\vee$ and the fact that $f(0) = 0$. This shows that the increments of the Lévy process are infinitely divisible (id) with canonical pdf $p_U = p_{id}$, which corresponds to the observation of the innovation through a unit rectangular window (see Proposition 4.12). Likewise, we find that the joint characteristic function of $U_1 = \langle \beta_0^\vee(\cdot - k_1), w \rangle = u[k_1]$ and $U_2 = \langle \beta_0^\vee(\cdot - k_2), w \rangle = u[k_2]$ for any $|k_1 - k_2| \geq 1$ factorizes as

$$\begin{aligned}
\widehat{p}_{U_1,U_2}(\omega_1, \omega_2) &= \widehat{\mathscr{P}}_w(\omega_1\beta_0^\vee(\cdot - k_1) + \omega_2\beta_0^\vee(\cdot - k_2)) \\
&= \widehat{p}_U(\omega_1)\,\widehat{p}_U(\omega_2),
\end{aligned}$$

where $\widehat{p}_U$ is given by (8.4). Here, we have used the fact that the supports of $\beta_0^\vee(\cdot - k_1)$ and $\beta_0^\vee(\cdot - k_2)$ are disjoint, together with the independence at every point of w (Proposition 4.11). This implies that the random variables U_1 and U_2 are independent for all pairs of distinct indices (k_1, k_2), which proves that the sequence of Lévy increments $\{u[k]\}_{k\in\mathbb{Z}}$ is i.i.d. with pdf $p_U = p_{\mathrm{id}}$. The bottom line is that the decoupling of the samples of a Lévy process afforded by (8.1) is perfect and that this transformation is reversible.

The alternative is to expand the Lévy process in a wavelet basis that is matched to the operator $\mathrm{L} = \mathrm{D}$. To that end, we revisit the Haar analysis of Section 1.3.4 by applying the generalized wavelet framework of Section 6.5.3 with $d = 1$ and (scalar) dilation $\mathbf{D} = 2$. The required ingredient is the $\mathrm{D}^*\mathrm{D}$-interpolator with respect to the grid $2^{i-1}\mathbb{Z}$: $\varphi_{\mathrm{int},i-1}(t) = \beta_{(0,0)}(t/2^{i-1} + 1)$, which is a symmetric triangular function of unit height and support $[-2^{i-1}, 2^{i-1}]$ (piecewise-linear B-spline). The specialized form of (6.54) then yields the *derivative-like wavelets*

$$\psi_{i,k}(t) = 2^{i/2-1}\mathrm{D}^*\varphi_{\mathrm{int},i-1}(t - 2^{i-1}k)$$
$$= \mathrm{D}^*\phi_i(t - 2^{i-1}k),$$

where $\phi_i(t) = 2^{i/2-1}\varphi_{\mathrm{int},i-1}(t) \propto \phi_0(t/2^i)$ is the normalized triangular kernel at resolution level i and $k \in \mathbb{Z} \setminus 2\mathbb{Z}$. The difference with the classical Haar-wavelet formulas (1.19) and (1.20) is that the polarity is reversed (since $\mathrm{D}^* = -\mathrm{D}$) and the location index k restricted to the set of odd integers. Apart from this change in indexing convention, required by the general (multidimensional) formulation, the two sets of basis functions are equivalent (see Figure 1.4). Next, we recall that the Lévy process can be represented as $s = \mathrm{I}_0 w$, where I_0 is the integral operator defined by (5.4). This allows us to express its Haar wavelet coefficients as

$$V_i[k] = \langle \psi_{i,k}, s \rangle$$
$$= \langle \mathrm{D}^*\phi_i(\cdot - 2^{i-1}k), \mathrm{I}_0 w \rangle$$
$$= \langle \mathrm{I}_0^* \mathrm{D}^*\phi_i(\cdot - 2^{i-1}k), w \rangle \qquad \text{(by duality)}$$
$$= \langle \phi_i(\cdot - 2^{i-1}k), w \rangle, \qquad \text{(left-inverse property of } \mathrm{I}_0^*\text{)}$$

which is very similar to (8.3), except that the rectangular smoothing kernel $\beta_0^\vee$ is replaced by a triangular one. By considering the continuous version $V_i(t) = \langle \phi_i(\cdot - t), w \rangle$ of the wavelet transform at scale i, we then invoke Proposition 7.2 to show that $V_i(t)$ is stationary with characteristic functional $\widehat{\mathscr{P}}_{V_i}(\varphi) = \widehat{\mathscr{P}}_w(\phi_i * \varphi)$. Moreover, since the smoothing kernels $\phi_i(\cdot - 2^{i-1}k)$ for i fixed and odd indices k are not overlapping, we deduce that the sequence of Haar-wavelet coefficients $\{V_i[k]\}_{k\in\mathbb{Z}\setminus 2\mathbb{Z}}$ is i.i.d. with characteristic function $\widehat{p}_{V_i}(\omega) = \mathbb{E}\{e^{j\omega\langle\phi_i, w\rangle}\} = \widehat{\mathscr{P}}_w(\omega\phi_i)$.

If we now compare the wavelet situation for $i = 1$ with that of the Lévy increments, we observe that the wavelet analysis involves one more layer of smoothing of the innovation with β_0 (since $\phi_1 = \varphi_{\mathrm{int},0} = \beta_0 * \beta_0^\vee$), which slightly complicates the statistical calculations. For $i > 1$, there is an additional coarsening effect which has some interesting statistical implications (see Section 9.8).

While the smoothing effect on the innovation is qualitatively the same in both scenarios, there are fundamental differences too. In the wavelet case, the underlying discrete transform is orthogonal, but the coefficients are not fully decoupled because of the unavoidable inter-scale dependencies, as we shall see in Section 8.4.1. Conversely, the finite-difference transform (8.1) is optimal for decoupling Lévy processes, but the representation is not orthogonal. In this chapter, we show that this dual way of expanding signals remains applicable for the complete family of sparse processes under mild conditions on the whitening operator L. While the consideration of higher-order operators makes us gain in generality, the price to pay is a slight loss in the decoupling ability of the transforms, because longer basis functions (and B-splines) tend to overlap more. Those limitations notwithstanding, we shall see that the derivation of the transform-domain statistics is in all points analogous to the Lévy scenario. In addition, we shall provide the tools for assessing the residual dependencies that are unavoidable when the random processes are non-Gaussian.

8.2   Extended theoretical framework

Fundamentally, a generalized stochastic process s is only observable indirectly through its inner products $X_n = \langle \varphi_n, s \rangle$ with some family $\{\varphi_n\}$ of test functions. Conversely, any linear measurement of a conventional or generalized stochastic process s is expressible as $Y = \langle \psi, s \rangle$ with some suitable kernel $\psi \in \mathscr{S}'(\mathbb{R}^d)$. We also recall that a generalized stochastic process on $\mathscr{S}'(\mathbb{R}^d)$ is completely characterized by its characteristic functional $\widehat{\mathscr{P}}_s(\varphi) = \mathbb{E}\{e^{j\langle \varphi, s \rangle}\}$ with $\varphi \in \mathscr{S}(\mathbb{R}^d)$ (see Minlos–Bochner Theorem 3.9). The combination of these two ingredients suggests a simple, powerful mechanism for the determination of the characteristic function of the random variable $Y = \langle \psi, s \rangle$ by formal substitution of $\varphi = \omega \psi$ in the characteristic functional of s. This leads to

$$\widehat{p}_Y(\omega) = \mathbb{E}\{e^{j\omega Y}\} = \mathbb{E}\{e^{j\omega \langle \psi, s \rangle}\} \qquad \text{(by definition)}$$
$$= \mathbb{E}\{e^{j\langle \omega \psi, s \rangle}\} \qquad \text{(by linearity)}$$
$$= \widehat{\mathscr{P}}_s(\omega \psi), \qquad \text{(by identification)}$$

where ψ is fixed and $\omega \in \mathbb{R}$ plays the role of the Fourier-domain variable. The same approach is extendable to the determination of the joint pdf of any finite collection of linear measurements $\{Y_n = \langle \psi_n, s \rangle\}_{n=1}^N$ (see Proposition 3.10).

This is very nice in principle, except that most of the kernels ψ_n considered in this chapter are non-smooth and almost systematically outside of $\mathscr{S}(\mathbb{R}^d)$. To properly handle this issue, we need to extend the domain of $\widehat{\mathscr{P}}_s$ to a larger class of (generalized) functions, as explained in the next two sections.

8.2.1   Discretization mechanism: sampling vs. projections

For practical purposes, it is desirable to represent a stochastic process s by a finite number of coefficients that can be stored and processed in a computer. The discretization methods can be divided in two categories. The first is a representation of s in terms of values that are sampled uniformly, which is formally described as

$$s[\mathbf{k}] = \langle \delta(\cdot - \mathbf{k}), s \rangle.$$

The implicit assumption here is that the process s has a sufficient degree of continuity for its sample values to be well defined. Since $s = L^{-1}w$, it is actually sufficient that the operator L^{-1} satisfies some mild regularity conditions. The simplest scenario is when L^{-1} is LSI with its impulse response ρ_L belonging to some function space $\mathcal{X} \subseteq L_1(\mathbb{R}^d)$. Then,

$$s[\mathbf{k}] = \langle \delta(\cdot - \mathbf{k}), s \rangle = \langle \delta(\cdot - \mathbf{k}), \rho_L * w \rangle = \langle \rho_L^{\vee}(\cdot - \mathbf{k}), w \rangle.$$

Now, if the characteristic functional of the noise admits a continuous extension from $\mathcal{S}(\mathbb{R}^d)$ to $\mathcal{X}$ (specifically, $\mathcal{X} = \mathcal{R}(\mathbb{R}^d)$ or $\mathcal{X} = L_p(\mathbb{R}^d)$, as shown in Section 8.2.2), we obtain the characteristic function of $s[\mathbf{k}]$ by replacing $\varphi \in \mathcal{X}$ by $\omega \rho_L^{\vee}(\cdot - \mathbf{k})$ in $\widehat{\mathscr{P}}_w(\varphi)$. The continuity and positive definiteness of $\widehat{\mathscr{P}}_w$ ensure that the sample values of the process have a well-defined probability distribution (by Bochner's Theorem 3.7) so that they can be interpreted as conventional random variables. The argument carries over to the non-shift-invariant scenario as well, by replacing the samples by their generalized increments. This is equivalent to sampling the *generalized increment process* $u = L_d s$, which is stationary by construction (see Proposition 7.4).

The second discretization option is to expand s in a suitable basis whose expansion coefficients (or projections) are given by

$$Y_n = \langle \psi_n, s \rangle,$$

where the ψ_n are appropriate analysis functions (typically, wavelets). The argument concerning the existence of the underlying expansion coefficients as conventional random variables is based on the following manipulation:

$$Y_n = \langle \psi_n, s \rangle = \langle \psi_n, L^{-1}w \rangle = \langle L^{-1*}\psi_n, w \rangle.$$

The condition for admissibility, which is essentially the same as in the previous scenario, is that $\phi_n = L^{-1*}\psi_n \in \mathcal{X}$, subject to the constraint that $\widehat{\mathscr{P}}_w$ admits an extension that is continuous and positive definite over $\mathcal{X}$. This is consistent with the notion of L-admissible wavelets (Definition 6.7), which asks that $\psi = L^*\phi$ with $\phi \in \mathcal{X} \subseteq L_1(\mathbb{R}^d)$.

8.2.2 Analysis of white noise with non-smooth functions

Having motivated the necessity of extending the domain of $\widehat{\mathscr{P}}_w(\varphi)$ to the largest possible class of functions $\mathcal{X}$, we now proceed with the mathematics. In particular, it is crucial to remove the stringent smoothness requirement associated with $\mathcal{S}(\mathbb{R}^d)$ (infinite order of differentiability), which is never met by the analysis functions used in practice. To that end, we extend the basic continuity result of Theorem 4.8 to the class of functions with rapid decay, which can be done without restriction on the Lévy exponent f. This takes care, in particular, of the cases where $\varphi = \psi_n$ is bounded with compact support, which is typical for an Nth-order B-spline or a wavelet basis function in 1-D.

PROPOSITION 8.1   *Let f be a Lévy–Schwartz exponent as specified in Theorem 4.8. Then, the Lévy noise functional*

$$\widehat{\mathscr{P}_w}(\varphi) = \exp\left(\int_{\mathbb{R}^d} f\big(\varphi(r)\big)\, dr\right)$$

is continuous and positive definite over $\mathscr{R}(\mathbb{R}^d)$.

The proof is the same as that of Theorem 4.8. It suffices to replace all occurrences of $\mathscr{S}(\mathbb{R}^d)$ by $\mathscr{R}(\mathbb{R}^d)$ and to restate the pointwise inequalities of the proof in the "almost everywhere" sense. The reason is that the only property that matters in the proof is the decay of φ, while its smoothness is irrelevant to the argument.

As we show next, reducing the constraints on the decay of the analysis functions is possible too, by imposing additional conditions on f.

THEOREM 8.2   *If f is a p-admissible Lévy exponent (see Definition 4.4) for some $p \geq 1$, then the Lévy-noise functional*

$$\widehat{\mathscr{P}_w}(\varphi) = \exp\left(\int_{\mathbb{R}^d} f\big(\varphi(r)\big)\, dr\right)$$

is continuous and positive definite over $L_p(\mathbb{R}^d)$.

Proof   Since the exponential function is continuous with $\exp(0) = 1$, it is sufficient to show that the functional

$$F(\varphi) = \log \widehat{\mathscr{P}_w}(\varphi) = \int_{\mathbb{R}^d} f\big(\varphi(r)\big)\, dr$$

is continuous over $L_p(\mathbb{R}^d)$ and conditionally positive definite of order one (see Definition 4.6). To that end, we first observe that $F(\varphi)$ is well defined for all $\varphi \in L_p(\mathbb{R}^d)$ with

$$|F(\varphi)| \leq \int_{\mathbb{R}^d} \big|f\big(\varphi(r)\big)\big|\, dr \leq C\|\varphi\|_p^p,$$

due to the p-admissibility of f. This assumption also implies that $|\omega| |f'(\omega)| \leq C|\omega|^p$, which translates into

$$|f(\omega_2) - f(\omega_1)| = \left|\int_{\omega_1}^{\omega_2} f'(\omega)\, d\omega\right|$$

$$\leq C\left|\int_{\omega_1}^{\omega_2} \omega^{p-1}\, d\omega\right| \leq C' \max(|\omega_1|^{p-1}, |\omega_2|^{p-1})\, |\omega_2 - \omega_1|$$

$$\leq C'(|\omega_1|^{p-1} + |\omega_2 - \omega_1|^{p-1})\, |\omega_2 - \omega_1|,$$

where we have used the fact that $\max(a, b) \leq |a| + |b - a|$. Next, we consider a convergent sequence $\{\varphi_n\}_{n=1}^\infty$ in $L_p(\mathbb{R}^d)$ whose limit is denoted by φ. We then have that

$$\left| F(\varphi_n) - F(\varphi) \right| \leq C' \int_{\mathbb{R}^d} \left(|\varphi(r)|^{p-1} |\varphi_n(r) - \varphi(r)| + |\varphi_n(r) - \varphi(r)|^p \right) dr$$

$$\leq C' \left(\|\varphi\|_p^{p-1} \|\varphi_n - \varphi\|_p + \|\varphi_n - \varphi\|_p^p \right),$$

$$\text{(by Hölder's inequality with } q = \tfrac{p}{p-1})$$

where the right-hand side converges to zero as $\lim_{n \to \infty} \|\varphi_n - \varphi\|_p = 0$, which proves the continuity of F on $L_p(\mathbb{R}^d)$. The second part of the statement is a direct consequence of the conditional positive definiteness of f. Indeed, for every choice $\varphi_1, \ldots, \varphi_N \in L_p(\mathbb{R}^d)$, $\xi_1, \ldots, \xi_N \in \mathbb{C}$, and $N \in \mathbb{Z}^+$, we have that

$$\sum_{m=1}^N \sum_{n=1}^N F(\varphi_m - \varphi_n) \xi_m \overline{\xi}_n = \int_{\mathbb{R}^d} \underbrace{\sum_{m=1}^N \sum_{n=1}^N f(\varphi_m(r) - \varphi_n(r)) \xi_m \overline{\xi}_n}_{\geq 0} \, dr \geq 0$$

subject to the constraint $\sum_{n=1}^N \xi_n = 0$.   □

8.3   Generalized increments for the decoupling of sample values

We recall that the generalized increment process associated with the stochastic process $s = \mathrm{L}^{-1}w$ is defined as

$$u(r) = \mathrm{L_d}s(r) = \sum_{k \in \mathbb{Z}^d} d_\mathrm{L}[k]s(r - k),$$

where $\mathrm{L_d}$ is the discrete-domain version of the operator L. It is a finite-difference-like operator that involves a suitable sequence of weights $d_\mathrm{L} \in \ell_1(\mathbb{Z}^d)$.

The motivation for computing $u = \mathrm{L_d}s$ is to partially decouple s and to remove its non-stationary components in the event that the inverse operator L^{-1} is not shift-invariant. The implicit requirement for producing stationary increments is that the continuous-domain composition of the operators $\mathrm{L_d}$ and L^{-1} is shift-invariant with impulse response $\beta_\mathrm{L} \in L_1(\mathbb{R}^d)$ (see Proposition 7.4). Our next result provides a general criterion for checking that this condition is met. It ensures that the scheme is applicable whenever the operator L is spline-admissible (see Definition 6.2) with associated generalized B-spline β_L, the construction of which is detailed in Section 6.4. The key, of course, is in the selection of the appropriate "localization" operator $\mathrm{L_d}$.

PROPOSITION 8.3   *Let* $\rho_\mathrm{L} = \mathcal{F}^{-1}\{1/\widehat{L}(\omega)\} \in \mathscr{S}'(\mathbb{R}^d)$ *be the Green's function of a spline-admissible operator* L *with frequency response* $\widehat{L}(\omega)$ *and associated B-spline*

$\beta_L = L_d \rho_L \in L_1(\mathbb{R}^d)$. Then, the operator $L^{-1} : \varphi(r) \mapsto \int_{\mathbb{R}^d} h(r, r')\varphi(r')\, dr'$ is a valid right inverse of L if its kernel is of the form

$$h(r, r') = L^{-1}\{\delta(\cdot - r')\}(r) = \rho_L(r - r') + p_0(r; r'),$$

where $p_0(r; r')$ with r' fixed is included in the null space $\mathcal{N}_L$ of L. Moreover, we have that

$$L_d L^{-1}\varphi = \beta_L * \varphi \tag{8.5}$$
$$L^{-1*}L_d^*\varphi = \beta_L^\vee * \varphi \tag{8.6}$$

for all $\varphi \in \mathscr{S}(\mathbb{R}^d)$.

Proof The equivalent kernel (see kernel Theorem 3.1) of the composed operator LL^{-1} is

$$\begin{aligned} LL^{-1}\{\delta(\cdot - r')\} &= L\{h(\cdot, r')\} \\ &= L\{\rho_L(\cdot - r') + p_0(\cdot; r')\} \\ &= L\{\rho_L(\cdot - r')\} + \underbrace{L\{p_0(\cdot; r')\}}_{0} \\ &= \delta(\cdot - r'), \end{aligned}$$

which proves that $LL^{-1} = \mathrm{Id}$, and hence that L^{-1} is a valid right inverse of L. Next, we consider the composed operator $L_d L^{-1}$ and apply the same procedure to show that

$$L_d L^{-1}\{\delta(\cdot - r')\} = \beta_L(\cdot - r'),$$

where we have used the property that $L_d p_0(\cdot; r') = 0$ since $\mathcal{N}_L$ is included in the null space of L_d (see (6.24) and text immediately below). It follows that

$$L_d L^{-1}\{\varphi\} = \int_{\mathbb{R}^d} \beta_L(\cdot - r')\varphi(r')\, dr' = \beta_L * \varphi,$$

which is a well-defined convolution operation since $\beta_L \in L_1(\mathbb{R}^d)$. Finally, Equation (8.6), which involves the corresponding left inverse L^{-1*} of L^*, is simply the adjoint of the latter relation. $\square$

A nice illustration of the first part of Proposition 8.3 is the specific form (7.51) of the kernel associated with fractional Brownian motion and its non-Gaussian variants (see Section 7.5.2).

As far as generalized B-splines are concerned, the proposition implies that the formal definition $\beta_L = L_d L^{-1}\delta$ is invariant to the actual choice of the right-inverse operator L^{-1} when $\mathcal{N}_L \cap \mathscr{S}'(\mathbb{R}^d)$ is non-trivial. This general form is an extension of the usual definition $\beta_L = L_d \rho_L$ (see (6.25)) that involves the (unique) shift-invariant right inverse of L whose impulse response is ρ_L.

From now on, we shall therefore assume that the inverse operator L^{-1} that acts on the innovation w fulfills the conditions in Proposition 8.3. We can then apply (8.5) to derive the convolution formula

$$u = L_d s = \beta_L * w. \tag{8.7}$$

Since this relation is central to our argument, we want to detail the explicit steps of its derivation. Specifically, for all $\varphi \in \mathscr{S}(\mathbb{R}^d)$, we have that

$$\langle \varphi, u \rangle = \langle \varphi, L_d s \rangle = \langle \varphi, L_d L^{-1} w \rangle$$
$$= \langle L^{-1*} L_d^* \varphi, w \rangle \qquad \text{(by duality)}$$
$$= \langle \beta_L^{\vee} * \varphi, w \rangle. \qquad \text{(from (8.6))}$$

This in turn implies that the characteristic functional of $u = L_d s$ is given by

$$\widehat{\mathscr{P}_u}(\varphi) = \widehat{\mathscr{P}_w}(\beta_L^{\vee} * \varphi). \qquad (8.8)$$

Consequently, the resulting generalized increment process is stationary even when the original process $s = L^{-1} w$ is not, which happens when L^{-1} fails to be shift-invariant. The intuition for this property is that L_d annihilates the signal components that are in the null space of L. Indeed, the form of the corrected left-inverse operators of Chapter 5 $\big($e.g., the integrator of (5.4)$\big)$ suggests that the non-stationary part of the signal corresponds to the null-space components (exponential polynomials) that are added to the solution during the inversion to fulfill the boundary conditions when the underlying SDE is unstable. The result also indicates that the decoupling effect will be strongest when the convolution kernel $\beta_L(r)$ is the most localized and closest to an impulse. The challenge is therefore to select L_d such that β_L is the most concentrated around the origin. This is precisely what the designer of "good" B-splines tries to achieve, as we saw in Chapter 6.

The final step is the discretization where $u = L_d s$ is sampled uniformly, which yields the sequence of generalized increments

$$u[k] = L_d s(r)|_{r=k} = (d_L * s)[k].$$

The practical interest of the last formula is that $\{u[k]\}_{k \in \mathbb{Z}^d}$ can be computed from the sampled values $s[k] = \langle \delta(\cdot - k), s \rangle, k \in \mathbb{Z}^d$, of the initial process s via a simple discrete convolution with d_L (digital filtering). In most cases, the mapping from $s[\cdot]$ to $u[\cdot]$ can also be inverted, as we saw in the case of the Lévy process $\big($see (8.2)$\big)$.

8.3.1   First-order statistical characterization

To determine the pointwise statistical description of the generalized increment process u, we consider the random variable $U = \langle \beta_L^{\vee}, w \rangle$. Its characteristic function $\widehat{p}_U(\omega)$ is obtained by plugging $\varphi = \omega \delta$ into the characteristic functional (8.8). In this way, we are able to specify its first-order pdf via the inverse Fourier transformation

$$p_U(x) = \int_{\mathbb{R}} e^{\int_{\mathbb{R}^d} f\left(\omega \beta_L^{\vee}(r)\right) dr} e^{-j\omega x} \frac{d\omega}{2\pi}.$$

This results in an id pdf with the modified Lévy exponent

$$f_{\beta_L}(\omega) = \int_{\mathbb{R}^d} f\left(\omega \beta_L(-r)\right) dr$$
$$= \int_{\mathbb{R}^d} f\left(\omega \beta_L(r)\right) dr. \qquad \text{(by change of variable)}$$

For instance, in the case of a symmetric-alpha-stable (SαS) innovation with $f(\omega) = -\frac{1}{\alpha!}|s_0\omega|^\alpha$ and dispersion parameter s_0, we find that p_U is SαS as well, with new dispersion parameter $s_U = s_0\|\beta_L\|_{L_\alpha}$, where $\|\beta_L\|_{L_\alpha}$ is the (pseudo) L_α-norm of the B-spline. In particular, for $\alpha = 2$ this shows that the generalized increments of a Gaussian process are Gaussian as well, with a variance that is proportional to the squared L_2-norm of the B-spline associated with the whitening operator L.

8.3.2   Higher-order statistical dependencies

As already mentioned, the decoupling afforded by L_d is not perfect. In the spatial (or temporal) domain, the remaining convolution of w with β_L induces dependencies over the spatial neighborhood that intersects with the support of the B-spline. When this support is greater than the sampling step ($T = 1$), this introduces residual coupling among the elements of $u[\cdot]$. This effect can be accounted for exactly by considering a K-point neighborhood $\Omega_K(k_0)$ of k_0 on the Cartesian lattice $\mathbb{Z}^d$ and by defining the corresponding K-dimensional vector $\mathbf{u}[k_0] = (U_1, \ldots, U_K)$ whose components are the values $u[k]$ with $k \in \Omega_K(k_0)$. The Kth-order joint pdf of $\mathbf{u}[k_0]$ is denoted by $p_{(U_1:U_K)}(\mathbf{u})$. Since the generalized-increment process $u = L_d s$ is stationary, $p_{(U_1:U_K)}(\mathbf{u})$ does not depend on the (absolute) location k_0. We determine its K-dimensional Fourier transform (Kth-order characteristic function) by making the substitution $\varphi = \sum_{k \in \Omega_K} \omega_k \delta(\cdot - k)$ in (8.8), which yields

$$\widehat{p}_{(U_1:U_K)}(\boldsymbol{\omega}) = \mathbb{E}\{e^{j\langle\boldsymbol{\omega},\mathbf{u}\rangle}\} = \exp\left(\int_{\mathbb{R}^d} f\left(\sum_{k \in \Omega_K} \omega_k \beta_L^\vee(r - k)\right) dr\right),$$

where $\boldsymbol{\omega}$ is the K-dimensional frequency variable with components ω_k where $k \in \Omega_K = \Omega_K(0)$. In addition, when the innovation fulfills the second-order conditions of Proposition 4.15, the autocorrelation function of the generalized-increment process is given by

$$R_{L_d s}(r) = \mathbb{E}\left\{u(\cdot + r)\overline{u(\cdot)}\right\} = \sigma_w^2(\beta_L * \overline{\beta_L^\vee})(r). \tag{8.9}$$

Hence, the correlation between the discretized increments is simply

$$\mathbb{E}\left\{u[k_1]\overline{u[k_2]}\right\} = \sigma_w^2(\beta_L * \overline{\beta_L^\vee})(k_1 - k_2).$$

This points once more to the importance of selecting the discrete operator L_d such that the support of $\beta_L = L_d \rho_L$ is minimal or decays as fast as possible when a compact support is not achievable.

This analysis clearly shows that the support of the B-spline governs the range of dependency of the generalized increments with the property that $u[k_1]$ and $u[k_2]$ are independent whenever $|k_1 - k_2| \geq \mathrm{support}(\beta_L)$. In particular, this implies that the sequence $u[\cdot]$ is i.i.d. if and only if $\mathrm{support}(\beta_L) \leq 1$, which is precisely the case for the first-order B-splines β_α with $\alpha \in \mathbb{C}$, which go hand-in-hand with the Lévy ($\alpha = 0$) and AR(1) processes.

8.3.3 Generalized increments and stochastic difference equations

To illustrate the procedure, we now focus on the extended family of generalized Lévy processes s_α of Section 7.4.2. These are solutions of ordinary differential equations with rational transfer functions specified by their poles $\boldsymbol{\alpha} = (\alpha_1, \ldots, \alpha_N)$ and zeros $\boldsymbol{\gamma} = (\gamma_1, \ldots, \gamma_M)$ with $M < N$. The extension over the classical CARMA framework is that the poles can be arbitrarily located in the complex plane so that the underlying system may be unstable. The associated B-spline is given by $\bigl($see (7.36)$\bigr)$

$$\beta_{\mathrm{L}}(t) = q_M(\mathrm{D})\beta_{\boldsymbol{\alpha}}(t) = \int_{\mathbb{R}} e^{\mathrm{j}\omega t} q_M(\mathrm{j}\omega) \prod_{n=1}^{N} \left(\frac{1 - e^{\alpha_n - \mathrm{j}\omega}}{\mathrm{j}\omega - \alpha_n} \right) \frac{\mathrm{d}\omega}{2\pi}, \tag{8.10}$$

where $q_M(\zeta) = b_M \prod_{m=1}^{M}(\zeta - \gamma_m)$ and $\beta_{\boldsymbol{\alpha}}$ is the exponential B-spline defined by (7.34). The generalized-increment process is then given by

$$u_{\boldsymbol{\alpha}}(t) = \Delta_{\boldsymbol{\alpha}} s_{\boldsymbol{\alpha}}(t) = (\beta_{\mathrm{L}} * w)(t), \tag{8.11}$$

where $\Delta_{\boldsymbol{\alpha}}$ is the finite-difference operator defined by (7.35). A more explicit characterization of $\Delta_{\boldsymbol{\alpha}}$ is through its weights $d_{\boldsymbol{\alpha}}$, whose representation in the z-transform domain is

$$D_{\boldsymbol{\alpha}}(z) = \sum_{k=0}^{N} d_{\boldsymbol{\alpha}}[k] z^{-k} = \prod_{n=1}^{N} (1 - e^{\alpha_n} z^{-1}). \tag{8.12}$$

By sampling (8.11) at the integers, we obtain

$$u_{\boldsymbol{\alpha}}[k] = u_{\boldsymbol{\alpha}}(t)|_{t=k} = \sum_{n=0}^{N} d_{\boldsymbol{\alpha}}[n] s_{\boldsymbol{\alpha}}[k - n]. \tag{8.13}$$

In light of this relation, it is tempting to investigate whether it is possible to decorrelate $u_{\boldsymbol{\alpha}}[\cdot]$ even further in order to describe $s_{\boldsymbol{\alpha}}[\cdot]$ through a discrete ARMA-type model. Ideally, we would like to come up with an equivalent discrete-domain innovation model that is easier to exploit numerically than the defining stochastic differential equation (7.19). To that end, we perform the spectral factorization of the sampled augmented B-spline kernel

$$B_{\mathrm{L}}(z) = \sum_{k=-N}^{N} \beta_{\overline{\mathrm{L}}^*}(k) z^{-k} = B_{\mathrm{L}}^+(z) B_{\mathrm{L}}^-(z), \tag{8.14}$$

where $B_{\mathrm{L}}^+(z) = \sum_{k=0}^{N-1} b_{\mathrm{L}}^+[k] z^{-k} = B_{\mathrm{L}}^-(z^{-1})$ specifies a causal finite-impulse-response (FIR) filter of size N. The crucial point for the argument below is that $B_{\mathrm{L}}^+(e^{\mathrm{j}\omega})$ $\bigl($or, equivalently, $B_{\mathrm{L}}(e^{\mathrm{j}\omega})\bigr)$ is non-vanishing. To that end, we invoke the Riesz-basis property of an admissible B-spline (see Definition 6.8) together with (6.19), which implies that

$$B_{\mathrm{L}}(e^{\mathrm{j}\omega}) = \sum_{n \in \mathbb{Z}} |\widehat{\beta}_{\mathrm{L}}(\omega + 2\pi n)|^2 \geq A^2 > 0.$$

PROPOSITION 8.4 (Stochastic difference equation)   *The sampled version $s_\alpha[\cdot]$ of a generalized second-order Lévy process with pole vector α and associated B-spline β_L satisfies the discrete ARMA-type whitening equation*

$$\sum_{n=0}^{N} d_\alpha[n] s_\alpha[k-n] = \sum_{m=0}^{N-1} b_L^+[m] e[k-m],$$

where d_α and b_L^+ are defined by (8.12) and (8.14), respectively. The driving term $e[k]$ is a discrete stationary white noise ("white" meaning fully decorrelated or with a flat power spectrum). However, $e[k]$ is a valid innovation sequence with i.i.d. samples only if the corresponding continuous-domain process is Gaussian, or, in full generality (i.e., in the non-Gaussian case), if it is a first-order Markov or Lévy-type process with $N = 1$.

Proof   Since $|B_L^+(e^{j\omega})| = \sqrt{B_L(e^{j\omega})}$ is non-vanishing and is a trigonometric polynomial of $e^{j\omega}$ whose roots are inside the unit circle, we have the guarantee that the inverse filter whose frequency response is $\frac{1}{B_L^+(e^{j\omega})}$ is causal-stable. It follows that $\Phi_e(e^{j\omega}) = \sigma_w^2 \frac{\sum_{n\in\mathbb{Z}} |\widehat{\beta}_L(\omega+2\pi n)|^2}{B_L(e^{j\omega})} = \sigma_w^2$, which proves the first part of the statement. As for the second part, we recall that decorrelation is equivalent to independence in the Gaussian case only. In the non-Gaussian case, the only way to ensure independence is by restricting ourselves to a first-order process, which results in an AR(1)-type equation with $e[n] = u_{\alpha_1}[n]$. Indeed, since the corresponding B-spline is a size 1, $u_{\alpha_1}[\cdot]$ is i.i.d., which implies that $p_U\left(u_{\alpha_1}[k] \,\middle|\, \{u_{\alpha_1}[k-m]\}_{m\in\mathbb{Z}^+}\right) = p_U(u_{\alpha_1}[k])$. This is equivalent to $s_{\alpha_1}[\cdot]$ having the Markov property since $p_S\left(s_{\alpha_1}[k] \,\middle|\, \{s_{\alpha_1}[k-m]\}_{m\in\mathbb{Z}^+}\right) = p_U(u_{\alpha_1}[k]) = p_S\left(s_{\alpha_1}[k] \,\middle|\, s_{\alpha_1}[k-1]\right)$.   □

8.3.4   Discrete whitening filter

The specification of a discrete-domain whitening filter that fully decorrelates the sampled version $s[\cdot]$ of a second-order stochastic process $s(\cdot)$ is actually feasible in all generality under the assumption that the operator L is spline-admissible. However, it is only really useful for Gaussian processes, since this is the only scenario where decorrelation is synonymous with statistical independence. The multidimensional frequency response of the corresponding discrete whitening filter is

$$\widehat{L}_G(\omega) = \frac{\widehat{L}_d(\omega)}{\sqrt{\sum_{n\in\mathbb{Z}^d} |\widehat{\beta}_L(\omega+2\pi n)|^2}}. \tag{8.15}$$

The Riesz-basis property of the B-spline ensures that the denominator of (8.15) is non-vanishing, so that we may invoke Wiener's lemma (Theorem 5.13) to show that the filter is stable. Generally, the support of its impulse response is infinite, unless the support of the B-spline is unity (Markov property).

8.3.5   Robust localization

We end our discussion of generalized increments with a description of a robust variant that is obtained by replacing the canonical localization operator L_d by some shorter filter $\tilde{L}_d$. This option is especially useful for decoupling fractal-type processes that are associated with fractional whitening operators whose discrete counterparts have an infinite support. The use of ordinary finite-difference operators, in particular, is motivated by Theorem 7.7 because of their stationarizing effect.

The guiding principle is to select a localization filter $\tilde{L}_d$ that has a compact support and is associated with some "augmented" operator $\tilde{L} = L_0 L$, where L_0 is a suitable differential operator. The natural candidate for $\tilde{L}$ is a (non-fractional) differential operator of integer order $\tilde{\gamma} \geq \gamma$ whose null space is identical to that of L and whose B-spline $\beta_{\tilde{L}}$ has a compact support. This latter function is given by

$$\beta_{\tilde{L}} = \tilde{L}_d \rho_{\tilde{L}},$$

where $\rho_{\tilde{L}} = \mathscr{F}^{-1}\{1/\widehat{\tilde{L}}(\omega)\}$ is the Green's function of $\tilde{L}$ whose Fourier symbol is $\widehat{\tilde{L}}(\omega) = \widehat{L}_0(\omega)\widehat{L}(\omega)$. The subsequent application of L_0 then yields the smoothing kernel

$$\phi(r) = L_0 \beta_{\tilde{L}}(r) \tag{8.16}$$
$$= \tilde{L}_d L_0 \rho_{\tilde{L}}(r) = \tilde{L}_d \rho_L(r) \tag{8.17}$$
$$= \sum_k d_{\tilde{L}}[k] \rho_L(r - k).$$

This shows that ϕ is a (finite) linear combination of the integer shifts of ρ_L (the Green's function of the original operator L), and hence is a cardinal L-spline (see Definition 6.4).

To describe the decoupling effect of $\tilde{L}_d$ on the signal $s = L^{-1}w$, we observe that (8.17) is equivalent to $\phi = \tilde{L}_d L^{-1}\delta$, which yields

$$\tilde{u} = \tilde{L}_d s = \tilde{L}_d L^{-1} w$$
$$= \phi * w.$$

This equation is the same as (8.7), except that we have replaced the original B-spline β_L by the smoothing kernel defined by (8.16). The procedure is acceptable whenever $\phi \in L_p(\mathbb{R}^d)$, and its decay at infinity is comparable to that of β_L. We call such a scheme a *robust localization* because its qualitative effect is the same as that of the canonical operator L_d. For instance, we can rely on the results of Chapter 9 to show that the statistical distributions of $\tilde{u}$ and u have the same global properties (sparsity pattern). The price to pay is a slight loss in decoupling power because the localization of ϕ is worse than that of β_L, itself being the best that can be achieved within the given spline space (B-spline property).

To access the remaining dependencies, it is instructive to determine the corresponding autocorrelation function

$$R_{\tilde{u}}(r) = \mathbb{E}\left\{\tilde{u}(\cdot + r)\overline{\tilde{u}(\cdot)}\right\} = \sigma_w^2 (L_0 \overline{L}_0^*)\{\beta_{\tilde{L}} * \overline{\beta}_{\tilde{L}}^{\vee}\}(r) \tag{8.18}$$

that primarily depends upon the size of the augmented-order B-spline $\beta_{\tilde{L}}$. Now, in the special case where L_0 is an all-pass operator with $|\widehat{L}_0(\omega)|^2 = 1$, we have that $|\widehat{\beta}_{\tilde{L}}(\omega)|^2 = |\widehat{\beta}_L(\omega)|^2$ so that the autocorrelation functions of $u = L_d s$ and $\tilde{u} = \tilde{L}_d s$ are identical. This implies that the decorrelation effect of the localization filters L_d and $\tilde{L}_d$ are equivalent, which justifies replacing one by the other.

Example of fractional derivatives

The fractional derivatives ∂_τ^γ, which are characterized in Proposition 5.6, are scale-invariant by construction. Since the family is complete, it implicitly specifies the broadest class of 1-D self-similar processes that are solutions of fractional SDEs (see Section 7.5). Here, we shall focus on the symmetric versions of these derivatives with $\tau = 0$ and Fourier multiplier $\widehat{L}(\omega) = |\omega|^\gamma$. The corresponding symmetric fractional B-splines of degree $\alpha = (\gamma - 1)$ are specified by the Fourier-domain formula (see [UB00])

$$\widehat{\beta_0^\alpha}(\omega) = \frac{\left|1 - e^{-j\omega}\right|^{\alpha+1}}{|\omega|^{\alpha+1}}. \tag{8.19}$$

To determine the time-domain counterpart of this equation, we rely on a generalized version of the binomial expansion, which is due to Thierry Blu.

THEOREM 8.5 (Blu's generalized binomial expansion)   *Let $u, v \in \mathbb{R}$ with $u + v > -\frac{1}{2}$. Then, for any $z = e^{j\theta}$ on the unit circle,*

$$(1 + z)^u (1 + z^{-1})^v = \sum_{k \in \mathbb{Z}} \binom{u + v}{u + k} z^{-k},$$

with generalized binomial coefficients

$$\binom{u}{v} = \frac{u!}{u! \, (u - v)!} = \frac{\Gamma(u + 1)}{\Gamma(v + 1) \, \Gamma(u - v + 1)}.$$

We then apply Theorem 8.5 with $u = v = \frac{\alpha+1}{2}$ to expand the numerator of (8.19) and compute the inverse Fourier transform (see Table A.1). This yields

$$\beta_0^\alpha(t) = \beta_{\partial_0^{\alpha+1}}(t) = \sum_{k \in \mathbb{Z}} d_{\alpha,0}[k] \rho_{\alpha,0}(t - k), \tag{8.20}$$

where

$$d_{\alpha,0}[k] = (-1)^k \binom{\alpha + 1}{\frac{\alpha+1}{2} + k} \tag{8.21}$$

and

$$\rho_{\alpha,0}(t) = \mathscr{F}^{-1}\left\{ \frac{1}{|\omega|^{\alpha+1}} \right\}(t) = \begin{cases} \frac{(-1)^{n+1}}{\pi(2n)!} t^{2n} \log|t|, & \text{for } \alpha = 2n \in 2\mathbb{N} \\[2mm] \frac{-1}{2\Gamma(\alpha+1) \sin(\frac{\pi}{2}\alpha)} |t|^\alpha, & \text{for } \alpha \in \mathbb{R}^+ \setminus 2\mathbb{N}. \end{cases}$$

Observe that $\rho_{\alpha,0}$ is the Green's function of the fractional-derivative operator $\partial_0^{\alpha+1}$, while the $d_{\alpha,0}[k]$ are the coefficients of the corresponding (canonical) localization filter. The simplest instance occurs for $\alpha = 1$, where (8.20) reduces to

$$\beta_0^1(t) = \tfrac{1}{2}|t + 1| - |t| + \tfrac{1}{2}|t - 1|,$$

which is the triangular function supported in $[-1, 1]$ (symmetric B-spline of degree one) shown in Figure 8.1c. In general, however, the fractional B-splines $\beta_0^\alpha(t)$ with $\alpha \in \mathbb{R}^+$ are not compactly supported, unless α is an odd integer. In fact, they can be shown to decay like $O(1/t^{\alpha+2})$, which is a behavior that is characteristic of fractional operators.

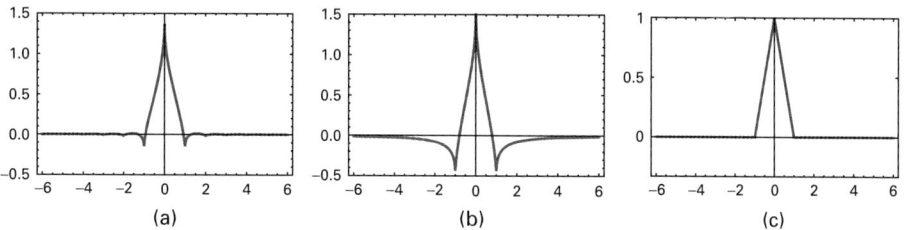

Figure 8.1 Comparison of kernels related to the symmetric fractional derivative $\partial_0^{1/2+1}$ with Fourier multipler $|\omega|^{3/2}$. (a) Fractional B-spline $\beta_0^{1/2}(t)$, (b) Localization kernel $\phi_{1/2,1}(t) = \partial_0^{1/2}\beta_0^1(t)$, (c) Augmented-order B-spline $\beta_0^1(t)$.

As far as the implementation of the localization operator L_d is concerned, the most favorable scenario is when $\alpha = 2n + 1$ is odd, in which case $d_{2n+1,0}[\cdot]$ is a (modulated) binomial sequence of length $2n + 2$ that corresponds to the $(2n + 2)$th centered finite-difference operator. Otherwise, $d_{\alpha,0}[\cdot]$ has an infinite length and decays like $O(1/|k|^{\alpha+2})$, which makes the numerical implementation of the filter impractical. This suggests switching to a robust localization where $d_{\alpha,0}$ is replaced by $d_{2n+1,0}$, with $(2n - 1) < \alpha \leq (2n + 1)$.

To be specific, we now consider the case $0 < \alpha \leq 1$ with $L = \partial_0^{\alpha+1}$ and the augmented-order operator $\tilde{L} = D^2 = \partial_0^{1-\alpha}\partial_0^{\alpha+1}$. The corresponding smoothing kernel is then given by

$$\phi_{\alpha,1}(t) = \partial_0^{1-\alpha}\beta_0^1(t) = \frac{1}{\Gamma(\alpha + 1)\sin\left(\frac{\pi}{2}\alpha\right)}\left(\tfrac{1}{2}|t + 1|^\alpha - |t|^\alpha + \tfrac{1}{2}|t - 1|^\alpha\right).$$

This function decays like $O(1/|t|^{-\alpha+2})$ since the finite-difference operator asymptotically acts as a second-order derivative. The relevant functions for $\alpha = 1/2$ are shown in Figure 8.1. The fractional-derivative operator $\partial_0^{1/2}$ is a non-local operator. Its application to the compactly supported B-spline β_0^1 produces a kernel with algebraically decaying tails. While $\beta_0^{1/2}$ and $\phi_{1/2,1}$ are qualitatively similar, the former function is better localized with smaller secondary lobes, which reflects the superior decoupling performance of the canonical scheme. Yet, this needs to be balanced against the fact that the robust version uses a three-tap filter (second-order finite difference), as opposed to the solution (8.21) of infinite support dictated by the theory.

8.4 Wavelet analysis

The other option for uncoupling the information is to analyze the signal $s(r)$ using wavelet-like functions. The implicit assumption for the following properties is that we have a real-valued wavelet basis available that is matched to the operator L. Specifically, the structure of the transform must be such that the basis functions $\psi_{i,k}^{(n)}$ at scale i are translated versions of a fixed number N_0 (typically, $N_0 = 1$ or $N_0 = \det(\mathbf{D}) - 1$) of normalized "mother" wavelets of the form

$$\psi_i^{(n)}(\mathbf{r}) = \mathrm{L}^* \phi_i^{(n)}(\mathbf{r}),$$

where the $\phi_i^{(n)}$ with $n = 1, \dots, N_0$ are scale-dependent smoothing kernels whose width is proportional to the scale $a_i = \det(\mathbf{D})^{i/d}$, where $\mathbf{D}$ is the underlying dilation matrix (e.g., $\mathbf{D} = 2\mathbf{I}$ for a standard dyadic scale progression). In the traditional wavelet transform, L is scale-invariant and the wavelets at resolution i are dilated versions of the primary ones at level $i = 0$ with $\phi_i^{(n)}(\mathbf{r}) \propto \phi_0^{(n)}(\mathbf{D}^{-i}\mathbf{r})$.

Here, for simplicity of notation, we shall focus on the general operator-like design of Section 6.5 with $N_0 = 1$, which has the advantage of involving the single mother wavelet

$$\psi_i(\mathbf{r}) = \mathrm{L}^* \phi_i(\mathbf{r}),$$

while the complete set of basis functions is represented by

$$\begin{aligned} \psi_{i,k} &= \mathrm{L}^* \phi_i(\cdot - \mathbf{D}^{i-1}k) \\ &= \mathrm{L}^* \phi_{i,k} \end{aligned} \tag{8.22}$$

with $i \in \mathbb{Z}$ and $k \in \mathbb{Z}^d \setminus \mathbf{D}\mathbb{Z}^d$. The technical assumption for the wavelet coefficients $Y_i[k] = \langle \psi_{i,k}, s \rangle$ to be well defined is that $\mathrm{L}^{-1*}\psi_{i,k} = \phi_{i,k} \in \mathscr{X}$, where $\mathscr{X} = \mathscr{R}(\mathbb{R}^d)$ or, in the event that f is p-admissible, $\mathscr{X} = L_p(\mathbb{R}^d)$ with $p \in [1, 2]$.

8.4.1   Wavelet-domain statistics

The key observation is that the wavelet coefficients $Y_{i,k} = \langle \psi_{i,k}, s \rangle$ of the random signal s at a given scale i can be obtained by first correlating the signal with the wavelet $\psi_i = \mathrm{L}^* \phi_i$ (continuous wavelet transform) and sampling thereafter. Similar to the case of the generalized increments, this has a stationarizing and decoupling effect.

PROPOSITION 8.6 (Wavelet-domain pdf)   *Let $v_i(\mathbf{r}) = \langle \psi_i(\cdot - \mathbf{r}), s \rangle$ with $\psi_i = \mathrm{L}^* \phi_i$ be the ith channel of the continuous wavelet transform of the generalized (stationary or non-stationary) stochastic process $s = \mathrm{L}^{-1}w$, where w is an innovation process with Lévy exponent f. Then, v_i is a generalized stationary process with characteristic functional $\widehat{\mathscr{P}}_{v_i}(\varphi) = \widehat{\mathscr{P}}_w(\phi_i * \varphi)$, where $\widehat{\mathscr{P}}_w$ is defined by (4.13). Moreover, the characteristic function of the (discrete) wavelet coefficient $Y_i[k] = \langle \psi_{i,k}, s \rangle = v_i(\mathbf{D}^{i-1}k)$ is given by $\widehat{p}_{Y_i}(\omega) = \widehat{\mathscr{P}}_w(\omega \phi_i) = e^{f_{\phi_i}(\omega)}$ and is infinitely divisible with Lévy exponent*

$$f_{\phi_i}(\omega) = \int_{\mathbb{R}^d} f\big(\omega \phi_i(\mathbf{r})\big) \, d\mathbf{r}.$$

Proof   Recalling that $s = \mathrm{L}^{-1}w$, we get

$$\begin{aligned} v_i(\mathbf{r}) &= \langle \psi_i(\cdot - \mathbf{r}), s \rangle = \langle \mathrm{L}^* \phi_i(\cdot - \mathbf{r}), \mathrm{L}^{-1}w \rangle \\ &= \langle \mathrm{L}^{-1*}\mathrm{L}^* \phi_i(\cdot - \mathbf{r}), w \rangle \\ &= \langle \phi_i(\cdot - \mathbf{r}), w \rangle = (\phi_i^\vee * w)(\mathbf{r}), \end{aligned}$$

where we have used the fact that L^{-1*} is a valid (continuous) left inverse of L^*. Since $\phi_i \in \mathscr{X}$, we can invoke Proposition 7.2 to prove the first part.

For the second part, we also use the fact that $Y_{i,k} = \langle \psi_{i,k}, s \rangle = \langle \phi_{i,k}, w \rangle$. Based on the definition of the characteristic functional $\widehat{\mathscr{P}}_w(\varphi) = \mathbb{E}\{e^{j\langle \varphi, w \rangle}\}$, we then obtain

$$
\begin{aligned}
\widehat{p}_{Y_i}(\omega) &= \mathbb{E}\{e^{j\omega Y_{i,k}}\} = \mathbb{E}\{e^{j\omega \langle \phi_{i,k}, w \rangle}\} \\
&= \mathbb{E}\{e^{j\omega \langle \phi_i, w \rangle}\} = \mathbb{E}\{e^{j\langle \omega\phi_i, w \rangle}\} \qquad \text{(by stationarity and linearity)} \\
&= \widehat{\mathscr{P}}_w(\omega\phi_i) = \exp\left(\int_{\mathbb{R}^d} f\big(\omega\phi_i(r)\big)\, dr\right),
\end{aligned}
$$

where we have inserted the explicit form given in (4.13). The result then follows by identification. Since $\widehat{\mathscr{P}}_w : \mathscr{X} \to \mathbb{C}$ is a continuous, positive definite functional with $\widehat{\mathscr{P}}_w(0) = 1$, we conclude that the function f_{ϕ_i} is conditionally positive of order one (by Schoenberg's correspondence Theorem 4.7) so that it is a valid Lévy exponent (see Definition 4.1). This proves that the underlying pdf is infinitely divisible (by Theorem 4.1). $\qquad \square$

We determine the joint characteristic function of any two wavelet coefficients $Y_1 = \langle \psi_{i_1,k_1}, s \rangle$ and $Y_2 = \langle \psi_{i_2,k_2}, s \rangle$ with indices (i_1, k_1) and (i_2, k_2) using a similar technique.

PROPOSITION 8.7 (Wavelet dependencies)   *The joint characteristic function of the wavelet coefficients $Y_1 = Y_{i_1}[k_1] = \langle \psi_{i_1,k_1}, s \rangle$ and $Y_2 = Y_{i_2}[k_2] = \langle \psi_{i_2,k_2}, s \rangle$ of the generalized stochastic process s in Proposition 8.6 is given by*

$$
\widehat{p}_{Y_1,Y_2}(\omega_1, \omega_2) = \exp\left(\int_{\mathbb{R}^d} f\big(\omega_1\phi_{i_1,k_1}(r) + \omega_2\phi_{i_2,k_2}(r)\big)\, dr\right),
$$

where f is the Lévy exponent of the innovation process w. The coefficients are independent if the kernels ϕ_{i_1,k_1} and ϕ_{i_2,k_2} have disjoint support. Their correlation is given by

$$
\mathbb{E}\{Y_1 Y_2\} = \sigma_w^2 \langle \phi_{i_1,k_1}, \phi_{i_2,k_2} \rangle \tag{8.23}
$$

under the assumption that the variance $\sigma_w^2 = -f''(0)$ of w is finite.

Proof   The first formula is obtained by substitution of $\varphi = \omega_1\psi_{i_1,k_1} + \omega_2\psi_{i_2,k_2}$ in $\mathbb{E}\{e^{j\langle \varphi, s \rangle}\} = \widehat{\mathscr{P}}_w(L^{-1*}\varphi)$ and simplification using the left-inverse property of L^{-1*}. The statement about independence follows from the exponential nature of the characteristic function and the property that $f(0) = 0$, which allows for the factorization of the characteristic function when the support of the kernels are distinct (independence of the noise at every point). The correlation formula is obtained by direct application of (7.7) with $\varphi_1 = \psi_{i_1,k_1} = L^*\phi_{i_1,k_1}$ and $\varphi_2 = \psi_{i_2,k_2} = L^*\phi_{i_2,k_2}$. $\qquad \square$

These results provide a complete characterization of the statistical distribution of sparse stochastic processes in some matched-wavelet domain. They also indicate that the representation is intrinsically sparse since the transformed-domain statistics are infinitely divisible. Practically, this translates into the wavelet-domain pdfs having heavier tails than a Gaussian (unless the process is Gaussian) – see the general argument and results of Chapter 9.

It should be noted, however, that the quality of the decoupling depends strongly on the spread of the smoothing kernels ϕ_i, which should be chosen to be maximally localized for best performance. In the case of a first-order system (see the example in Section 6.3.3 and the wavelets in Figure 6.3d), the basis functions for i fixed are not overlapping, which implies that the wavelet coefficients within a given scale are independent. This is not so across scales because of the cone-shaped region where the support of the kernels ϕ_{i_1} and ϕ_{i_2} overlap. Incidentally, the inter-scale correlation of wavelet coefficients is often exploited in practice to improve coding performance [Sha93] and signal reconstruction by imposing joint sparsity constraints [CNB98].

8.4.2   Higher-order wavelet dependencies and cumulants

To describe higher-order interactions, we use K-dimensional multi-index vectors of the form $\boldsymbol{n} = (n_1, \ldots, n_K)$ whose entries n_k are non-negative integers. We then define the following multi-index operations and operators:

- Sum of components $|\boldsymbol{n}| = \sum_{k=1}^{K} n_k = n$
- Factorial $\boldsymbol{n}! = n_1! n_2! \cdots n_K!$ with the convention that $0! = 1$
- Exponentiation of a vector $\boldsymbol{z} = (z_1, \ldots, z_K) \in \mathbb{C}^K$: $\boldsymbol{z}^{\boldsymbol{n}} = z_1^{n_1} \cdots z_K^{n_K}$
- Higher-order partial derivative of a function $f(\boldsymbol{\omega})$, $\boldsymbol{\omega} = (\omega_1, \ldots, \omega_K) \in \mathbb{R}^K$:

$$\partial^{\boldsymbol{n}} f(\boldsymbol{\omega}) = \frac{\partial^{|\boldsymbol{n}|} f(\boldsymbol{\omega})}{\partial \omega_1^{n_1} \cdots \partial \omega_K^{n_K}}.$$

The notation allows for the concise description of the multinomial theorem given by

$$\left(\sum_{k=1}^{K} \mathrm{j}\omega_k \right)^n = \sum_{|\boldsymbol{n}|=n} \frac{n!}{\boldsymbol{n}!} \mathrm{j}^n \boldsymbol{\omega}^{\boldsymbol{n}},$$

which involves a summation over $\binom{n+K-1}{K-1}$ distinct monomials of the form $\boldsymbol{\omega}^{\boldsymbol{n}} = \omega_1^{n_1} \cdots \omega_K^{n_K}$ with $n_1 + \cdots + n_K = n$. It also yields a compact formula for the Nth-order Taylor series of a multidimensional function $f(\boldsymbol{\omega})$ with well-defined derivatives up to order $N+1$,

$$f(\boldsymbol{\omega}_0 + \boldsymbol{\omega}) = \sum_{n=|\boldsymbol{n}|=0}^{N} \frac{\partial^{\boldsymbol{n}} f(\boldsymbol{\omega}_0)}{\boldsymbol{n}!} \boldsymbol{\omega} - \boldsymbol{\omega}_0^{\boldsymbol{n}} + O(\|\boldsymbol{\omega} - \boldsymbol{\omega}_0\|^{N+1}). \tag{8.24}$$

Let $\widehat{p}_{(X_1 : X_K)}(\boldsymbol{\omega})$, $\boldsymbol{\omega} \in \mathbb{R}^K$, be the multidimensional characteristic function associated with the Kth-order joint pdf $p_{(X_1 : X_K)}(\boldsymbol{x})$ whose polynomial moments are all assumed to be finite. Then, if the function $f(\boldsymbol{\omega}) = \log \widehat{p}_{(X_1 : X_K)}(\boldsymbol{\omega})$ is well defined, we can write the full multidimensional Taylor series expansion

$$f(\boldsymbol{\omega}) = \log \widehat{p}_{(X_1 : X_K)}(\boldsymbol{\omega}) = \sum_{n=|\boldsymbol{n}|=0}^{\infty} \frac{\partial^{\boldsymbol{n}} f(0)}{\boldsymbol{n}!} \boldsymbol{\omega}^{\boldsymbol{n}}$$

$$= \sum_{n=|\boldsymbol{n}|=0}^{\infty} \underbrace{(-\mathrm{j})^n \partial^{\boldsymbol{n}} f(0)}_{\kappa_{\boldsymbol{n}}} \frac{\mathrm{j}^n \boldsymbol{\omega}^{\boldsymbol{n}}}{\boldsymbol{n}!}, \tag{8.25}$$

where the internal summation is through all multi-indices whose sum is $|n| = n$. By definition, the *cumulants* of $p_{(X_1:X_K)}(x)$ are the coefficients of this expansion, so that

$$\kappa_n = (-\mathrm{j})^{|n|} \partial^n f(0).$$

The interest of these quantities is that they are in one-to-one relation with the (multidimensional) moments of $p_{(X_1:X_K)}$ defined by

$$m_n = \int_{\mathbb{R}^K} x^n p_{(X_1:X_K)}(x) \, \mathrm{d}x = \int_{\mathbb{R}^K} x_1^{n_1} \cdots x_K^{n_K} p_{(X_1:X_K)}(x) \, \mathrm{d}x,$$

which also happen to be the coefficients of the Taylor-series expansion of $\widehat{p}_{(X_1:X_K)}(\omega)$ around $\omega = 0$. Since id laws are specified in terms of their Lévy exponent f, it is often easier to determine their cumulants than their moments. Another advantage of cumulants is that they provide a direct measure of the deviation from a Gaussian, whose cumulants are zero for orders greater than two (since a Gaussian Lévy exponent is necessarily quadratic).

PROPOSITION 8.8 (Higher-order wavelet dependencies) *Let $\{(i_k, k_k)\}_{k=1}^K$ be a set of indices and $\{Y_k = \langle \psi_{i_k,k_k}, s \rangle\}_{k=1}^K$ with $\psi_{i_k,k_k} = \mathrm{L}^* \phi_{i_k,k_k}$ be the corresponding wavelet coefficients of the generalized stochastic process s in Proposition 8.6. Then, the joint characteristic function of $(Y_1, \ldots, Y_K)$ is given by*

$$\widehat{p}_{(Y_1:Y_K)}(\omega_1, \ldots, \omega_K) = \exp\left(\int_{\mathbb{R}^d} f\big(\omega_1 \phi_{i_1,k_1}(r) + \cdots + \omega_K \phi_{i_K,k_K}(r)\big) \, \mathrm{d}r \right),$$

where f is the Lévy exponent of the innovation process w. The coefficients are independent if the kernels ϕ_{i_k,k_k} with $k = 1, \ldots, K$ have disjoint support. The wavelet cumulants with multi-index n are given by

$$\kappa_n\{Y_1 : Y_K\} = \kappa_n \int_{\mathbb{R}^d} \phi_{i_1,k_1}^{n_1}(r) \cdots \phi_{i_K,k_K}^{n_K}(r) \, \mathrm{d}r, \tag{8.26}$$

under the assumption that the innovation cumulants of order $n = \sum_{k=1}^K n_k$,

$$\kappa_n = (-\mathrm{j})^n \left. \frac{\partial^n f(\omega)}{\partial \omega^n} \right|_{\omega=0},$$

are finite.

Proof Since $Y_k = \langle \psi_{i_k,k_k}, \mathrm{L}^{-1} w \rangle = \langle \phi_{i_k,k_k}, w \rangle$, the first part is obtained by direct substitution of $\varphi = \omega_1 \phi_{i_1,k_1} + \cdots + \omega_K \phi_{i_K,k_K}$ in the characteristic functional of the innovation $\mathbb{E}\{e^{\langle \varphi, w \rangle}\} = \widehat{\mathscr{P}}_w(\varphi) = \exp\left(\int_{\mathbb{R}} f\big(\varphi(r)\big) \, \mathrm{d}r \right)$. To prove the second part, we start from the Taylor-series expansion of f, which reads

$$f(\omega) = \sum_{n=0}^{\infty} \kappa_n \frac{(\mathrm{j}\omega)^n}{n!},$$

where the κ_n are the cumulants of the canonical innovation pdf p_{id}. Next, we consider the multidimensional wavelet Lévy exponent $f_Y(\omega) = \log \widehat{p}_{(Y_1:Y_K)}(\omega)$ and expand it as

$$
\begin{aligned}
f_Y(\omega) &= \int_{\mathbb{R}^d} f\left(\omega_1 \phi_{i_1,k_1}(r) + \cdots + \omega_K \phi_{i_K,k_K}(r)\right) dr \\
&= \int_{\mathbb{R}^d} \sum_{n=0}^{\infty} \kappa_n \frac{\left(j\omega_1 \phi_{i_1,k_1}(r) + \cdots + j\omega_K \phi_{i_K,k_K}(r)\right)^n}{n!} \, dr \quad \text{(1-D Taylor expansion of } f) \\
&= \int_{\mathbb{R}^d} \sum_{n=0}^{\infty} \kappa_n \frac{1}{n!} \sum_{n=|n|} \frac{n!}{n!} \left(j\omega_1 \phi_{i_1,k_1}(r)\right)^{n_1} \cdots \left(j\omega_K \phi_{i_K,k_K}(r)\right)^{n_K} dr
\end{aligned}
$$

(Multinomial expansion of inner sum)

$$
= \sum_{n=|n|=0}^{\infty} \frac{j^n \omega^n}{n!} \kappa_n \underbrace{\int_{\mathbb{R}^d} \phi_{i_1,k_1}^{n_1}(r) \cdots \phi_{i_K,k_K}^{n_K}(r) \, dr}.
$$

The formula for the cumulants of $(Y_1, \ldots, Y_K)$ is then obtained by identification with (8.25). □

We note that, for $N = 2$, we recover Proposition 8.8. For instance, under the second-order hypothesis, we have that

$$
\kappa_1 = -j f'(0) = 0
$$

$$
\kappa_2 = (-j)^2 f''(0) = \sigma_w^2,
$$

so that (8.26) with $n = (1, 1)$ and $n = 2$ is equivalent to the wavelet-correlation formula (8.23).

Finally, we mention that the cumulant formula in Proposition 8.8 is directly transposable to the generalized increments of Section 8.3 by merely replacing ϕ_{i_k,k_k} by $\beta_L^{\vee}(\cdot - k_k)$ in (8.26) in accordance with the identity $u[k_k] = \langle \beta_L^{\vee}(\cdot - k_k), w \rangle$. In fact, the result of Proposition 8.8 applies for any linear transform of s that involves a collection of basis functions $\{\psi_k\}_{k=1}^K$ such that $\phi_k = L^{-1*}\psi_k \in \mathcal{R}(\mathbb{R}^d)$, and more generally $\phi_k \in L_p(\mathbb{R}^d)$ when f is p-admissible.

8.5 Optimal representation of Lévy and AR(1) processes

We conclude the chapter with a detailed investigation of the effect of such signal decompositions on first-order processes. We are especially interested in the evaluation of performance for data reduction and the comparison with the "optimal" solutions provided by the Karhunen–Loève transform (KLT) and independent-component analysis (ICA). While ICA is usually determined empirically by running a suitable algorithm, the good news is that it can be worked out analytically for this particular class of signal models and used as gold standard.

The Gaussian AR(1) model is of special relevance since it has been put forward in the past to justify two popular source-encoding algorithms: linear predictive coding (LPC) and DCT-based transform coding [JN84]. LPC, on the one hand, is used for voice compression in the GSM standard for mobile phones (2G cellular network). It is also part of the FLAC lossless audio codec. The DCT, on the other hand, was introduced as an approximation of the KLT of an AR(1) process [Ahm74]. It forms the core of the widely used JPEG method of lossy compression for digital pictures. Our primary interest here is to investigate the extent to which these classical tools of signal processing remain relevant when switching to the non-Gaussian regime.

8.5.1 Generalized increments and first-order linear prediction

The generalized CAR(1) processes of interest to us are solutions of the first-order SDE (see Section 7.3.4)

$$(D - \alpha_1 \mathrm{Id})s(t) = w(t)$$

with a single scalar coefficient $\alpha_1 \in \mathbb{R}$. The underlying system is causal-stable for $\alpha_1 < 0$ and results in a stationary output s, which may be Gaussian or not, depending on the type of innovation w. The unstable scenario $\alpha_1 \to 0$ corresponds to the Lévy processes described in Section 7.4.1.

If we sample $s(\cdot)$ at the integers, we obtain the discrete AR(1) process $s[\cdot]$ that satisfies the first-order difference equation

$$s[k] - a_1 s[k-1] = u[k] \tag{8.27}$$

with $a_1 = \mathrm{e}^{\alpha_1}$. From the result in Section 8.3.1 and Proposition 8.4, we know that $u[\cdot]$ is i.i.d. with an infinitely divisible distribution characterized by the (modified) Lévy exponent

$$f_U(\omega) = \int_{\mathbb{R}} f\big(\omega \beta_{\alpha_1}(t)\big)\, \mathrm{d}t = \int_0^1 f\big(\omega \mathrm{e}^{\alpha_1 t}\big)\, \mathrm{d}t.$$

Here, $\beta_{\alpha_1}(t) = \mathbb{1}_{[0,1)}(t)\mathrm{e}^{\alpha_1 t}$ is the exponential B-spline associated with the first-order whitening operator $\mathrm{L} = \mathrm{D} - \alpha_1 \mathrm{Id}$ (see Section 6.3) and f is the exponent of the continuous domain innovation w.

To make the link with predictive coding and classical estimation theory, we observe that $\tilde{s}[k] = a_1 s[k-1]$ is the best (minimum-error) linear predictor of $s[k]$ given the past of the signal $\{s[k-n]\}_{n=1}^{\infty}$. This suggests interpreting the generalized increments $u[k] = s[k] - \tilde{s}[k]$ as prediction errors.

For the benefit of the reader, we recall that the main idea behind linear predictive coding (LPC) is to sequentially transmit the prediction error $u[k]$, rather than the sample values of the signal, which are inherently correlated. One also typically uses higher-order AR models to better represent the effect of the sound-production system. A refinement for real-world signals is to re-estimate the prediction coefficients from time to time in order to readapt the model to the data.

8.5.2   Vector-matrix formulation

The common practice in signal processing is to describe finite-dimensional signal transforms within the framework of linear algebra. To that end, one considers a series of
N consecutive samples $X_n = s[n]$ that are concatenated into the signal vector $\mathbf{x} = (X_1, \ldots, X_N)$, which, from now on, will be treated as a multivariate random variable.
One also imposes some boundary conditions for the signal values that are outside the
observation window, such as, for instance, the periodic extension $s[k] = s[k + N]$. Another example is $s[0] = 0$, which is consistent with the definition of a Lévy process.

In the present scenario, the underlying signal is specified by the discrete AR(1) innovation model (8.27) and the Lévy exponent f_U of its driving term $u[\cdot]$. The transcription
of this model in matrix-vector form is

$$\mathbf{L}\mathbf{x} = \mathbf{u},$$

where $\mathbf{u} = (U_1, \ldots, U_N)$ with $U_n = u[n]$ and $\mathbf{L}$ is a Toeplitz matrix with non-zero entries
$[\mathbf{L}]_{n,n} = 1$ and $[\mathbf{L}]_{n-1,n} = a_1 = e^{\alpha_1}$.

Assuming that $\mathbf{L}$ is invertible,[1] we then solve this linear system of equations, which
yields

$$\mathbf{x} = \mathbf{L}^{-1}\mathbf{u}, \tag{8.28}$$

where $\mathbf{u}$ is i.i.d., in direct analogy with the continuous-domain representation of signal
$s = \mathrm{L}^{-1}w$. Since (8.28) is a special instance of the general finite-dimensional innovation
model, we can refer to the formulas of Section 4.3 for the complete multivariate statistical description. For instance, the joint characteristic function of the signal is given by
(4.12) with $\mathbf{A} = \mathbf{L}^{-1}$ and $f = f_U$.

8.5.3   Transform-domain statistics

The primary motivation for applying a linear transform to $\mathbf{x}$ is to produce the equivalent
signal representation

$$\mathbf{y} = (Y_1, \ldots, Y_N) = \mathbf{T}\mathbf{x} \tag{8.29}$$

whose components Y_n can be processed or analyzed individually, which is justifiable
when the Y_n are (approximately) decoupled. A transform is thereby characterized by
an $(N \times N)$ matrix $\mathbf{T} = [\mathbf{t}_1 \cdots \mathbf{t}_N]^T$ which is assumed to be invertible. If no further
constraint is imposed, then the optimal transform for the present model is $\mathbf{T} \propto \mathbf{L}^{-1}$,
which results in a perfect decoupling, as in LPC. However, there are many applications
such as denoising and coding where it is important to preserve the norm of the signal, so
that one constrains the transform matrix $\mathbf{T}$ to be orthonornal. In what follows, we shall
describe and compare the solutions that are available for that purpose and quantify the
penalty that is incurred by imposing the orthonormality condition.

---

[1] This property is dependent upon a proper choice of discrete boundary conditions (e.g., $s[0] = 0$).

But, before that, let us determine the transform-domain statistics of the signal in order to qualify the effect of $\mathbf{T}$. Generally, if we know the Nth-order pdf of $\mathbf{x}$, we can readily deduce the joint pdf of the transform-domain coefficients $\mathbf{y} = \mathbf{Tx}$ as

$$p_{(Y_1:Y_N)}(\mathbf{y}) = \frac{1}{|\det(\mathbf{T})|} p_{(X_1:X_N)}(\mathbf{T}^{-1}\mathbf{y}). \tag{8.30}$$

The Fourier equivalent of this relation is

$$\widehat{p}_{(Y_1:Y_N)}(\boldsymbol{\omega}) = \widehat{p}_{(X_1:X_N)}(\mathbf{T}^T\boldsymbol{\omega}),$$

where $\widehat{p}_{(Y_1:Y_N)}$ and $\widehat{p}_{(X_1:X_N)}$ are the characteristic functions of $\mathbf{y}$ and $\mathbf{x}$, respectively. In the case of the innovation model (8.28), we can be completely explicit and obtain closed formulas, including the complete multivariate characterization of the transform-domain cumulants. To that end, we rewrite (8.29) as $\mathbf{y} = \mathbf{Au}$, where

$$\mathbf{A} = \mathbf{TL}^{-1} = [\mathbf{a}_1 \cdots \mathbf{a}_N]^T = [\mathbf{b}_1 \cdots \mathbf{b}_N] \tag{8.31}$$

is the composite matrix that combines the effect of noise shaping (innovation model) and the linear transformation of the data. The row vectors of $\mathbf{A}$ are $\mathbf{a}_m = (a_{m,1}, \ldots, a_{m,N})$ with $a_{m,n} = [\mathbf{A}]_{m,n}$, while its column vectors are denoted by $\mathbf{b}_n = (a_{1,n}, \ldots, a_{N,n})$.

PROPOSITION 8.9 *Let* $\mathbf{y} = \mathbf{Au}$, *where* $\mathbf{A} = [\mathbf{b}_1 \cdots \mathbf{b}_N]$ *is an invertible matrix of size N and $\mathbf{u}$ is i.i.d. with an infinite-divisible pdf with Lévy exponent f_U. Then, the joint characteristic function of $\mathbf{y} = (Y_1, \ldots, Y_N)$ is given by*

$$\widehat{p}_{(Y_1:Y_N)}(\boldsymbol{\omega}) = \exp\left(\sum_{n=1}^{N} f_U([\mathbf{A}^T\boldsymbol{\omega}]_n)\right)$$

$$= \exp\left(\sum_{n=1}^{N} f_U(\langle \mathbf{b}_n, \boldsymbol{\omega}\rangle)\right). \tag{8.32}$$

The corresponding multivariate cumulants with multi-index $\boldsymbol{m} = (m_1, \ldots, m_N)$ are

$$\kappa_{\boldsymbol{m}}\{Y_1 : Y_N\} = \kappa_m\{U\} \sum_{n=1}^{N} a_{1,n}^{m_1} \cdots a_{N,n}^{m_N} \tag{8.33}$$

$$= \kappa_m\{U\} \sum_{n=1}^{N} \mathbf{b}_n^{\boldsymbol{m}}, \tag{8.34}$$

where $a_{m,n} = [\mathbf{A}]_{m,n} = [\mathbf{b}_n]_m$ *and* $\kappa_m\{U\} = (-j)^m \partial^m f_U(0)$ *is the (scalar) cumulant of order $m = \sum_{n=1}^{N} m_n$ of the innovation. Finally, the marginal distribution of $Y_n = \langle \mathbf{t}_n, \mathbf{x}\rangle = \langle \mathbf{a}_n, \mathbf{u}\rangle$ is infinitely divisible with Lévy exponent*

$$f_{Y_n}(\omega) = \sum_{m=1}^{N} f_U(a_{n,m}\omega). \tag{8.35}$$

Proof Since $\mathbf{u} = (U_1, \ldots, U_N)$ is i.i.d. with characteristic function $\widehat{p}_U(\omega) = e^{f_U(\omega)}$, we have that

$$\widehat{p}_{(U_1:U_N)}(\boldsymbol{\omega}) = \prod_{n=1}^N e^{f_U(\omega_n)} = \exp\left(\sum_{n=1}^N f_U(\omega_n)\right).$$

Moreover,

$$\widehat{p}_{(Y_1:Y_N)}(\boldsymbol{\omega}) = \mathbb{E}\left\{e^{j\langle \mathbf{y},\boldsymbol{\omega}\rangle}\right\} = \mathbb{E}\left\{e^{j\langle \mathbf{Au},\boldsymbol{\omega}\rangle}\right\}$$

$$= \mathbb{E}\left\{e^{j\langle \mathbf{u},\mathbf{A}^T\boldsymbol{\omega}\rangle}\right\} = \widehat{p}_{(U_1:U_N)}(\mathbf{A}^T\boldsymbol{\omega}).$$

Combining these two formulas yields (8.32).

The second part is obtained by adapting the proof of Proposition 8.8. In essence, the idea is to replace the integral over $\mathbb{R}^d$ by a sum over n, and the basis functions by the row vectors of $\mathbf{A}$.

As for the last statement, the characteristic function $\widehat{p}_{Y_n}(\omega)$ is obtained by the replacement of $\boldsymbol{\omega}$ by $\omega \mathbf{e}_n$ in (8.32), where $\mathbf{e}_n$ is the nth canonical unit vector. Indeed, setting one of the frequency-domain variables to zero is equivalent to performing the corresponding marginalization (integration) of the joint pdf. We thereby obtain

$$\widehat{p}_{Y_n}(\omega) = \exp\left(\sum_{m=1}^N f_U(a_{n,m}\omega)\right),$$

whose exponent is the sought-after quantity. Implicit to this result is the fact that the infinite-divisibility property is preserved when performing linear combination of id variables. Specifically, let f_1 and f_2 be two valid Lévy exponents. Then, it is not hard to see that $f(\omega) = f_1(a_1\omega) + f_2(a_2\omega)$, where $a_1, a_2 \in \mathbb{R}$ are arbitrary constants, is a valid Lévy exponent as well (in reference to Definition 4.1). □

Proposition 8.9 in conjunction with (8.31) provides us with the complete characterization for the transform-domain statistics. To illustrate its usage, we shall now deduce the expression for the transform-domain covariances. Specifically, the covariance between Y_{n_1} and Y_{n_2} is given by the second-order cumulant with multi-index $\mathbf{m} = \mathbf{e}_{n_1} + \mathbf{e}_{n_2}$, whose expression (8.33) with $m = 2$ simplifies to

$$\mathbb{E}\left\{\left(Y_{n_1} - \mathbb{E}\{Y_{n_1}\}\right)\left(Y_{n_2} - \mathbb{E}\{Y_{n_2}\}\right)\right\} = \kappa_2\{U\}\sum_{n=1}^N a_{n_1,n}a_{n_2,n}$$

$$= \sigma_0^2\,\langle \mathbf{a}_{n_1}, \mathbf{a}_{n_2}\rangle, \tag{8.36}$$

where $\sigma_0^2 = -f_U''(0)$ is the variance of the innovation and where $\mathbf{a}_{n_1}$ and $\mathbf{a}_{n_2}$ are the n_1th and n_2th row vectors of the matrix $\mathbf{A} = [\mathbf{a}_1 \cdots \mathbf{a}_N]^T$, respectively. In particular, for $n_1 = n_2 = n$, we find that the variance of Y_n is given by

$$\mathrm{Var}\{Y_n\} = \sigma_0^2\|\mathbf{a}_n\|^2.$$

Covariance matrix

An equivalent way of expressing second-order moments is to use covariance matrices. Specifically, the covariance matrix of the random vector $\mathbf{x} \in \mathbb{R}^N$ is defined as

$$\mathbf{C}_X = \mathbb{E}\left\{ (\mathbf{x} - \mathbb{E}\{\mathbf{x}\})\,(\mathbf{x} - \mathbb{E}\{\mathbf{x}\})^T \right\}. \tag{8.37}$$

This is a $(N \times N)$ symmetric, positive definite matrix. A standard manipulation then yields

$$\begin{aligned}
\mathbf{C}_Y &= \mathbf{T}\mathbf{C}_X\mathbf{T}^T \\
&= \mathbf{A}\mathbf{C}_U\mathbf{A}^T = \sigma_0^2\,\mathbf{A}\mathbf{A}^T,
\end{aligned} \tag{8.38}$$

where $\mathbf{A}$ is defined by (8.31). The reader can easily verify that this result is equivalent to (8.36). Likewise, the second-order transcription of the innovation model is

$$\begin{aligned}
\mathbf{C}_X &= \mathbf{L}^{-1}\mathbf{C}_U\mathbf{L}^{-1T} \\
&= \sigma_0^2\,(\mathbf{L}^T\mathbf{L})^{-1}.
\end{aligned} \tag{8.39}$$

Differential entropy

A final important theoretical quantity is the differential entropy of the random vector $\mathbf{x} - (X_1, \ldots, X_N)$, which is defined as

$$\begin{aligned}
H_{(X_1:X_N)} &= \mathbb{E}\left\{ -\log p_{(X_1:X_N)}(\mathbf{x}) \right\} \\
&= -\int_{\mathbb{R}^N} p_{(X_1:X_N)}(\mathbf{x})\,\log p_{(X_1:X_N)}(\mathbf{x})\,\mathrm{d}\mathbf{x}.
\end{aligned} \tag{8.40}$$

For instance, the differential entropy of the N-dimensional multivariate Gaussian pdf with mean $\mathbf{x}_0$ and covariance matrix $\mathbf{C}_X$,

$$p_{\text{Gauss}}(\mathbf{x}) = \frac{1}{\sqrt{(2\pi)^N \det(\mathbf{C}_X)}}\,\exp\left(-\tfrac{1}{2}(\mathbf{x} - \mathbf{x}_0)^T \mathbf{C}_X^{-1}(\mathbf{x} - \mathbf{x}_0) \right),$$

is given by

$$\begin{aligned}
H(p_{\text{Gauss}}) &= \frac{1}{2}\,(N + N\log(2\pi) + \log\det(\mathbf{C}_X)) \\
&= \frac{1}{2}\,\log\left((2\pi e)^N \det(\mathbf{C}_X) \right).
\end{aligned} \tag{8.41}$$

The special relevance of this expression is that the Gaussian distribution is known to have the maximum differential entropy among all densities with a given covariance $\mathbf{C}_X$. This leads to the inequality

$$-H_{(X_1:X_N)} + \frac{1}{2}\,\log\left((2\pi e)^N \det(\mathbf{C}_X) \right) \le 0, \tag{8.42}$$

where the quantity on the left is called the *negentropy*.

Table 8.1 Table of differential entropies.[a]

Probability density function $p(x)$	Differential entropy $\left(-\int_{\mathbb{R}} p(x)\log p(x)\,dx\right)$
Gaussian $\dfrac{1}{2\pi\sigma^2}e^{-x^2/(2\sigma^2)}$	$\frac{1}{2}\left(1+\log(2\pi\sigma^2)\right)$
Laplace $\dfrac{1}{2\lambda}e^{-|x|/\lambda}$	$1+\log(2\lambda)$
Cauchy $\dfrac{s_0}{\pi}\dfrac{1}{s_0^2+x^2}$	$\log(4\pi s_0)$
Student $\dfrac{1}{B\left(r,\frac{1}{2}\right)}\left(\dfrac{1}{x^2+1}\right)^{r+\frac{1}{2}}$	$\left(r+\frac{1}{2}\right)\left(\psi\left(r+\frac{1}{2}\right)+\psi(r)\right)+\log\left(\sqrt{\frac{r}{2}}B\left(r,\frac{1}{2}\right)\right)$

[a] $B(r,s)$ and $\psi(z)=\frac{d}{dz}\log\Gamma(z)$ are Euler's beta and digamma functions, respectively (see Appendix C).

To quantify the effect of the linear transformation $\mathbf{T}$, we calculate the entropy of (8.30), which, upon change of variables, yields

$$H_{(Y_1:Y_N)} = -\int_{\mathbb{R}^N} p_{(Y_1:Y_N)}(\pmb{x})\log p_{(Y_1:Y_N)}(\pmb{x})\,d\pmb{x}$$
$$= H_{(X_1:X_N)} - \log|\det(\mathbf{T})| \tag{8.43}$$
$$= H_{(U_1:U_N)} - \log|\det(\mathbf{A})|$$
$$= N\cdot H_U - \log|\det(\mathbf{T})| + \log|\det(\mathbf{L})|, \tag{8.44}$$

where $H_U = -\int_{\mathbb{R}} p_U(x)\log p_U(x)\,dx$ is the differential entropy of the (scalar) innovation. Equation (8.43) implies that the differential entropy is invariant to orthonormal transformations (since $\det(\mathbf{T})=1$), while (8.44) shows that it is primarily determined by the probability law of the innovation. Some specific formulas for the differential entropy of id laws are given in Table 8.1.

8.5.4 Comparison of orthogonal transforms

While the obvious candidate for expanding sparse AR(1) processes is the operator-like transform of Section 6.3.3, we shall also consider two classical data analysis methods – principal-component analysis (PCA) and independent-component analysis (ICA) – which can be readily transposed to our framework.

Discrete Karhunen–Loève transform

The discrete KLT is the model-based version of PCA. It relies on prior knowledge of the covariance matrix of the signal $\left(\text{see } (8.37)\right)$. Specifically, the KLT matrix $\mathbf{T}_{\text{KLT}} = [\mathbf{h}_1 \cdots \mathbf{h}_N]^T$ is build from the eigenvectors $\mathbf{h}_n$ of $\mathbf{C}_X$, with the eigenvalues $\lambda_1 \geq \cdots \geq \lambda_N$ being ordered by decreasing magnitude. This results in a representation that is perfectly decorrelated, with

$$\mathbf{C}_{\text{KLT}} = \mathbf{T}_{\text{KLT}}\mathbf{C}_X\mathbf{T}_{\text{KLT}}^T = \text{diag}(\lambda_1, \ldots, \lambda_N).$$

This implies that the solution is optimal in the Gaussian scenario, where decorrelation is equivalent to independence. PCA is essentially the same technique, except that it replaces $\mathbf{C}_X$ by a scatter matrix that is estimated empirically from the data.

In addition to the decorrelation property, the KLT minimizes any criterion of the form (see [Uns84, Appendix A])

$$R(\mathbf{T}) = \sum_{n=1}^{N} G\big(\mathrm{Var}\{Y_n\}\big), \tag{8.45}$$

where $G : \mathbb{R}^+ \rightarrow \mathbb{R}$ is an arbitrary, continuous, monotonically decreasing, convex (or increasing concave) function and $\mathrm{Var}\{Y_n\} = \mathbf{t}_n^T \mathbf{C}_X \mathbf{t}_n$ is the variance of $Y_n = \langle \mathbf{t}_n, \mathbf{x} \rangle$.

Under the finite-variance hypothesis, the covariance matrix of the AR(1) process is given by (8.39), where $\mathbf{L}^{-1}$ is the inverse of the prediction filter and σ_0^2 is the variance of the discrete innovation $\mathbf{u}$. There are several instances of the model for which the KLT can be determined analytically, based on the fact that the eigenvectors of $\sigma_0^2 (\mathbf{L}^T \mathbf{L})^{-1}$ are the same as those of $(\mathbf{L}^T \mathbf{L})$. Specifically, $\mathbf{h}$ is an eigenvector of $\mathbf{C}_X$ with eigenvalue λ if and only if

$$(\mathbf{L}^T \mathbf{L}) \mathbf{h} = \frac{\sigma_0^2}{\lambda} \mathbf{h}.$$

For the AR(1) model, $\mathbf{L}^T \mathbf{L}$ is tridiagonal. This can then be converted into a set of second-order difference equations that may be solved recursively. In particular, for $\alpha_1 = 0$ (Lévy process), the eigenvectors for $n = 1, \ldots, N$ correspond to the sinusoidal sequences

$$h_n[k] = \frac{2}{\sqrt{2N+1}} \sin\left(\pi \frac{2n-1}{2N+1} k\right). \tag{8.46}$$

Depending on the boundary conditions, one can obtain similar formulas for the eigenvectors when $\alpha_1 \neq 0$. The bottom line is that the solutions are generally sinusoids, with minor variations in phase and frequency. This is consistent with the fact that the correlation matrix $\mathbf{C}_X$ is very close to being circular, and that circular matrices are diagonalized by the discrete Fourier transform (DFT). Another "universal" transform that provides an excellent approximation of the KLT of an AR(1) process (see [Ahm74]) is the discrete cosine transform (DCT) whose basis vectors for $n = 1, \ldots, N$ are

$$g_n[k] = \frac{2}{\sqrt{2N}} \cos\left(\frac{\pi n}{N} \left(k - \frac{1}{2}\right)\right). \tag{8.47}$$

A key property is that the DCT is asymptotically optimal in the sense that its performance is equivalent to that of the KLT when the block size N tends to infinity. Remarkably, this is a result that holds for the complete class of wide-sense-stationary processes [Uns84], which may explain why this transform performs so well in practice.

Independent-component analysis

The decoupling afforded by the KLT emphasizes decorrelation, which is not necessarily appropriate when the underlying model is non-Gaussian. Instead, one would rather like to obtain a representation that maximizes statistical independence, which is the goal pursued by *independent-component analysis* (ICA). Unlike the KLT, there is no single ICA but a variety of numerical solutions that differ in terms of the criterion being optimized [Com94]. The preferred measure of independence is *mutual information* (MI), with the caveat that it is often difficult to estimate reliably from the data. Here, we take advantage of our analytic framework to bypass this estimation step, which is the empirical part of ICA.

Specifically, let

$$\mathbf{z} = (Z_1, \ldots, Z_N) = \mathbf{T}_{ICA}\mathbf{x}$$

with $\mathbf{T}_{ICA}^T \mathbf{T}_{ICA} = \mathbf{I}$. By definition, the mutual information of the random vector $(Z_1, \ldots, Z_N)$ is given by

$$I(Z_1, \ldots, Z_N) = \left(\sum_{n=1}^{N} H_{Z_n}\right) - H_{(Z_1:Z_N)} \geq 0, \tag{8.48}$$

where $H_{Z_n} = -\int_{\mathbb{R}} p_{Z_n}(z) \log p_{Z_n}(z)\, dz$ is the differential entropy of the component Z_n (which is computed from the marginal distribution p_{Z_n}), and $H_{(Z_1:Z_N)}$ is the Nth-order differential entropy of $\mathbf{z}$ (see (8.40)). The relevance of this criterion is that $I(Z_1, \ldots, Z_N) = 0$ if and only if the variables $Z_1, \ldots, Z_N$ are statistically independent. The other important property is that $H_{(Z_1:Z_N)} = H_{(Y_1:Y_N)} = H_{(X_1:X_N)}$, meaning that the joint differential entropy does not depend upon the choice of transformation as long as $|\det(\mathbf{T})| = 1$ (see (8.43)). Therefore, the ICA transform that minimizes $I(Z_1, \ldots, Z_N)$ subject to the orthonormality constraint is also the one that minimizes the sum of the transform-domain entropies. We call this optimal solution the *min-entropy ICA*.

The implicit understanding with ICA is that the components of $\mathbf{z}$ are (approximately) independent so that it makes good sense to approximate the joint pdf of $\mathbf{z}$ by the product of the marginals given by

$$p_{(Z_1:Z_N)}(z) \approx \prod_{n=1}^{N} p_{Z_n}(z_n) = q_{(Z_1:Z_N)}(z). \tag{8.49}$$

As it turns out, the min-entropy ICA minimizes the Kullback–Leibler divergence between $p_{(Z_1:Z_N)}$ and its separable approximation $q_{(Z_1:Z_N)}$. Indeed, the Kullback–Leibler divergence between two N-dimensional probability density functions p and q is defined as

$$\mathbb{D}(p\|q) = \int_{\mathbb{R}^N} \log\left(\frac{p(z)}{q(z)}\right) p(z)\, dz.$$

In the present context, this simplifies to

$$\mathbb{D}\left(p_{(Z_1:Z_N)} \parallel \prod_{n=1}^{N} p_{Z_n}\right) = -H_{(Z_1:Z_N)} - \int_{\mathbb{R}^N} p_{(Z_1:Z_N)}(\mathbf{z}) \sum_{n=1}^{N} \log p_{Z_n}(z_n)\, d\mathbf{z}$$

$$= -H_{(Z_1:Z_N)} - \sum_{n=1}^{N} \int_{\mathbb{R}} p_{Z_n}(z) \log p_{Z_n}(z)\, dz$$

$$= -H_{(Z_1:Z_N)} - \sum_{n=1}^{N} H_{Z_n} = I(Z_1, \ldots, Z_N) \ge 0,$$

with equality to zero if and only if $p_{(Z_1:Z_N)}(\mathbf{z})$ and its product approximation on the middle/left-hand-side of (8.49) are equal.

In the special case where the innovation follows an $S\alpha S$ distribution with $f_U(\omega) = -|s_0\omega|^\alpha$, we can derive a form of the entropy criterion that is directly amenable to numerical optimization. Indeed, based on Proposition 8.9, we find that the characteristic function of Y_n is given by

$$\widehat{p}_{Y_n}(\omega) = \exp\left(\sum_{m=1}^{N} -|s_0 a_{n,m}\omega|^\alpha\right)$$

$$= \exp\left(-s_0^\alpha \|\mathbf{a}_n\|_{\ell_\alpha}^\alpha |\omega|^\alpha\right) = e^{-|s_n\omega|^\alpha}, \tag{8.50}$$

for which we deduce that Y_n is $S\alpha S$ with dispersion parameter

$$s_n = s_0 \|\mathbf{a}_n\|_{\ell_\alpha} = s_0 \left(\sum_{m=1}^{N} |a_{n,m}^\alpha|^\alpha\right)^{\frac{1}{\alpha}}.$$

This implies that p_{Y_n} is a rescaled version of p_U so that its entropy is given by

$$H_{Y_n} = H_U - \log \|\mathbf{a}_n\|_{\ell_\alpha}.$$

Hence, minimizing the mutual information (or, equivalently, the sum of the entropies of the transformed coefficients) is equivalent to the optimization of

$$I(\mathbf{T}; \alpha) = -\sum_{n=1}^{N} \log \|\mathbf{a}_n\|_{\ell_\alpha}$$

$$= -\sum_{n=1}^{N} \frac{1}{\alpha} \log \left(\sum_{m=1}^{N} \left|[\mathbf{TL}^{-1}]_{n,m}^\alpha\right|^\alpha\right). \tag{8.51}$$

Based on (8.48), (8.44), and the property that $|\det(\mathbf{L})| = 1$, we can actually verify that (8.51) is the exact formula for the mutual information $I(Y_1, \ldots, Y_N)$. In particular, for $\alpha = 2$ (Gaussian case), we find that

$$I(\mathbf{T}; 2) = -\sum_{n=1}^{N} \frac{1}{2} \log \|\mathbf{a}_n\|_2^2 = -\sum_{n=1}^{N} \frac{1}{2} \log \text{Var}\{Y_n\},$$

which is a special instance of (8.45). It follows that, for $\alpha = 2$, ICA is equivalent to the KLT, as expected.

Experimental results and discussion

To quantify the performance of the various transforms, we considered a series of signal vectors of size $N = 64$, which are sampled versions of Lévy processes and CAR(1) signals with SαS innovations. The motivation for this particular setup is twofold. First, the use of a stable excitation lends itself to a complete analytical treatment due to the preservation of the SαS property (see (8.50)). Second, the model offers direct control over the degree of sparsity via the adjustment of $\alpha \in (0, 2]$. The classical Gaussian configuration is achieved with $\alpha = 2$, while the distribution becomes more and more heavy-tailed as α decreases with $p_{Y_n}(x) = O(1/x^{\alpha+1})$ (see Appendix C).

Our quality index is the average mutual information of the transform-domain coefficients given by $\frac{1}{N}I(Y_1, \ldots, Y_N) \geq 0$. A value of zero indicates that the transform coefficients are completely independent. The baseline transformation is the identity, which is expected to yield the worst results. The optimal performance is achieved by ICA, which is determined numerically for each brand of signals based on the optimization of (8.51).

In Figure 8.2, we compare the performance of the DCT, the Haar wavelet transform, and the ICA gold standard for the representation of Lévy processes (with $a_1 = e^0 = 1$). We verified that the difference in performance between the DCT and the KLT associated with Brownian motion was marginal so that the DCT curve is truly representative of the best that is achievable within the realm of "classical" sinusoidal transforms. Interestingly, the Haar wavelet transform is indistinguishable from the optimal solution for values of α below 1, and better than the DCT up to some transition point. The change of regime of the DCT for larger values of α is explained by the property that it converges to the optimal solution for $\alpha = 2$, the single point at which ICA and the KLT are equivalent.

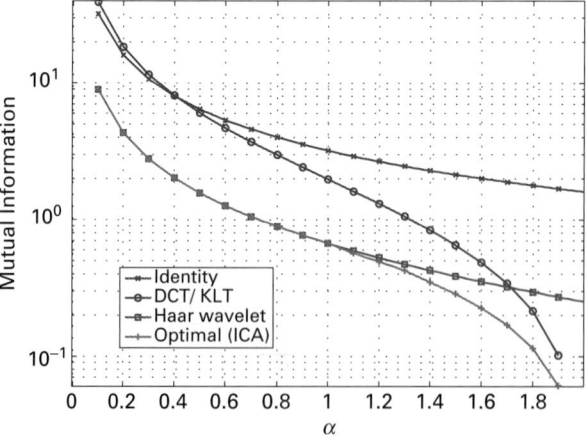

Figure 8.2 Decoupling ability of various transforms of size $N = 64$ for the representation of SαS Lévy processes as a function of the stability/sparsity index α. The criterion is the average mutual information. The Gaussian case $\alpha = 2$ corresponds to a Brownian motion.

We also examined the basis functions of ICA and found that they were very similar to Haar wavelets. In particular, it appeared that the ICA training algorithm would almost systematically uncover the $N/2$ shorter Haar wavelets, which is not overly surprising since these are basis vectors also shared by the discrete whitening filter **L**.

Remarkably, this statistical model supports the (quasi)-optimality of wavelets. Since mutual information is in direct relation to the transform-domain entropies which are predictors of coding performance, these results also provide an explanation of the superiority of wavelets for the M-term approximation of Lévy processes reported in Section 1.3.4 (see Figure 1.5).

The graph in Figure 8.3 provides the matching results for AR(1) signals with $a_1 = 0.9$. Also included is a comparison between the Haar transform and the operator-like wavelet transform that is matched to the underlying AR(1) model. The conclusions are essentially the same as before. Here, too, the DCT closely replicates the performance of the KLT associated with the Gaussian brand of these processes. While the Haar transform is superior to the DCT for the sparser processes (small values of α), it is generally outperformed by the operator-like wavelet transform, which confirms the relevance of applying matched basis functions.

The finding that a fixed set of basis functions (operator-like wavelet transform) is capable of essentially replicating the performance of ICA is good news for applications. Indeed, the computational cost of the wavelet algorithm is $O(N)$, as opposed to $O(N^2)$ for ICA, not to mention the price of the training procedure (iterative optimization) which is even more demanding ($O(N^2)$ per iteration of a gradient-based scheme). A further conceptual advantage is that the operator-like wavelets are known in analytic form (see Section 6.3.3), while ICA can, at best, only be determined numerically by running a suitable optimization algorithm.

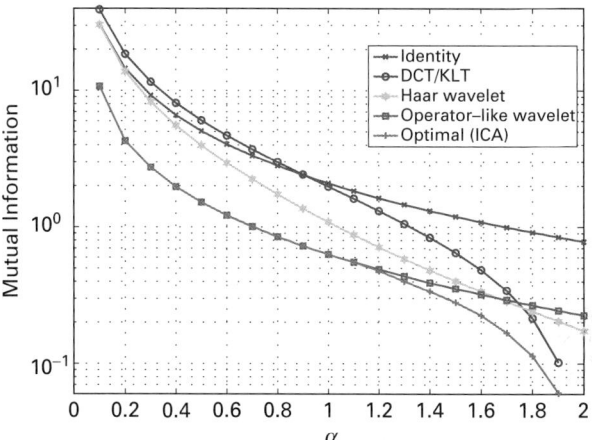

Figure 8.3 Decoupling ability of various transforms for the representation of stable AR(1) signals of size $N = 64$ with correlation $a_1 = 0.9$ as a function of the stability/sparsity index α. The criterion is the average mutual information. The DCT is known to be asymptotically optimal in the Gaussian case ($\alpha = 2$).

8.6   Bibliographical notes

Sections 8.1–8.4

The property that finite differences decouple Lévy processes is well known – in fact, it is the starting point of the definition of such "additive" processes [Lév65]. By contrast, the observation that Haar wavelets have the same kind of ability (on a scale-by-scale basis) is much more recent [UTS14].

A crucial theoretical aspect is the extension of the domain of the characteristic functional that was carried out in Section 8.2. The proof of Theorem 8.2 is adapted from [UTS14, Theorem 3], while more general results for arbitrary Lévy exponents can be found in [Fag14].

The material in Section 8.3 is an extension of the results presented in [UTAK14]. In that respect, we note that the correspondence between continuous-time and discrete-time ARMA models (the Gaussian part of Proposition 8.4) is a classical result in the theory of Gaussian stationary processes [Doo90]. The localization of the fractional-derivative operators in Section 8.3.5 is intimately linked to the construction of fractional B-splines with the help of the generalized binomial expansion (see [BU03, UB00]).

The theoretical results on wavelet-domain statistics in Section 8.4 are an extension of [UTS14, Section V.D]. In particular, the general cumulant formulas (Proposition 8.8) are new, to the best of our knowledge.

Section 8.5

The use of linear transforms for the decorrelation of signals is a classical topic in signal and image processing [Pra91, JN84, Jai89]. The DCT was introduced by Ahmed *et al.* as a practical substitute for the KLT of an AR(1) process [Ahm74]. As part of his thesis work, Unser proved its asymptotic equivalence with the KLT for the complete class of Gaussian stationary processes [Uns84]. The AR(1) model has frequently been used for comparing linear transforms using various performance metrics derived from the Gaussian hypothesis [PAP72, HP76, Jai79, JN84]. While such measures clearly point to the superiority of sinusoidal transforms, they lose their relevance in the context of sparse processes. The derivation of the KLT of a Lévy process (see (8.46)) can be found in [KPAU13, Appendix II].

Two classical references on ICA are [Com94, HO00]. In practice, ICA is determined from the data based on the minimization of a suitable contrast that favors independence or, by extension, non-Gaussianity/sparsity. There is some empirical evidence of a link between ICA and wavelets. The first is a famous experiment by Olshausen and Field, who computed ICA from a large collection of natural image patches and pointed out the similarity between the extracted factors and a directional wavelet/Gabor analysis [OF96a]. In 1999, Cardoso and Donoho reported a numerical experiment involving a (non-Gaussian) sawtooth process that resulted in ICA basis functions that were remarkably similar to wavelets [Car99]. The characterization of ICA for $S\alpha S$ processes that is presented in Section 8.5.4, and the demonstration of the connection with operator-like wavelets, are based on the more recent work of Pad and Unser [PU13].

The source for the calculation of the differential entropies in Table 8.1 is [LR78].

9　Infinite divisibility and transform-domain statistics

As we saw in Chapter 8, we have at our disposal two primary tools to analyze/ characterize sparse signals: (1) the construction of the generalized-increment process by application of the discrete version of the whitening operator (localization), and (2) a wavelet analysis using operator-like basis functions. The goal shared by the two approaches is to approximately recover the innovations by numerical inversion of the signal-generation model. While the localization option appears to give the finest level of control, the wavelet option is interesting as well because of its greater robustness to modeling errors. Indeed, performing a wavelet transform (especially if it is orthogonal) is a much stabler operation than taking a discrete derivative, which tends to amplify high-frequency perturbations.

In this chapter, we take advantage of our functional framework to gain a better understanding of the true nature of the transform-domain statistics. Our investigation revolves around the fact that the marginal pdfs are infinitely divisible (id) and that they can be determined explicitly via the analysis of a common white noise (innovation). We make use of this result to examine a number of relevant properties of the underlying id distributions and to show that they are, for the most part, invariant to the actual shape of the analysis window. This implies, among other things, that the qualitative effect of operator-like filtering is the same in both analysis scenarios, in the sense that the sparsity pattern of the innovation is essentially preserved.

The chapter is organized as follows. In Section 9.1, we interpret as spectral mixing the observation of white Lévy noise through some general analysis function $\varphi \in L_p(\mathbb{R}^d)$. We prove that the resulting pdf is infinitely divisible under mild conditions on φ and f. The key idea is that the pdf of $X_\varphi = \langle \varphi, w \rangle$ is completely characterized by its modified Lévy exponent

$$f_\varphi(\omega) = \int_{\mathbb{R}^d} f\big(\omega\varphi(r)\big)\, dr, \tag{9.1}$$

where f is the Lévy exponent of the continuous-domain innovations. This also translates into a correspondence between the Lévy density $v_\varphi(a)$ of f_φ and the canonical $v(a)$ that characterizes the innovation (see Theorem 4.2).

The central part of the chapter is devoted to the demonstration that f_φ (resp., v_φ) preserves the primary features of f (resp. v) and by the same token those of the underlying pdf. The properties of unimodality, self-decomposability, and stability are covered in

Sections 9.2, 9.3, and 9.4, respectively, while the characterization of the decay and the determination of the moments are carried out in Sections 9.5–9.6. The conclusion that follows is that the shape of the analysis window does not fundamentally change the nature of the transform-domain pdfs. This is good news for practical applications since it allows us to do some relatively robust modeling by sticking to a particular family of id distributions (such as the Student, symmetric gamma, or SαS laws in Table 4.1) whose dispersion and decay parameters can be tuned to fit the statistics of a given type of signal. These findings also suggest that the transform-domain statistics are only mildly dependent upon the choice of a particular wavelet basis as long as the analysis wavelets are matched to the whitening operator of the process in the sense that $\psi_i = L^* \phi_i$ with $\phi_i \in L_p(\mathbb{R}^d)$.

In the case where the operator is scale-invariant (fractal process), we can be more specific and obtain a precise characterization of the evolution of the wavelet statistics across scale. Our mathematical analysis hinges upon the semigroup properties of id laws, which are reviewed in Section 9.7. These results are then used in Section 9.8 to show that the wavelet-domain pdfs are ruled by a diffusion-like equation. This allows us to prove that the wavelet-domain pdfs converge to Gaussians – or, more generally, to stable distributions – as the scale gets coarser.

9.1   Composition of id laws, spectral mixing, and analysis of white noise

The fundamental aspect of infinite divisibility is that the property is conserved through basic convolution-based compositions of probability density functions. Specifically, let p_{X_1} and p_{X_2} be two id pdfs. Then, the id property is maintained for

$$p_Y(x) = \left(\tfrac{1}{|s_1|} p_{X_1}\left(\tfrac{\cdot}{s_1}\right) * \tfrac{1}{|s_2|} p_{X_2}\left(\tfrac{\cdot}{s_2}\right) \right)(x) \Leftrightarrow \widehat{p}_Y(\omega) = \widehat{p}_{X_1}(s_1\omega) \cdot \widehat{p}_{X_2}(s_2\omega),$$ where p_Y

can be interpreted as the pdf of the linear combination $Y = s_1 X_1 + s_2 X_2$ of independent random variables X_1 and X_2. Indeed, the resulting Lévy triplet is $(s_1 m_1 + s_1 m_2, s_1^2 b_1 + s_2^2 b_2, \tfrac{1}{s_1} v_1(\tfrac{\cdot}{s_1}) + \tfrac{1}{s_2} v_2(\tfrac{\cdot}{s_2}))$, where (m_i, b_i, v_i), $i = 1, 2$ are the Lévy triplets of p_{X_1} and p_{X_2}. Along the same lines, we have the implication that, for any finite $\tau \geq 0$,

$$f(\omega) \text{ (p-)admissible} \Rightarrow \tau f(\omega) \text{ is a (p-)admissible Lévy exponent.}$$

The Fourier-domain interpretation here is $\widehat{p}_{X_\tau}(\omega) = \left(\widehat{p}_{X_1}(\omega) \right)^\tau$, which translates into a τ-fold (fractional) convolution of p_{X_1}.

In this work, we rely heavily on the property that the composition of id probability laws can undergo a limit process by considering a mixing that involves a continuum of independent random contributions (integration of white noise). For instance, we have already emphasized in Section 4.4.3 that the observation of a white Lévy noise through a rectangular window results in a reference random variable $X_{\text{id}} = \langle \text{rect}, w \rangle$ with a canonical id pdf $p_{\text{id}}(x)$ (see Proposition 4.12). The important question that we are addressing here is to describe the effect of applying a non-rectangular analysis window, as in (2.4) or (2.1).

The first step is to determine the pdf of the random variable $X_\varphi = \langle \varphi, w \rangle$ for an arbitrary φ. We have already alluded to the fact that p_{X_φ} is infinitely divisible. Indeed, the characteristic function of X_φ is obtained by making the substitution $\varphi \to \omega\varphi$ in the characteristic functional (4.13) of the Lévy noise, which yields

$$\widehat{p}_{X_\varphi}(\omega) = \widehat{\mathscr{P}_w}(\omega\varphi) = e^{f_\varphi(\omega)},$$

where f_φ is defined by (9.1).

The next result shows that f_φ is indeed an admissible Lévy exponent – that is, a conditional, positive definite function of order one admitting a Lévy–Khintchine decomposition with some modified Lévy density v_φ.

THEOREM 9.1   *Let $f(\omega)$ be a valid Lévy exponent with Lévy triplet $(b_1, b_2, v(a))$ subject to the constraint that $\int_{\mathbb{R}} \min(a^2, |a|^p) v(a)\, da < \infty$ for some $p \in [0, 2]$. Then, for any $\varphi \in L_p(\mathbb{R}^d) \cap L_2(\mathbb{R}^d) \cap L_1(\mathbb{R}^d)$, $f_\varphi(\omega) = \int_{\mathbb{R}^d} f(\omega\varphi(r))\, dr$ is an admissible Lévy exponent with Gaussian parameter*

$$b_{2,\varphi} = b_2 \|\varphi\|_{L_2}^2$$

and modified Lévy density

$$v_\varphi(a) = \int_{\Pi_\varphi} \frac{1}{|\varphi(r)|} v(a/\varphi(r))\, dr, \tag{9.2}$$

where $\Pi_\varphi \subseteq \mathbb{R}^d$ denotes the domain over which $\varphi(r) \neq 0$. Moreover, $f_\varphi(\omega)$ (resp., v_φ) satisfies the same type of admissibility condition as $f(\omega)$ (resp., v).

Note that the restriction $\varphi \in L_1(\mathbb{R}^d)$ is only required for the non-centered scenarios, where $\int_{\mathbb{R}} av(a)\, da \neq 0$ and $b_1 \neq 0$. The corresponding Lévy parameter $b_{1,\varphi}$ then depends upon the type of Lévy–Khintchine formula. In the case of the fully-compensated representation (4.5), we have $b_{1,\varphi} = b_1'' \int_{\mathbb{R}^d} \varphi(r)\, dr$.

LEMMA 9.2   *Let $v(a) \geq 0$ be a valid Lévy density such that $\int_{\mathbb{R}} \min(a^2, |a|^p) v(a)\, da < \infty$ for some $p \in [0, 2]$. Then, for any given $\varphi \in L_p(\mathbb{R}^d) \cap L_2(\mathbb{R}^d)$, the modified density $v_\varphi(a)$ specified by (9.2) is Lévy-admissible with*

$$\int_{\mathbb{R}} \min(a^2, |a|^p) v_\varphi(a)\, da < 2\|\varphi\|_{L_2}^2 \int_{|a|<1} a^2 v(a)\, da + 2\|\varphi\|_{L_p}^p \int_{|a|\geq 1} |a|^p v(a)\, da. \tag{9.3}$$

If, in addition, $\int_{\mathbb{R}} |a|^p v(a)\, da < \infty$ for some $p \geq 0$, then the result holds for any $\varphi \in L_p(\mathbb{R}^d)$ and v_φ satisfies the inequality

$$\int_{\mathbb{R}} |a|^p v_\varphi(a)\, da < 2\|\varphi\|_{L_p}^p \int_{\mathbb{R}} |a|^p v(a)\, da. \tag{9.4}$$

The limit case $p = 0$ is covered by using the convention that $L_0(\mathbb{R}^d)$ is the space of bounded, compactly supported functions with $\|\varphi\|_{L_0}^0 = \int_{\Pi_\varphi} dr$, where $\Pi_\varphi = \{r : \varphi(r) \neq 0\}$.

Proof of Lemma 9.2. $v_\varphi(a)$ is non-negative by construction. To prove that it satisfies the required bound, we rewrite it as

$$v_\varphi(a) = \int_{-\infty}^{\infty} \frac{1}{|\theta|} v(a/\theta)\, \mu_\varphi(d\theta),$$

where μ_φ is the measure describing the amplitude distribution of $\varphi(r)$ within the range $\theta \in \mathbb{R}$, with zero contribution at $\theta = 0$ (to avoid dividing by zero). For further convenience, we also define the measure $\mu_{|\varphi|}$ that specifies the amplitude distribution of $|\varphi(r)|$. To check the finiteness of $\int_{|a|>1} |a|^p v_\varphi(a)\, da$, we first consider the contribution I_1 of the positive values

$$I_1 = \int_1^\infty |a|^p v_\varphi(a)\, da = \int_1^\infty \int_{-\infty}^\infty \frac{|a|^p}{|\theta|} v(a/\theta)\, \mu_\varphi(d\theta)\, da$$

$$= \int_{-\infty}^\infty \int_1^\infty \frac{|a|^p}{|\theta|} v(a/\theta)\, da\, \mu_\varphi(d\theta)$$

$$= \int_0^\infty \int_{1/\theta}^\infty |a'\theta|^p v(a')\, da'\, \mu_\varphi(d\theta)$$

$$\leq \int_0^\infty \left(\int_{\min(1,1/\theta)}^1 |a'\theta|^p v(a')\, da' + \int_1^\infty |a'\theta|^p v(a')\, da' \right) \mu_\varphi(d\theta),$$

where we are relying on Tonelli's theorem to interchange the integrals and where we have made the reverse change of variable $a' = a/\theta$. The crucial step is to note that $a'\theta \geq 1$ for the range of values within the first inner integral, which yields

$$I_1 \leq \int_0^\infty \left(\int_{\min(1,1/\theta)}^1 |a'\theta|^2 v(a')\, da' + \int_1^\infty |a'\theta|^p v(a')\, da' \right) \mu_\varphi(d\theta)$$

$$\leq \int_{-\infty}^\infty \theta^2 \mu_\varphi(d\theta) \int_0^1 a^2 v(a)\, da + \int_{-\infty}^\infty |\theta|^p \mu_\varphi(d\theta) \int_1^\infty a^p v(a)\, da.$$

Proceeding in the same fashion for the negative values and recalling that $\int_\mathbb{R} |\theta|^p \mu_\varphi(d\theta) = \|\varphi\|_{L_p}^p$, we find that

$$\int_{|a|\geq 1} |a|^p v_\varphi(a)\, da \leq \|\varphi\|_{L_2}^2 \int_{|a|<1} a^2 v(a)\, da + \|\varphi\|_{L_p}^p \int_{|a|\geq 1} |a|^p v(a)\, da < \infty,$$

where we have used $\int_\mathbb{R} \min(a^2, |a|^p) v(a)\, da = \int_{|a|<1} a^2 v(a)\, da + \int_{|a|\geq 1} |a|^p v(a)\, da$. As for the quadratic part (I_2) of the admissibility condition, we consider the integral

$$I_2 = \int_0^1 a^2 v_\varphi(a)\, da$$

$$= \int_0^1 \int_{-\infty}^\infty \frac{a^2}{\theta} v(a/\theta)\, \mu_\varphi(d\theta)\, da$$

$$= \int_0^\infty \int_0^{1/\theta} (a'\theta)^2 v(a')\, da'\, \mu_\varphi(d\theta)$$

with the change of variable $a' = a/\theta$. Since $a' < 1/\theta$ within the bound of the inner integral, we have

$$I_2 \leq \int_0^\infty \left(\int_0^1 (a'\theta)^2 v(a') \, da' + \int_1^{\max(1,1/\theta)} |a'\theta|^p v(a') \, da' \right) \mu_\varphi(d\theta)$$

$$\leq \int_{-\infty}^\infty \theta^2 \mu_\varphi(d\theta) \int_0^1 a^2 v(a) \, da + \int_{-\infty}^\infty |\theta|^p \mu_\varphi(d\theta) \int_1^\infty a^p v(a) \, da.$$

Using the same technique for the negative values, we get

$$\int_{|a|<1} a^2 v_\varphi(a) \, da \leq \|\varphi\|_{L_2}^2 \int_{|a|<1} a^2 v(a) \, da + \|\varphi\|_{L_p}^p \int_{|a|\geq 1} |a|^p v(a) \, da < \infty,$$

which is then combined with the previous result to yield (9.3). The announced L_p inequality is obtained in a similar fashion without the necessity of splitting the integrals in subparts. $\qquad\square$

LEMMA 9.3   *Let $v(a) \geq 0$ be a Lévy density such that $\int_\mathbb{R} \min(a^2, |a|^p) v(a) \, da < \infty$ for some $p \in [0,2]$. Then, the non-Gaussian Lévy exponent*

$$g(\omega) = \int_{\mathbb{R}\setminus\{0\}} \left(e^{\mathrm{j} a\omega} - 1 - \mathrm{j} a\omega \mathbb{1}_{|a|<1}(a) \right) v(a) \, da$$

is bounded by

$$|g(\omega)| \leq C_2 |\omega|^2 + C_q |\omega|^q,$$

with $C_2 = \int_{|a|<1} |a|^2 v(a) \, da$, $q = \min(1,p)$, and $C_q = 2 \int_{|a|\geq 1} |a|^q v(a) \, da$.

Proof of Lemma 9.3.   We rewrite $g(\omega)$ as

$$g(\omega) = \int_{|a|<1} \left(e^{\mathrm{j} a\omega} - 1 - \mathrm{j} a\omega \right) v(a) \, da + \int_{|a|\geq 1} \left(e^{\mathrm{j} a\omega} - 1 \right) v(a) \, da.$$

Next, we observe that the condition in the statement implies that

$$\int_{|a|>1} |a|^q v(a) \, da < \infty$$

for all $0 \leq q \leq p$, and, in particular, $q = \min(1,p)$. We then consider the inequalities

$$\left| e^{\mathrm{j}x} - 1 - \mathrm{j}x \right| \leq x^2$$

and

$$\left| e^{\mathrm{j}x} - 1 \right| \leq \min(|x|, 2)$$

$$\leq 2 \min(|x|, 1)$$

$$\leq 2|x|^q,$$

where the restriction $q \leq 1$ ensures that $|x|^q \geq |x|$ for $|x| \leq 1$. By combining these elements, we construct the upper bound

$$
\begin{aligned}
|g(\omega)| &\leq \int_{|a|<1} \left| e^{ja\omega} - 1 - ja\omega \right| v(a)\, da + \int_{|a|\geq 1} \left| e^{ja\omega} - 1 \right| v(a)\, da \\
&\leq \int_{|a|<1} |a\omega|^2 v(a)\, da + 2 \int_{|a|\geq 1} |\omega a|^q v(a)\, da \\
&\leq |\omega|^2 \int_{|a|<1} |a|^2 v(a)\, da + 2|\omega|^q \int_{|a|\geq 1} |a|^q v(a)\, da.
\end{aligned}
$$

$\square$

Proof of Theorem 9.1. First, we use Lemma 9.3 together with the Lévy–Khintchine formula (4.3) to show that the modified Lévy exponent f_φ is a well-defined function of ω. This is achieved by establishing the upper bound

$$
\left| f_\varphi(\omega) \right| \leq A_1 |\omega| + A_2 |\omega|^2 + A_q |\omega|^q, \tag{9.5}
$$

where $q = \min(1,p)$, $A_1 = |b_1'|\,\|\varphi\|_{L_1}$, $A_2 = \left(\frac{b_2}{2} + \int_{|a|<1} |a|^2 v(a)\, da \right) \|\varphi\|_{L_2}^2$ and $A_3 = 2 \int_{|a|\leq 1} |a|^q v(a)\, da\, \|\varphi\|_{L_q}^q$.

To lay out the technique of proof, we first assume that $\int_{|a|>1} |a|\, v(a)\, da < \infty$ and consider the Lévy–Khintchine representation (4.5) of f. This yields

$$
\begin{aligned}
f_\varphi(\omega) &= jb_1'' \int_{\mathbb{R}^d} \varphi(r)\omega\, dr - b_2 \int_{\mathbb{R}^d} |\varphi(r)|^2 \frac{\omega^2}{2}\, dr + g_\varphi(\omega) \\
&= j\, b_1'' \underbrace{\int_{\mathbb{R}^d} \varphi(r)\, dr}_{b_{1,\varphi}} \omega - \underbrace{b_2 \|\varphi\|_{L_2}^2}_{b_{2,\varphi}} \frac{\omega^2}{2} + g_\varphi(\omega),
\end{aligned}
$$

where

$$
g_\varphi(\omega) = \int_{\mathbb{R}^d} \int_{\mathbb{R}\setminus\{0\}} \left(e^{ja\varphi(r)\omega} - 1 - ja\varphi(r)\omega \right) v(a)\, da\, dr,
$$

with the property that $g_\varphi(0) = 0$. Next, we identify $v_\varphi(a)$ by making the change of variable $a'(r) = a\varphi(r)$, while restricting the domain of integration to the subregion of $\mathbb{R}^d$ over which the argument is non-zero:

$$
\begin{aligned}
g_\varphi(\omega) &= \int_{\Pi_\varphi} \int_{\mathbb{R}\setminus\{0\}} \left(e^{ja'\omega} - 1 - ja'\omega \right) \left(\frac{1}{|\varphi(r)|} v\!\left(a'/\varphi(r)\right) \right) da'\, dr \\
&= \int_{\mathbb{R}\setminus\{0\}} \left(e^{ja'\omega} - 1 - ja'\omega \right) \underbrace{\int_{\Pi_\varphi} \frac{1}{|\varphi(r)|} v\!\left(a'/\varphi(r)\right) dr}_{v_\varphi(a')} da',
\end{aligned}
$$

where the interchange of integrals is legitimate thanks to (9.5) (by Fubini). Lemma 9.2 ensures that v_φ is admissible in accordance with Definition 4.3.

The scenario $\int_{|a|>1} |a|\, v(a)\, da = \infty$ is trickier because it calls for a more careful compensation of the singularity of the Lévy density. The classical Lévy–Khintchine formula (4.4) leads to an integral of the form

$$g_\varphi'(\omega) = \int_{\mathbb{R}^d} \int_{\mathbb{R}\setminus\{0\}} \left(e^{ja\varphi(r)\omega} - 1 - ja\varphi(r)\omega\, h\big(a, \varphi(r)\big) \right) v(a)\, da\, dr$$

with $h\big(a, \varphi(r)\big) = \mathbb{1}_{|a|<1}(a)$. It turns out that the exact form of the compensation $h\big(a, \varphi(r)\big)$ is not important as long as it stabilizes the integral by introducing an appropriate linear bias which results in a modification of the constant b_1'. Instead of the canonical solution, we propose an alternative regularization with $h\big(a, \varphi(r)\big) = \mathbb{1}_{|a\varphi(r)|<1}(a)$, which is compatible with the change of variable $a' = a/\varphi(r)$. The rationale is that this particular choice is guaranteed to lead to a convergent integral, as a consequence of Lemma 9.2, so that the remainder of the proof is the same as in the previous case: change of variable and interchange of integrals justified by Fubini's theorem. This also leads to some modified constant $b_{1,\varphi}'$ which is necessarily finite since both $g_\varphi'(\omega)$ and $f_\varphi(\omega)$ are bounded.   □

To carry out the proof of Lemma 9.2, we have exploited the fact that the integration of f against a function φ amounts to a spectral mixing $f(\omega; \mu_\varphi) = \int_{\mathbb{R}} f(\theta\omega)\, \mu_\varphi(d\theta)$. This results in an admissible Lévy exponent provided that the pth absolute moment of the measure μ is finite: $\int_{\mathbb{R}} |\theta|^p\, \mu_\varphi(d\theta) < \infty$. The equivalence with $f_\varphi(\omega)$ is obtained by defining $\mu_\varphi((-\infty, \theta]) = \mathrm{Meas}\{r : \varphi(r) < \theta \text{ and } \varphi(r) \neq 0\}$, which specifies the amplitude distribution of $\varphi(r)$ as r ranges over $\mathbb{R}^d$. To gain further insight into the mixing process, we like to view the Lebesgue integral $\int_{\mathbb{R}} f(\theta\omega)\, \mu_\varphi(d\theta)$ as the limit of a sequence of Lévy exponents $f_N(\omega) = \sum_{n=1}^{N} f(\theta_{n,N}\omega)\tau_n$, each corresponding to a characteristic function of the form $e^{f_N(\omega)} = \prod_{n=1}^{N} \widehat{p}_X(s_n\omega)^{\tau_n}$ with $s_n = \theta_{n,N}$ and $\tau_n = \mu_\varphi\big([\theta_{n-1,N}, \theta_{n,N})\big)$. The latter is interesting because it provides a convolutional interpretation of the mixing process, and also because it shows that all that matters is the amplitude distribution of φ, and not its actual spatial structure.

COROLLARY 9.4 (Symmetric Lévy exponents)   *Let f be an admissible Lévy exponent and let $\varphi \in L_p(\mathbb{R})$ for $p \geq 1$ be a function such that its amplitude distribution (or histogram) μ_φ is symmetric. Then, $f_\varphi(\omega) = \int_{\mathbb{R}^d} f\big(\omega\varphi(r)\big)\, dr = \int_{\mathbb{R}} f(\theta\omega)\, \mu_\varphi(d\theta)$ is a valid symmetric Lévy exponent that admits the canonical representation (4.4) with modified Lévy parameters $\big(b_\varphi, v_\varphi(a) = v_\varphi(-a)\big)$, as specified in Theorem 9.1. Conversely, if f is symmetric to start with, then f_φ stays symmetric for any φ, irrespective of its amplitude distribution.*

Proof   Based on the Lévy–Khintchine representation of f and by relying on Fubini's theorem to justify the interchange of integrals, we get

$$f_\varphi(\omega) = -\frac{b_2}{2}\omega^2 \underbrace{\int_{\mathbb{R}^d} \theta^2\, \mu_\varphi(d\theta)}_{\|\varphi\|^2} + \int_{\mathbb{R}} \int_{\mathbb{R}} \big(e^{ja\theta\omega} - 1 - ja\theta\omega \big) v(a)\, da\, \mu_\varphi(d\theta)$$

$$= -\frac{b_2}{2}\omega^2 \|\varphi\|^2 + \int_{\mathbb{R}} \int_0^\infty 2\big(\cos(a\theta\omega) - 1 \big)\, \mu_\varphi(d\theta) v(a)\, da,$$

where we have made use of the symmetry assumption $\mu_\varphi(E) = \mu_\varphi(-E)$. Since the above formula is symmetric in ω and $f_\varphi(\omega)$ is a valid Lévy exponent (by Theorem 9.1), we can invoke Corollary 4.3, which yields the desired result. The converse part of Corollary 9.4 is obvious.   □

The practical relevance of Corollary 9.4 is that we can restrict ourselves to a symmetric model of noise without any loss of generality as long as the analysis function (typically, a wavelet) has a symmetric amplitude distribution μ_φ. This is equivalent to all odd moments of μ_φ being zero, including the mean.

One of the cornerstones of our formulation is that the mixing (white-noise integration) does not fundamentally affect the key properties of the Lévy exponent, as will be made clear in what follows.

9.2   Class C and unimodality

DEFINITION 9.1   An id distribution is said to be of *class* C if its Lévy density $v(a)$ is unimodal with mode 0; more precisely, if $v'(a) \le 0$ for $a > 0$, and $v'(a) \ge 0$ for $a < 0$.

This is again a property that is conserved through basic composition: linear combination of random variables and Fourier-domain exponentiation (fractional convolution of pdf). Less obvious is the fact that the class-C property is a necessary condition for the family of distributions p_{X_τ} with $\widehat{p}_{X_\tau}(\omega) = (\widehat{p}_{X_1}(\omega))^\tau$ to be unimodal for all $\tau \in \mathbb{R}^+$ [Wol78]. Conversely, if p_{X_1} is symmetric and of class C, then the whole family p_{X_τ} is symmetric unimodal. The class-C unimodality equivalence, however, is only partial. In particular, there exist non-symmetric distributions of class C that are not unimodal, and some symmetric unimodal id distributions that are not in the class C for which $p_{X_\tau}(x)$ is unimodal only for some subrange of τ [Wol78, Section 3]. The key result for our purpose is as follows.

COROLLARY 9.5   *If the Lévy density of an id distribution is symmetric and unimodal, then its pdf $p_X(x) = \int_\mathbb{R} \exp\left(-\frac{b_2}{2}\omega^2 + \int_\mathbb{R} \cos(a\omega)v(a)\, da\right) e^{-jx\omega} \frac{d\omega}{2\pi}$ is symmetric and unimodal as well. Moreover, the property is preserved through the white-noise integration process with some arbitrary bounded function $\varphi \in L_2(\mathbb{R}^d)$.*

Proof   The first part is a restatement of a classical result in the theory of infinitely divisible distributions [Wol78, Theorem 1]. As for the second part, we invoke Theorem 9.1, which ensures that the modified Lévy exponent v_φ given by (9.2) is admissible. Since $v(a) = v(-a)$, v_φ is automatically symmetric as well. We prove its unimodality by computing the derivative for $a > 0$ as

$$v'_\varphi(a) = \frac{d}{da} \int_{\Pi_\varphi} \frac{1}{|\varphi(r)|} v(a/|\varphi(r)|) \, dr$$

$$= \int_{\Pi_\varphi} \frac{1}{|\varphi(r)|^2} v'(a/|\varphi(r)|) \, dr \le 0,$$

which follows from the condition $v'(a) \le 0$ for $a > 0$ (unimodality of $v(a)$).   □

The pleasing observation is that all the examples of id distributions in Table 4.1 are in this category.

PROPOSITION 9.6 (Alternative statement)   *Let $f(\omega) = f(-\omega)$ be a real-valued admissible Lévy exponent of class C associated with a symmetric unimodal distribution and φ a d-dimensional function such that $\|\varphi\|_{L_p} < \infty$ for $p \in \mathbb{R}^+$. Then, $f_\varphi(\omega) = \int_{\mathbb{R}^d} f(\omega\varphi(r)) \, dr$ is a valid Lévy exponent that retains the symmetry and unimodality properties.*

The interest of the above is more in the constructive proof given below than in the statement, which is slightly less general than Corollary 9.5.

Proof of Proposition 9.6   The result is obtained as a corollary of two theorems in [GK68]: (i) the convolution of two symmetric [1] unimodal distributions is unimodal, and (ii) if a sequence of unimodal distributions converges to a distribution, then the limit function is unimodal as well. The idea is then to view the Lebesgue integral as the limit as n goes to infinity of the sequence of sums $\sum_{k=1}^{n^2-1} f(\omega s_{k,n}) \mu_{k,n}$, where $\mu_{k,n}$ is the measure of the set $E_{k,n} = \{r : \frac{k}{n} \leq |\varphi(r)| < \frac{k+1}{n}\}$ and $s_{k,n} = \arg\min_{\varphi(r):r\in E_{n,k}} |f(\omega\varphi(r))|$, recalling that the Lévy exponent f is continuous. Each individual term corresponds to a characteristic function $e^{\mu_{k,n} f(\omega s_{k,n})} = (e^{f(\omega s_{k,n})})^{\mu_{k,n}}$ (Fourier-domain convolutional factor) that is id (thanks to the rescaling and exponentiation property), symmetric, and unimodal. Finally, we rely on the admissibility condition and Lebesgue's dominated-convergence theorem to show that the sequence converges to the limit $f_\varphi(\omega) = f_\varphi(-\omega)$ which specifies a valid symmetric id distribution.   □

This proof also suggests that the class-C property is the tightest possible condition in the mixing scenario with arbitrary φ. Indeed, class C plus symmetry is necessary and sufficient for $p_{X_\tau}(x)$ to be unimodal for all $\tau \in \mathbb{R}^+$ [Wol78].

Another interesting class of id distributions are those that are *completely monotonic*. We recall that a function (or density) $q(x)$ is completely monotonic on $\mathbb{R}^+$ if it is of class C^∞ with alternating derivatives, so that

$$(-1)^n \frac{d^n q(x)}{dx^n} \geq 0 \text{ for } n \in \mathbb{N}, x \geq 0.$$

Thanks to Bernstein's theorem, the symmetric version of completely monotonic distributions can be expressed as mixtures of Laplacians; that is,

$$q(x) = \int_0^{+\infty} \frac{1}{2} \lambda e^{-\lambda|x|} p_Z(\lambda) \, d\lambda \tag{9.6}$$

for some mixing density $p_Z(\lambda)$ on $\mathbb{R}^+$. By making the change of variable $\theta = 1/\lambda$ (scale of the exponential distribution), we may also express $q(x)$ as

$$q(x) = \int_0^{+\infty} \frac{1}{2\theta} e^{-|x|/\theta} p_{1/Z}(\theta) \, d\theta = \int_0^\infty p_Y(x|\theta) p_{1/Z}(\theta) \, d\theta$$

[1] While translating Gnedenko and Kolmogorov's book into English, K. L. Chung realized that Lapin's theorem on the convolution of unimodal distributions, which does not impose the symmetry condition, is generally not true. The result for symmetric distributions goes back to Wintner in 1936 (see [Wol78] for a historical account).

with $p_Z(\lambda)\,\mathrm{d}\lambda = p_{1/Z}(\theta)\,\mathrm{d}\theta$, where $p_Y(x|\theta)$ is the Laplace distribution with standard deviation θ. The probabilistic interpretation of the above expansion is that of an exponential scale mixture: $q = p_X$ is the pdf of the ratio $X = Y/Z$ of two independent random variables Y and Z with pdfs p_Y (standardized Laplacian with $\lambda = 1$) and p_Z (under the constraint that Z is positive), respectively.

Complete monotonicity is one of the few simple criteria that ensures that a distribution is infinitely divisible [SVH03, Theorem 10.1]. Moreover, one can readily show that a symmetric completely monotonic distribution is log-convex, since

$$\frac{\mathrm{d}^2 \log q(x)}{\mathrm{d}x^2} = \frac{q''(x)}{q(x)} - \left(\frac{q'(x)}{q(x)}\right)^2 \geq 0. \tag{9.7}$$

Indeed, based on the canonical form (9.6), we have that

$$q'(x)^2 = \left(\int_0^{+\infty} -\lambda e^{-\lambda|x|}\frac{\lambda}{2}p_Z(\lambda)\,\mathrm{d}\lambda\right)^2$$
$$\leq \left(\int_0^{+\infty} e^{-\lambda|x|}\frac{\lambda}{2}p_Z(\lambda)\,\mathrm{d}\lambda\right)\left(\int_0^{+\infty} \lambda^2 e^{-\lambda|x|}\frac{\lambda}{2}p_Z(\lambda)\,\mathrm{d}\lambda\right) = q(x)q''(x),$$

where we have made use of the Cauchy–Schwarz inequality applied to the inner product $\langle u, v\rangle_Z = \mathbb{E}_Z\{\frac{\lambda}{2}uv\}$ with $u(\lambda) = e^{-\frac{1}{2}\lambda|x|}$ and $v(\lambda) = -\lambda e^{-\frac{1}{2}\lambda|x|}$.

Note that, unlike monotonicity, complete monotonicity is generally not preserved through spectral mixing.

A related property at the other end of the spectrum is log-concavity. It is equivalent to the convexity of the log-likelihood potential $\Phi_X(x) = -\log p_X(x)$, which is advantageous for optimization purposes and for designing MAP estimators (see Chapter 10). The id laws in Table 4.1 that are log-concave are the Gaussian, Laplace, and hyperbolic secant distributions. It should be kept in mind, however, that most id distributions are not log-concave since the property is incompatible with a slower-than-exponential decay. The limit case is the symmetric unimodal Laplace distribution, which is both log-concave and log-convex.

9.3   Self-decomposable distributions

DEFINITION 9.2   A random variable X is said to be *self-decomposable* if, for every $\lambda \in (0,1)$, $X = \lambda X + X_\lambda$ in law, where X and X_λ are independent random variables with pdf p_X and p_{X_λ}, respectively. The corresponding p_X (resp., $\widehat{p}_X$) is also said to be self-decomposable.

The self-decomposability of X is therefore equivalent to the requirement that its characteristic function $\widehat{p}_X(\omega) = \mathbb{E}\{e^{\mathrm{j}\omega X}\}$ can be factorized as

$$\widehat{p}_X(\omega) = \widehat{p}_X(\lambda\omega) \cdot \widehat{p}_{X_\lambda}(\omega), \tag{9.8}$$

where $\widehat{p}_{X_\lambda}(\omega)$ is a valid characteristic function for any $\lambda \in (0,1)$.

All the examples of id distributions in Table 4.1 are self-decomposable. For instance, in the case of the symmetric gamma (sym gamma) distribution, we have that

$$\widehat{p}_\lambda(\omega) = \frac{\widehat{p}_X(\omega)}{\widehat{p}_X(\lambda\omega)} = \left(\lambda^2 + (1 - \lambda^2)\frac{1}{1 + \omega^2}\right)^r.$$

The latter is of the form $(\widehat{p}_0(\omega))^r$, where $p_0(x) = \lambda^2\delta(x) + (1 - \lambda^2)\frac{1}{2}e^{-|x|}$ is a valid (id) pdf for any $\lambda \in (0, 1)$. The key properties of self-decomposable distributions are [GK68, SVH03]:

- All self-decomposable distributions are infinitely divisible and of class C.
- Self-decomposability is preserved through basic composition: multiplication, rescaling, and exponentiation of characteristic functions.
- A self-decomposable distribution is necessarily unimodal [SVH03, Theorem 6.23]. In particular, if p_X is symmetric, then $\max\{p_X\} = p_X(0)$ and $p'_X(x) \leq 0$ for $x > 0$.
- If $\widehat{p}_X(\omega)$ is self-decomposable then $\widehat{p}_{X_\lambda}(\omega) = \frac{\widehat{p}_X(\omega)}{\widehat{p}_X(\lambda\omega)}$ is self-decomposable as well for any $\lambda \in (0, 1)$. The converse is true with the weaker requirement that $\widehat{p}_{X_\lambda}(\omega)$ need only be a valid characteristic function.
- A distribution is self-decomposable if and only if its Lévy density v is such that $av(a)$ is non-increasing on $(-\infty, 0)$ and on $(0, +\infty)$ [SVH03, Theorem 6.12].

The last property provides a simple operational criterion for testing self-decomposability when the Lévy density is known explicitly. The requirement is that $v(a)$ should either be zero or decrease monotonically at least as fast as $1/a$.

PROPOSITION 9.7 *Let $f(\omega)$ be a (p)-admissible Lévy exponent associated with a self-decomposable distribution and let $\varphi \in L_p(\mathbb{R}^d)$. Then, $f_\varphi(\omega) = \int_{\mathbb{R}^d} f(\omega\varphi(r)) \, dr$ is an admissible Lévy exponent that retains the self-decomposability property.*

Proof First, we invoke Theorem 9.1, which ensures that f_φ is a proper Lévy exponent to start with. The self-decomposability condition (9.8) is equivalent to

$$f(\omega) = f(\lambda\omega) + f_\lambda(\omega),$$

where $f_\lambda(\omega)$ is a valid Lévy exponent for any $\lambda \in (0, 1)$. Inserting φ and taking the integral on both sides gives

$$f_\varphi(\omega) = \int_{\mathbb{R}^d} f(\lambda\omega\varphi(r)) \, dr + \int_{\mathbb{R}^d} f_\lambda(\omega\varphi(r)) \, dr$$
$$= f_\varphi(\lambda\omega) + f_{\lambda,\varphi}(\omega),$$

where $f_{\lambda,\varphi}(\omega) = \int_{\mathbb{R}^d} f_\lambda(\omega\varphi(r)) \, dr$. Finally, since $|f_\lambda(\omega)| \leq |f(\omega)| + |f(\lambda\omega)|$, we are guaranteed that the integration with respect to φ yields an acceptable Lévy exponent. This implies that both $e^{f_\varphi(\omega)}$ and $e^{f_{\lambda,\varphi}(\omega)}$ are valid characteristic functions, which establishes the self-decomposability property. ☐

9.4   Stable distributions

Stability is the most stringent distributional property in the chain since all stable distributions are self-decomposable and hence unimodal of class C.

DEFINITION 9.3   A random variable X is called (strictly) *stable* if, for every $n \in \mathbb{Z}^+$, there exists c_n such that $X = c_n(X_1 + \cdots + X_n)$ in law, where the X_i are i.i.d. with pdf $p_X(x)$.

The definition clearly implies infinite divisibility. On the side of the Lévy exponent, the condition translates into the homogeneity requirement $f(a\omega) = a^\alpha f(\omega)$, with $\alpha \in (0, 2]$ being the stability index. One can then exploit this property to get the complete parameterization of the family of stable distributions. If we add the symmetry constraint, then the class of homogeneous Lévy exponents reduces to $f(\omega) \propto -|\omega|^\alpha$. These exponents are associated with the symmetric-alpha-stable (SαS) laws whose characteristic function is specified by

$$\widehat{p}(\omega; \alpha, s_0) = \mathrm{e}^{-|s_0\omega|^\alpha} = \mathrm{e}^{f_\alpha(s_0\omega)}$$

with $\alpha \in (0, 2]$ and $s_0 \in \mathbb{R}^+$.

The fundamental point is that stability is preserved through linear combinations, as a consequence of the definition. This remains true when we switch to a continuum by integrating stable white noise against some analysis function φ.

PROPOSITION 9.8   *Let $f(\omega) = -|s_0\omega|^\alpha$ be a Lévy exponent associated with an SαS distribution of order α. Then, the SαS property is preserved for $\int_{\mathbb{R}^d} f(\omega\varphi(r))\, dr$, provided that $\|\varphi\|_{L_\alpha} < \infty$.*

Specifically, we have that

$$\int_{\mathbb{R}^d} f_\alpha\big(s_0\omega\varphi(r)\big)\, dr = \int_{\mathbb{R}^d} -|s_0\,\omega\varphi(r)|^\alpha\, dr$$

$$= -\underbrace{\left(s_0^\alpha \int_{\mathbb{R}^d} |\varphi(r)|^\alpha\, dr\right)}_{s_0'^\alpha} |\omega|^\alpha\, dr$$

$$= -|s_0'\omega|^\alpha$$

with $s_0' = s_0\|\varphi\|_{L_\alpha}$.

The result also holds in the non-symmetric scenario where the distributions are parameterized by (α, β, τ). It then requires the additional assumption that $\varphi \in L_1(\mathbb{R})$ (for the non-zero-mean case) and $\int_{\mathbb{R}^d} |\varphi(r)| \log |\varphi(r)|\, dr$ (in the special asymmetric case where $\alpha = 1$ and $\beta \neq 0$).

Stable laws play a crucial role in the generalized formulation of the central-limit theorem, the standard (finite-variance) case being $\alpha = 2$ (Gaussian distribution).

9.5   Rate of decay

A fundamental property of infinitely divisible distributions is that they cannot decay faster than a Gaussian (heavy-tail behavior). The range of behaviors (from faster to slower decay) can be categorized as follows:

- Gaussian with $p_{\mathrm{id}}(x) = \mathrm{e}^{-O(|x|^2|)}$
- Supra-exponential with $p_{\mathrm{id}}(x) = \mathrm{e}^{-O(|x|\,\log(|x|+1))}$
- Exponential with $p_{\mathrm{id}}(x) = \mathrm{e}^{-O(|x|)}$ (e.g., the Laplace distribution)
- Exponential polynomial with $p_{\mathrm{id}}(x) = O(x^{\theta_1} \mathrm{e}^{-\theta_2(|x|)})$ for some $\theta_1 > 0$ and $\theta_2 > 0$ (e.g., the beta and Meixner distributions)
- Algebraic or inverse polynomial with $p_{\mathrm{id}}(x) = O(1/|x|^\theta)$ (e.g., SαS and Student) for some $\theta > 1$.

The most convenient way of probing decay is through the determination of some generalized moments of the form $m_{X,\theta} = \int_{\mathbb{R}} g_\theta(x) p_X(x)\, \mathrm{d}x = \mathbb{E}\{g_\theta(X)\}$, where $g_\theta(x) = g_\theta(-x)$ denotes some suitable family of symmetric functions that are increasing away from the origin. The typical choices for g_θ are $\mathrm{e}^{\theta|x|}$ (increasing exponential with $\theta > 0$), $|x|^{\theta_1} \mathrm{e}^{\theta_2 x}$ (exponential polynomial with $\theta_1 \geq 0, \theta_2 > 0$), and $|x|^\theta$ (monomial with $\theta > 0$). We also recall that the conventional (polynomial) moments with $g_n(x) = x^n$ and n integer can be readily obtained from the Taylor series of the characteristic function $\mathrm{e}^{f(\omega)}$ at the origin, since

$$\mathbb{E}\{X^n\} = \int_{\mathbb{R}} a^n p_X(a)\, \mathrm{d}a = \frac{1}{j^n} \left. \frac{\mathrm{d}^n \widehat{p}_X(\omega)}{\mathrm{d}\omega^n} \right|_{\omega=0}. \tag{9.9}$$

An interesting theoretical aspect that is further developed in Section 9.6 is that these quantities are in direct correspondence with the polynomial moments of the Lévy density v, which are closely linked to the cumulants of p_X. In particular, we have the following relations for the mean and variance of the distribution:

$$\mathbb{E}\{X\} = \int_{\mathbb{R}} x p_X(x)\, \mathrm{d}x = b_1' - \int_{|a|\geq 1} a v(a)\, \mathrm{d}a$$

$$\mathrm{Var}\{X\} = \int_{\mathbb{R}} (x - \mathbb{E}\{X\})^2 p_X(x)\, \mathrm{d}x = b_2 + \int_{\mathbb{R}} a^2 v(a)\, \mathrm{d}a.$$

In order to characterize the decay of the distribution, it is actually sufficient to check whether a given type of generalized moment is finite and to determine the range of exponents θ over which $\mathbb{E}\{g_\theta(X)\}$ is not well defined (i.e., infinite). Once more, the good news is that this information can be deduced directly from the Lévy density $v(a)$ [Sat94].

DEFINITION 9.4   A weighting function $g(x)$ on $\mathbb{R}$ is called *submultiplicative* if there exists a constant $C \geq 1$ such that

$$g(x + y) \leq C g(x) g(y) \text{ for } x, y \in \mathbb{R},$$

and if it is bounded and measurable on any compact subset.

THEOREM 9.9 (Kruglov [Kru70])   *For an id distribution p_X with Lévy density v, we have the following equivalences for the existence of generalized moments:*

$$\int_{\mathbb{R}} |x|^\theta p_X(x)\, dx < \infty \quad \Leftrightarrow \quad \int_{|a|>1} |a|^\theta v(a)\, da < \infty$$

$$\int_{\mathbb{R}} |x|^{\theta_1} e^{\theta_2 |x|} p_X(x)\, dx < \infty \quad \Leftrightarrow \quad \int_{|a|>1} |a|^{\theta_1} e^{\theta_2 |a|} v(a)\, da < \infty$$

$$\int_{\mathbb{R}} g_\theta(x) p_X(x)\, dx < \infty \quad \Leftrightarrow \quad \int_{|a|>1} g_\theta(a) v(a)\, da < \infty,$$

for any family of submultiplicative weighting functions g_θ.

The first two equivalences are deduced as special cases of the last one by considering the weighting sequences $g_\theta(x) = \max(1, |x|^\theta)$ and $g_{\theta_1,\theta_2}(x) = \max(1, |x|^{\theta_1}) e^{\theta_2 x}$, which are submultiplicative for $\theta, \theta_2 > 0$ and $\theta_1 \geq 0$.

As an application of Theorem 9.9, we can restate the Lévy–Schwartz admissibility condition in Theorem 4.8 as

$$\mathbb{E}\{|\langle \mathbb{1}_{[0,1)}, w \rangle|^\epsilon\} < \infty \text{ for some } \epsilon > 0. \tag{9.10}$$

The key here is that $X_{\mathrm{id}} = \langle \mathbb{1}_{[0,1)}, w \rangle$ (the canonical observation of the innovation w) is an id random variable whose characteristic function is $\widehat{p}_{\mathrm{id}}(\omega) = e^{f(\omega)}$ (see Proposition 4.12).

The direct implication is that p_X will have the same decay behavior as v, keeping in mind that it can be no better than supra-exponential (e.g., when v is compactly supported), unless we are in the purely Gaussian scenario with $v(a) = 0$. This also means that, from a decay point of view, supra-exponential and compact support are to be placed in the same category.

The fundamental reason for the equivalences in Theorem 9.9 is that the corresponding types of decay (exponential vs. polynomial) are preserved through convolution. In the latter case, this can be quantified quite precisely using a generalized version of the Young inequality for weighted L_p-spaces (see [AG01]):

$$\|g_\theta (h * q)\|_{L_p} \leq C_\theta \|g_\theta h\|_{L_1} \cdot \|g_\theta q\|_{L_p}$$

for any $p \geq 1$ and any family of submultiplicative weighting functions $g_\theta(x)$. The canonical example of weighting function that is used to characterize algebraic decay is $g_\alpha(x) = (1 + |x|)^\alpha$ with $\alpha \in \mathbb{R}^+$, which is submultiplicative with constant $C_\alpha = 1$.

In light of Theorem 9.9, our next result implies that the rate of polynomial decay and the moment-related properties of p_X are preserved through the white-noise integration process.

PROPOSITION 9.10   *Let v be a Lévy density with a p_0th order of algebraic decay at infinity for some $p_0 \geq 1$ (or better) and φ a function such that $\|\varphi\|_{L_p}^p = \int_{\mathbb{R}^d} |\varphi(r)|^p\, dr < \infty$ for all $p \geq p_0 - 1$, with the formal convention that $\|\varphi\|_{L_0}^0 = \int_{\Pi_\varphi} dr$. Then, the transformed density v_φ given by (9.2) will retain the decay of v and belong to the same* *general class (e.g., finite-variance vs. infinite-variance density, Poisson, Gaussian or alpha-stable). In particular, its pth-order absolute moment for any $p \geq 0$ is given by*

$$m_{v_\varphi,p} = \int_{\mathbb{R}} |a|^p v_\varphi(a)\, \mathrm{d}a = \|\varphi\|_{L_p}^p m_{v,p} \tag{9.11}$$

and is well defined whenever $m_{v,p}$, the pth-order absolute moment of v, is bounded. Likewise, the integration against φ will preserve the finiteness of all absolute moments of the underlying distribution, including those for $p \leq 2$.

Proof For the admissibility of v_φ, we refer to Theorem 9.1. To show that the integrand in (9.2) is well defined when $\varphi(r)$ tends to zero and that v_φ has the same decay properties at infinity as v, we consider the bound $v(a) < C|a|^{-p_0}$ with $p_0 \geq 1$, which implies that $\frac{1}{|\varphi(r)|} v(a/\varphi(r)) \leq C|\varphi(r)|^{p_0-1}|a|^{-p_0}$. It follows that $v_\varphi(a) \leq C\|\varphi\|_{L_p}^p |a|^{-p_0}$ with $p = p_0 - 1$, meaning that the rate of decay of v is preserved. To check the integrability of v_φ, we make the reverse change of variable $a = \varphi(r)a'$ and obtain

$$\int_{\mathbb{R}} \int_{\Pi_\varphi} \frac{1}{|\varphi(r)|} v(a/\varphi(r))\, \mathrm{d}r\, \mathrm{d}a = \int_{\mathbb{R}} \int_{\Pi_\varphi} v(a')\, \mathrm{d}r\, \mathrm{d}a'$$
$$= \int_{\Pi_\varphi} \mathrm{d}r \int_{\mathbb{R}} v(a')\, \mathrm{d}a' = \|\varphi\|_{L_0}^0 \int_{\mathbb{R}} v(a')\, \mathrm{d}a'.$$

Likewise, we can readily show that $\int_{\mathbb{R}} |a|^p v_\varphi(a)\, \mathrm{d}a = \|\varphi\|_{L_p}^p \int_{\mathbb{R}} |a|^p v(a)\, \mathrm{d}a$, assuming that these latter quantities are well defined. For the absolute moments of order $p < 2$, we can refer to Lemma 9.2 and the conservation of the admissibility condition $\int_{\mathbb{R}} \min(a^2, |a|^p) v_\varphi(a)\, \mathrm{d}a < \infty$. $\qquad\square$

In particular, when v is of the Poisson type, we have that $\int_{\mathbb{R}} v(a)\, \mathrm{d}a = \lambda < +\infty$, and it follows that v_φ will be of the Poisson type as well, provided that φ is compactly supported. The fact that the Poisson parameter λ increases in proportion to the support of φ is consistent with the window intersecting more Dirac impulses (components of the white Poisson noise).

We shall now see that a relation similar to (9.11) holds for the standard integer-order moments of v_φ and v; for $p > 2$, these actually happen to be the cumulants of the underlying probability distributions.

9.6   Lévy exponents and cumulants

The fact is that, with id distributions, it is often simpler to work with cumulants than with ordinary moments, the idea being that cumulants are to the Lévy exponent what moments are to the characteristic function.

Specifically, let $\widehat{p}_X(\omega) = \int_{\mathbb{R}} p_X(x) e^{\mathrm{j}\omega x}\, \mathrm{d}x$ be the characteristic function of a pdf (not necessarily id) whose moments $m_p = \int_{\mathbb{R}} x^p p_X(x)\, \mathrm{d}x < \infty$ are assumed to be well defined for any $p \in \mathbb{N}$. The boundedness of the moments implies that $\widehat{p}_X(\omega)$ (the conjugate

Fourier transform of p_X) is infinitely differentiable at the origin. Hence, it is possible to represent $\widehat{p}_X$, as well as its logarithm, by the Taylor series expansions

$$\widehat{p}_X(\omega) = \sum_{p=0}^{\infty} \frac{(j\omega)^p}{p!} m_p, \tag{9.12}$$

$$\log \widehat{p}_X(\omega) = \sum_{n=1}^{\infty} \frac{(j\omega)^n}{n!} \kappa_n, \tag{9.13}$$

where the second formula provides the definition of the κ_n, which are the cumulants of p_X. This clearly shows that p_X is uniquely characterized by its moments (by way of its characteristic function), or, equivalently, in terms of its cumulants by way of its cumulant generating function $\log \widehat{p}_X(\omega)$. Another equivalent way of expressing this correspondence is

$$\kappa_n\{X\} = \frac{1}{j^n} \frac{d^n \log \widehat{p}_X(\omega)}{d\omega^n} \bigg|_{\omega=0},$$

which is the direct counterpart of (9.9). By equating the Taylor series of the exponential of the right-hand sum in (9.13) with (9.12), one can derive a direct relation between the moments and the cumulants

$$m_p = \sum_{n=0}^{p-1} \binom{p-1}{n} m_n \kappa_{p-n},$$

with $\kappa_0 = 0$ and $m_0 = 1$. There is also a converse version of this formula. In practice, however, it is often more convenient to compute the cumulants from the centered moments $\mu_p = \mathbb{E}\{(x - \mathbb{E}\{x\})^p\}$ of the distribution. For reference purposes, we provide the relevant formulas up to order 6:

$$\kappa_2 = \mu_2$$
$$\kappa_3 = \mu_3$$
$$\kappa_4 = \mu_4 - 3\mu_2^2$$
$$\kappa_5 = \mu_5 - 10\mu_3\mu_2$$
$$\kappa_6 = \mu_6 - 15\mu_4\mu_2 - 10\mu_3^2 + 30\mu_2^3,$$

while $\kappa_1 = m_1$ is simply the mean of the distribution.

In the case of interest to us where p_X is id, $\log \widehat{p}_X(\omega) = f(\omega)$, which translates into

$$\kappa_n\{X\} = \frac{1}{j^n} \frac{d^n f(\omega)}{d\omega^n} \bigg|_{\omega=0} = \begin{cases} b_1' - \int_{|a| \geq 1} a v(a)\, da, & n = 1 \\ b_2 + \int_{\mathbb{R}} a^2 v(a)\, da, & n = 2 \\ \int_{\mathbb{R}} a^n v(a)\, da, & n > 2. \end{cases}$$

This relation is fundamental for it relates the cumulants of the id distribution to the moments of the Lévy density. It also implies that $\kappa_4\{X\} > 0$ whenever $v(a) \neq 0$, which translates into the property that all non-Gaussian id distributions are *leptokurtic*. This is consistent with id distributions being more peaky than a Gaussian around the mean and exhibiting fatter tails.

A further theoretical motivation for using cumulants is that they provide a direct measure of the deviation from Gaussianity since the cumulants of a Gaussian are necessarily zero for $n > 2$ (because $\log \widehat{g}_\sigma(\omega) = -\frac{\sigma^2}{2}\omega^2$).

A final practical advantage of cumulants is that they offer a convenient means of quantifying – and possibly inverting – the effect of the white-noise integration process.

PROPOSITION 9.11 *Let f be an admissible Lévy exponent and let $\varphi \in L_p(\mathbb{R})$ for all $p \geq 1$. Then, the cumulants of $f_\varphi(\omega) = \int_{\mathbb{R}^d} f(\omega\varphi(r))\,dr$ are related to those of $f(\omega)$ by*

$$\kappa_{n,\varphi} = \kappa_n \int_{\mathbb{R}^d} (\varphi(r))^n \, dr.$$

Proof   We start by writing the Taylor series expansion of f as

$$f(\omega) = \sum_{n=1}^\infty \kappa_n \frac{(j\omega)^n}{n!},$$

where the quantities κ_n are, by definition, the cumulants of the density function $p_X(x) = \mathscr{F}^{-1}\{e^{f(\omega)}\}(x)$. Next, we replace ω by $\omega\varphi(r)$ and integrate over $\mathbb{R}^d$, which gives

$$\int_{\mathbb{R}^d} f(\omega\varphi(r))\,dr = \sum_{n=1}^\infty \kappa_n \frac{(j\omega)^n}{n!} \int_{\mathbb{R}^d} (\varphi(r))^n \, dr = f_\psi(\omega).$$

The last step is to equate the above expression to the expansion of the cumulant generating function f_φ

$$f_\varphi(\omega) = \sum_{n=1}^\infty \kappa_{n,\varphi} \frac{(j\omega)^n}{n!},$$

which yields the desired result.   □

A direct implication is that the odd-order cumulants of $\langle \varphi, w \rangle$ are zero whenever the analysis kernel φ has a symmetric amplitude distribution. This results in a modified Lévy exponent f_φ that is real-valued symmetric, which is consistent with Corollary 9.4. In particular, this condition is fulfilled when the analysis kernel has an axis of antisymmetry; that is, when there exists r_0 such that $\varphi(r_0 + r) = -\varphi(r_0 - r), \forall r \in \mathbb{R}^d$.

As for the non-zero even-order cumulants, we can use the relation

$$\kappa_{2m} = \frac{\kappa_{2m,\varphi}}{\|\varphi\|_{L_{2m}(\mathbb{R}^d)}^{2m}}$$

to recover the cumulants of w from the moments of the observed random variable $\langle \varphi, w \rangle$.

9.7   Semigroup property

It turns out that every id pdf is embedded in a semigroup that plays a central role in the classical theory of Lévy processes. These convolution semigroups define natural families of id pdfs such as the sym gamma and Meixner distributions that are extending the Laplace and hyperbolic secant distributions, respectively. In Section 9.8, we put the semigroup property to good use to characterize the behavior of the wavelet-domain statistics across scales.

Let us recall the exponentiation property: if p_{X_1} is id with characteristic function $e^{f(\omega)}$ then p_{X_τ} with $\widehat{p}_{X_\tau}(\omega) = e^{\tau f(\omega)} = \left(\widehat{p}_{X_1}(\omega)\right)^\tau$ is id as well for any $\tau \in \mathbb{R}^+$. A direct implication is the convolution relation that relates the pdfs at scale $\tau + \tau_0$ and τ_0 with $\tau > 0$, expressed as

$$p_{X_{\tau+\tau_0}}(x) = \left(p_{X_\tau} * p_{X_{\tau_0}}\right)(x).$$

This suggests that the family of pdfs $\{p_{X_\tau} : \tau \in [0, \infty)\}$ is endowed with a semigroup-like structure. To specify the semigroup properties and spell out the implications, we introduce the family of linear convolution operators Q_τ with $\tau \geq 0$,

$$Q_\tau q(x) = \left(p_{X_\tau} * q\right)(x)$$

for any $q \in L_1(\mathbb{R})$. Clearly, the family $\{Q_\tau : \tau \geq 0\}$ is bounded over $L_1(\mathbb{R})$ with $\|Q_\tau\| = \sup_{\|q\|_{L_1}=1} \|Q_\tau q\|_{L_1} \leq 1$ (as a consequence of Young's inequality) and is such that: (1) $Q_{\tau_1} Q_{\tau_2} = Q_{\tau_1+\tau_2}$ for $\tau_1, \tau_2 \in [0, \infty)$, (2) $Q_0 = \text{Id}$, and (3) $\lim_{\tau \downarrow 0} \|Q_\tau q - q\|_{L_1} = 0$ for any $q \in L_1(\mathbb{R})$. It therefore satisfies all the properties of a *strongly continuous contraction semigroup*. Such semigroups are entirely characterized by their *infinitesimal generator* G, which is defined as

$$Gq(x) = \lim_{\tau \downarrow 0} \frac{Q_\tau q(x) - q(x)}{\tau}. \tag{9.14}$$

Based on this generator, the members of the group can be represented via the exponential map

$$Q_\tau = e^{\tau G} = \text{Id} + \tau G + \frac{1}{2!} \tau^2 G^2 + \cdots .$$

Likewise, we may also write

$$\lim_{\Delta\tau \to 0} \frac{p_{X_{\tau+\Delta\tau}}(x) - p_{X_\tau}(x)}{\Delta\tau} = G p_{X_\tau}(x), \tag{9.15}$$

which implies that $p_{X_\tau}(x) = p(x, \tau)$ is the solution of the partial differential equation

$$\frac{\partial}{\partial\tau} p(x, \tau) = G p(x, \tau),$$

with initial condition $p(x, 0) = p_{X_0}(x) = \delta(x)$. In the present case where $Q_\tau = e^{\tau G}$ is shift-invariant, we have a direct correspondence with the frequency response $e^{\tau f(\omega)}$. By transposing Definition (9.14) into the Fourier domain, we identify G as the LSI operator specified by

$$Gq(x) = \lim_{\tau \downarrow 0} \int_{\mathbb{R}} \widehat{q}(\omega) \frac{e^{\tau f(\omega)} - 1}{\tau} e^{-j\omega x} \frac{d\omega}{2\pi}$$

$$= \int_{\mathbb{R}} \widehat{q}(\omega) f(\omega) e^{-j\omega x} \frac{d\omega}{2\pi}, \tag{9.16}$$

where $\widehat{q}(\omega)$ is the (conjugate) Fourier transform of $q(x)$. The fact that the Lévy exponent $f(\omega)$ is the frequency response of G has some pleasing consequences for the interpretation of the PDE that rules the evolution of $p_{X_\tau}(x)$ as a function of τ.

9.7.1 Gaussian case

When $f(\omega) = -\frac{b_2}{2}\omega^2$, the operator G boils down to a second derivative along x. This results in the diffusion-type evolution equation

$$\frac{\partial}{\partial \tau}p(x,\tau) = \frac{b_2}{2}\frac{\partial^2}{\partial x^2}p(x,\tau),$$

where $\frac{\partial^2}{\partial x^2} = \Delta_x$ is the 1D equivalent of the Laplacian. The solution is a Gaussian with standard deviation $\sqrt{\tau b_2}$.

9.7.2 SαS case

When $f(\omega) = -s_0^\alpha|\omega|^\alpha$, the operator is proportional to the fractional Laplacian of order $\alpha/2$, denoted by $(-\Delta_x)^{\alpha/2}$. Hence, we can write down the fractional diffusion equation

$$\frac{\partial}{\partial \tau}p(x,\tau) = -s_0^\alpha(-\Delta_x)^{\alpha/2}p(x,\tau),$$

which is quite similar to the Gaussian case, except that the impulse response of G is no longer a point distribution. Here, the solution is an SαS distribution of order α and dispersion $s_\tau = \tau^{1/\alpha}s_0$.

9.7.3 Compound-Poisson case

Under the assumption that $v = \lambda p_A \in L_1(\mathbb{R})$ and $b_1 = b_2 = 0$, we express the Lévy exponent as $f(\omega) = \int_{\mathbb{R}}(e^{j a \omega} - 1)v(a)\,da = \lambda(\widehat{p}_A(\omega) - \widehat{p}_A(0))$, where $\widehat{p}_A(\omega) = \int_{\mathbb{R}}p_A(x)e^{j\omega x}\,dx$ is the characteristic function of $p_A(x) \geq 0$ with the normalization constraint $\widehat{p}_A(0) = 1$. It follows that the generator G is a filter characterized by the impulse response $g(x) = \overline{\mathscr{F}}^{-1}\{f\}(x) = \lambda p_A(x) - \lambda\delta(x)$. The corresponding evolution equation is

$$\frac{\partial}{\partial \tau}p(x,\tau) = \lambda\big(p_A * p(\cdot,\tau)\big)(x) - \lambda p(x,\tau)$$

and involves a convolution with the Poisson amplitude distribution p_A.

9.7.4 General iterated-convolution interpretation

By following up on the Poisson example, we propose to describe the action of G in full generality as a convolution with some impulse response $g(x) = G\{\delta\}(x)$. This leads to the equivalent form of the evolution equation

$$\frac{\partial}{\partial \tau}p(x,\tau) = \big(g * p(\cdot,\tau)\big)(x), \tag{9.17}$$

which is quite attractive from an engineering perspective. In the case where $f(\omega) = -b_2 \frac{\omega^2}{2} + \int_{\mathbb{R}} (e^{j a \omega} - 1 - j a \omega) \, v(a) \, da$, we obtain the explicit form of the impulse response by the formal inverse (conjugate) Fourier transformation

$$g(x) = G\{\delta\}(x) = \frac{b_2}{2} \delta''(x) + \int_{\mathbb{R}} \big(\delta(x - a) - \delta(x) + a\delta'(x)\big) v(a) \, da,$$

where δ' and δ'' are the first and second derivatives of the Dirac impulse. Some further simplification is possible if we can split the second component of $g(x)$ into parts, although this requires some special care because all non-Poisson Lévy densities are singular around the origin. A first simplification occurs when $\int_{\mathbb{R}} |a| v(a) \, da < \infty$, which allows us to pull δ' out of the integral with its weight being $m_1 = \int_{\mathbb{R}} a v(a) \, da$. To bypass the singularity issue, we consider the sequence of non-singular Lévy densities $v_{1/n}(a) = v(a)$ for $|a| > 1/n$ and zero otherwise, which converges to $v(a)$ as n goes to infinity. Using the fact that $v_{1/n}$ is Lebesgue-integrable (as a consequence of the admissibility condition), we can perform a standard (conjugate) Fourier inversion, which yields

$$g(x) = -m_1 \delta'(x) + \frac{b_2}{2} \delta''(x) + \lim_{n \to \infty} \left(v_{1/n}(x) - \delta(x) \int_{|a| < 1/n} v(a) \, da \right),$$

with the limit component $\lim_{n \to \infty} v_{1/n}(x) = v(x)$ being the original Lévy density. The above representation is enlightening because the impulse response is now composed of two terms: a distribution that is completely localized at the origin (linear combination of Dirac impulse and its derivatives up to order two) plus a smoothing component that converges to the initial Lévy density v. The Dirac correction is actually crucial because it converts an essentially lowpass filter (convolution with $v(x) \geq 0$) into a highpass one which is consistent with the requirement that $f(0) = 0$.

We can also use this result to describe the evolution of p_{X_τ} for some small increment $\Delta \tau$, like

$$p_{X_{\tau + \Delta \tau}}(x) = \overline{\mathscr{F}}^{-1} \left\{ \widehat{p}_{X_\tau}(\omega) \left(1 + \Delta \tau f(\omega) + \frac{1}{2!} \Delta \tau^2 f^2(\omega) + \cdots \right) \right\}(x)$$

$$= p_{X_\tau}(x) + \Delta \tau \, (g * p_{X_\tau})(x) + \frac{1}{2!} \Delta \tau^2 (g * g * p_{X_\tau})(x) + O(\Delta \tau^3),$$

where $g = G\{\delta\}$ is the impulse response of the semigroup generator. Since g is representable as the sum of a point distribution concentrated at the origin and a Lévy density v, this partly explains why p_{X_τ} will be endowed with the properties of v that are preserved through convolution; in particular, the rate of decay (exponential vs. polynomial), symmetry, and unimodality.

9.8   Multiscale analysis

The semigroup property is especially relevant for describing the evolution across scale of the wavelet-domain statistics of a sparse stochastic process. The premise for the validity of such an analysis is that the whitening operator L is scale-invariant of order γ and that the mother wavelet ψ can be written as

$$\psi(\mathbf{r}) = L^* \phi(\mathbf{r}),$$

where $\phi \in L_1(\mathbb{R}^d)$ is a suitable "smoothing" kernel. We also assume that the wavelets are normalized to have a constant L_2 norm across scales. In the framework of the continuous wavelet transform, the wavelet at scale $a > 0$ is therefore given by

$$
\begin{aligned}
\psi_a(\mathbf{r}) &= a^{-d/2} \psi(\mathbf{r}/a) \\
&= a^{-d/2} L^* \{\phi(\cdot)\}(\mathbf{r}/a) \\
&= L^* \left\{ a^{-d/2+\gamma} \phi(\cdot/a) \right\}(\mathbf{r}) \\
&= L^* \phi_a(\mathbf{r}) \quad\quad\quad (9.18)
\end{aligned}
$$

with

$$\phi_a(\mathbf{r}) = a^{\gamma - d/2} \phi(\mathbf{r}/a), \quad\quad\quad (9.19)$$

where we have made use of the scale-invariance property of L (see Definition 5.2 in Chapter 5). The modified Lévy exponent at scale a can therefore be determined to be

$$
\begin{aligned}
f_{\phi_a}(\omega) &= \int_{\mathbb{R}^d} f\left(\omega a^{\gamma - d/2} \phi(\mathbf{r}/a)\right) d\mathbf{r} \\
&= \int_{\mathbb{R}^d} f\left(\omega a^{\gamma - d/2} \phi(\mathbf{r}')\right) |a|^d \, d\mathbf{r}' \quad\quad \text{(change of variable } \mathbf{r}' = \mathbf{r}/a) \\
&= a^d f_\phi\left(a^{\gamma - d/2} \omega\right). \quad\quad\quad (9.20)
\end{aligned}
$$

Thus, we see that there are two mechanisms at play that determine the evolution of the wavelet distribution across scale. The first is a simple change of amplitude[2] of the wavelet coefficients with their standard deviation being divided by the factor $a^{\gamma - d/2}$ that multiplies ω in (9.20). The second is the multiplication of the Lévy exponent by $\tau = a^d$, which induces the kind of convolution semigroup investigated in Section 9.7.

9.8.1 Scale evolution of the pdf

To make the link with the previous results on semigroups explicit, we introduce the family of id pdfs

$$p_{\mathrm{id}}(x; \tau, f_\phi) = \mathscr{F}^{-1}\left\{ e^{\tau f_\phi(\omega)} \right\}(x)$$

---

[2] Let $Y = bX$ where X is a random variable with pdf p_X and b a fixed scaling constant. Then, $p_Y(x) = |b|^{-1} p_X(x/b)$, $\mathrm{Var}(Y) = b^2 \mathrm{Var}(X)$, and $\hat{p}_Y(\omega) = \hat{p}_X(b\omega)$. In the present case, X is id with $\hat{p}_X(\omega) = \exp(f_\phi(\omega))$.

that are tied to the smoothing kernel ϕ and indexed by the parameter $\tau \in \mathbb{R}^+$. Next, we consider the random variables associated with the wavelet coefficients of the stochastic process $s(r)$ at some fixed location r_0 and scale a:

$$V_a = \langle \psi_a(\cdot - r_0), s \rangle = \langle \phi_a(\cdot - r_0), w \rangle.$$

Since the Lévy noise w is stationary, it follows that V_a has an id distribution with modified Lévy exponent f_{ϕ_a}, as specified by (9.20). This allows us to express p_{V_a}, the pdf of the wavelet coefficients V_a, as

$$p_{V_a}(x) = p_{\mathrm{id}}(x; a^d, f_\phi(b\cdot))$$
$$= |b|^{-1} p_{\mathrm{id}}(x/b; a^d, f_\phi)$$

with $b = a^{\gamma - d/2}$. Instead of rescaling the argument of f_ϕ or the pdf, we can also consider the renormalized wavelet coefficients $Y_a = V_a/a^{\gamma - d/2}$ whose distribution is characterized by

$$p_{Y_a}(x) = p_{\mathrm{id}}(x; a^d, f_\phi),$$

which indicates that the evolution across scale is part of the same extended id family. This connection allows us to transpose the results of Section 9.7 into the wavelet domain and to infer the corresponding evolution equation. These findings are summarized as follows.

PROPOSITION 9.12   *Let s be a sparse process with scale-invariant whitening operator* L *of order* γ *and Lévy exponent f. The continuous wavelet transform at scale a of s is given by* $v_a(r) = \langle \psi_a(\cdot - r), s \rangle$ *with the wavelet* $\psi = \mathrm{L}^* \phi$ *being* L-*compatible. Then,* $v_a(r)$ *is stationary for all a > 0 and its first-order pdf* p_{V_a} *is infinitely divisible with Lévy exponent given by (9.20).*

Moreover, let p_{Y_τ} *denote the pdf of the scale-normalized wavelet coefficients* $y_\tau = a^{d/2 - \gamma} v_a$ *with* $\tau = a^d$. *Then,* $p_{Y_\tau}(x) = p_Y(x; \tau)$ *satisfies the differential evolution equation*

$$\frac{\partial}{\partial \tau} p_Y(x, \tau) = \big(g_\phi * p_Y(\cdot, \tau)\big)(x), \tag{9.21}$$

where g_ϕ *is the generalized function that is the inverse (conjugate) Fourier transform of* $f_\phi(\omega) = \int_{\mathbb{R}^d} f(\omega \phi(r)) \, dr$.

9.8.2   Scale evolution of the moments

Equation (9.21) suggests that the evolution of the wavelet moments $\mathbb{E}\{V_a^p\}$ is dependent upon two effects: the first is the dilation of the pdf by $a^{\gamma - d/2}$, which translates into a simple moment proportionality factor $(a^{q(\gamma - d/2)})$, while the second is the fractional convolution (multiplication of the Lévy exponent by a^d), which induces some additional spreading of the pdf. For the case where p is an integer, we are able to derive an explicit formula by considering cumulants rather than moments.

Since we know the wavelet-domain Lévy exponent $f_{\phi_a}(\omega) = \log \widehat{p}_{V_a}(\omega)$, we apply the technique of Section 9.6 to determine the evolution of the cumulants across scale. We obtain

$$
\begin{aligned}
\kappa_n\{V_a\} &= \frac{1}{j^n} \left.\frac{d^n \log \widehat{p}_{V_a}(\omega)}{d\omega^n}\right|_{\omega=0} \\
&= \frac{1}{j^n} \left.\frac{d^n a^d f_\phi\left(a^{\gamma-d/2}\omega\right)}{d\omega^n}\right|_{\omega=0} \\
&= a^d \left(a^{\gamma-d/2}\right)^n \frac{1}{j^n} \left.\frac{d^n f_\phi(\omega)}{d\omega^n}\right|_{\omega=0} \qquad \text{(chain rule of differentiation)} \\
&= a^d a^{n(\gamma-d/2)} \kappa_n\{V_1\}.
\end{aligned} \tag{9.22}
$$

In the case of the variance (i.e., $\kappa_2\{V_a\} = \mathrm{Var}\{V_a\}$), this simplifies to

$$
\mathrm{Var}\{V_a\} = a^{2\gamma} \mathrm{Var}\{V_1\}. \tag{9.23}
$$

Not too surprisingly, the result is compatible with the scaling law of the variance of the wavelet coefficients of a fractional Brownian field: $\mathrm{Var}\{V_a\} = \sigma_0^2 a^{2H+d}$, where $H = \gamma - \frac{d}{2}$ is the Hurst exponent of the process. The latter corresponds to the special Gaussian version of the theory with $\kappa_n\{V_a\} = 0$ for $n > 2$.

The implication of (9.22) is that the evolution of the wavelet cumulants across scale is linear in a log-log plot. Specifically, we have that

$$
\log_a \kappa_n\{V_a\} = \log_a \kappa_n\{V_1\} + d + n(\gamma - d/2),
$$

which suggests a simple regression scheme for estimating γ from the moments of the wavelet coefficients of a self-similar process.

Based on (9.22), we relate the evolution of the kurtosis to the scale as

$$
\eta_4(a) = \frac{\kappa_4\{V_a\}}{\kappa_2^2\{V_a\}} = a^{-d}\eta_4(1). \tag{9.24}
$$

This implies that the kurtosis (if initially well defined) converges to zero as the scale gets coarser. We also see that the rate of convergence is universal (independent of the order γ) and that it is faster in higher dimensions.

The normalization ratio for the other cumulants is given by

$$
\eta_m(a) = \frac{\kappa_m\{V_a\}}{\kappa_2^{m/2}\{V_a\}} = \frac{1}{a^{d(m/2-1)}}\eta_m(1), \tag{9.25}
$$

which shows that the relative contributions of the higher-order cumulants (with $m > 2$) tend to zero as the scale increases. This implies that the limit distribution converges to a Gaussian under the working hypothesis that the moments (or cumulants) of p_{id} are well defined. This asymptotic behavior happens to be a manifestation of a generalized version of the central-limit theorem, the idea being that the dilation of the observation window has an effect that is equivalent to the summation of an increasing number of i.i.d. random contributions.

9.8.3   Asymptotic convergence to a Gaussian/stable distribution

Let us first investigate the case of an $S\alpha S$ innovation model, which is quite instructive and also fundamental for the asymptotic theory. The corresponding form of the Lévy exponent is $f(\omega) = -|s_0\omega|^\alpha$ with $\alpha \in (0, 2)$ and dispersion parameter $s_0 \in \mathbb{R}^+$. By substituting this expression in (9.20), we get

$$f_{\phi_a}(\omega) = - \int_{\mathbb{R}^d} \left| s_0 a^{\gamma - d/2} \omega \phi(r) \right|^\alpha a^d \, dr$$

$$= - \left| s_0 a^{\gamma - d/2} a^{d/\alpha} \omega \right|^\alpha \int_{\mathbb{R}^d} |\phi(r)|^\alpha \, dr$$

$$= - \left| s_{a,\phi} \, \omega \right|^\alpha, \tag{9.26}$$

where

$$s_{a,\phi} = s_0 a^{\gamma - d/2 + d/\alpha} \|\phi\|_{L_\alpha}. \tag{9.27}$$

This confirms the fact that the stability property is conserved in the wavelet domain. In the particular case of a Gaussian process with $\alpha = 2$ and $s_0 = \sigma_0/\sqrt{2}$, we obtain $s_{a,\phi} = s_0 a^\gamma \|\phi\|_{L_2}$, which is compatible with (9.23).

The remarkable aspect is that the combination of (9.26) and (9.27) specifies the limit distribution of the wavelet pdf for a sufficiently large under very general conditions, where $\alpha \le 2$ is the critical exponent of the underlying distribution. The parameter s_0 is related to the αth moment of the canonical pdf, where α is the largest possible exponent for this moment to be finite, the standard case being $\alpha = 2$.

Since the derivation of this kind of limit result for $\alpha < 2$ (generalized version of the central-limit theorem) is rather technical, we concentrate now on the finite-variance case ($\alpha = 2$), which can be established under rather general conditions using a basic Taylor-series argument.

For simplicity, we consider a centered scenario with modified Lévy triplet $b_{1,\phi} = 0$, $b_{2,\phi} = b_2 \|\phi\|_{L_2}^2$, and v_ϕ as specified by (9.2) such that $\int_{\mathbb{R}} t v_\phi(t) \, dt = 0$. Note that these conditions are automatically satisfied when the variance of the innovations is finite and f is real-valued symmetric (see Corollary 9.4). Due to the finite-variance hypothesis, we have that $m_{2,\phi} = \int_{\mathbb{R}} t^2 v_\phi(t) \, dt < \infty$. This allows us to write the Taylor series expansion of $f_{\phi_a}(\omega/a^\gamma)$ as

$$f_{\phi_a}(\omega/a^\gamma) = a^d f_\phi\left(a^{\gamma - d/2} \omega/a^\gamma\right)$$

$$= a^d f_\phi\left(a^{-d/2}\omega\right)$$

$$= -\frac{b_{2,\phi}\omega^2}{2} - \frac{m_{2,\phi}}{2}\omega^2 + O(a^{-d}\omega^4),$$

which corresponds to the Lévy exponent of the normalized variable $Z_a = V_a/a^\gamma$. This implies that: (1) the variance of Z_a is given by $\mathbb{E}\{Z_a^2\} = b_{2,\phi} + m_{2,\phi} = \text{Var}\{V_1\}$ and is independent of a, and (2) $\lim_{a \to +\infty} f_{\phi_a}(\omega/a^\gamma) = -\frac{b_{2,\phi} + m_{2,\phi}}{2}\omega^2$, which indicates that the limit distribution of Z_a is a centered Gaussian (central-limit theorem).

Practically, this translates into the Gaussian approximation of the pdf of the wavelet coefficients given by

$$p_{V_a}(x) \approx \text{Gauss}\big(0, a^{2\gamma} \text{Var}\{V_1\}\big),$$

which becomes more and more accurate as the scale a increases. We note the simplicity of the asymptotic model and the fact that it is consistent with (9.23) which specifies the general evolution of the variance across scale.

9.9   Notes and pointers to the literature

While most of the results in this chapter are specific to sparse stochastic processes, the presentation relies heavily on standard results from the theory of infinitely divisible laws. Much of this theory gravitates around the one-to-one relation that exists between the pdf (p_{id}) and the Lévy density (v) that appears in the Lévy–Khintchine representation (4.2) of the exponent $f(\omega)$, the general idea being to translate the properties from one domain to the other. Two classical references on this subject are the textbooks by Sato [Sat94] and Steutel and Van Harn [SVH03].

The novel twist here is that the observation of white noise through an analysis window φ results in a modified Lévy exponent f_φ and hence a modified Lévy density v_φ, as specified by (9.2). The technical aspect is to make sure that both f_φ and v_φ are admissible, which is the topic of Section 9.1. Determining the properties of the pdf of $X_\varphi = \langle \varphi, w \rangle$ then reduces to the investigation of v_φ, which can be carried out using the classical tools of the theory of id laws. An alternative proof of Theorem 9.1 as well as some complementary results on infinite divisibility and tail decay are reported in [AU14].

Many of the basic concepts such as stability and self-decomposability go back to the groundwork of Lévy [Lév25, Lév54]. The general form of Theorem 9.9 is due to Kruglov [Kru70] (see also [Sat94, pp. 159–160]), but there are antecedents from Ramachandran [Ram69] and Wolfe [Wol71].

The use of semigroups and potential theory is a fruitful approach to the study of continuous-time Markov processes, including Lévy processes. The concept appears to have been pioneered by William Feller [Fel71, see Chapters 9 and 10]. Suggested readings on this fascinating topic are [Sat94, Chapter 8], [App09, Chapter 3], and [Jac01]. The transposition of these tools in Section 9.8 for characterizing the evolution of the wavelet statistics across scale is new, to the best of our knowledge; an expanded version of this material will be published elsewhere.

10 Recovery of sparse signals

In this chapter, we apply the theory of sparse stochastic processes to the reconstruction of signals from noisy measurements. The foundation of the approach is the specification of a corresponding (finite-dimensional) Bayesian framework for the resolution of ill-posed inverse problems. Given some noisy measurement vector $\mathbf{y} \in \mathbb{R}^M$ produced by an imaging or signal acquisition device (e.g., optical or X-ray tomography, magnetic resonance), the problem is to reconstruct the unknown object (or signal) s as a d-dimensional function of the space-domain variable $\mathbf{r} \in \mathbb{R}^d$ based on the accurate physical modeling of the imaging process (which is assumed to be linear).

The non-standard aspect here is that the reconstruction problem is stated in the continuous domain. A practical numerical scheme is obtained by projecting the solution onto some finite-dimensional reconstruction space. Interestingly, the derived MAP estimators result in optimization problems that are very similar to the variational formulations that are in use today in the field of biomaging, including Tikhonov regularization and ℓ_1-norm minimization. The proposed framework provides insights of a statistical nature and also suggests novel computational schemes and solutions.

The chapter is organized as follows. In Section 10.1, we present a general method for the discretization of a linear inverse problem in a shift-invariant basis. The corresponding finite-dimensional statistical characterization of the signal is obtained by suitable "projection" of the innovation model onto the reconstruction space. This information is then used in Section 10.2 to specify the maximum a posteriori (MAP) reconstruction of the signal. We also develop an iterative optimization scheme that alternates between a classical linear reconstructor and a shrinkage estimator that is specified by the signal prior. In Section 10.3, we apply these techniques to the reconstruction of biomedical images. After reviewing the physical principles of image formation, we derive practical MAP estimators for the deconvolution of fluorescence micrographs, for the reconstruction of magnetic resonance images, and for X-ray computed tomography. We present illustrative examples and discuss the connections with several reconstruction algorithms currently in favor. In Section 10.4, we investigate the extent to which such variational methods approximate the minimum-mean-square-error (MMSE) solution for the simpler problem of signal denoising. To that end, we present a direct algorithm for the MMSE denoising of Lévy processes that is based on belief propagation. This optimal solution is then used as reference for assessing the performance of non-Gaussian MAP estimators.

10.1 Discretization of linear inverse problems

The proposed discretization approach is inspired by the classical method of resolution of PDEs using finite elements. The underlying principle is to reconstruct a continuously defined approximation of the original signal that lives in some finite-dimensional subspace spanned by basis functions located on a reconstruction grid with step size h. The advantages of this strategy are twofold. First, it provides a clean analytical solution to the discretization problem via the specification of basis functions (splines or wavelets). Second, it offers the same type of error control as finite-element methods: in other words, one can make the discretization error arbitrarily small by selecting a reconstruction grid that is sufficiently fine.

To discretize the problem, one represents the unknown signal s as a weighted sum of basis functions $s(r) = \sum_{k \in \Omega} s[k]\beta_k(r)$ with $\text{card}(\Omega) = K$, and specifies the linear (noise-free) forward model $\mathbf{y}_0 = \mathbf{Hs}$ with $\mathbf{s} = (s[k])_{k \in \Omega}$, where the $(M \times K)$ system matrix $\mathbf{H}$ accounts for the image-formation physics. The reconstruction problem then essentially boils down to inverting this system of equations, which is typically very large and ill-conditioned (underdetermined or unstable). Another confounding factor is that the measurements are often corrupted by measurement noise. In practice, the ill-posedness of the reconstruction problem is dealt with by introducing regularization constraints that favor certain types of solutions.

Here, we assume that the underlying signal $s(r)$ is a realization of a sparse stochastic process and take advantage of the continuous-domain innovation model $\mathrm{L}s = w$ to specify the joint pdf $p_S(\mathbf{s})$ of the discrete representation of the signal. We then use this prior information to derive solutions to the reconstruction problem that are optimal in some well-defined statistical sense. We emphasize the MAP solution (i.e., the signal that best explains the measured data) and propose some practical reconstruction algorithms. We will also show that MMSE estimators are accessible under certain conditions (e.g., Gaussianity or Markov property).

10.1.1 Shift-invariant reconstruction subspace

Since the original specification of the problem is in the continuous domain, its proper discretization requires a scheme by which the solution can be expressed in some reconstruction space with minimal loss of information. For practical purposes, we would also like to associate the discrete representation of the signal with its equidistant samples and control the quality of the discretization through the use of a simple resolution parameter: the sampling step h. This is achieved within the context of generalized sampling using "shift-invariant" reconstruction spaces [Uns00].

To simplify the presentation and to set aside the technicalities associated with boundary conditions, we start by considering signals that are defined over the infinite domain $\mathbb{R}^d$. Our reconstruction space at resolution h is then defined as

$$V_h = \left\{ s_h(r) = \sum_{k \in \mathbb{Z}^d} c_h[k]\beta\left(\frac{r - hk}{h}\right) : c_h[k] \in \ell_\infty(\mathbb{Z}^d), r \in \mathbb{R}^d \right\},$$

which involves a family of shifted basis functions specified by a single B-spline-like generator $\beta(r)$. The basis functions are rescaled to match the resolution (dilation by h) and positioned on the reconstruction grid $h\mathbb{Z}^d$. For computational reasons, it is important that the signal representation in terms of its coefficients $c_h[k]$ be stable and unambiguous (Riesz-basis property). Moreover, for the discretization procedure to be acceptable, we require that the error between the original signal $s(r)$ and its projection onto V_h vanishes as the sampling step h tends to zero. These properties are fulfilled if and only if β satisfies the following conditions, which are standard in sampling and approximation theory [Uns00]:

(1) Riesz-basis condition: for all $\omega \in \mathbb{R}^d, 0 < A \le \sum_{n \in \mathbb{Z}^d} |\hat{\beta}(\omega + 2\pi n)|^2 \le B < \infty$
where $\hat{\beta} = \mathscr{F}\{\beta\}$ is the d-dimensional Fourier transform of β (see Section 6.2.3).
(2) L_p-stability for $p \ge 1$: $\sum_{k \in \mathbb{Z}^d} |\beta(r - k)| < \infty$, for all $r \in \mathbb{R}^d$
(3) Partition of unity: $\sum_{k \in \mathbb{Z}^d} \beta(r - k) = 1$, for all $r \in \mathbb{R}^d$.

We note that these conditions are met by the polynomial B-splines. These functions are known to offer the best cost–quality tradeoff among the known families of interpolation kernels [TBU00]. The computational cost is typically proportional to the support of a basis function, while the quality is determined by its approximation order (asymptotic rate of decay of the approximation error). The approximation order of a B-spline of degree n is $(n + 1)$, which is the maximum that is achievable with a support of size $n + 1$ [BTU01].

In practice, it makes good sense to choose β compactly supported so that condition (2) is automatically satisfied. Under such a hypothesis, we can extend the representation for sequences $c_h[\cdot]$ with polynomial growth, which may be required for handling non-stationary signals such as Lévy processes and fractals.

Our next ingredient is a biorthogonal (generalized) function $\tilde{\beta}$ such that

$$\langle \beta(\cdot - k), \tilde{\beta}(\cdot - k') \rangle = \delta_{k-k'}, \tag{10.1}$$

where $\tilde{\beta}$ is not necessarily included in V_h. Based on this biorthogonality property, we specify a signal projector onto the reconstruction space V_h as

$$P_{V_h} s(r) = \sum_{k \in \mathbb{Z}^d} \langle \frac{1}{h^d} \tilde{\beta} \left(\frac{\cdot - hk}{h} \right), s \rangle \, \beta \left(\frac{r - hk}{h} \right).$$

Using (10.1), it is not hard to verify that P_{V_h} is idempotent (i.e., $P_{V_h} P_{V_h} s = P_{V_h} s$) and hence a valid (linear) projection operator. Moreover, the partition of unity guarantees that the error of approximation decays like $\|s - P_{V_h} s\|_{L_p} = O(h)$ (or even faster if β has an order of approximation greater than one), so that the discretization error becomes negligible for h sufficiently small.

To simplify the notation, we shall assume from here on that $h = 1$ and that the control of the discretization error is adequate. (If not, we can decrease the sampling step further, which may also be achieved through an appropriate rescaling of the system of coordinates in which r is expressed.) Hence, our signal reconstruction model is

$$s_1(r) = P_{V_1} s(r) = \sum_{k \in \mathbb{Z}^d} s[k] \beta(r - k) \tag{10.2}$$

with

$$s[k] = \langle \tilde{\beta}(\cdot - k), s \rangle.$$

The main point here is that $s_1(r)$ – the discretized version of $s(r)$ – is uniquely described by its expansion coefficients $(s[k])_{k \in \mathbb{Z}^d}$, so that we can reformulate the reconstruction problem in terms of those quantities. The clear advantage of working with such a hybrid representation is that it is statistically tractable (countable set of parameters) and yet continuously defined so that it lends itself to a proper modeling of the signal-acquisition/measurement process.

Discrete innovation and its statistical characterization

In Chapter 8, we have shown how to obtain the discrete-domain counterpart of the innovation model $Ls = w$ by applying to the samples of the signal the discrete version L_d of the whitening operator L. In the present scenario, where the continuous-domain process s is prefiltered by $\tilde{\beta}^\vee$ prior to sampling, we construct a variant of the generalized increments as

$$u[k] = L_d(\tilde{\beta}^\vee * s)(r)\Big|_{r=k} = (d_L * s)[k],$$

where the right-hand convolution between d_L and $s[\cdot]$ is discrete. Specifically, d_L is the discrete-domain impulse response of L_d (i.e., $L_d\{\delta\}(r) = \sum_{k \in \mathbb{Z}^d} d_L[k]\delta(r - k)$), while the signal coefficients are given by $s[k]$ in (10.2). Based on the defining properties $s = L^{-1}w$ and $L_d L^{-1}\delta = \beta_L$, where β_L is the B-spline associated with L, we express the *discrete innovation u* as

$$u[k] = \langle \tilde{\beta}(\cdot - k), L_d L^{-1} w \rangle = \langle \tilde{\beta}(\cdot - k), \beta_L * w \rangle$$
$$= \langle (\beta_L^\vee * \tilde{\beta})(\cdot - k), w \rangle.$$

This implies that u is stationary and that its statistical properties are in direct relation with those of the white Lévy noise w (continuous-domain innovation). Recalling that w is specified by its characteristic form $\widehat{\mathscr{P}}_w(\varphi) = \mathbb{E}\{e^{j\langle \varphi, w \rangle}\}$, we obtain the joint characteristic function of the innovation values in a K-point neighborhood, $\mathbf{u} = (u[k])_{k \in \Omega_K}$, by direct substitution of $\varphi(r) = \sum_{k \in \Omega_K} \omega_k(\tilde{\beta} * \beta_L^\vee)(r - k)$, where $\boldsymbol{\omega} = (\omega_k)_{k \in \Omega_K}$ is the corresponding Fourier-domain indexing vector. Specifically, $\widehat{p}_U(\boldsymbol{\omega})$, which is the (conjugate) Fourier transform of the joint pdf $p_U(\mathbf{u})$, is given by

$$\widehat{p}_U(\boldsymbol{\omega}) = \widehat{\mathscr{P}}_w \left(\sum_{k \in \Omega_K} \omega_k (\tilde{\beta} * \beta_L^\vee)(\cdot - k) \right), \tag{10.3}$$

where $\widehat{\mathscr{P}}_w(\varphi)$ is determined by its Lévy exponent f through Equation (4.13).

Since L_d is the best discrete approximation of L, we can expect the samples of u to be approximately decoupled. Yet, in light of (10.3), we also see that the quality of the decoupling is dependent upon the extent of the kernel $\tilde{\beta} * \beta_L^{\vee}$, which should be as concentrated as possible. Since β_L has minimal support by design, we can improve the situation only by taking $\tilde{\beta}$ to be a point distribution. Practically, this suggests the choice of a discretization model that uses basis functions that are interpolating with $\beta = \varphi_{int}$. Indeed, the interpolation condition is equivalent to

$$\langle \varphi_{int}, \delta(\cdot - k) \rangle = \delta_k$$

so that the corresponding biorthogonal analysis function is $\tilde{\beta}(r) = \delta(r)$. This is the solution that minimizes the overlap of the basis functions in (10.3).

Decoupling simplification

To obtain an innovation-domain description that is more directly applicable in practice, we make the decoupling simplification $\widehat{p}_U(\boldsymbol{\omega}) \approx \prod_{k \in \Omega_K} \widehat{p}_U(\omega_k)$, which is equivalent to assuming that the discrete innovation sequence $u[\cdot]$ is i.i.d. This means that the Kth-order joint pdf of the discrete innovation can be factorized as

$$p_U(\mathbf{u}) = \prod_{k \in \Omega_K} p_U(u[k]), \qquad (10.4)$$

with the following explicit formula for its first-order pdf:

$$p_U(x) = \overline{\mathscr{F}}^{-1} \left\{ \widehat{\mathscr{P}}_w \left(\omega(\tilde{\beta} * \beta_L^{\vee}) \right) \right\}(x). \qquad (10.5)$$

Note that the latter comes as a special case of (10.3) with $K = 1$. The important practical point is that $p_U(x)$ is infinitely divisible, with modified Lévy exponent

$$\log \widehat{p}_U(\omega) = f_{\tilde{\beta}*\beta_L^{\vee}}(\omega) = \int_{\mathbb{R}^d} f\left(\omega(\tilde{\beta} * \beta_L^{\vee})(r) \right) dr. \qquad (10.6)$$

This makes it fall within the general family of distributions investigated in Chapter 9 (by setting $\varphi = \beta * \beta_L^{\vee}$). Equivalently, we can write the corresponding log-likelihood function

$$-\log(p_U(\mathbf{u})) = \sum_{k \in \Omega_K} \Phi_U(u[k]) \qquad (10.7)$$

with $\Phi_U(u) = -\log p_U(u)$.

10.1.2　Finite-dimensional formulation

To turn the formulation of Section 10.1.1 into a practical scheme that is numerically useful, we take a finite section of the signal model (10.2) by restricting ourselves to a subset of K basis functions with $k \in \Omega$ over some region of interest (ROI). In practice, one often introduces problem-specific boundary conditions (e.g., truncation of the support of the signal, periodization, mirror symmetric extension) that can be enforced by suitable modification of the basis functions that intersect the boundaries of the ROI. The corresponding signal representation then reads

$$s_1(\boldsymbol{r}) = \sum_{k \in \Omega} s[\boldsymbol{k}]\beta_{\boldsymbol{k}}(\boldsymbol{r}), \tag{10.8}$$

where $\beta_{\boldsymbol{k}}(\boldsymbol{r})$ is the basis function corresponding to $\beta(\boldsymbol{r} - \boldsymbol{k})$ in (10.2) up to the modifications at the boundaries. This model is specified by the K-dimensional signal vector $\mathbf{s} = (s[\boldsymbol{k}])_{\boldsymbol{k} \in \Omega}$ that is related to the discrete innovation vector $\mathbf{u} = (u[\boldsymbol{k}])_{\boldsymbol{k} \in \Omega}$ by

$$\mathbf{u} = \mathbf{Ls},$$

where $\mathbf{L}$ is the $(K \times K)$ matrix representation of L_{d}, the discrete version of the whitening operator L.

The general form of a linear, continuous-domain measurement model is

$$y_m = \int_{\mathbb{R}^d} s_1(\boldsymbol{r})\eta_m(\boldsymbol{r}) \, \mathrm{d}\boldsymbol{r} + n[m] = \langle s_1, \eta_m \rangle + n[m], \quad (m = 1, \ldots, M)$$

with sampling/imaging functions $\{\eta_m\}_{m=1}^M$ and additive measurement noise $n[\cdot]$. The sampling function $\eta_m(\boldsymbol{r})$ represents the spatio-temporal response of the mth detector of an imaging/acquisition device. For instance, it can be a 3-D point-spread function in deconvolution microscopy, a line integral across a 2-D or 3-D specimen in computed tomography, or a complex exponential (or a localized version thereof) in the case of magnetic resonance imaging. This measurement model is conveniently described in matrix-vector form as

$$\mathbf{y} = \mathbf{y}_0 + \mathbf{n} = \mathbf{Hs} + \mathbf{n}, \tag{10.9}$$

where

- $\mathbf{y} = (y_1, \ldots, y_M)$ is the M-dimensional measurement vector
- $\mathbf{s} = (s[\boldsymbol{k}])_{\boldsymbol{k} \in \Omega}$ is the K-dimensional signal vector
- $\mathbf{H}$ is the $(M \times K)$ system matrix with

$$[\mathbf{H}]_{m,k} = \langle \eta_m, \beta_k \rangle = \int_{\mathbb{R}^d} \eta_m(\boldsymbol{r})\beta_k(\boldsymbol{r}) \, \mathrm{d}\boldsymbol{r} \tag{10.10}$$

- $\mathbf{y}_0 = \mathbf{Hs}$ is the noise-free measurement vector
- $\mathbf{n}$ is an additive i.i.d. noise component with pdf p_N.

The final ingredient is the set of biorthogonal analysis functions $\{\tilde{\beta}_k\}_{k \in \Omega}$, which are such that $\langle \tilde{\beta}_k, \beta_{k'} \rangle = \delta_{k-k'}$, with $k, k' \in \Omega$. We recall that the biorthogonality with the synthesis functions $\{\beta_k\}_{k \in \Omega}$ is essential for determining the projection of the stochastic process $s(\boldsymbol{r})$ onto the reconstruction space. The other important point is that the choice $\tilde{\beta} = \delta$, which corresponds to interpolating basis functions, facilitates the determination of the joint pdf $p_U(\mathbf{u})$ of the discrete innovation vector $\mathbf{u}$ from (10.3) or (10.4).

The reconstruction task is now to recover the unknown signal vector $\mathbf{s}$ given the noisy measurements $\mathbf{y}$. The statistical approaches to such problems are all based on the determination of the posterior distribution $p_{S|Y}$, which depends on the prior distribution p_S and the underlying noise model. Using Bayes' rule, we have that

$$p_{S|Y}(\mathbf{s}|\mathbf{y}) = \frac{p_{Y|S}(\mathbf{y}|\mathbf{s})p_S(\mathbf{s})}{p_Y(\mathbf{y})} = \frac{p_N(\mathbf{y} - \mathbf{Hs})p_S(\mathbf{s})}{p_Y(\mathbf{y})}$$

$$= \frac{1}{Z}p_N(\mathbf{y} - \mathbf{Hs})p_S(\mathbf{s}),$$

where the proportionality factor Z is not essential to the estimation procedure because it only depends on the input data $\mathbf{y}$, which is a known quantity. We also note that Z can be recalculated by imposing the normalization constraint $\int_{\mathbb{R}^K} p_{S|Y}(\mathbf{s}|\mathbf{y})\,d\mathbf{s} = 1$, which is the way it is handled in message-passing algorithms (see Section 10.4.2). The next step is to introduce the discrete innovation variable $\mathbf{u} = \mathbf{Ls}$, whose pdf $p_U(\mathbf{u})$ has been derived explicitly. If the linear mapping between $\mathbf{u}$ and $\mathbf{s}$ is one-to-one, [1] we clearly have that

$$p_S(\mathbf{s}) \propto p_U(\mathbf{Ls}).$$

Using this relation together with the decoupling simplification (10.4), we find that

$$p_{S|Y}(\mathbf{s}|\mathbf{y}) \propto p_N(\mathbf{y} - \mathbf{Hs})p_U(\mathbf{Ls}) \approx p_N(\mathbf{y} - \mathbf{Hs}) \prod_{k \in \Omega} p_U([\mathbf{Ls}]_k), \qquad (10.11)$$

where p_U is specified by (10.6) and solely depends on the Lévy exponent f of the continuous-domain innovation w and the B-spline kernel $\beta_L = L_d L^{-1}\delta$ associated with the whitening operator L.

In the standard additive white Gaussian noise scenario (AWGN), we find that

$$p_{S|Y}(\mathbf{s}|\mathbf{y}) \propto \exp\left(-\frac{\|\mathbf{y} - \mathbf{Hs}\|^2}{2\sigma^2}\right) \prod_{k \in \Omega} p_U([\mathbf{Ls}]_k),$$

where σ^2 is the variance of the discrete measurement noise.

Conceptually, the best solution to the reconstruction problem is the MMSE estimator, which is given by the mean of the posterior distribution

$$\mathbf{s}_{\mathrm{MMSE}}(\mathbf{y}) = \mathbb{E}\{\mathbf{s}|\mathbf{y}\}.$$

The estimate $\mathbf{s}_{\mathrm{MMSE}}(\mathbf{y})$ is optimal in that it is closest (in the mean-square sense) to the (unknown) noise-free signal $\mathbf{s}$; i.e., $\mathbb{E}\{|\mathbf{s}_{\mathrm{MMSE}}(\mathbf{y}) - \mathbf{s}|^2\} = \min\{\mathbb{E}\{|\tilde{\mathbf{s}}(\mathbf{y}) - \mathbf{s}|^2\}\}$ among all signal estimators $\tilde{\mathbf{s}}(\mathbf{y})$. The downside of this estimator is that it is difficult to compute in practice, except for special cases such as those discussed in Sections 10.2.2 and 10.4.2.

---

[1] A similar proportionality relation can be established for the cases where $\mathbf{L}$ has a non-empty null space via the imposition of the boundary conditions of the SDE. While the rigorous formulation can be carried out, it is often not worth the effort because it will only result in a very slight modification of the solution such that $s_1(r)$ satisfies some prescribed boundary conditions (e.g., $s_1(0) = 0$ for a Lévy process), which are artificial anyway. The pragmatic approach, which better suits real-world applications, is to ignore this technical issue by adopting the present stationary formulation, and to let the optimization algorithm adjust the null-space component of the signal to produce the solution that is maximally consistent with the data.

10.2   MAP estimation and regularization

Among the statistical estimators that incorporate prior information, the most popular is the MAP solution, which extracts the mode of the posterior distribution. Its use can be justified by the fact that it produces the signal estimate that best explains the observed data. While MAP does not necessarily yield the estimator with the best average performance, it has the advantage of being tractable numerically.

Here, we make use of the prior information that the continuous-domain signal s satisfies the innovation model $Ls = w$, where w is a white Lévy noise. The finite-dimensional transposition of this model (under the decoupling simplification) is that the discrete innovation vector $\mathbf{Ls} = \mathbf{u}$ can be assumed to be i.i.d.,[2] where $\mathbf{L}$ is the matrix counterpart of the whitening operator L. For a given set of noisy measurements $\mathbf{y} = \mathbf{Hs} + \mathbf{n}$ with AWGN of variance σ^2, we obtain the MAP estimator through the maximization of (10.11). This results in

$$s_{\mathrm{MAP}} = \arg\min_{s \in \mathbb{R}^K} \left(\frac{1}{2} \|\mathbf{y} - \mathbf{Hs}\|_2^2 + \sigma^2 \sum_{k \in \Omega} \Phi_U([\mathbf{Ls}]_k) \right), \tag{10.12}$$

with $\Phi_U(x) = -\log p_U(x)$, where $p_U(x)$ is given by (10.5). Observe that the cost functional in (10.12) has two components: a data term $\frac{1}{2}\|\mathbf{y} - \mathbf{Hs}\|_2^2$ that enforces the consistency between the data and the simulated, noise-free measurements $\mathbf{Hs}$, and a second regularization term that favors likeliest solutions in reference to the prior stochastic model. The balancing factor is the variance σ^2 which amplifies the influence of the prior information as the data get noisier. The specificity of the present formulation is that the potential function is given by the log-likelihood of the infinitely divisible random variable U, which has strong theoretical implications, as discussed in Sections 10.2.1 and 10.2.3.

For the time being, we observe that the general form of the estimator (10.12) is compatible with the standard variational approaches used in signal processing. The three cases of interest that correspond to valid id (infinitely divisible) log-likelihood functions (see Table 4.1) are:

(1) Gaussian: $p_U(x) = \frac{1}{\sqrt{2\pi}\sigma_0} e^{-x^2/(2\sigma_0^2)}$ $\quad\Rightarrow\quad$ $\Phi_U(x) = \frac{1}{2\sigma_0^2}x^2 + C_1$

(2) Laplace: $p_U(x) = \frac{\lambda}{2} e^{-\lambda|x|}$ $\quad\Rightarrow\quad$ $\Phi_U(x) = \lambda|x| + C_2$

(3) Student: $p_U(x) = \dfrac{1}{B\left(r, \frac{1}{2}\right)} \left(\dfrac{1}{x^2+1}\right)^{r+\frac{1}{2}}$ $\quad\Rightarrow\quad$ $\Phi_U(x) = \left(r + \frac{1}{2}\right)\log(x^2+1) + C_3,$

where the constants C_1, C_2, and C_3 can be ignored since they do not affect the solution. The first quadratic potential leads to the classical Tikhonov regularizer, which yields a stabilized linear solution. The second absolute-value potential produces an ℓ_1-type regularizer; it is the preferred solution for solving deterministic compressed-sensing and sparse-signal-recovery problems. If L is a first-order derivative operator,

---

[2] The underlying discrete innovation sequence $u[k] = L_d s(r)|_{r=k}$ is stationary and therefore identically distributed. It is also independent for Markov and/or Gaussian processes, but only approximatively so otherwise. To justify the decoupling simplification, we like to invoke the minimum-support property of the B-spline $\beta_L = L_d L^{-1} \delta \in L_1(\mathbb{R}^d)$, where L_d is the discrete counterpart of the whitening operator L.

then (10.12) maps into total-variation (TV) regularization, which is widely used in applications [ROF92]. The third log-based potential is interesting as well because it relates to
the limit on an ℓ_p-relaxation scheme when p tends to zero [WN10]. The latter has been
proposed by several authors as a practical "debiasing" method for improving the sparsity of the solution of a compressed-sensing problem [CW08]. The connection between
log and ℓ_p norm relaxation is provided by the limit

$$\log x^2 = \lim_{p \to 0} \frac{x^{2p} - 1}{p},$$

which is compatible with the Student prior for $x^2 \gg 1$.

10.2.1   Potential function

In the present Bayesian framework, the potential function $\Phi_U(x) = -\log p_U(x)$ is determined by the Lévy exponent $f(\omega)$ of the continuous-domain innovation w or, equivalently, by the canonical noise pdf $p_{id}(x)$ in Proposition 4.12. Specifically, $p_U(x)$ is
infinitely divisible with modified Lévy exponent $f_{\tilde{\beta}*\beta_L^\vee}(\omega)$ given by (10.7). While the
exact form of $p_U(x)$ also depends on the B-spline kernel β_L, a remarkable aspect of
the theory is that its global characteristics remain qualitatively very similar to those of
$p_{id}(x)$. Indeed, based on the analysis of Chapter 9 with $\varphi = \tilde{\beta} * \beta_L^\vee \in L_p(\mathbb{R}^d)$, we can
infer the following properties:

- If $p_{id}(x)$ is symmetric and unimodal, then the same is true for $p_U(x)$ (by
  Corollary 9.5). These are very desirable properties for they ensure that $\Phi_U(x) = \Phi_U(|x|)$ is symmetric and increases monotonically away from zero. The preser-
  vation of these features is important because most image-processing practitioners
  would be reluctant to apply a regularization scheme that does not fulfill these basic
  monotonicity constraints and cannot be interpreted as a penalty.
- In general, $p_U(x)$ will not be Gaussian unless $p_{id}(x)$ is Gaussian to start with, in which
  case $\Phi_U(x)$ is quadratic.
- If $p_{id}(x)$ is stable (e.g., SαS) then the property is preserved for $p_U(x)$ (by Proposition 9.8). This corresponds to a sparse scenario as soon as $\alpha \neq 2$.
- In general, $p_U(x)$ will display the same sparsity patterns as $p_{id}(x)$. In particular, if
  $p_{id}(x) = O(1/|x|^p)$ with $p > 1$ (heavy-tailed behavior), then the same holds true for
  $p_U(x)$. Likewise, its pth-moment will be finite if and only if $\int_{\mathbb{R}} |x|^p p_{id}(x)\, dx < \infty$
  (see Proposition 9.10).
- The qualitative behavior of $p_U(x)$ around the origin will be the same as that of $p_{id}(x)$,
  regarding properties such as symmetry and order of differentiability (or lack thereof).

An important remark concerning the last point is in order. We have already seen
that compound-Poisson distributions exhibit a Dirac delta impulse at the origin and
that the Poisson property is preserved through the noise-integration process when β_L
is compactly supported (set $p = 0$ in Proposition 9.10). Since taking the logarithm
of $\delta(x)$ is not an admissible operation, we cannot define a proper potential function
in such cases (finite rate of innovation scenario). Another way to put this is that the

Table 10.1 Asymptotic behavior of the potential function $\Phi_X(x)$ for the infinitely divisible distributions in Table 4.1.[a]

$p_X(x)$	$\Phi_X(x) = -\log p_X(x)$ as $x \to 0$	$\Phi_X(x)$ as $x \to \pm\infty$	Smooth	Convex
Gaussian	$a_0 + \dfrac{x^2}{2\sigma^2}$	$a_0 + \dfrac{x^2}{2\sigma^2}$	Yes	Yes
Laplace ($\lambda \in \mathbb{R}^+$)	$a_0 + \lambda|x|$	$a_0 + \lambda|x|$	No	Yes
Sym gamma $r \in \mathbb{R}^+$	$\begin{cases} \log(a_0' + a_r'|x|^{2r-1} + O(x^2)), & r < 3/2 \\ a_0 + \dfrac{x^2}{4r-6} + O(|x|^{\min(4,2r-1)}), & r > 3/2 \end{cases}$	$b_0 + |x| - (r-1)\log|x|$	No	No
Hyperbolic secant	$a_0 + \dfrac{\pi^2 x^2}{8\sigma_0^2} + O(x^4)$	$-\log\sigma_0 + \dfrac{\pi}{2\sigma_0}|x|$	Yes	Yes
Meixner $r, s \in \mathbb{R}^+$	$a_0 + \dfrac{\psi^{(1)}(r/2)}{4s^2}x^2 + O(x^4)$	$b_0 + \dfrac{\pi}{2s}|x| - (r-1)\log|x|$	Yes	No
Cauchy $s \in \mathbb{R}^+$	$a_0 + \frac{x^2}{s^2} + O(x^4)$	$b_0 - \log s + 2\log|x|$	Yes	No
Sym Student $r \in \mathbb{R}^+$	$a_0 + \left(r + \frac{1}{2}\right)x^2 + O(x^4)$	$b_0 + (2r+1)\log|x|$	Yes	No
SαS, $\alpha \in (0,2)$, $s \in \mathbb{R}^+$	$a_0 + \frac{\Gamma(\frac{3}{\alpha})}{2s^2\Gamma(\frac{1}{\alpha})}x^2 + O(x^4)$	$b_0 - \alpha\log s + (\alpha+1)\log|x|$	Yes	No

[a] $\Gamma(z)$ and $\psi^{(1)}(r)$ are Euler's gamma and first-order polygamma functions, respectively (see Appendix C).

compound-Poisson MAP estimator would correspond to $u = 0$ because the probability of getting zero is overwhelmingly larger than that of observing any other value.

These limitations notwithstanding, we can rely on the properties of infinitely divisible laws to make some general statements about the asymptotic form of the potential function. First, $\Phi_U(x)$ will typically exhibit a Gaussian (i.e., quadratic) behavior near the origin. Indeed, when $p_U(x)$ is symmetric and twice-differentiable at the origin, we can write the Taylor series of the potential as

$$\Phi_U(x) = \Phi_U(0) + \frac{\Phi_U''(0)}{2}x^2 + O(|x|^4),\qquad(10.13)$$

with

$$\Phi_U''(0) = -\left.\frac{d^2\log p_U(x)}{dx^2}\right|_{x=0} = -\frac{p_U''(0)}{p_U(0)}.$$

By contrast, unless U is Gaussian, the rate of growth of Φ_U decreases progressively away from the origin and becomes less than quadratic. Indeed, by using the tail properties of id distributions, one can prove that a non-Gaussian id potential cannot grow faster than $x\log(x)$ as $x \to \infty$. A more typical asymptotic trend is $O(x)$ when $p_U(x)$ has exponential decay, or $O(\log(x))$ when $p_U(x)$ has algebraic decay. These behaviors are exemplified in Table 10.1. For the precise specification of these potential functions, including the derivation of their asymptotics, we refer to Appendix C. In particular, we note that the families of sym gamma and Meixner distributions give rise to an intermediate range of behaviors with potential functions falling in between the linear solution (Laplace and hyperbolic secant) and the log form that is characteristic of the heavier-tailed laws (Student and SαS).

10.2.2   LMMSE/Gaussian solution

Before addressing the general MAP estimation problem, it is instructive to investigate the Gaussian scenario, which admits a closed-form solution. Moreover, under the Gaussian hypothesis, we can compute a perfectly decoupled representation by using a modified discrete filter $\mathbf{L}_G$ that performs an exact whitening[3] of the signal in the discrete domain (see Section 8.3.4). In the present setting, the Gaussian MAP estimator is the minimizer of the quadratic cost functional

$$\mathscr{C}_2(\mathbf{s}, \mathbf{y}) = \frac{1}{2}\|\mathbf{y} - \mathbf{H}\mathbf{s}\|_2^2 + \sigma^2 \frac{1}{2}\|\mathbf{L}_G\mathbf{s}\|_2^2$$

$$\mathbf{s}_{\mathrm{MAP}}(\mathbf{y}) = \arg\min_{\mathbf{s}\in\mathbb{R}^K} \mathscr{C}_2(\mathbf{s}, \mathbf{y})$$

with $\mathbf{L}_G = \mathbf{C}_{ss}^{-1/2}$, where $\mathbf{C}_{ss} = \mathbb{E}\{\mathbf{s}\mathbf{s}^T\}$ is the $(K \times K)$ symmetric covariance matrix of the signal. The implicit assumption here is that $\mathbf{C}_{ss}$ is invertible and $\mathbb{E}\{\mathbf{s}\} = \mathbf{0}$ (zero-mean signal). The gradient of $\mathscr{C}_2(\mathbf{s}, \mathbf{y})$ is given by

$$\frac{\partial\mathscr{C}_2(\mathbf{s}, \mathbf{y})}{\partial\mathbf{s}} = -\mathbf{H}^T(\mathbf{y} - \mathbf{H}\mathbf{s}) + \sigma^2\mathbf{L}_G^T\mathbf{L}_G\mathbf{s},$$

and the MAP estimator is found by equating it to zero. This yields the classical linear solution

$$\mathbf{s}_{\mathrm{MAP}}(\mathbf{y}) = \left(\mathbf{H}^T\mathbf{H} + \sigma^2\mathbf{L}_G^T\mathbf{L}_G\right)^{-1}\mathbf{H}^T\mathbf{y} \tag{10.14}$$

$$= \left(\mathbf{H}^T\mathbf{H} + \sigma^2\mathbf{C}_{ss}^{-1}\right)^{-1}\mathbf{H}^T\mathbf{y}.$$

Note that the term $\sigma^2\mathbf{C}_{ss}^{-1}$, which is bounded from above and below (due to the finite-variance and invertibility hypotheses), acts as a stabilizer. It ensures that the matrix inverse in (10.14) is well defined, and hence that the solution is unique.

Alternatively, it is also possible to derive the LMMSE estimator (or Wiener filter) which takes the standard form

$$\mathbf{s}_{\mathrm{LMMSE}}(\mathbf{y}) = \mathbf{C}_{ss}\mathbf{H}^T\left(\mathbf{H}\mathbf{C}_{ss}\mathbf{H}^T + \mathbf{C}_{nn}\right)^{-1}\mathbf{y}, \tag{10.15}$$

where $\mathbf{C}_{nn} = \sigma^2\mathbf{I}$ is the $(N \times N)$ covariance matrix of the noise. We note that the LMMSE solution is also valid for the non-Gaussian, finite-variance scenarios: it provides the MMSE solution among all linear estimators, irrespective of the type of signal model.

---

[3] We recommend the use of the classical discrete whitening filter $\mathbf{L}_G = \mathbf{C}_{ss}^{-1/2}$ as a substitute for $\mathbf{L}$ in the Gaussian case because it results in an exact formulation. However, we do not advise one to do so for non-Gaussian models because it may induce undesirable long-range dependencies, the difficulty being that decorrelation alone is no longer synonymous with statistical independence.

It is well known from estimation theory that the Gaussian MAP and LMMSE solutions are equivalent. This can be seen by considering the following sequence of equivalent matrix identities:

$$\mathbf{H}^T \mathbf{H} \mathbf{C}_{ss} \mathbf{H}^T + \sigma^2 \mathbf{C}_{ss}^{-1} \mathbf{C}_{ss} \mathbf{H}^T = \mathbf{H}^T \mathbf{H} \mathbf{C}_{ss} \mathbf{H}^T + \sigma^2 \mathbf{H}^T$$

$$\left(\mathbf{H}^T \mathbf{H} + \sigma^2 \mathbf{C}_{ss}^{-1} \right) \mathbf{C}_{ss} \mathbf{H}^T = \mathbf{H}^T \left(\mathbf{H} \mathbf{C}_{ss} \mathbf{H}^T + \sigma^2 \mathbf{I} \right)$$

$$\mathbf{C}_{ss} \mathbf{H}^T \left(\mathbf{H} \mathbf{C}_{ss} \mathbf{H}^T + \sigma^2 \mathbf{I} \right)^{-1} = \left(\mathbf{H}^T \mathbf{H} + \sigma^2 \mathbf{C}_{ss}^{-1} \right)^{-1} \mathbf{H}^T,$$

where have used the hypothesis that the covariance matrix $\mathbf{C}_{ss}$ is invertible.

The availability of the closed-form solution (10.14) or (10.15) is nice conceptually, but it is not necessarily applicable for large-scale problems because the system matrix is too large to be stored in memory and inverted explicitly. The usual numerical approach is to solve the corresponding system of linear equations iteratively using the conjugate-gradient (CG) method. The convergence speed of CG can often be improved by applying some problem-specific preconditioner. A particularly favorable situation is when the matrix $\mathbf{H}^T \mathbf{H} + \sigma^2 \mathbf{L}_G^T \mathbf{L}_G$ is block-Toeplitz (or circulant) and is diagonalized by the Fourier transform. The signal reconstruction can then be computed very efficiently with the help of an FFT-based inversion. This strategy is applicable for the basic flavors of deconvolution, computed tomography, and magnetic resonance imaging. It also makes the link with the classical methods of Wiener filtering, filtered backprojection, or backprojection filtering, which result in direct image reconstruction.

10.2.3 Proximal operators

The optimization problem (10.12) becomes more challenging when the potential function Φ_U is non-quadratic. To get a better handle on the effect of such a regularization, it is instructive to consider a simplified scalar version of the problem. Conceptually, this is equivalent to treating the components of the problem as if they were decoupled. To that end, we define the proximal operator with weight σ^2 as

$$\text{prox}_{\Phi_U}(y; \sigma^2) = \arg \min_{u \in \mathbb{R}} \frac{1}{2} |y - u|^2 + \sigma^2 \Phi_U(u), \tag{10.16}$$

which is tied to the underlying stochastic model. Since $\Phi_U(u) = -\log p_U(u)$ is bounded from below and cannot grow faster that $O(|u|^2)$, the solution of (10.16) exists for any $y \in \mathbb{R}$. When the minimizer is not unique, we can establish arbitrary preference (for instance, pick the smallest solution) and specify $\tilde{u} = \text{prox}_{\Phi_U}(y; \sigma^2)$ as the proximal estimator of y. This function returns the closest approximation of y, subject to the penalty induced by $\sigma^2 \Phi_U(u)$. In practice, σ^2 is fixed and the mapping $\tilde{u} = \text{prox}_{\Phi_U}(y; \sigma^2)$ can be specified either analytically or, at least, numerically, in the form of a 1-D lookup table that maps y to $\tilde{u}$ (shrinkage function).

The proximal operator induces a perturbation of the identity map that is more or less pronounced, depending upon the magnitude of σ^2 and the characteristics of $\Phi_U(u)$. At the points u where is $\Phi_U(u)$ is differentiable, $\tilde{u}$ satisfies

$$-y + \tilde{u} + \sigma^2 \Phi'_U(\tilde{u}) = 0, \tag{10.17}$$

where $\Phi'_U(u) = \frac{d\Phi_U(u)}{du}$. In particular if $\Phi_U(u)$ is twice-differentiable with $1 + \sigma^2\Phi''_U(u) \geq 0$, then the above mapping is one-to-one, which implies that $\tilde{u}(y)$ is the inverse function of $y(\tilde{u}) = \text{prox}^{-1}_{\Phi_U}(\tilde{u}; \sigma^2) = \tilde{u} + \sigma^2\Phi'_U(\tilde{u})$. The quantity $(-\Phi'_U(u)) = \frac{p'_U(u)}{p_U(u)}$ is sometimes called the *score*. It is related to the Fisher information $\mathbb{E}\{|\Phi'_U(U)|^2\} \geq 1/\text{Var}\{U\}$, which is minimum in the Gaussian case.

Equation (10.17) can be used to derive the closed-form representation of the two better-known examples of proximal maps,

$$T_1(y) = \text{prox}_{\Phi_1}(y; \sigma^2) = \text{sign}(y)\left(|y| - \lambda\sigma^2\right)_+ \tag{10.18}$$

$$T_2(y) = \text{prox}_{\Phi_2}(y; \sigma^2) = \frac{\sigma_0^2}{\sigma_0^2 + \sigma^2}y, \tag{10.19}$$

with $\Phi_1(u) = \lambda|u|$ (Laplace law) and $\Phi_2(u) = |u|^2/(2\sigma_0^2)$ (Gaussian). The first is a soft-threshold (shrinkage operator) and the second a linear scaling (scalar Wiener filter). While not all id laws lend themselves to such an analytic treatment, we can nevertheless determine the asymptotic form of their proximal operator. For instance, if Φ_U is symmetric and twice-differentiable at the origin, which is the case for most[4] examples in Table 10.1, then

$$\text{prox}_{\Phi_U}(y; \sigma^2) = \frac{y}{1 + \sigma^2\Phi''_U(0)} \quad \text{as } y \to 0.$$

The result is established by using a basic first-order Taylor series argument: $\Phi'_U(u) = \Phi''_U(0)u + O(u^2)$. The required slope parameter is

$$\Phi''_U(0) = -\frac{p''_U(0)}{p_U(0)} = \frac{\int_{\mathbb{R}} \omega^2\hat{p}_U(\omega)\,d\omega}{\int_{\mathbb{R}} \hat{p}_U(\omega)\,d\omega}, \tag{10.20}$$

which is computable from the moments of the characteristic function.

To determine the larger-scale behavior, one has to distinguish between the (non-Gaussian) intermediate scenarios, where the asymptotic trend of the potential is predominantly linear (e.g., Laplace, hyperbolic secant, sym gamma, and Meixner families), and the distributions with algebraic decay, where it is logarithmic (Student and SαS stable), as made explicit in Table 10.1. (This information can readily be extracted from the special-function formulas in Appendix C.) For the first exponential and sub-exponential category, we have that $\lim_{u\to\infty} \Phi'_U(u) = b_1 + O(1/u)$ with $b_1 > 0$, which implies that

$$\text{prox}_{\Phi_U}(y; \sigma^2) \sim y - \sigma^2 b_1 \quad \text{as } y \to +\infty.$$

This corresponds to a shrinkage-type estimator.

The limit behavior in the second heavy-tailed category is $\lim_{u\to\infty} \Phi'_U(u) = b_2/u$ with $b_2 > 0$, which translates into the asymptotic identity-like behavior

$$\text{prox}_{\Phi_U}(y; \sigma^2) \sim y \quad \text{as } y \to \infty.$$

[4] Only the Laplace law and its sym gamma variants with $r < 3/2$ fail to meet the differentiability requirement.

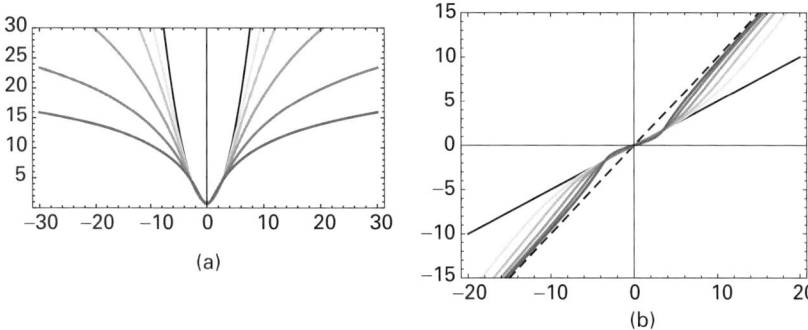

Figure 10.1 Normalized Student potentials for $r = 2, 4, 8, 16, 32$ (dark to light) and corresponding proximity maps. (a) Student potentials $\Phi_{\text{Student}}(y)$ with unit signal variance. (b) Shrinkage operator $\text{prox}_{\Phi_{\text{Student}}}(y; \sigma^2)$ for $\sigma^2 = 1$.

Some examples of normalized potential functions for symmetric Student distributions with increasing asymptotic decay are shown in Figure 10.1, together with their corresponding proximity maps for a unit signal-to-noise ratio. The proximity maps fall nicely in between the best linear (or ℓ_2) solution (solid black line) and the identity map (dashed black line). Since the Student distributions are maximal at $y = 0$, symmetric, and infinitely differentiable, the thresholding functions are linear around the origin – as predicted by the theory – and remarkably close to the pointwise Wiener (LMMSE) solution. For larger values of the input, the estimation progressively switches to an identity, which is consistent with the algebraic decay of the Student pdfs, the transition being faster for the heavier-tailed distributions (r small).

In summary, the proximal maps associated with id distributions have two predominant regimes: (1) a linear, Gaussian mode (attenuation) around the origin and (2) a shrinkage mode as one moves away from the origin. The amount of asymptotic shrinkage depends on the rate of decay of the pdf – it converges to zero (identity map) when the decay is algebraic, as opposed to exponential.

10.2.4 MAP estimation

We have now all the elements in hand to derive a generic MAP estimation algorithm. The basic idea is to consider the innovation vector $\mathbf{u}$ as an auxiliary variable and to reformulate the MAP estimation as a constrained optimization problem,

$$\mathbf{s}_{\text{MAP}} = \arg\min_{\mathbf{s} \in \mathbb{R}^K} \left(\frac{1}{2}\|\mathbf{y} - \mathbf{H}\mathbf{s}\|_2^2 + \sigma^2 \sum_{k \in \Omega} \Phi_U\big(u[k]\big) \right) \text{ subject to } \mathbf{u} = \mathbf{L}\mathbf{s}.$$

Instead of addressing the constrained optimization heads-on, we impose the constraint by adding a quadratic penalty term $\frac{\mu}{2}\|\mathbf{L}\mathbf{s} - \mathbf{u}\|_2^2$ with μ sufficiently large. To make the scheme more efficient, we rely on the augmented-Lagrangian (AL) method. The corresponding AL functional is

$$\mathcal{L}_{\mathcal{A}}(\mathbf{s}, \mathbf{u}, \boldsymbol{\alpha}) = \frac{1}{2}\|\mathbf{y} - \mathbf{H}\mathbf{s}\|_2^2 + \sigma^2 \sum_{k \in \Omega} \Phi_U\big(u[k]\big) - \boldsymbol{\alpha}^T(\mathbf{L}\mathbf{s} - \mathbf{u}) + \frac{\mu}{2}\|\mathbf{L}\mathbf{s} - \mathbf{u}\|_2^2,$$

where the vector $\boldsymbol{\alpha} = (\alpha_k)_{k \in \Omega}$ represents the Lagrange multipliers for the desired constraint.

To find the solution, we handle the optimization task sequentially according to the alternating-direction method of multipliers (ADMM) [BPCE10]. Specifically, we consider the unknown variables $\mathbf{s}$ and $\mathbf{u}$ in succession, minimizing $\mathcal{L}_{\mathscr{A}}(\mathbf{s}, \mathbf{u}, \boldsymbol{\alpha})$ with respect to each of them, while keeping $\boldsymbol{\alpha}$ and the other one fixed. This is combined with an update of $\boldsymbol{\alpha}$ to refine the current estimate of the Lagrange multipliers. By using the index k to denote the iteration number, the algorithm cycles through the following steps until convergence:

$$\mathbf{s}^{k+1} \leftarrow \arg \min_{\mathbf{s} \in \mathbb{R}^N} \mathcal{L}_{\mathscr{A}}(\mathbf{s}, \mathbf{u}^k, \boldsymbol{\alpha}^k)$$

$$\boldsymbol{\alpha}^{k+1} = \boldsymbol{\alpha}^k - \mu(\mathbf{L}\mathbf{s}^{k+1} - \mathbf{u}^k)$$

$$\mathbf{u}^{k+1} \leftarrow \arg \min_{\mathbf{u} \in \mathbb{R}^N} \mathcal{L}_{\mathscr{A}}(\mathbf{s}^{k+1}, \mathbf{u}, \boldsymbol{\alpha}^{k+1}).$$

We now look into the details of the optimization. The first step amounts to the minimization of the quadratic form in $\mathbf{s}$ given by

$$\mathcal{L}_{\mathscr{A}}(\mathbf{s}, \mathbf{u}^k, \boldsymbol{\alpha}^k) = \frac{1}{2}\|\mathbf{y} - \mathbf{H}\mathbf{s}\|_2^2 - (\mathbf{L}\mathbf{s} - \mathbf{u}^k)^T \boldsymbol{\alpha}^k + \frac{\mu}{2}\|\mathbf{L}\mathbf{s} - \mathbf{u}^k\|_2^2 + C_1, \qquad (10.21)$$

where $C_1 = C_1(\mathbf{u}^k)$ is a constant that does not depend on $\mathbf{s}$. By setting the gradient of the above expression to zero, as in

$$\frac{\partial \mathcal{L}_{\mathscr{A}}(\mathbf{s}, \mathbf{u}^k, \boldsymbol{\alpha}^k)}{\partial \mathbf{s}} = -\mathbf{H}^T(\mathbf{y} - \mathbf{H}\mathbf{s}) - \mathbf{L}^T(\boldsymbol{\alpha}^k - \mu(\mathbf{L}\mathbf{s} - \mathbf{u}^k)) = \mathbf{0}, \qquad (10.22)$$

we obtain the intermediate linear estimate,

$$\mathbf{s}^{k+1} = \left(\mathbf{H}^T\mathbf{H} + \mu\mathbf{L}^T\mathbf{L}\right)^{-1}\left(\mathbf{H}^T\mathbf{y} - \mathbf{z}^{k+1}\right), \qquad (10.23)$$

with $\mathbf{z}^{k+1} = \mathbf{L}^T(\boldsymbol{\alpha}^k + \mu\mathbf{u}^k)$. Remarkably, this is essentially the same result as the Gaussian solution (10.14) with a slight adjustment of the data term and regularization strength. Note that the condition $\text{Ker}(\mathbf{H}) \cap \text{Ker}(\mathbf{L}) = \{\mathbf{0}\}$ is required for this linear solution to be well defined and unique.

To justify the update of the Lagrange multipliers, we note that (10.22) can be rewritten as

$$\frac{\partial \mathcal{L}_{\mathscr{A}}(\mathbf{s}, \mathbf{u}^k, \boldsymbol{\alpha}^k)}{\partial \mathbf{s}} = -\mathbf{H}^T(\mathbf{y} - \mathbf{H}\mathbf{s}) - \mathbf{L}^T\boldsymbol{\alpha}^{k+1} = \mathbf{0},$$

which is consistent with the global optimality conditions

$$\mathbf{L}\mathbf{s}^\star - \mathbf{u}^\star = \mathbf{0},$$

$$-\mathbf{H}^T(\mathbf{y} - \mathbf{H}\mathbf{s}^\star) - \mathbf{L}^T\boldsymbol{\alpha}^\star = \mathbf{0},$$

where $(\mathbf{s}^\star, \mathbf{u}^\star, \boldsymbol{\alpha}^\star)$ is a stationary point of the optimization problem. This ensures that $\boldsymbol{\alpha}^{k+1}$ gets closer to the correct vector of Lagrange multipliers $\boldsymbol{\alpha}^\star$ as $\mathbf{u}^k$ converges to $\mathbf{u}^\star = \mathbf{L}\mathbf{s}^\star$.

As for the third step, we define $\tilde{u}[k] = [\mathbf{L}\mathbf{s}^{k+1}]_k$ and rewrite the AL criterion as

$$\mathscr{L}_{\mathscr{A}}(\mathbf{s}^{k+1}, \mathbf{u}, \boldsymbol{\alpha}^{k+1}) = C_2 + \sum_{k \in \Omega} \sigma^2 \Phi_U\big(u[k]\big) + \alpha_k u[k] + \frac{\mu}{2}\,(\tilde{u}[k] - u[k])^2$$

$$= C_3 + \sum_{k \in \Omega} \frac{\mu}{2}\left(\left(\tilde{u}[k] - \frac{\alpha_k}{\mu}\right) - u[k]\right)^2 + \sigma^2 \Phi_U\big(u[k]\big),$$

where C_2 and C_3 are constants that do not depend on $\mathbf{u}$. This shows that the optimization problem is decoupled and that the update can be obtained by direct application of the proximal operator (10.16) in a coordinate-wise fashion, so that

$$\mathbf{u}^{k+1} = \text{prox}_{\Phi_U}\left(\mathbf{L}\mathbf{s}^{k+1} - \tfrac{1}{\mu}\boldsymbol{\alpha}^{k+1}; \frac{\sigma^2}{\mu}\right). \qquad (10.24)$$

A few remarks are in order. While there are other possible numerical approaches to the present MAP-estimation problem, the proposed algorithm is probably the simplest to deploy because it makes use of two very basic modules: a linear solver (akin to a Wiener filter) and a model-specific proximal operator that can be implemented as a pointwise non-linearity. The approach provides a powerful recipe for improving on some prior linear solution by reapplying the solver sequentially and embedding it into a proper computational loop. The linear solver needs to be carefully engineered because it has a major impact on the efficiency of the method. The adaptation to a given sparse stochastic model amounts to a simple adjustment of the proximal map (lookup table).

The ADMM is guaranteed to converge to the global optimum when the cost functional is convex. Unfortunately, convexity is not necessarily observed in our case (see Table 10.1). But, since each step of the algorithm involves an exact minimization, the cost functional is guaranteed to decrease so that the method remains applicable in non-convex situations. There is the risk, though, that it gets trapped into a local optimum. A way around the difficulty is to consider a warm start that may be obtained by running the ℓ_2 (Gauss) or ℓ_1 (Laplace) version of the method.

10.3 MAP reconstruction of biomedical images

In this section, we apply the MAP-estimation paradigm to the reconstruction of biomedical images. We concentrate on three imaging modalities: deconvolution microscopy, magnetic resonance imaging (MRI), and X-ray computed tomography (CT). In each case, we briefly recall the underlying linear image-formation model and then proceed with the determination of the system matrix $\mathbf{H}$ in accordance with the discretization principle presented in Section 10.1. An important practical aspect is the selection of an appropriate reconstruction basis. This selection is guided by approximation-theoretic and computational considerations. The outcome for each modality is a generic iterative reconstruction algorithm whose regularization parameters (whitening/regularization operator, potential function) can be tuned to the characteristics of the underlying class of biomedical images. Since our primary intent here is illustrative, we have fixed the regularization operator to the most popular choice in the field (i.e., magnitude of the gradient) and are presenting practical reconstruction results that highlight the influence of the potential function.

10.3.1 Scale-invariant image model and common numerical setup

A well-documented observation is that natural images tend to exhibit a Fourier spectrum that is mostly isotropic and roughly decaying like $1/\|\omega\|^{\gamma}$ with $1/2 \leq \gamma \leq 2$. The same holds true for many biomedical images, in both 2-D and 3-D. This is consistent with the idea of rotation- and scale-invariance since the objects and elementary structures in an image can appear at arbitrary orientations and magnifications. In the biomedical context, it can be argued that natural growth often induces fractal-like patterns whose appearance is highly irregular despite the simplicity of the underlying generative rules. Prominent examples of such structures are (micro)vascular networks, dendritic trees, trabecular bone, and cellular scaffolds. The other important observation is that the wavelet-domain statistics of images are typically non-Gaussian, with a small proportion of large coefficients (typically corresponding to contours) and the majority being close to zero (in smooth image regions). This is the primary reason why sparsity has become such an important topic in signal and image processing (see Introduction).

Our justification for the present model-based approach is that these qualitative properties of images are consistent with the stochastic models investigated in Chapter 7 if we take the whitening operator to be $(-\Delta)^{\gamma/2}$, the fractional Laplacian of order γ. While it could make sense to fit such a model very precisely and apply the corresponding (fractional) isotropic localization operator to uncouple the information, we prefer to rely on a robust scheme that involves shorter filters, as justified by the analysis of Section 8.3.5. A natural refinement over the series of coordinate-wise differences suggested by Theorem 7.7 is to consider the magnitude of the discrete gradient. This makes the regularization rotation-invariant without adding to the computational cost, and provides an efficient scheme for obtaining a robust estimate of the absolute value of the discrete increment process. Indeed, working with the absolute value is legitimate as long as the potential function is symmetric, which is always the case for the priors used in practice.

Before moving to the specifics of the various imaging modalities, we briefly describe the common numerical setup used for the experiments. Image acquisition is simulated numerically by applying the forward model (matrix $\mathbf{H}$) to the noise-free image $\mathbf{s}$ and subsequently adding white Gaussian noise of variance σ^2. The images are then reconstructed from the noisy measurements $\mathbf{y}$ by applying the algorithm described in Section 10.2.4, with $\mathbf{L}$ being the discrete gradient operator. The latter is implemented using forward finite differences while the corresponding $\mathbf{L}^T\mathbf{L}$ is the (scalar) discrete Laplacian. Our discretization assumes that the images are extended using periodic boundary conditions. This lends itself to the use of FFT-based techniques for the implementation of the matrix multiplications required by the linear step of the algorithm: Equation (10.23) or its corresponding gradient updates.

For each set of measurements, we are presenting three reconstruction results to compare the effect of the primary types of potential functions Φ_U.

- Gaussian prior: $\Phi_{\text{Gauss}}(x) = Ax^2$. In this case, the algorithm implements a classical linear reconstruction. This solution is representative of the level of performance of the reconstruction methods that are currently in use and that do not impose any sparsity on the solution.
- Laplace prior: $\Phi_{\text{Laplace}}(x) = B|x|$. This configuration is at the limit of convexity and imposes some medium level of sparsity. It corresponds to an ℓ_1-type minimization and is presently very popular for the recovery of sparse signals in the context of compressed sensing.
- Student prior: $\Phi_{\text{Student}}(x) = C\log(x^2 + \epsilon)$. This log-like penalty is non-convex and allows for the kind of heavy-tail behavior associated with the sparsest processes.

The regularization constants A, B, C are optimized for each experiment by comparison of the solution with the noise-free reference (oracle) to obtain the highest possible SNR. The proximal step of the algorithm described by (10.24) is adapted to handle the discrete gradient operator by merely shrinking its magnitude. This is the natural vectorial extension dictated by Definition (10.16). The shrinkage function for the Laplace prior is a soft-threshold (see (10.18)), while Student's solution with ϵ small is much closer to a hard threshold and favors sparser solutions. The reconstruction is initialized in a systematic fashion: the solution of the Gaussian estimator is used as initialization for the Laplace estimator and the output of the Laplace estimator is used as initial guess for Student's estimator. The parameter for Student's estimator is set to $\epsilon = 10^{-7}$.

10.3.2 Deconvolution of fluorescence micrographs

The signal of interest in fluorescence microscopy is the 3-D spatial density of the fluorescent labels that are embedded in a sample [VAVU06]. A molecular label is a fluorophore which, upon illumination at a specific wavelength, has the ability to re-emit light at another (typically) longer wavelength. A notable representative is the green fluorescence protein (GFP) that has its primary excitation peak at 395 nm and its emission peak at 509 nm.

Physical model of a diffraction-limited microscope

In a standard wide-field microscope, the fluorophores are excited by applying a uniform beam of light. The optical system includes an excitation filter that is tuned to the excitation wavelength of the fluorophore and an emission filter that collects the re-emitted light. Due to the random nature of photon re-emission, each point in the sample contributes independently to the light intensity in the image space, so that the overall system is linear. The microscope makes use of lenses to obtain a magnified view of the object. When the optical axis is properly aligned (paraxial configuration), the lateral [5] translation of a point source produces a corresponding (magnified) shift in the image plane. If, on the other hand, the point source moves axially out of focus, one observes a progressive spreading of the light (blurring), the effect being stationary with respect to its lateral

---

[5] In microscopy, "lateral" refers to the x-y focal plane, while "axial" refers to the z coordinate (depth), which is along the optical axis.

position. This implies that the fluorescence microscope acts as a linear shift-invariant system. It can therefore be described by the convolution equation

$$g(x,y,z) = (h_{3D} * s)(x,y,z),$$ (10.25)

where h_{3D} is the 3-D impulse response, also known as the *point-spread function* (PSF) of the microscope. A high-performance microscope can be modeled as a perfect aberration-free optical system whose only limiting factor is the finite size of the pupil of the objective. The PSF is then given by

$$h_{3D}(x,y,z) = I_0 \left| p_\lambda \left(\frac{x}{M}, \frac{y}{M}, \frac{z}{M^2} \right) \right|^2,$$ (10.26)

where I_0 is a constant gain, M is the magnification factor, and p_λ is the coherent diffraction pattern of an ideal point source with emission wavelength λ that is induced by the pupil. The square modulus accounts for the fact that the quantity measured at the detector is the (incoherent) light intensity, which is the energy of the electric field. Also note that the rescaling is not the same along the z dimension. The specific form of p_λ, as provided by the Fraunhofer theory of diffraction, is

$$p_\lambda(x,y,z) = \int_{\mathbb{R}^2} P(\omega_1, \omega_2) \exp\left(j2\pi z \frac{\omega_1^2 + \omega_2^2}{2\lambda f_0^2} \right) \exp\left(-j2\pi \frac{x\omega_1 + y\omega_2}{\lambda f_0} \right) d\omega_1 \, d\omega_2,$$ (10.27)

where f_0 is the focal length of the objective and $P(\omega_1, \omega_2) = \mathbb{1}_{\|\omega\| < R_0}$ is the pupil function. The latter is an indicator function that describes the circular aperture of radius R_0 in the so-called Fourier plane. The ratio R_0/f_0 is a good predictor of the *numerical aperture* [6] (NA), the optical parameter that is used in microscopy to specify the resolution of an objective through Abbe's law.

The 3-D PSF given by (10.26) is shown in Figure 10.2. We observe that h_{3D} is circularly symmetric with respect to the origin $(x,y) = (0,0)$ in the planes perpendicular to the optical axis z and that it exhibits characteristic diffraction rings. It is narrowest in the focal x-y plane with $z = 0$. The focal spot in Figure 10.2a is the Airy pattern that determines the lateral resolution of the microscope (see (10.29) and the discussion below). By contrast, the PSF is significantly broader in the axial direction. It also spreads out linearly along the lateral dimension as one moves away from the focal plane. The latter represents the effect of defocusing, with the external cone-shaped envelope in Figure 10.2c being consistent with the simplified behavior predicted by ray optics. This shows that, besides the fundamental limit on the lateral resolution that is imposed by the pupil function, the primary source of blur in wide-field microscopy is along the optical axis and is due to the superposition of the light contributions coming from the neighboring planes that are out of focus. The good news is that these effects can be partly compensated through the use of 3-D deconvolution techniques. In practical deconvolution

---

[6] The precise definition is $NA = n \sin\theta$, where n is the index of refraction of the operating medium (e.g., 1.0 for air, 1.33 for pure water, and 1.51 for immersion oil) and θ is the half-angle of the cone of light entering the objective. The small-angle approximation for a normal use in air is $NA \approx R_0/f_0$. A finer analysis that takes into account the curvature of the lens shows that this simple ratio formula remains accurate even at large numerical apertures in a well-corrected optical system.

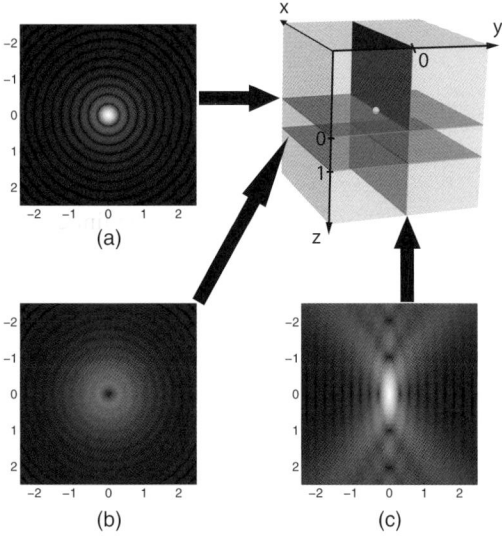

Figure 10.2 Visualization of the 3-D PSF of a wide-field microscope in a normalized coordinate system. (a) Cut through the lateral x-y plane with $z = 0$ (in focus). (b) Cut through a lateral x-y plane with $z = 1$ (out of focus). (c) Cut through the axial x-z plane with $y = 0$.

microscopy, one typically acquires a focal series of images (called a z-stack) which are then deconvolved in 3-D by using a PSF that is derived either experimentally, through the imaging of fluorescent nano beads, or theoretically $\big($see (10.26) and (10.27)$\big)$, based on the optical parameters of the microscope (e.g., NA, λ, M). Note that there are also optical solutions, such as confocal or light-sheet microscopy, for partially suppressing the out-of-focus light, but that these require more sophisticated instrumentation and longer acquisition times. These modalities may also benefit from deconvolution, but to a lesser extent.

Here, for simplicity, we shall concentrate on the simpler 2-D scenario where one is imaging a thin horizontal specimen s whose fluorescence emitters are confined to the focal plane $z = 0$. The image formation then simplifies to the purely 2-D convolution

$$g(x, y) = (h_{2D} * s)(x, y), \tag{10.28}$$

where the PSF is now given by

$$h_{2D}(x, y) = I_0 \left| 2 \frac{J_1(r/r_0)}{r/r_0} \right|^2 \tag{10.29}$$

with $r = \sqrt{x^2 + y^2}$, $r_0 = \frac{\lambda f_0}{2\pi R_0}$, and $J_1(r)$ is the first-order Bessel function. This characteristic pattern is called the *Airy disc*, while $J_1(r)/r$ is the inverse Fourier transform of a circular pupil of unit diameter (see (10.27) with $z = 0$).

The Fourier transform of the 2-D PSF of a microscope is called the *modulation transfer function*. Because of the squaring in (10.29), it is equal to the Fourier-domain autocorrelation function of the circular pupil function $P(\omega_1, \omega_2)$. The calculation of this convolution yields

$$
\hat{h}_{2D}(\boldsymbol{\omega}) = \begin{cases} \frac{2}{\pi}\left(\arccos\left(\frac{\|\boldsymbol{\omega}\|}{\omega_0}\right) - \frac{\|\boldsymbol{\omega}\|}{\omega_0}\sqrt{1 - \left(\frac{\|\boldsymbol{\omega}\|}{\omega_0}\right)^2}\right), & \text{for } 0 \le \|\boldsymbol{\omega}\| < \omega_0 \\[2mm] 0, & \text{otherwise,} \end{cases} \tag{10.30}
$$

where

$$
\omega_0 = \frac{2R_0}{\lambda f_0} = \frac{\pi}{r_0} \approx \frac{2\mathrm{NA}}{\lambda}
$$

is the Rayleigh frequency. This shows that a microscope is a radially symmetric low-pass filter whose cutoff frequency ω_0 imposes a fundamental limit on resolution. A Shannon-type sampling argument suggests that the ultimate resolution is r_0, assuming that one is able to deconvolve the image. Alternatively, one can apply Rayleigh's space-domain criterion, which stipulates that the minimum distance at which two point sources can be separated is when the first diffraction minimum coincides with the maximum response of the second source. Since the function $(J_1(r)/r)^2$ reaches its first zero at $r = 3.8317$, this corresponds to a resolution limit at $d_{\mathrm{Rayleigh}} = 0.61\lambda \times (f_0/R_0)$. This is consistent with Abbe's celebrated formula for the diffraction limit of a microscope: $d_{\mathrm{Abbe}} = \lambda/(2\mathrm{NA})$. The better objectives have a large numerical aperture with the current manufacturing limit being $\mathrm{NA} < 1.45$ (with oil immersion). This puts the resolution limit to about one-third the wavelength λ (or, one-half the wavelength for the more typical value of $\mathrm{NA} = 1$). Sometimes in the literature, the PSF is approximated by a 2-D isotropic Gaussian. The standard deviation that provides the closest fit with the physical model (10.29) is $\sigma_0 = 0.42\lambda(f_0/R_0) = 0.84/\omega_0$.

Discretization

From now on, we assume that $\omega_0 \le \pi$ so that we can sample the data on the integer grid while meeting the Nyquist criterion. The corresponding analysis functions, which are indexed by $\boldsymbol{m} = (m_1, m_2)$, are therefore given by

$$
\eta_{\boldsymbol{m}}(x, y) = h_{2D}(x - m_1, y - m_2).
$$

In order to discretize the system, we select a sinc basis $\{\mathrm{sinc}(\boldsymbol{x} - \boldsymbol{k})\}_{\boldsymbol{k} \in \mathbb{Z}^2}$ with

$$
\mathrm{sinc}(x, y) = \mathrm{sinc}(x)\mathrm{sinc}(y),
$$

where $\mathrm{sinc}(x) = \sin(\pi x)/(\pi x)$. The entries of the system matrix in (10.9) are then obtained as

$$
\begin{aligned}
[\mathbf{H}]_{\boldsymbol{m},\boldsymbol{k}} &= \langle \eta_{\boldsymbol{m}}, \mathrm{sinc}(\cdot - \boldsymbol{k}) \rangle \\
&= \langle h_{2D}(\cdot - \boldsymbol{m}), \mathrm{sinc}(\cdot - \boldsymbol{k}) \rangle \\
&= (\mathrm{sinc} * h_{2D})(\boldsymbol{m} - \boldsymbol{k}) = h_{2D}(\boldsymbol{m} - \boldsymbol{k}).
\end{aligned}
$$

In effect, this is equivalent to constructing the system matrix from the samples of the PSF since h_{2D} is already band-limited as a result of the imaging physics (diffraction-limited microscope).

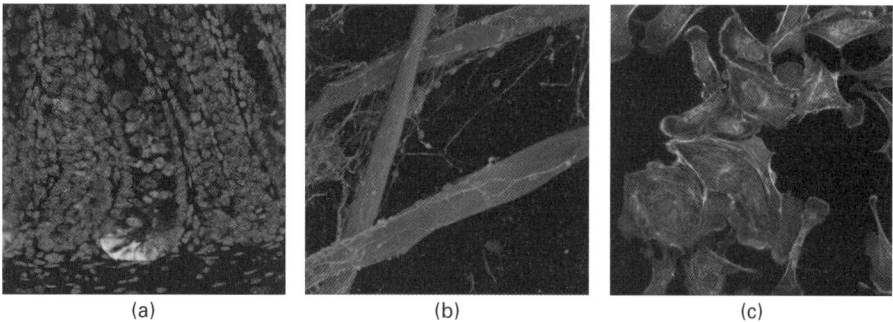

(a)   (b)   (c)

Figure 10.3 Images used in deconvolution experiments. (a) Stem cells surrounded by goblet cells. (b) Nerve cells growing around fibers. (c) Artery cells.

An important aspect for the implementation of the signal-recovery algorithm is that **H** is a discrete convolution matrix which is diagonalized by the discrete Fourier transform. The same is true for the regularization operator **L** as well as for any linear combination, product, or inverse of such convolution matrices. This allows us to convert (10.23) to a simple Fourier-domain multiplication which yields a fast and direct implementation of the linear step of the algorithm. The computational cost is essentially that of two FFTs (one forward and one backward Fourier transform).

Experimental results

The reference data are provided by the three microscopic images in Figure 10.3, which display different types of cells. The input images of size (512×512) are blurred with a Gaussian PSF of support (9×9) and standard deviation $\sigma_0 = 4$ to simulate the effect of a wide-field microscope with a low-NA objective. The measurements are degraded with additive white Gaussian noise so as to meet some prescribed blurred SNR (BSNR), defined as BSNR = var(**Hs**)$/\sigma^2$.

For deconvolution, the algorithm is run for a maximum of 500 iterations, or until the absolute relative error between the successive iterates is less than 5×10^{-6}. The results are summarized in Table 10.2. The first observation is that the standard linear deconvolution (MAP estimator based on a Gaussian prior) performs remarkably well for the image in Figure 10.3a, which is heavily textured. The MAP estimator based on the Laplace prior, on the other hand, yields the best performance for images having sharp edges with a moderate amount of texture, such as those in Figures 10.3b,c. This confirms the general claim that it is possible to improve the reconstruction performance through the promotion of sparse solutions. However, as the application of the Student prior to images typically encountered in microscopy demonstrates, exaggeration in the enforcement of sparsity is a distinct risk. Finally, we note that the Gaussian and Laplace versions of the algorithm are compatible with the methods commonly used in the field; for instance, ℓ_2-Tikhonov regularization [PMC93] and ℓ_1/TV regularization [DBFZ$^+$06].

10.3.3   Magnetic resonance imaging

Magnetic resonance refers to the property of atomic nuclei in a static magnetic field to absorb and re-emit electromagnetic radiation. This energy is re-emitted at a resonance

Table 10.2 Deconvolution performance of MAP estimators based on different prior distributions. The best results are shown in boldface.

	BSNR (dB)	*Estimation performance (SNR in dB)*		
		Gaussian	Laplace	Student
Stem cells	20	**14.43**	13.76	11.86
	30	**15.92**	15.77	13.15
	40	**18.11**	**18.11**	13.83
Nerve cells	20	13.86	**15.31**	14.01
	30	15.89	**18.18**	15.81
	40	18.58	**20.57**	16.92
Artery cells	20	14.86	**15.23**	13.48
	30	16.59	**17.21**	14.92
	40	18.68	**19.61**	15.94

frequency that is proportional to the strength of the magnetic field. The basic idea of magnetic resonance imaging (MRI) is to induce a space-dependent variation of the frequency of resonance by imposing spatial magnetic gradients. The specimen is then excited by applying pulsed radio waves that cause the nuclei (or spins) in the specimen to produce a rotating magnetic field detectable by the receiving coil(s) of the scanner.

Here, we shall focus on 2-D MRI, where the excitation is confined to a single plane. In effect, by applying a proper sequence of magnetic gradient fields, one is able to sample the (spatial) Fourier transform of the spin density $s(r)$ with $r \in \mathbb{R}^2$. Specifically, the mth (noise-free) measurement is given by

$$\widehat{s}(\omega_m) = \int_{\mathbb{R}^2} s(r) e^{-j\langle \omega_m, r \rangle} \, dr,$$

where the sampling occurs according to some predefined k-space trajectory (the convention in MRI is to use $k = \omega_m$ as the spatial frequency variable). This is to say that the underlying analysis functions are the complex exponentials $\eta_m(r) = e^{-j\langle \omega_m, r \rangle}$.

The basic problem in MRI is then to reconstruct $s(r)$ based on the partial knowledge of its Fourier coefficients which are also corrupted by noise. While the reconstruction in the case of a dense Cartesian sampling amounts to a simple inverse Fourier transform, it becomes more challenging for other trajectories, especially as the sampling density decreases.

For simplicity, we discretize the forward model by using the same sinc basis functions as for the deconvolution problem of Section 10.3.2. This results in the system matrix

$$[\mathbf{H}]_{m,n} = \langle \eta_m, \text{sinc}(\cdot - n) \rangle$$
$$= \langle e^{-j\langle \omega_m, \cdot \rangle}, \text{sinc}(\cdot - n) \rangle = e^{-j\langle \omega_m, n \rangle}$$

Table 10.3 MR reconstruction performance of MAP estimators based on different prior distributions.

	Radial lines	Estimation performance (SNR in dB)		
		Gaussian	Laplace	Student
Wrist	20	8.82	**11.8**	5.97
	40	11.30	**14.69**	13.81
Angiogram	20	4.30	9.01	**9.40**
	40	6.31	14.48	**14.97**

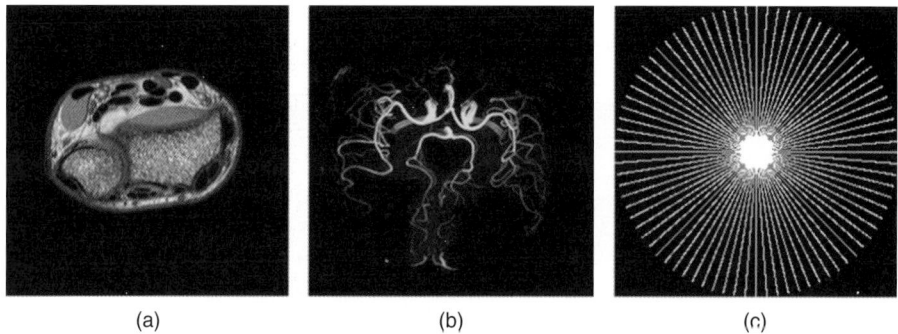

(a)   (b)   (c)

Figure 10.4 Data used in MR reconstruction experiments. (a) Cross section of a wrist. (b) Angiography image. (c) k-space sampling pattern along 40 radial lines.

under the assumption that $\|\boldsymbol{\omega}_m\|_\infty \le \pi$. The clear advantage of using the sinc basis is that $\mathbf{H}$ reduces to a discrete Fourier-like matrix, with the caveat that the frequency sampling is not necessarily uniform.

A convenient feature of this imaging model is that the matrix $\mathbf{H}^T\mathbf{H}$ is circulant so that the linear iteration step of the algorithm can be computed in exact form using the FFT.

Experimental results

To illustrate the method, we consider the reconstruction of the two MR images of size (256×256) shown in Figure 10.4: a cross section of a wrist and a MR angiogram. The Fourier-domain measurements are simulated using the type of radial sampling pattern shown in Figure 10.4c. The reconstruction algorithm is run with the same stopping criteria as in Section 10.3.2. The reconstruction results for two sampling scenarios are quantified in Table 10.3.

The first observation is that the estimator based on the Laplace prior generally outperforms the Gaussian solution, which corresponds to the traditional type of linear reconstruction. The Laplace prior is a clear winner for the wrist image, which has sharp edges and some amount of texture. While this is similar to the microscopy scenario, the tendency appears to be more systematic for MRI: we were unable to find a single MR scan in our database for which the Gaussian solution performs best. Yet, the supremacy of the ℓ_1 solution is not universal, as illustrated by the reconstruction of the angiogram, for

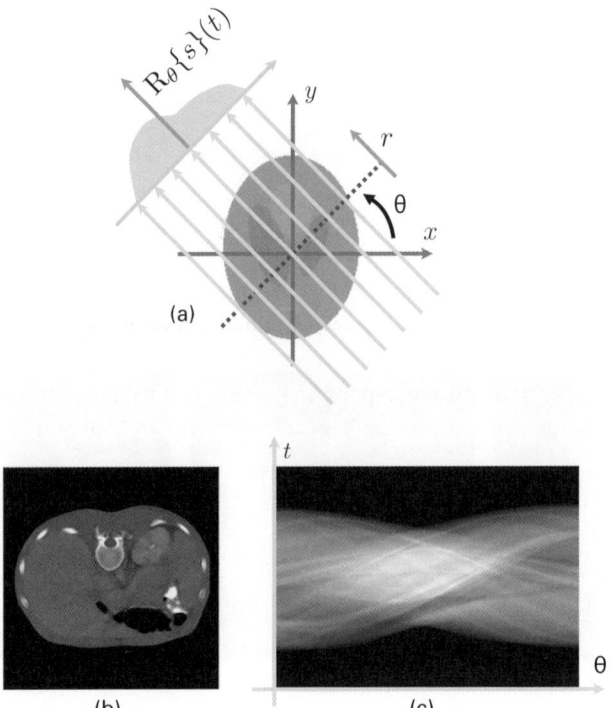

Figure 10.5 X-ray tomography and the Radon transform. (a) Imaging geometry. (b) 2-D reconstruction of a tomogram. (c) Its Radon transform (sinogram).

which the Student prior yields the best results, because this image is inherently sparse and composed of piecewise-smooth components. Similarly to the microscopy modality, we note that the present MAP formulation is compatible with the deterministic schemes used for practical MRI reconstruction; in particular, the methods that rely on total variation [BUF07] and log-based regularization [TM09], which are in direct correspondence with the Laplace and Student priors, respectively.

10.3.4   X-ray tomography

X-ray computed tomography (CT) aims at the reconstruction of an object (X-ray absorption map) from its projections (or line integrals) taken along different directions. The projections are obtained by integrating the function along a set of parallel rays. To specify the imaging geometry, we refer to Figure 10.5a. Letting $x = (x, y) \in \mathbb{R}^2$ be the spatial coordinates of the input function, we have that

$$x = t\theta + r\theta^{\perp},$$

where $\theta = (\cos\theta, \sin\theta)$ is the vector defining the t axis and $\theta^{\perp} = (-\sin\theta, \cos\theta)$ is the unit vector that points along the direction of integration (r axis). The mathematical model of conventional CT is based on the Radon transform, which is an invertible operator from $L_2(\mathbb{R}^2) \to L_2(\mathbb{R} \times [-\pi, \pi])$. It is defined as

$$R_\theta\{s(\boldsymbol{x})\}(t) = \int_\mathbb{R} s(t\boldsymbol{\theta} + r\boldsymbol{\theta}^\perp)\mathrm{d}r \qquad (10.31)$$

$$= \int_{\mathbb{R}^2} s(\boldsymbol{x})\delta(t - \langle\boldsymbol{x}, \boldsymbol{\theta}\rangle)\,\mathrm{d}\boldsymbol{x}, \qquad (10.32)$$

with index variables $t \in \mathbb{R}$ and $\theta \in [-\pi, \pi]$. An example of a Radon transform corresponding to a cross section of the human thorax is shown in Figure 10.5c. The resulting family of functions $g_\theta(t) = R_\theta\{s\}(t)$ is called the *sinogram*, owing to the property that the trajectory of a point $(x_0, y_0) = (R_0 \cos \phi_0, R_0 \sin \phi_0)$ in Radon space is a sinusoidal curve given by $t_0(\theta) = x_0 \cos\theta + y_0 \sin\theta = R_0 \cos(\theta - \phi_0)$.

In practice, the measurements correspond to the sampled values of the Radon transform of the absorption map $s(\boldsymbol{x})$ at a series of points $(t_m, \theta_m), m = 1, \ldots, M$. From (10.32), we deduce that the analysis functions are

$$\eta_m(\boldsymbol{x}) = \delta(t_m - \langle\boldsymbol{x}, \boldsymbol{\theta}_m\rangle),$$

which represent a series of idealized lines in $\mathbb{R}^2$ perpendicular to $\boldsymbol{\theta}_m = (\cos\theta_m, \sin\theta_m)$.

Discretization
For discretization purposes, we represent the absorption distribution as the weighted sum of separable B-spline-like basis functions:

$$s(\boldsymbol{x}) = \sum_{\boldsymbol{k}} s[\boldsymbol{k}]\beta(\boldsymbol{x} - \boldsymbol{k}),$$

with $\beta(\boldsymbol{x}) = \beta(x)\beta(y)$, where $\beta(x)$ is a suitable symmetric kernel (typically, a polynomial B-spline of degree n). The constraint here is that β ought to have a short support to reduce computations, which rules out the use of the sinc basis.

In order to determine the system matrix, we need to compute the Radon transform of the basis functions. The properties of the Radon transform that are helpful for that purpose are

(1) Projected translation-invariance

$$R_\theta\{\varphi(\cdot - \boldsymbol{x}_0)\}(t) = R_\theta\{\varphi\}(t - \langle\boldsymbol{x}_0, \boldsymbol{\theta}\rangle) \qquad (10.33)$$

(2) Pseudo-distributivity with respect to convolution

$$R_\theta\{\varphi_1 * \varphi_2\}(t) = (R_\theta\{\varphi_1\} * R_\theta\{\varphi_2\})(t) \qquad (10.34)$$

(3) Fourier central-slice theorem

$$\int_\mathbb{R} R_\theta\{\varphi\}(t)\mathrm{e}^{-\mathrm{j}\omega t}\,\mathrm{d}t = \widehat{\varphi}(\boldsymbol{\omega})|_{\boldsymbol{\omega}=\omega\boldsymbol{\theta}}. \qquad (10.35)$$

The first result is obtained by a simple change of variable in (10.32). The second is a direct consequence of (10.35). The Fourier central-slice theorem states that the 1-D Fourier transform of the projection of φ at angle θ is equal to a corresponding central cut of the 2-D Fourier transform of the function. The result is easy to establish for $\theta = 0$, in which case the t and x axes coincide. For this particular configuration, we have that

$$\int_{\mathbb{R}} R_0\{\varphi\}(x)e^{-j\omega x} \, dx = \int_{\mathbb{R}} \left(\int_{\mathbb{R}} \varphi(x,y) \, dy \right) e^{-j\omega x} \, dx$$

$$= \int_{\mathbb{R}^2} \varphi(x,y)e^{-j\omega x} \, dx \, dy$$

$$= \widehat{\varphi}(\omega,0), \qquad\qquad\qquad \text{(by definition)}$$

where the interchange of integrals in the second line requires that $\varphi \in L_1(\mathbb{R}^2)$ (Fubini). The angle-dependent formula (10.35) is obtained by rotating the system of coordinates and invoking the rotation property of the Fourier transform.

Next we show that the Radon transform of the basis functions can be obtained through the convolution of two rescaled 1-D kernels.

PROPOSITION 10.1   *The Radon transform of the separable function $\varphi(\mathbf{x} - \mathbf{x}_0)$ where $\varphi(\mathbf{x}) = \varphi_1(x)\varphi_2(y)$ is given by*

$$R_\theta\{\varphi(\cdot - \mathbf{x}_0)\}(t) = \varphi_\theta(t - t_0),$$

where $t_0 = \langle \mathbf{x}_0, \boldsymbol{\theta} \rangle$ and

$$\varphi_\theta(t) = \left(\frac{1}{|\cos\theta|}\varphi_1\left(\frac{\cdot}{\cos\theta}\right) * \frac{1}{|\sin\theta|}\varphi_2\left(\frac{\cdot}{\sin\theta}\right) \right)(t),$$

with the convention that

$$\lim_{a\to 0} \frac{1}{|a|}\varphi\left(\frac{x}{a}\right) = \delta(x)\left(\int_{\mathbb{R}} \varphi(x) \, dx \right).$$

Proof   Since φ is separable, its 2-D Fourier transform is given by

$$\widehat{\varphi}(\boldsymbol{\omega}) = \widehat{\varphi}_1(\omega_1)\,\widehat{\varphi}_2(\omega_2),$$

where $\widehat{\varphi}_1$ and $\widehat{\varphi}_2$ are the 1-D Fourier transforms of φ_1 and φ_2, respectively. The Fourier central-slice theorem then implies that

$$\widehat{\varphi_\theta}(\omega) = \widehat{R_\theta\varphi}(\omega) = \widehat{\varphi}_1(\omega\cos\theta)\,\widehat{\varphi}_2(\omega\sin\theta).$$

Next, we note that $\widehat{\varphi}_1(\omega\cos\theta)$ and $\widehat{\varphi}_2(\omega\sin\theta)$ are the 1-D Fourier transforms of $\frac{1}{|\cos\theta|}\varphi_1\left(\frac{t}{\cos\theta}\right)$ and $\frac{1}{|\sin\theta|}\varphi_2\left(\frac{t}{\sin\theta}\right)$, respectively. The final result then follows from (10.33) and the property that the Fourier-domain product maps into a time-domain convolution.   □

This allows us to write the entries of the system matrix as

$$[\mathbf{H}]_{m,k} = \langle \delta(t_m - \langle\cdot, \boldsymbol{\theta}_m\rangle), \beta(\cdot - \mathbf{k}) \rangle$$

$$= R_{\theta_m}\{\beta(\cdot - \mathbf{k})\}(t_m) = \beta_{\theta_m}(t_m - \langle \mathbf{k}, \boldsymbol{\theta}_m \rangle),$$

where $\beta_{\theta_m}(t)$ is the projection of $\beta(\mathbf{x}) = \beta(x)\beta(y)$ along the direction θ_m, as specified in Proposition 10.1.

We shall now apply the result of Proposition 10.1 to determine the Radon transform of a symmetric tensor-product polynomial spline of degree n. The relevant 1-D formula for $\beta(x)$ is

$$\beta^n(x) = \sum_{k=0}^{n+1} (-1)^k \binom{n+1}{k} \frac{\left(x - k + \frac{n+1}{2}\right)_+^n}{n!},$$

which is the recentered version of (1.11). Next, by making use of the distributivity of convolution and the relation

$$\frac{t_+^{n_1}}{n_1!} * \frac{t_+^{n_2}}{n_2!} = \frac{t_+^{n_1+n_2+1}}{(n_1 + n_2 + 1)!},$$

we find that

$$R_\theta \left\{\beta^n(x)\beta^n(y)\right\}(t) =$$

$$\sum_{k=0}^{n+1}\sum_{k'=0}^{n+1}(-1)^{k+k'} \binom{n+1}{k}\binom{n+1}{k'} \frac{\left(t + \left(\frac{n+1}{2} - k\right)\cos\theta + \left(\frac{n+1}{2} - k'\right)\sin\theta\right)_+^{2n+1}}{|\cos\theta|^{n+1} \, |\sin\theta|^{n+1} \, (2n+1)!},$$

$$(10.36)$$

which provides an explicit formula for the Radon transform of a B-spline of any degree n for $\theta \neq 0, \pm\frac{\pi}{2}, \pm\pi$ (when the projection is along a coordinate axis, the Radon transform is simply $\beta^n(t)$). This result is a special case of the general box-spline calculus described in [ENU12].

For the present experiments, we select $n = 1$. This corresponds to piecewise bilinear basis functions whose Radon transforms are the (non-uniform) cubic splines specified by (10.36) for $\theta \neq 0, \pm\frac{\pi}{2}, \pm\pi$, or simple triangle functions otherwise. The Radon profiles are stored in a lookup table to speed up computations. In essence, the forward matrix $\mathbf{H}$ amounts to a "standard" projection with angle-dependent interpolation weights given by β_θ, while $\mathbf{H}^T$ is the corresponding backprojection. For a parallel geometry, their computation complexity is $O(N \times M_\theta \times (n+1))$ where N is the number of pixels (or B-spline coefficients) of the reconstructed image, M_θ the number of distinct angular directions, and n the degree of the B-spline.

Experimental results
We consider the two images shown in Figure 10.6. The first is the Shepp–Logan (SL) phantom of size (256×256), while the second is a real CT reconstruction of the cross section of the lung of size (750×750). In the simulations of the forward model, we use a standard parallel geometry with an angular sampling that is matched to the size of the images. Specifically, the projections are taken along $M_\theta = 180, 360$ equiangular directions for the lung image and $M_\theta = 120, 180$ directions for the SL phantom. The measurements are degraded with Gaussian noise with a signal-to-noise ratio of 20 dB.

For the reconstruction, we solve the quadratic minimization problem (10.21) iteratively by using 50 conjugate-gradient (inner) iterations. The reconstruction results are reported in Table 10.4.

Table 10.4 Reconstruction results of X-ray computed tomography using different estimators.

	Directions	*Estimation performance (SNR in dB)*		
		Gaussian	Laplace	Student
SL phantom	120	16.8	17.53	**18.76**
SL phantom	180	18.13	18.75	**20.34**
Lung	180	**22.49**	21.52	21.45
Lung	360	**24.38**	22.47	22.37

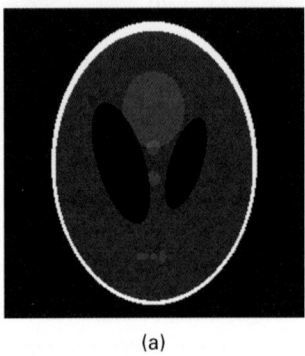
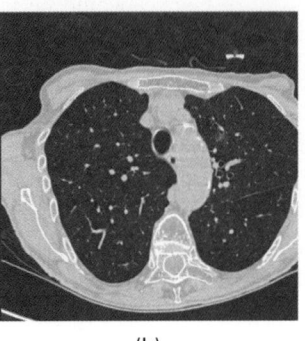

(a)   (b)

Figure 10.6 Images used in X-ray tomographic reconstruction experiments. (a) The Shepp-Logan (SL) phantom. (b) Cross section of a human lung.

We observe that the imposition of the strong level of sparsity brought by Student priors is advantageous for the SL phantom. This is not overly surprising given that the SL phantom is an artificial construct composed of piecewise-constant regions (ellipses). For the realistic lung image (true CT), we find that the Gaussian solution outperforms the others. Similarly to the deconvolution and MRI problems, the present MAP estimators are in line with the Tikhonov-type [WLLL06] and TV [XQJ05] reconstructions used for X-ray CT.

10.3.5   Discussion

During our investigation of real image-reconstruction problems, we have highlighted the similarity between deterministic sparsity-promoting methods and MAP estimators for sparse stochastic processes. The experiments we conducted with different imaging modalities confirm the importance of sparse modeling in the reconstruction of biomedical images. We found that imposing a medium level of sparsity, as afforded by the Laplace prior (ℓ_1-norm minimization), is beneficial in most instances. Heavier-tailed priors are available too, but they are helpful only for a limited class of images that are inherently sparse. At the other end of the palette is the "classical" linear type of reconstruction (Gaussian prior), which performs remarkably well for images whose content is more diffuse/textured or when the inverse problem is well conditioned. This confirms that the efficiency of a potential function depends strongly on the type of image being considered. In our model, this is related to the Lévy exponent of the underlying continuous-domain innovation process w, which is in direct relationship with the signal prior.

As far as the relevance of the underlying model is concerned, we like to view the present set of techniques and continuous-domain stochastic models as a conceptual framework for deriving and refining state-of-the-art algorithms in a principled fashion. The reassuring aspect is that the approach gives support to several algorithms that are presently used in the field.

The next step, of course, would be to determine how to best fit the model to the data. However, the inherent difficulty with this Bayesian view of the problem is that there is actually no guarantee that (non-Gaussian) MAP estimation performs best for the class of signals for which it is designed. There is even evidence that a slight model mismatch (e.g., modification of the MAP criterion) can be beneficial in some instances (see Section 10.4.3 for explicit illustrations of this statement).

The current challenge is to take full advantage of the statistical model and to find a proper way of constraining the solution. One possible approach is to specify reconstruction methods that are (sub)optimal in the MMSE sense for particular classes of stochastic processes. While it is still not clear how this can be achieved in full generality, Section 10.4 demonstrates how to proceed for the simpler signal-denoising scenario where the system matrix is the identity.

10.4   The quest for the minimum-error solution

In the Bayesian framework where the prior distribution of the signal is known, the optimal (MMSE) reconstruction is the conditional expectation of the signal given the measurements. Unfortunately, the direct computation of the MMSE solution, which is specified by an N-dimensional integral, is intractable numerically for the (non-Gaussian) cases of interest to us. This is to be contrasted with the previous MAP formulation which translates into the "Gibbs energy" minimization problem (10.12) that can be solved numerically using standard optimization techniques. Since the algorithms favored by practitioners are based on similar variational principles, a key issue is to characterize their degree of (sub)optimality and, in the case of deficiency of the MAP criterion, to understand how the energy functional should be modified in order to improve the quality of the reconstruction.

In this section, we investigate the problem of the denoising of Lévy processes for which questions regarding optimality can be answered to a large extent. Specifically, we shall define the corresponding MMSE signal estimator, derive a computational solution based on belief propagation, and use the latter as gold standard to assess the performance of the primary types of MAP estimators previously considered.

10.4.1   MMSE estimators for first-order processes

We now focus on the problem of the recovery of non-Gaussian AR(1) and Lévy processes from their noisy sampled values. The corresponding statistical measurement model is

$$p_{(Y_1:Y_N|X_1:X_N)}(\mathbf{y}|\mathbf{x}) = \prod_{n=1}^{N} p_{Y|X}(y_n|x_n),$$

which assumes that the noise contributions are independent and characterized by the conditional pdf $p_{Y|X}$. For instance, in the case of AWGN, we have that $p_{Y|X}(y|x) = g_\sigma(y-x)$, where g_σ is a centered Gaussian distribution with standard deviation σ.

Given the measurements $\mathbf{y} = (y_1, \ldots, y_N)$, the problem is to recover the unknown signal vector $\mathbf{x} = (x_1, \ldots, x_N)$ based on the knowledge that the latter is a realization of a sparse first-order process of the type characterized in Section 8.5.2. This prior information is summarized by the stochastic difference equation

$$u_n = x_n - a_1 x_{n-1},$$

where (u_n) is an i.i.d. sequence with infinitely divisible pdf p_U, with the implicit convention that $x_0 = 0$ (or, alternatively, $x_0 = x_N$ if we are applying circular boundary conditions). This model covers the cases of the non-Gaussian AR(1) processes (when $|a_1| < 1$) and of the Lévy processes for $a_1 = 1$. The posterior distribution of the signal is therefore given by

$$p_{(X_1:X_N|Y_1:Y_N)}(\mathbf{x}|\mathbf{y}) = \frac{1}{Z} \prod_{n=1}^{N} p_{Y|X}(y_n|x_n) \prod_{n=1}^{N} p_U \big(\underbrace{x_n - a_1 x_{n-1}}_{u_n}\big), \tag{10.37}$$

where Z is a proper normalization constant. We can then formally specify the optimal signal estimate as

$$\mathbf{x}_{MMSE}(\mathbf{y}) = \mathbb{E}\{\mathbf{x}|\mathbf{y}\} = \int_{\mathbb{R}^N} \mathbf{x}\, p_{(X_1:X_N|Y_1:Y_N)}(\mathbf{x}|\mathbf{y})\, d\mathbf{x}. \tag{10.38}$$

This is to be contrasted with the MAP estimator, which is defined as

$$\mathbf{x}_{MAP}(\mathbf{y}) = \arg\max_{\mathbf{x}\in\mathbb{R}^N} \{p_{(X_1:X_N|Y_1:Y_N)}(\mathbf{x}|\mathbf{y})\}. \tag{10.39}$$

For completeness, we now briefly show that the conditional mean (10.38) minimizes the mean-square estimation error among all signal estimators. An estimator $\tilde{\mathbf{x}} = \tilde{\mathbf{x}}(\mathbf{y})$ is a specific function of the measurement vector $\mathbf{y}$ and its performance is measured by the (conditional) mean-square estimation error

$$\mathbb{E}\{(\mathbf{x}-\tilde{\mathbf{x}})^2|\mathbf{y}\} = \int_{\mathbb{R}^N} (\mathbf{x}-\tilde{\mathbf{x}})^2 p_{(X_1:X_N|Y_1:Y_N)}(\mathbf{x}|\mathbf{y})\, d\mathbf{x}. \tag{10.40}$$

Since $\mathbf{y}$ is fixed, we can minimize this expression by annihilating its partial derivatives with respect to $\tilde{\mathbf{x}}$. This gives

$$\frac{\partial\, \mathbb{E}\{(\mathbf{x} - \tilde{\mathbf{x}})^2 | \mathbf{y}\}}{\partial \tilde{\mathbf{x}}} = -\int_{\mathbb{R}^N} 2(\mathbf{x} - \tilde{\mathbf{x}}) p_{(X_1 : X_N | Y_1 : Y_N)}(\mathbf{x}|\mathbf{y})\, d\mathbf{x} = 0$$

$$\Rightarrow \quad \tilde{\mathbf{x}}_{\text{opt}} = \int_{\mathbb{R}^N} \mathbf{x}\, p_{(X_1 : X_N | Y_1 : Y_N)}(\mathbf{x}|\mathbf{y})\, d\mathbf{x},$$

which proves that (10.38) is the MMSE solution. The implicit assumption here is that $\int_{\mathbb{R}^N} |\mathbf{x}|^n p_{X|Y}(\mathbf{x}|\mathbf{y})\, d\mathbf{x} < \infty$ for $n = 1, 2$ so that (10.40) is well defined and so that we can safely differentiate under the integral sign (by Lebesgue's dominated-convergence theorem).

Finally, we note that the MMSE estimator provided by (10.38) has the following properties:

- It is unbiased with $\mathbb{E}\{\mathbf{x}_{\text{MMSE}}(\mathbf{y})\} = \mathbb{E}\{\mathbf{x}\}$.
- It satisfies the statistical "orthogonality" principle

$$\mathbb{E}\left\{ \tilde{\mathbf{x}}(\mathbf{y}) \left(\mathbf{x} - \mathbf{x}_{\text{MMSE}}(\mathbf{y})\right)^T \right\} = 0$$

for any estimator $\tilde{\mathbf{x}}(\mathbf{y}) : \mathbb{R}^N \to \mathbb{R}^N$ with $\mathbb{E}\{\|\tilde{\mathbf{x}}(\mathbf{y})\|^2\} < \infty$ that is a (non-linear) function of the measurements.
- It is typically non-linear unless $\mathbf{y}$ and $\mathbf{x}$ are jointly Gaussian.

10.4.2 Direct solution by belief propagation

The first approach that we consider is an explicit calculation of (10.38) based on a recursive evaluation of the required integrals. The algorithm relies on *belief propagation* (BP). BP is a general graph-based technique for computing the marginal distributions of a high-dimensional posterior pdf that admits a decomposition into a product of simple low-order factors [KFL01]. It operates by passing messages along the edges of a factor graph (see Figure 10.7). The role of these messages is twofold: (1) to perform the summation (or integration) of the factors of the pdf with respect to a given node variable $\left(\text{e.g., } x_n \text{ in (10.37)}\right)$ and (2) to progressively combine the factors into more global entities by forming appropriate products.

Example: estimation of a three-point Lévy process
The best way of explaining BP is to detail a simple example. To that end, we consider the signal $\mathbf{x} = (x_1, x_2, x_3)$ corresponding to three consecutive samples of a Lévy process. The posterior distribution of the signal given the vector of noisy measurements $\mathbf{y} = (y_1, y_2, y_3)$ is obtained from (10.37) with $N = 3$ and $a_1 = 1$. The complete factorized expression, which is represented by the factor graph in Figure 10.7, is

$$p_{(X_1 : X_3 | Y_1 : Y_3)}(\mathbf{x}|\mathbf{y}) \propto p_U(x_1)\, p_{Y|X}(y_1|x_1)\, p_U(x_2 - x_1) \times$$
$$p_{Y|X}(y_2|x_2)\, p_U(x_3 - x_2)\, p_{Y|X}(y_3|x_3). \tag{10.41}$$

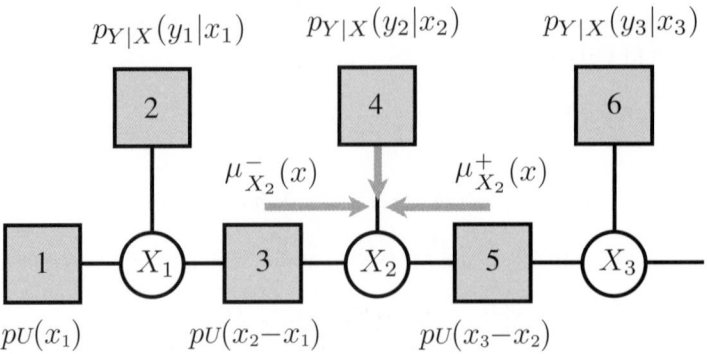

Figure 10.7 Factor-graph representation of the posterior distribution (10.41) or (10.37) in greater generality. The boxed nodes represent the factors of the pdf and the circled nodes the unknown variables. The presence of an edge between a factor and a variable node indicates that the latter variable is active within the factor. The functions $\mu_{X_2}^-(x)$ and $\mu_{X_2}^+(x)$ represent the beliefs at the variable node X_2; they condense all the statistical information coming from the left and right of the graph, respectively.

The crucial step for exact Bayesian inference is to marginalize $p_{(X_1:X_N|Y_1:Y_N)}(\mathbf{x}|\mathbf{y})$ with respect to x_n. In short, we need to integrate the posterior pdf over all other variables. For instance, in the case of x_2, we get

$$p_{(X_2|Y_1:Y_3)}(x_2|\mathbf{y}) = \int_{\mathbb{R}}\int_{\mathbb{R}} p_{(X_1:X_3|Y_1:Y_3)}(\mathbf{x}|\mathbf{y})\, dx_1\, dx_3$$

$$\propto \underbrace{\int_{\mathbb{R}} \overbrace{p_U(x_1)}^{\mu_{X_1}^-(x_1)}\, p_{Y|X}(y_1|x_1)\, p_U(x_2 - x_1)\, dx_1}_{\mu_{X_2}^-(x_2)}$$

$$\cdot\, p_{Y|X}(y_2|x_2) \cdot \underbrace{\int_{\mathbb{R}} p_U(x_3 - x_2)\, p_{Y|X}(y_3|x_3) \cdot \overbrace{1}^{\mu_{X_3}^+(x_3)}\, dx_3}_{\mu_{X_2}^+(x_2)}\,.$$

To evaluate the marginal with respect to x_3, we take advantage of the previous integration over x_1 encoded in the "belief" function $\mu_{X_2}^-(x_2)$ and proceed as follows:

$$p_{(X_3|Y_1:Y_3)}(x_3|\mathbf{y}) = \int_{\mathbb{R}}\int_{\mathbb{R}} p_{(X_1:X_3|Y_1:Y_3)}(\mathbf{x}|\mathbf{y})\, dx_1\, dx_2$$

$$\propto \underbrace{\int_{\mathbb{R}} \mu_{X_2}^-(x_2)\, p_{Y|X}(y_2|x_2)\, p_U(x_3 - x_2)\, dx_2}_{\mu_{X_3}^-(x_3)} \cdot p_{Y|X}(y_3|x_3) \cdot \underbrace{1}_{\mu_{X_3}^+(x_3)}\,.$$

The emerging pattern is that the marginal distribution of the variable x_n can be expressed as a product of three terms:

$$p_{(X_n|Y_1:Y_N)}(x_n|\mathbf{y}) = \mu^-_{X_n}(x_n) \cdot p_{Y|X}(y_n|x_n) \cdot \mu^+_{X_n}(x_n),$$

where the so-called belief functions $\mu^-_{X_n}$ and $\mu^+_{X_n}$ condense the statistical information carried by the variables with indices below n and above n, respectively. This aggregation mechanism is summarized graphically with arrows in Figure 10.7.

BP for Lévy and non-Gaussian AR(1) processes
The fundamental idea for extending the scheme to a larger number of samples is that the beliefs $\mu^-_{X_n}(x)$ and $\mu^+_{X_n}(x)$ can be updated recursively. The algorithm below is a generalization that is applicable to the MMSE denoising of the broad class of Markov-1 signals. Since the underlying factor graph has no loops, it computes exact marginals and terminates after one forward and backward sweep of message passing.

- *Initialization.* Set

$$\mu^-_{X_1}(x) = p_U(x)$$
$$\mu^+_{X_N}(x) = 1$$

- *Forward message recursion.* For $n = 2$ to N, compute

$$\mu^-_{X_n}(x) \propto \int_{\mathbb{R}} \mu^-_{X_{n-1}}(z)\, p_{Y|X}(y_{n-1}|z)\, p_U(x - a_1 z)\, dz \qquad (10.42)$$

- *Backward message recursion.* For $n = (N-1)$ down to 1, compute

$$\mu^+_{X_n}(x) \propto \int_{\mathbb{R}} p_U(z - a_1 x)\, p_{Y|X}(y_{n+1}|z)\, \mu^+_{X_{n+1}}(z)\, dz \qquad (10.43)$$

- *Results.* For $n = 1$ to N, compute

$$p_{(X_n|Y_1:Y_N)}(x|\mathbf{y}) \propto \mu^-_{X_n}(x) \cdot p_{Y|X}(y_n|x) \cdot \mu^+_{X_n}(x)$$
$$[\mathbf{x}_{\text{MMSE}}]_n = \int_{\mathbb{R}} x\, p_{(X_n|Y_1:Y_N)}(x|\mathbf{y})\, dx. \qquad (10.44)$$

The symbol $\propto$ denotes a renormalization such that the resulting function integrates to one. The critical part of this algorithm is the evaluation of the convolution-like integrals (10.42) and (10.43). The scalar belief functions $\left(\mu^-_{X_n}(x), \mu^+_{X_n}(x)\right)_{n=1}^{N}$ that result from these calculations also need to be stored, which presupposes some form of discretization.

Fourier-based version of BP for Lévy processes
There are two practical reasons for transcribing the above BP estimation algorithm into the Fourier domain. The first is that closed-form expressions are not available for all infinitely divisible pdfs. The preferred mode of description is the characteristic function $\widehat{p}_U(\omega) = e^{f_U(\omega)}$, where f_U is the Lévy exponent of the innovation, as has been made

clear in Chapters 4 and 9. The second reason is that, for $a_1 = 1$ (Lévy process), (10.42) (as well as (10.43)) may be rewritten as the convolution

$$\bar{\mu}_{X_n}(x) \propto \int_{\mathbb{R}} g(z) p_U(x - z) \, dz = (p_U * g)(x) = \mathcal{F}^{-1}\{e^{f_U(\omega)} \widehat{g}(\omega)\}(x),$$

where $g(z) = \bar{\mu}_{X_{n-1}}(z) \, p_{Y|X}(y_{n-1}|z)$. In order to obtain a cost-effective implementation, we suggest evaluating the various formulas by relying only on products by switching back and forth between the "time" domain to compute $g(z)$ and the Fourier domain to evaluate the convolution. The corresponding algorithm is summarized below. For simplicity, we are assuming that the underlying pdfs are symmetric; therefore, the presence of complex conjugation and the differences in convention from the Fourier transform as used by statisticians are inconsequential.

- *Initialization.* Set

$$\widehat{\bar{\mu}}_{X_1}(\omega) = e^{f_U(\omega)}$$

$$\widehat{\mu}^+_{X_N}(\omega) = \delta(\omega)$$

- *Forward message recursion.* For $n = 2$ to N, compute

$$\widehat{g}(\omega) = \mathcal{F}_z\left\{p_{Y|X}(y_{n-1}|z) \, \mathcal{F}^{-1}\{\widehat{\bar{\mu}}_{X_{n-1}}\}(z)\right\}(\omega)$$

$$\widehat{\bar{\mu}}_{X_n}(\omega) \propto \widehat{g}(\omega) \, \widehat{p}_U(\omega)$$

- *Backward message recursion.* For $n = (N - 1)$ down to 1, compute

$$\widehat{g}(\omega) = \mathcal{F}_z\left\{p_{Y|X}(y_{n+1}|z) \, \mathcal{F}^{-1}\{\widehat{\mu}^+_{X_{n+1}}\}(z)\right\}(\omega)$$

$$\widehat{\mu}^+_{X_n}(\omega) \propto \widehat{p}_U(\omega) \, \widehat{g}(\omega)$$

- *Results.* For $n = 1$ to N, compute

$$\widehat{p}_{(X_n|Y_1:Y_N)}(\omega) \propto \mathcal{F}\left\{\mathcal{F}^{-1}\{\widehat{\bar{\mu}}_{X_n}\}(x) \, p_{Y|X}(y_n|x) \, \mathcal{F}^{-1}\{\widehat{\mu}^+_{X_n}\}(x)\right\}(\omega)$$

$$[\mathbf{x}_{\text{MMSE}}]_n = j \left. \frac{d\widehat{p}_{(X_n|Y_1:Y_N)}(\omega)}{d\omega} \right|_{\omega=0}. \tag{10.45}$$

The symbol $\propto$ denotes a renormalization such that the value of the Fourier transform at the origin is one. We have taken advantage of the moment-generating property of the Fourier transform to establish (10.45).

The conventional and Fourier-based versions of the BP algorithm yield the exact MMSE estimator for our problem. However, both involve continuous mathematics (integrals and/or Fourier transforms) and neither one can be implemented in the given form. The simplest and most practical generic solution is to represent the belief functions by their samples on a uniform grid with a sampling step that is sufficiently fine and to truncate their support while maintaining the error within an acceptable bound. Integrals are then approximated by Riemann sums and the Fourier transform is implemented using the FFT.

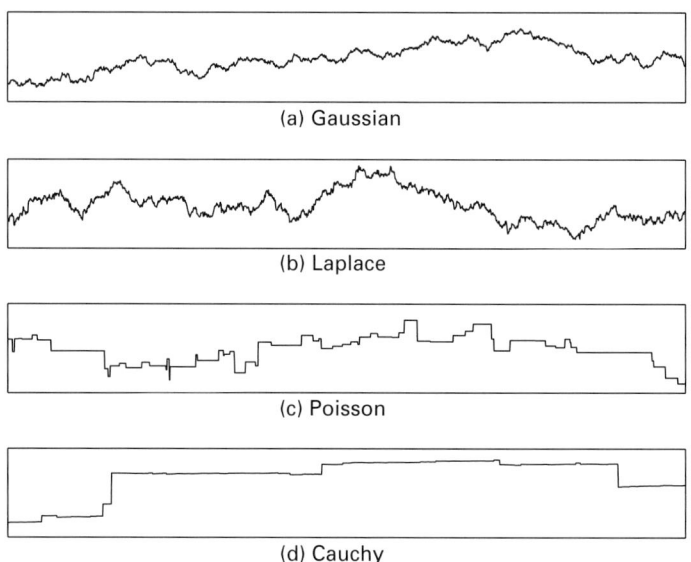

(a) Gaussian

(b) Laplace

(c) Poisson

(d) Cauchy

Figure 10.8 Examples of Lévy processes with increasing degree of sparsity. (a) Brownian motion (with Gaussian increments). (b) Lévy–Laplace motion. (c) Compound–Poisson process. (d) Lévy flight with Cauchy-distributed increments.

10.4.3   MMSE vs. MAP denoising of Lévy processes

We now present a series of denoising experiments with the goal of comparing the various estimation techniques introduced so far. We have considered signals associated with the four Lévy processes displayed in Figure 10.8. These were generated by discrete integration of sequences of i.i.d. increments $\big($see (8.2)$\big)$ with statistical distributions as indicated below.

- *Brownian motion.* (u_n) follows the standard Gaussian distribution $p_U(x) = g_1(x)$.
- *Laplace motion.* (u_n) follows a Laplace distribution with dispersion parameter $1/\lambda = 1$.
- *Compound-Poisson process.* (u_n) follows a compound-Poisson distribution with parameters $(\lambda = 0.6, p_A)$ such that $p_A = g_1$ is a standard Gaussian distribution and $\mathrm{Prob}(u_n = 0) = \mathrm{e}^{-\lambda} = 0.549$.
- *Lévy flight.* (u_n) follows a Cauchy distribution (SαS with $\alpha = 1$ or, equivalently, Student with $r = \frac{1}{2}$) with dispersion $s_0 = 1$.

The reader can refer to Table 4.1 for the precise definition of the underlying pdfs.

Each realization $\mathbf{x}$ of the signal was corrupted with additive white Gaussian noise of variance σ^2 to yield the measurement vector $\mathbf{y}$. Based on (10.12) with $\mathbf{H} = \mathbf{I}$ and $\mathbf{s} = \mathbf{x}$, the relevant MAP estimators are therefore given by

$$\mathbf{x}_{\mathrm{MAP}}(\mathbf{y}) = \arg\min_{\mathbf{x} \in \mathbb{R}^N} \left(\frac{1}{2}\|\mathbf{y} - \mathbf{x}\|_2^2 + \tau \sum_{n=1}^{N} \Phi_U\big([\mathbf{L}\mathbf{x}]_n\big) \right), \tag{10.46}$$

where $\tau \propto \sigma^2$ is a regularization parameter and Φ_U is one of the following potential functions:

- $\Phi_{\text{Gauss}}(x) = \frac{1}{2}|x|^2$ (Gaussian prior), which is known to yield the MMSE solution for Brownian motion and the best linear estimator otherwise (Wiener filter or LMMSE solution).
- $\Phi_{\text{Laplace}}(x) = |x|$ (Laplace prior), also termed TV, because it provides the signal estimate whose discrete *total variation* is minimum. The corresponding estimator, which involves the minimization of an ℓ_1-norm, is illustrative of the kind of recovery techniques used in compressed sensing.
- $\Phi_{\text{Student}}(x) = \log\left(|x|^2 + 1\right)$ (Student or Cauchy prior), an instance of a non-convex potential function that promotes sparsity even further than Φ_{Gauss}.

These MAP estimators are computed iteratively according to the generic procedure described in Section 10.2.4 using the LMMSE solution as warm start. The performance of an estimator $\tilde{x}(y)$ is measured by the signal-to-noise ratio improvement, which is given by

$$\Delta\text{SNR}(\tilde{x}, x) = 10\log_{10}\left(\frac{\|y - x\|^2}{\|\tilde{x}(y) - x\|^2}\right).$$

The signal length was set to $N = 100$. Each data point in a graph is an average resulting from 500 realizations of the denoising experiment. The MMSE estimator (Fourier-based implementation of belief propagation) relies on the correct prior signal model, while the regularization parameter τ of each of the other three estimators is kept constant for a given noise level, and adjusted to yield the lowest collective estimation error.

We show in the graph of Figure 10.9 the signal-to-noise improvement of the various algorithms for the denoising of Brownian motion. The first observation is that the results of the BP MMSE estimator and the Wiener filter (LMMSE=MAP Gaussian) are indistinguishable and that these methods perform best over the whole range of experimentation, in agreement with the theory. The worst results are obtained for TV regularization, while the Log penalty gives intermediate results. A possible explanation of the latter finding is that the Log potential is quadratic around the origin, so that it can replicate the behavior of Φ_2, but only over some limited range of input values.

A similar scenario is repeated in Figure 10.10 for the compound-Poisson process. We note that the corresponding MAP estimate is a constant signal since the probability of its increments being zero is overwhelmingly larger (Dirac mass at the origin) than any other acceptable value. This trivial estimator is excluded from the comparison. At low noise levels where the sparsity of the source dictates the structure, the performance of the TV estimator is very close to that of the MMSE denoiser, which can be considered as gold standard. Yet, the relative performance of the TV estimator deteriorates with increasing noise, so much so that TV ends being worst at the other end of the scale. One can observe a reverse trend for the LMMSE estimator, which progressively converges to the MMSE solution as the variance of the noise increases. Here, the explanation is that the statistics of the noisy signal is dominated by the Gaussian constituent, which is favorable to the LMMSE estimator.

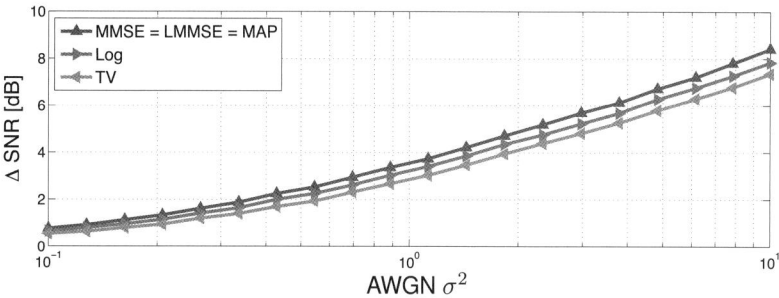

Figure 10.9 SNR improvement as a function of the level of noise for Brownian motion. The denoising methods by order of decreasing performance are: MMSE estimator (which is equivalent to the LMMSE and MAP estimators), Log regularization, and TV regularization.

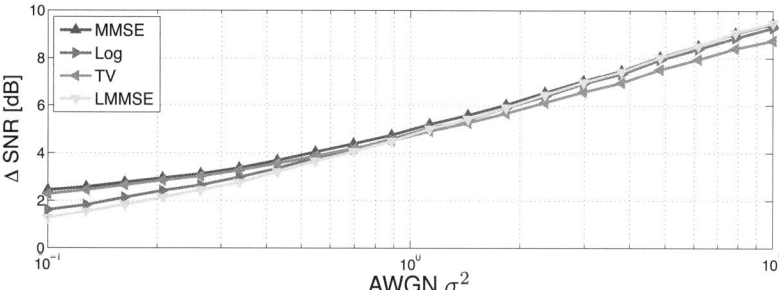

Figure 10.10 SNR improvement as a function of the level of noise for a piecewise-constant signal (compound-Poisson process). The denoising methods are: MMSE estimator, Log regularization, TV regularization, and LMMSE estimator.

The more challenging case of a Lévy flight, which is the sparsest process in the series, is documented in Figure 10.11. Here, the MAP estimator (Log potential) performs well over the whole range of experimentation. A possible explanation is that the corrupted signal looks sparse even at large noise powers since the convolution of a heavy-tailed pdf with a Gaussian remains heavy-tailed. The dominance of the non-Gaussian regime also explains why the LMMSE performs so poorly. The main limitation of the LMMSE algorithm is that it fails to preserve the sharp edges that are characteristic of this type of signal.

The last example, in Figure 10.12, is particularly telling because the results go against our initial expectation, especially at higher noise levels. While the MAP (TV) estimator performs well in the low-noise regime, it progressively falls behind all other estimators as the variance of the noise increases. Particularly surprising is the good behavior of the LMMSE algorithm, which matches the MMSE solution at higher noise levels. Apart from the MMSE denoiser, there is no single estimator that outperforms the others over the whole range of noise. The possible reason for the poor performance of MAP is that the underlying signal is at the very low end of sparsity, with its general appearance being rather similar to Brownian motion (see Figure 10.8a,b). This finding suggests that one should be cautious with the Bayesian justification of ℓ_1-norm minimization techniques

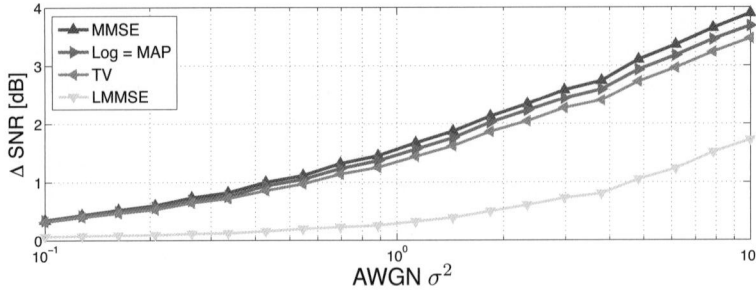

Figure 10.11 SNR improvement as a function of the level of noise for a Lévy flight with Cauchy-distributed increments. The denoising methods by order of decreasing performance are: MMSE estimator, MAP estimator (Log regularization), TV regularization, and LMMSE estimator.

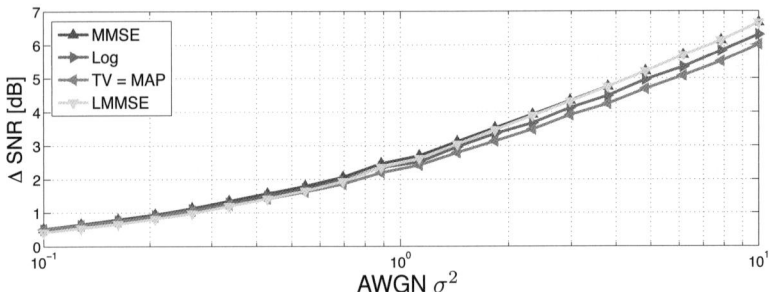

Figure 10.12 SNR improvement as a function of the level of noise for a Lévy process with Laplace-distributed increments. The denoising methods are: MMSE estimator, LMMSE estimator, MAP estimator (TV regularization), and Log regularization.

based on Laplace priors since MAP-TV does not necessarily perform so well for the model to which it is matched – in fact, often worse than the classical Wiener solution (LMMSE).

   While this series of experiments stands as a warning against a strict application of the MAP principle, it also shows that it is possible to specify variational estimators that approximate the MMSE solution well. The caveat is that the best performing potential is not necessarily the prior log-likelihood function associated with the probability model, which calls for further investigation.

10.5   Bibliographical notes

Image reconstruction is a classical topic in image processing and biomedical imaging [Nat84, Pra91]. In the variational formulation, the reconstruction task is cast as an optimization problem. The idea of including a special penalty term (*regularization*) to stabilize ill-posed problems can be traced back to Tikhonov [Tik63, TA77]. Classical Tikhonov regularization involves quadratic terms only; it results in a linear reconstruction (regularized least-squares solution) and is equivalent to *ridge regression* in statistics [HK70]. An informative discussion of such techniques in the context of image restoration and their connection with statistical modeling is given in [KV90]. The Bayesian interpretation of this type of algorithm suggests a natural refinement which is to combine the quadratic regularization with a (non-quadratic) log-likelihood data term that incorporates the knowledge of the statistics of the noise [Hun77, TH79]. Still, it remains the case that quadratic regularization has a tendency to weaken signal discontinuities, which has led researchers to look for alternatives. A significant step forward was the introduction of potential functions that have a better ability to preserve edges [GR92, CBFAB97]. Total variation is a popular instance of such a non-quadratic regularizer that has the important property of being convex [ROF92]. The connection between (Gibbs) energy minimization and MAP estimation is well understood and has been exploited to make the link with discrete Markov random fields [GG84, BS93].

Sections 10.1 and 10.2
While researchers have spent a significant effort on the discrete stochastic modeling of images for defining suitable non-Gaussian priors [GG84, BS93], there is little work on the use of continuous-domain stochastic models, except in the Gaussian case where the MMSE solution is linear and intimately linked to smoothing splines [KW70] and hybrid Wiener filtering [UB05b, RVDVBU08]. The primary reference for the model-based discretization procedure of Sections 10.1–10.2 is [BKNU13]. The generic algorithm presented in Section 10.2.4 is inspired by the work of Ramani and Fessler, who proposed an ADMM algorithm for the reconstruction of parallel MRI [RF11]. Other relevant works on iterative reconstruction algorithms are [WYYZ08, BT09a, BT09b, ABDF11]. See also [CP11] for the general definition and application of proximal operators.

Section 10.3
Self-similar probability models are commonly used as prior knowledge for image processing [PPLV02]. The property of scale-invariance is supported by empirical observations of the power spectrum of natural images [Fie87, RB94, OF96b, SO01]; it is also motivated by physics and biology [Man82, Rud97, MG01]. The non-Gaussian character of images is well documented too [SLSZ03]. The wavelet-domain histograms, in particular, are typically leptokurtotic (with a peak at the origin) and heavy-tailed [Fie87, Mal89, SLG02, PSWS03]. The same holds true for the pdfs of image derivatives, which are often exponential or even subexponential. Grenander and Srivastava have shown that this behavior can be induced by a simple generative model that involves the random superposition of a fixed collection of templates [GS01]. Mumford and Gidas have introduced a scale-invariant adaptation of this model that takes the form of a stochastic wavelet expansion with a random placement and scaling of the atoms [MG01]. This random wavelet model has the same phenomenological characteristics – infinite divisibility and wide-sense self-similarity – as the sparse stochastic processes being considered here. However, it does not lend itself as well to statistical inference because of the lack of an underlying innovation model.

The primary areas of application of deconvolution are astronomy and fluorescence microscopy, where the dominant source of noise is photon counting (Poisson statistics) [MKCC99, SPM02]. Deconvolution is particularly effective in wide-field fluorescence microscopy because it can be deployed in full 3-D [AS83]. The traditional deconvolution method is the Richardson–Lucy (RC) algorithm: a multiplicative gradient-based technique that maximizes a Poisson-likelihood criterion [Ric72, Luc74]. Note that the level of smoothing of RC is controlled through the number of iterations since the cost function does not include a regularization term. While this type of algorithm may seem primitive by modern standards, it is the workhorse of many applications, with some of its variants performing remarkably well [vKvVVvdV97]. A significant advantage of RC is that it results in a positive solution. In that respect, we note that the iterative algorithm of Carrington *et al.* optimizes a quadratic Tikhonov-type criterion subject to the positivity constraint [CLM$^+$95]. By contrast, the use of *unconstrained* penalized least-squares methods, as in Section 10.3.2, is slightly more academic, but typical of the evaluation of deconvolution algorithms [BT09b, BDF07]. The MAP formulation can also be refined by adopting the Poisson-likelihood term of RC and complementing it with a suitable penalty such as total variation [DBFZ$^+$06, FBD10].

Iterative reconstruction methods became relevant for MRI with the invention of parallel imaging (SENSE) by Pruessmann *et al.* [PWSB99]. The first-generation algorithms were linear, with the optimization being typically performed by the conjugate gradient method [Pru06]. Non-linear techniques based on TV-regularizer were introduced later as a natural extension [BUF07] and, more significantly, as one of the first practical demonstrations of the "compressed-sensing" (CS) methodology [LDP07].

X-ray computed tomography is a classical medical imaging modality that relies on the numerical inversion of the Radon transform [Nat84, Kal06]. Tomographic reconstruction is also of interest in biology for the imaging of small animals using microscanners [HT02] and for the determination of 3-D molecular structure from cryoelectron micrographs [Fra92, LFB05]. When the number of projections is large with a uniform angular distribution, the tomogram is reconstructed by filtered backprojection (FBP) [RL71, PSV09]. For less ideal acquisition conditions (e.g., noisy and/or missing data, non-even angular distribution of the projections), better results are obtained using iterative algorithms such as the algebraic reconstruction technique (ART) [GBH70]. ART is a basic linear reconstructor that computes the solution of a least-squares minimization problem [HL76]; it may also be interpreted as a Bayesian estimator with a simple Gaussian prior [HHLL79]. The recent trend is to consider non-quadratic regularizers imposed upon the gradient, with a marked preference for total variation [PBE01, SP08, RBFK11]. A precursor to this type of algorithm is Bouman and Sauer's MAP estimator that involves a Markov random-field model with generalized Gaussian priors [BS93]. The discretization of the tomography reconstruction problem using local basis functions (including square pixels and low-order B-splines) was pioneered by Hanson *et al.* [HW85]. Many iterative algorithms use isotropic band-limited basis functions, which facilitates the implementation of the forward model since the Radon transform of a basis function does not depend on the direction of integration [Lew90]. The spline-based discretization that is used in the present experiments is slightly more involved but has better approximation properties [ENU12].

Section 10.4

The connection between TV denoising and the MAP estimation of Lévy processes was pointed out in [UT11]. The direct solution for the MMSE denoising of Lévy processes, which is based on belief propagation, was proposed by Kamilov *et al.* [KPAU13]. For a general presentation of the methods of belief propagation and message passing, we refer to the articles of Loeliger *et al.* [KFL01, Loe04]. The paper by Amini *et al.* [AKBU13] provides the basic tools for the proper statistical formulation of signal recovery for higher-order sparse stochastic models. It also includes the type of experimental comparison presented in Section 10.4.3. The term "Lévy flight" was coined by Mandelbrot [Man82]. This stochastic model induces a chaotic behavior with random displacements interspersed with sudden jumps. It is characteristic of the path followed by birds and other animals when searching for food [VBB$^+$02].

Several authors have identified deficiencies of non-Gaussian MAP estimation techniques [Nik07, Gri11, SDFR13]. Conversely, Gribonval has shown that, in the AWGN scenario, there exists a penalized least-squares estimator with an appropriate penalty that is equivalent to the MMSE solution, and that this modified penalty may not coincide with the prior log-likelihood function associated with the underlying statistical model.

11 Wavelet-domain methods

A simple and surprisingly effective approach for removing noise in images is to expand the signal in an orthogonal wavelet basis, apply a soft-threshold to the wavelet coefficients, and reconstruct the "denoised" image by inverse wavelet transformation. The classical justification for the algorithm is that i.i.d. noise is spread out uniformly in the wavelet domain while the signal gets concentrated in a few significant coefficients (sparsity property) so that the smaller values can be primarily attributed to noise and easily suppressed.

In this chapter, we take advantage of our statistical framework to revisit such wavelet-based reconstruction methods. Our first objective is to present some alternative dictionary-based techniques for the resolution of general inverse problems based on the same stochastic models as in Chapter 10. Our second goal is to take advantage of the orthogonality of wavelets to get a deeper understanding of the effect of proximal operators while investigating the possibility of optimizing shrinkage/thresholding functions for better performance. Finally, we shall attempt to bridge the gap between operator-based regularization, as discussed in Sections 10.2–10.3, and the imposition of sparsity constraints in the wavelet domain. Fundamentally, this relates to the dichotomy between an analysis point of view of the problem (typically in the form of the minimization of an energy functional with a regularization term) vs. a synthesis point of view, where a signal is represented as a sum of elementary constituents (wavelets.)

The chapter is composed of two main parts. The first is devoted to inverse problems in general. Specifically, in Section 11.1 we apply our general discretization and modeling paradigm to the derivation of wavelet-domain MAP estimators for the resolution of linear inverse problems. One of the key differences from the innovation-based formulation of Chapter 10 is the presence of scale-dependent potential functions whose form is specified by the stochastic model. We then address practical issues in Section 11.2 with the presentation of the two primary iterative thresholding algorithms (ISTA and FISTA). These methods are illustrated with the deconvolution of fluorescence micrographs.

The second part of the chapter focuses on the denoising problem, with the aim of improving upon simple soft-thresholding and wavelet-domain MAP estimation. Section 11.3 presents a detailed investigation of shrinkage functions in relation to infinitely divisible laws, with the emphasis on pointwise estimators that are optimal in the MMSE sense. In Section 11.4, we show how the performance of wavelet denoising can be boosted even further through the use of redundant representations (tight wavelet frames). In particular, we describe the concept of consistent cycle spinning, which provides a conceptual bridge with the optimal estimation techniques of Section 10.4. We then close the circle in Section 11.4.4 by combining all ingredients – tight operator-like wavelet frames, MMSE shrinkage functions, and consistent cycle spinning – and present an iterative wavelet-based algorithm that converges empirically to the reference MMSE solution of Section 10.4.2.

11.1 Discretization of inverse problems in a wavelet basis

As an alternative to the shift-invariant formulation presented in Chapter 10, we may choose to discretize a linear inverse problem in a wavelet basis. To that end, we consider a biorthogonal wavelet system of the type investigated in Chapter 8 that is matched to the whitening operator L. The underlying signal representation is the wavelet counterpart of (10.2) in Section 10.1. It is given by

$$s_1(r) = \sum_{i=1}^{\infty} \sum_{k \in \mathbb{Z}^d \setminus \mathbf{D}\mathbb{Z}^d} v_i[k]\psi_{i,k}(r) = \sum_{k \in \mathbb{Z}^d} s[k]\beta_L(r-k), \qquad (11.1)$$

where the wavelet coefficients are obtained as

$$v_i[k] = \langle \tilde{\psi}_{i,k}, s \rangle.$$

The mathematical requirement is that the family of analysis/synthesis functions $(\tilde{\psi}_{i,k}, \psi_{i,k})$ forms a biorthonormal wavelet basis.

Observe that the central wavelet expansion in (11.1) excludes the finer-scale wavelet coefficients with $i < 1$, so that the signal approximation $s_1(r)$, which is the projection of $s(r)$ onto the reference space V_0, can also be represented as a linear combination of the integer shifts of the scaling function β_L.

The crucial ingredient for our formulation (see Section 6.5.3) is that the analysis wavelets are such that

$$\tilde{\psi}_{i,k}(r) = L^* \tilde{\phi}_i(r - \mathbf{D}^{i-1}k), \qquad (11.2)$$

where $\tilde{\phi}_i \in L_1(\mathbb{R}^d)$ is some suitable (possibly, scale-dependent) smoothing kernel and $\mathbf{D}$ is the dilation matrix that specifies the multiresolution decomposition. Recalling that $s = L^{-1}w$, this implies that

$$v_i[k] = \langle \tilde{\psi}_{i,k}, s \rangle = \langle L^* \tilde{\phi}_i(\cdot - \mathbf{D}^{i-1}k), L^{-1}w \rangle$$
$$= \langle \tilde{\phi}_i(\cdot - \mathbf{D}^{i-1}k), w \rangle,$$

so that it is possible to derive any finite-dimensional joint pdf of the wavelet coefficients $v_i[\cdot]$ by using the general white-noise analysis exposed in Chapters 8 and 9. In particular, Proposition 8.6 tells us that p_{V_i}, the pdf of the wavelet coefficients at scale i, is infinitely divisible with modified Lévy exponent $f_{\tilde{\phi}_i}(\omega) = \int_{\mathbb{R}^d} f(\omega\tilde{\phi}_i(r))\, dr$.

11.1.1   Specification of wavelet-domain MAP estimator

To obtain a practical reconstruction model, we adopt the same strategy as in Section 10.1.2: we truncate the signal over a spatial region Ω and introduce problem-specific boundary conditions that are enforced by suitable modifications of the basis functions. This yields the finite-dimensional signal model

$$s_1(r) = \sum_{i=1}^{I_{max}} \sum_{k \in \Omega_i} v_i[k]\psi_{i,k}(r) = \sum_{k \in \Omega} s[k]\beta_{L,k}(r), \qquad (11.3)$$

where Ω_i denotes the wavelet-domain index set corresponding to the region of interest (ROI) Ω. Note that the above expansion spans the same signal space as (10.8), provided that we select $\beta = \beta_L$ as the scaling function of the wavelet system $\{\psi_{i,k}\}$.

The signal in (11.3) is uniquely specified by an N-dimensional vector $\mathbf{v}$ of pooled wavelet coefficients $v_i[k], k \in \Omega_i, i = 1, \ldots, I_{max}$. The right-hand side of (11.3) also indicates that there is a linear, one-to-one correspondence between the sequence of wavelet coefficients $v_i[\cdot]$ and the discrete signal $s[\cdot]$. This mapping specifies the discrete wavelet transform which admits a fast filterbank implementation. In vector notation, this translates into

$$\mathbf{v} = \tilde{\mathbf{W}}\mathbf{s} \quad \Leftrightarrow \quad \mathbf{s} = \mathbf{W}\mathbf{v}$$

with $\mathbf{W} = \tilde{\mathbf{W}}^{-1}$, where the entries of the $(N \times N)$ wavelet matrices $\tilde{\mathbf{W}}$ and $\mathbf{W}$ are given by

$$[\tilde{\mathbf{W}}]_{(i,k),k'} = \langle \tilde{\psi}_{i,k}, \beta_{L,k'} \rangle$$
$$[\mathbf{W}]_{k',(i,k)} = \langle \tilde{\beta}_{L,k'}, \psi_{i,k} \rangle,$$

respectively. Also note that the wavelet basis is orthonormal if and only if $\tilde{\psi}_{i,k} = \psi_{i,k}$, which translates into $\tilde{\mathbf{W}} = \mathbf{W}^T$ being an orthonormal matrix; this latter property presupposes that the underlying scaling functions are orthogonal too.

With the above convention, we write the wavelet version of the measurement equation (10.9) as

$$\mathbf{y} = \mathbf{H}_{wav}\mathbf{v} + \mathbf{n},$$

with wavelet-domain system matrix $\mathbf{H}_{wav}$ whose entries are given by

$$[\mathbf{H}_{wav}]_{m,(i,k)} = \langle \eta_m, \psi_{i,k} \rangle, \qquad (11.4)$$

where η_m is the analysis function corresponding to the mth measurement. The link with (10.10) in Section 10.1.2 is $\mathbf{H}_{wav} = \mathbf{H}\mathbf{W}$ with the proper choice of analysis function $\tilde{\beta} = \tilde{\beta}_L$.

For the purpose of simplification and mathematical tractability, we now make the same kind of decoupling simplification as in Section 10.1.2, treating the wavelet components as if they were independent.[1] Using Bayes' rule, we get the corresponding

---

[1] While this approximation is legitimate within a given scale for sufficiently well-localized wavelets, it is less so between scales because the wavelet smoothing kernels $\tilde{\phi}_i$ and $\tilde{\phi}_{i'}$ typically overlap. (A more refined probabilistic model should take those inter-scale dependencies into consideration.)

expression for the posterior probability distribution as

$$p_{V|Y}(\mathbf{v}|\mathbf{y}) \propto \exp\left(-\frac{\|\mathbf{y} - \mathbf{H}_{\text{wav}}\mathbf{v}\|^2}{2\sigma^2}\right) p_V(\mathbf{v})$$

$$\approx \exp\left(-\frac{\|\mathbf{y} - \mathbf{H}_{\text{wav}}\mathbf{v}\|^2}{2\sigma^2}\right) \prod_{i=1}^{I_{\max}} \prod_{\mathbf{k}\in\Omega_i} p_{V_i}\big(v_i[\mathbf{k}]\big),$$

where p_{V_i} is the (conjugate) inverse Fourier transform of $\hat{p}_{V_i}(\omega) = e^{f_{\tilde{\phi}_i}(\omega)}$. By maximizing $p_{V|Y}$, we derive the wavelet-domain version of the MAP estimator

$$\mathbf{v}_{\text{MAP}}(\mathbf{y}) = \arg\min_{\mathbf{v}} \left\{\frac{1}{2}\|\mathbf{y} - \mathbf{H}_{\text{wav}}\mathbf{v}\|_2^2 + \sigma^2 \sum_i \sum_{\mathbf{k}\in\Omega_i} \Phi_{V_i}\big(v_i[\mathbf{k}]\big)\right\}, \qquad (11.5)$$

which is similar to (10.12), except that it now involves the series wavelet potentials

$$\Phi_{V_i}(x) = -\log p_{V_i}(x).$$

The specificity of the present MAP formulation is that the potential functions Φ_{V_i} are scale-dependent and tied to the Lévy exponent f of the continuous-domain innovation w. Since the pdfs p_{V_i} of the wavelet coefficients are infinitely divisible with Lévy exponent $f_{\tilde{\phi}_i}$, we can determine the exact form of the potentials as

$$\Phi_{V_i}(x) = -\log \int_{\mathbb{R}} \exp\big(f_{\tilde{\phi}_i}(\omega) - \mathrm{j}\omega x\big) \frac{\mathrm{d}\omega}{2\pi} \qquad (11.6)$$

with

$$f_{\tilde{\phi}_i}(\omega) = \int_{\mathbb{R}^d} f\big(\omega\tilde{\phi}_i(\mathbf{r})\big) \, \mathrm{d}\mathbf{r},$$

where $\tilde{\phi}_i$ is the wavelet smoothing kernel at resolution i in (11.2). Moreover, we can rely on the theoretical analysis of id potentials in Section 10.2.1, which remains valid in the wavelet domain, to extract the global characteristics of Φ_{V_i}. The general trend that emerges is that these characteristics are mostly insensitive to the exact shape of $\tilde{\phi}_i$, and hence to the choice of a particular wavelet basis.

11.1.2 Evolution of the potential function across scales

In a conventional wavelet analysis, the basis functions are dilated versions of a small number ($N_0 = \det(\mathbf{D}) - 1$, typically) of mother wavelets. For simplicity of notation, we consider the case of a single mother wavelet $\tilde{\psi}$ and a dyadic dilation matrix $\mathbf{D} = 2\mathbf{I}$. The analysis wavelets at scale $a = 2^i$ (or resolution i) are shifted versions of

$$\tilde{\psi}_i(\mathbf{r}) = 2^{-id/2}\tilde{\psi}(\mathbf{r}/2^i) = \mathrm{L}^*\tilde{\phi}_i(\mathbf{r}),$$

where L is scale-invariant of order γ and

$$\tilde{\phi}_i(\mathbf{r}) = 2^{i(\gamma-d/2)}\tilde{\phi}(\mathbf{r}/2^i)$$

in accordance with (9.19).

The assumption that the underlying signal $s(r)$ is (second-order) self-similar has direct repercussions on the form of the potentials and their evolution across scale. Based on the analysis in Section 9.8, we find that the wavelet-domain pdfs p_{V_i} are members of the same class. Specifically, their Lévy exponent at resolution i is given by

$$f_{\tilde{\phi}_i}(\omega) = 2^{id} f_{\tilde{\phi}}\left(2^{i(\gamma - d/2)}\omega\right). \tag{11.7}$$

It follows that the wavelet potential at resolution i can be written as

$$\Phi_{V_i}(x) = i \log b_1 + \Phi\left(\frac{x}{b_1^i}; 2^{id}\right) \tag{11.8}$$

with $b_1 = 2^{\gamma - d/2}$ and

$$\Phi(x, \tau) = -\log \overline{\mathscr{F}}^{-1}\left\{e^{\tau f_{\tilde{\phi}}(\omega)}\right\}(x).$$

The main point is that, up to a dilation by $(b_1)^i$, the wavelet potentials are part of the parametric family $\Phi(x, \tau)$, which corresponds to the natural semigroup extension of the wavelet pdf at scale $i = 0$.

Interestingly, we can also provide an iterated convolution interpretation of this result by considering the pdfs of the scale-normalized wavelet coefficients $z_i = v_i/(b_1)^i$. To see this, we express the characteristic function of z_i as

$$\widehat{p}_{Z_i}(\omega) = \widehat{p}_{V_i}(\omega/b_1^i) = \exp\left(2^{id} f_{\tilde{\phi}}(\omega)\right)$$

$$= \left(\widehat{p}_{Z_{i-1}}(\omega)\right)^{2^d}$$

$$= \left(\widehat{p}_{Z_0}(\omega)\right)^{2^{id}},$$

which indicates that p_{Z_i} is the 2^{id}-fold convolution of $p_{Z_0} = p_{V_0}$, which is itself the pdf of the wavelet coefficients at resolution 0. In 1-D, this translates into the recursive relation

$$p_{Z_i}(x) = \left(p_{Z_{i-1}} * p_{Z_{i-1}}\right)(x), \tag{11.9}$$

which we like to view as the probabilistic counterpart of the two-scale relation of the multiresolution theory of the wavelet transform. Incidentally, this iterated convolution relation also explains why p_{Z_i} spreads out and converges to a Gaussian as the scale increases. Observe that the effect is more pronounced in higher dimensions since the number of elementary convolution factors in the probabilistic two-scale relation grows exponentially with d.

11.2   Wavelet-based methods for solving linear inverse problems

Having specified the statistical reconstruction problem in a wavelet basis, we now describe numerical methods of solution. To that end, we consider the general optimization problem

$$\min_{s}\left\{\frac{1}{2}\|y - Hs\|_2^2 + \tau \Phi(W^T s)\right\}, \tag{11.10}$$

where $\boldsymbol{\Phi}(\mathbf{v}) = \sum_{n=1}^{N} \Phi_n(v_n)$ is a separable potential function and $\mathbf{W}^T = \mathbf{W}^{-1}$ an orthonormal transform matrix. The qualitative effect of the second term in (11.10) is to favor solutions that admit a sparse wavelet expansion; the strength of this "regularization" constraint is controlled by the parameter $\tau \in \mathbb{R}^+$. Clearly, the solution of (11.10) is equivalent to the MAP estimator (11.5) if we set $\tau = \sigma^2$ and $\Phi(\mathbf{v}) = \sum_i \sum_{k \in \Omega_i} \Phi_{V_i}(v_i[k])$.

While a possible approach for solving (11.10) is to apply the ADMM algorithm of Section 10.2.4 with the replacement of $\mathbf{L}$ by $\mathbf{W}^T$ and a slight adjustment for scale-dependent potentials, we shall present two alternative techniques (ISTA and FISTA) that capitalize on the orthogonality of the matrix $\mathbf{W}$. The second algorithm (FISTA) is a modification of the first one that results in faster convergence.

11.2.1 Preliminaries

To exploit the separability of the potential function $\boldsymbol{\Phi}$, we restate the reconstruction problem in terms of the wavelet coefficients $\mathbf{v} = (v_1, \ldots, v_N) = \mathbf{W}^T \mathbf{s}$ as the minimization of the cost functional

$$\mathscr{C}(\mathbf{v}) = \frac{1}{2} \|\mathbf{y} - \mathbf{H}_{\text{wav}} \mathbf{v}\|_2^2 + \tau \sum_{n=1}^{N} \Phi_n(v_n), \qquad (11.11)$$

where $\mathbf{H}_{\text{wav}} = \mathbf{H}\mathbf{W}$. In order to gain insights into the algorithmic components of ISTA, we first investigate two extreme cases for which the solution can be written down explicitly.

Least-squares estimation
For $\tau = 0$, the minimization of (11.11) reduces to a classical least-squares estimation problem, and there is no advantage in expressing the signal in terms of wavelets. The solution of the reconstruction problem is given by

$$\mathbf{s}_{\text{LS}} = (\mathbf{H}^T \mathbf{H})^{-1} \mathbf{H}^T \mathbf{y}$$

under the assumption that $\mathbf{H}^T \mathbf{H}$ is invertible. When the underlying matrix is too large to be inverted numerically, the corresponding linear system of equations is solved iteratively. The simplest iterative reconstruction method is the Landweber algorithm

$$\mathbf{s}^{k+1} = \mathbf{s}^k + \mu \, \mathbf{H}^T (\mathbf{y} - \mathbf{H}\mathbf{s}^k) \qquad (11.12)$$

with $\mu \in \mathbb{R}^+$, which progressively builds up the solution by applying a steepest-descent update. It is a first-order optimizer whose efficiency depends on the step size μ and the conditioning of $\mathbf{H}$. A classical result is that this iterative scheme will converge to the solution provided that $0 < \mu < 2/L$, where $L = \lambda_{\max}(\mathbf{H}^T \mathbf{H})$ is the spectral radius of the iteration matrix $\mathbf{A} = \mathbf{H}^T \mathbf{H}$.

Simple denoising problem

When both $\mathbf{s}$ and $\mathbf{y}$ are expressed in the wavelet basis and $\mathbf{H} = \mathbf{I}$, (11.10) reduces to a separable denoising problem. Specifically, by defining $\mathbf{z} = \mathbf{W}^T \mathbf{y}$, we get

$$\tilde{\mathbf{v}} = \arg\min_{\mathbf{v}} \left\{ \frac{1}{2} \|\mathbf{y} - \mathbf{W}\mathbf{v}\|_2^2 + \tau \mathbf{\Phi}(\mathbf{v}) \right\} \tag{11.13}$$

$$= \arg\min_{\mathbf{v}} \left\{ \frac{1}{2} \|\mathbf{z} - \mathbf{v}\|_2^2 + \tau \sum_{n=1}^N \Phi_n(v_n) \right\}, \tag{by Parseval}$$

so that

$$\tilde{\mathbf{v}} = \text{prox}_{\mathbf{\Phi}}(\mathbf{z}; \tau) = \begin{pmatrix} \text{prox}_{\Phi_1}(z_1; \tau) \\ \vdots \\ \text{prox}_{\Phi_N}(z_N; \tau) \end{pmatrix}, \tag{11.14}$$

where the definition of the underlying proximal operators (vectorial and scalar) is consistent with the formulation of Section 10.2.3. Hence, the solution $\tilde{\mathbf{v}}$ can be computed by applying a series of component-wise shrinkage/thresholding functions to the wavelet coefficients of $\mathbf{y}$. This is the model-based version of the standard denoising algorithm mentioned in the introduction. The relation between prox_{Φ_n} and the underlying probability model is investigated in more detail in Section 11.3. The bottom line is that these are scale-dependent non-linear maps (see examples in Figure 11.5) that can be precomputed and stored in a lookup table, which makes the denoising procedure very efficient.

11.2.2   Iterative shrinkage/thresholding algorithm

The idea behind the iterative shrinkage/thresholding algorithm (ISTA) is to solve (11.10) iteratively by alternatively switching between a simple Landweber update and a denoising step.

ISTA produces a sequence $\mathbf{v}^k$ that converges to the minimizer $\mathbf{v}^\star$ of (11.11) when $\mathbf{\Phi}$ is convex. At each step, it minimizes a simpler auxiliary cost $\mathscr{C}'(\mathbf{v}, \mathbf{v}^k)$ that depends on the current estimate $\mathbf{v}^k$. The design constraint is that $\mathscr{C}'(\mathbf{v}, \mathbf{v}^k) \geq \mathscr{C}(\mathbf{v}^k)$, with equality when $\mathbf{v} = \mathbf{v}^k$. This guarantees that the cost functional decreases monotonically with the iteration number k. The standard choice is

$$\mathscr{C}'(\mathbf{v}, \mathbf{v}^k) = \mathscr{C}(\mathbf{v}) + \underbrace{\frac{L}{2}\|\mathbf{v} - \mathbf{v}^k\|_2^2 - \frac{1}{2}\|\mathbf{H}_{\text{wav}}(\mathbf{v} - \mathbf{v}^k)\|_2^2}_{\geq 0}, \tag{11.15}$$

with L such that

$$L\|\mathbf{e}\|^2 \geq \|\mathbf{H}_{\text{wav}}\mathbf{e}\|^2,$$

for all $\mathbf{e} \in \mathbb{R}^N$. The critical value of L is $\lambda_{\max}(\mathbf{H}_{\text{wav}}^T \mathbf{H}_{\text{wav}})$, which is the same L as in the Landweber algorithm of Section 11.2.1 since $\mathbf{W}$ is unitary. The derivation of ISTA is based on the rewriting of (11.15) as

$$\mathscr{C}'(\mathbf{v}, \mathbf{v}^k) = \frac{L}{2}\|\mathbf{v} - \mathbf{z}^k\|_2^2 + \tau \mathbf{\Phi}(\mathbf{v}) + C_0(\mathbf{v}^k, \mathbf{y}), \tag{11.16}$$

where $C_0(\mathbf{v}^k, \mathbf{y})$ is a term that does not depend on $\mathbf{v}$ and where the auxiliary variable $\mathbf{z}^k$ is given by

$$
\begin{aligned}
\mathbf{z}^k &= \mathbf{v}^k + \tfrac{1}{L}\mathbf{H}_{\text{wav}}^T(\mathbf{y} - \mathbf{H}_{\text{wav}}\mathbf{v}^k) \\
&= \mathbf{W}^T\big(\mathbf{s}^k + \tfrac{1}{L}\mathbf{H}^T(\mathbf{y} - \mathbf{H}\mathbf{s}^k)\big).
\end{aligned}
\tag{11.17}
$$

The crucial point is that the minimization of (11.16) with respect to $\mathbf{v}$ is equivalent to the denoising problem (11.13). This implies that

$$
\arg\min_{\mathbf{v}\in\mathbb{R}^N}\mathscr{C}'(\mathbf{v}, \mathbf{v}^k) = \operatorname{prox}_{\Phi}\left(\mathbf{z}^k; \frac{\tau}{L}\right),
$$

which corresponds to a shrinkage/thresholding of the wavelet coefficients of the signal. The form of the update equation (11.17) is also highly suggestive, for it boils down to a Landweber iteration $\big($see (11.12)$\big)$ followed by a wavelet transform. The resulting ISTA is summarized in Algorithm 1.

Algorithm 1: ISTA solves $\mathbf{s}^{\star} = \arg\min_{\mathbf{s}}\left\{\frac{1}{2}\|\mathbf{y} - \mathbf{H}\mathbf{s}\|_2^2 + \tau\,\Phi(\mathbf{W}^T\mathbf{s})\right\}$

input: $\mathbf{A} = \mathbf{H}^T\mathbf{H}$, $\mathbf{a} = \mathbf{H}^T\mathbf{y}$, $\mathbf{s}^0$, τ, and L
set: $k \leftarrow 0$
repeat

    $\mathbf{s}^{k+1} \leftarrow \mathbf{s}^k + \frac{1}{L}(\mathbf{a} - \mathbf{A}\mathbf{s}^k)$ (Landweber step)

    $\mathbf{v}^{k+1} \leftarrow \operatorname{prox}_{\Phi}(\mathbf{W}^T\mathbf{s}^{k+1}; \frac{\tau}{L})$ (wavelet-domain denoising)

    $\mathbf{s}^{k+1} \leftarrow \mathbf{W}\mathbf{v}^{k+1}$ (inverse wavelet transform)

    $k \leftarrow k + 1$

until stopping criterion
return $\mathbf{s}^k$

The remarkable aspect is that this simple sequence of Landweber updates and wavelet-domain thresholding operations converges to the solution of (11.10). The only subtle point is that the strength of the thresholding (τ/L) is tied to the step size of the gradient update.

11.2.3 Fast iterative shrinkage/thresholding algorithm

While ISTA converges to a (possibly local) minimum, it may do so rather slowly since the amount of error reduction at each step is dictated by the Landweber update. The latter, which is a basic first-order technique, is known to be quite inefficient when the system matrix is poorly conditioned.

When Φ is convex, it is possible to characterize the convergence behavior of ISTA. Specifically, Beck and Teboulle [BT09b, Theorem 3.1] have shown that, for any $k > 1$,

$$
\mathscr{C}(\mathbf{v}_{\text{ISTA}}^k) - \mathscr{C}(\mathbf{v}^{\star}) \le \frac{L}{2k}\|\mathbf{v}_{\text{ISTA}}^k - \mathbf{v}^{\star}\|_2^2,
$$

which indicates that the cost function decreases linearly with the iteration number k.

In the same paper, these authors have proposed a refinement of the scheme, called the fast iterative shrinkage/thresholding algorithm (FISTA), which improves the rate of convergence by one order. This is achieved via a controlled over-relaxation that utilizes the previous iterates to produce a better guess for the next update. A possible implementation of FISTA is shown in Algorithm 2.

Algorithm 2: FISTA solves $\mathbf{s}^\star = \arg\min_{\mathbf{s}} \left\{ \frac{1}{2}\|\mathbf{y} - \mathbf{H}\mathbf{s}\|_2^2 + \tau \, \Phi(\mathbf{W}^T\mathbf{s}) \right\}$

    input: $\mathbf{A} = \mathbf{H}^T\mathbf{H}$, $\mathbf{a} = \mathbf{H}^T\mathbf{y}$, $\mathbf{s}^0$, τ and L

    set: $k \leftarrow 0$, $\mathbf{w}_0 \leftarrow \mathbf{W}\mathbf{s}^0$, $t_0 \leftarrow 0$;

    repeat

$$\mathbf{w}^{k+1} \leftarrow \text{prox}_\Phi \left(\mathbf{W}^T(\mathbf{s}^k + \frac{1}{L}(\mathbf{a} - \mathbf{A}\mathbf{s}^k)); \frac{\tau}{L} \right) \quad \text{(ISTA step)}$$

$$t_{k+1} \leftarrow \frac{1}{2}\left(1 + \sqrt{1 + 4t_k^2}\right)$$

$$\mathbf{v}^{k+1} \leftarrow \mathbf{w}^{k+1} + \frac{t_k - 1}{t_{k+1}}\left(\mathbf{w}^{k+1} - \mathbf{w}^k\right)$$

$$\mathbf{s}^{k+1} \leftarrow \mathbf{W}\mathbf{v}^{k+1}$$

$$k \leftarrow k + 1$$

    until stopping criterion

    return $\mathbf{s}^k$

The only difference from ISTA is the update of $\mathbf{v}^{k+1}$, which is an extrapolation of the two previous ISTA computations $\mathbf{w}^{k+1}$ and $\mathbf{w}^k$. The variable t_k controls the strength of the over-relaxation, which increases with k up to some asymptotic limit.

The theoretical justification for FISTA (see [BT09b, Theorem 4.4]) is that the scheme improves the convergence such that, for any $k > 1$,

$$\mathscr{C}(\mathbf{v}^k_{\text{FISTA}}) - \mathscr{C}(\mathbf{v}^\star) \leq \frac{2L}{(k+1)^2}\|\mathbf{v}^k_{\text{FISTA}} - \mathbf{v}^\star\|_2^2.$$

Practically, switching from a linear to a quadratic convergence rate can translate to a spectacular speed improvement over ISTA, with the advantage that this change of regime essentially comes for free. FISTA therefore constitutes the method of choice for wavelet-based regularization; it typically delivers state-of-the-art performance for the kind of large-scale optimization problems encountered in imaging.

11.2.4   Discussion of wavelet-based image reconstruction

Iterative shrinkage/thresholding algorithms can be applied to the reconstruction of images for a whole variety of biomedical imaging modalities in the same way as we saw in Section 10.3. For illustration purposes, we have applied ISTA and FISTA to the deconvolution of the fluorescence micrographs of Section 10.3.2. In order to mimic the regularizing effect of the gradient operator, we have selected 2-D Haar wavelets, which act qualitatively as (smoothed) first-order derivatives. We have also

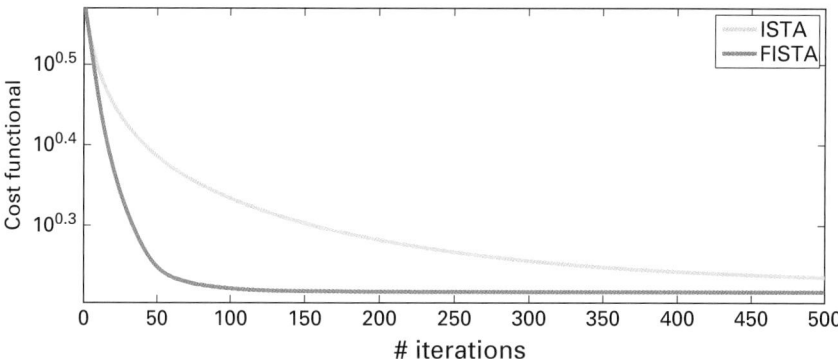

Figure 11.1 Comparison of the convergence properties of ISTA (light) and FISTA (dark) for the image in 11.2(b) as a function of the iteration index.

used the same type of potential functions: $\Phi_{\text{Gauss}}(x) = A_i|x|^2$, $\Phi_{\text{Laplace}}(x) = B_i|x|$, and $\Phi_{\text{Student}}(x) = C_i \log(x^2 + \epsilon)$, where A_i, B_i, and C_i are some proper scale-dependent constants. As in the previous experiments, the overall regularization strength τ was tuned for best performance (maximum SNR with respect to the reference). Here, we are presenting the results for the image of nerve cells (see Figure 10.3b) with the use of ℓ_1 wavelet-domain regularization.

The plot in Figure 11.1 documents the evolution of the cost functional (11.11) as a function of the iteration index for both ISTA and FISTA. It illustrates the faster convergence rate of FISTA, in agreement with Beck and Teboulle's prediction. As far as quality is concerned, a general observation is that the output of the basic version of the wavelet-based reconstruction algorithm is not on a par with the results of Section 10.3. The main problem (see Figure 11.2f) is that the reconstructed images suffer from artifacts (in the form of wavelet footprints) that are typically the consequence of the lack of shift-invariance of the wavelet representation. Fortunately, there is a simple remedy to correct for this effect via a mechanism called *cycle spinning*. [2] The approach is to randomly shift the signal back and forth during the course of iterations, which is equivalent to cycling through a family of shifted wavelet transforms, as will be described in Section 11.4.2. Incorporating cycle spinning in ISTA does not increase the computational cost but improves the SNR of the reconstruction significantly, as shown in Figure 11.2e. Hence, we end up with a result that is comparable in quality to the output of the MAP reconstruction algorithm of Section 10.2 (see Figure 11.2c). This trend persists with other images and across imaging modalities. Combining cycle spinning with FISTA is feasible as well, with the advantage that the convergence rate of the latter is typically superior to that of the ADMM technique.

While averaging across shifts appears to be essential for making wavelets competitive, we are left with the conceptual problem that the cycle-spun version of ISTA does not rigorously fit our statistical formulation. It converges to a solution that is not a strict

---

[2] Cycle spinning is used almost systematically for the wavelet-based reconstructions showcased in the literature. However, the method is rarely accounted for in the accompanying theory.

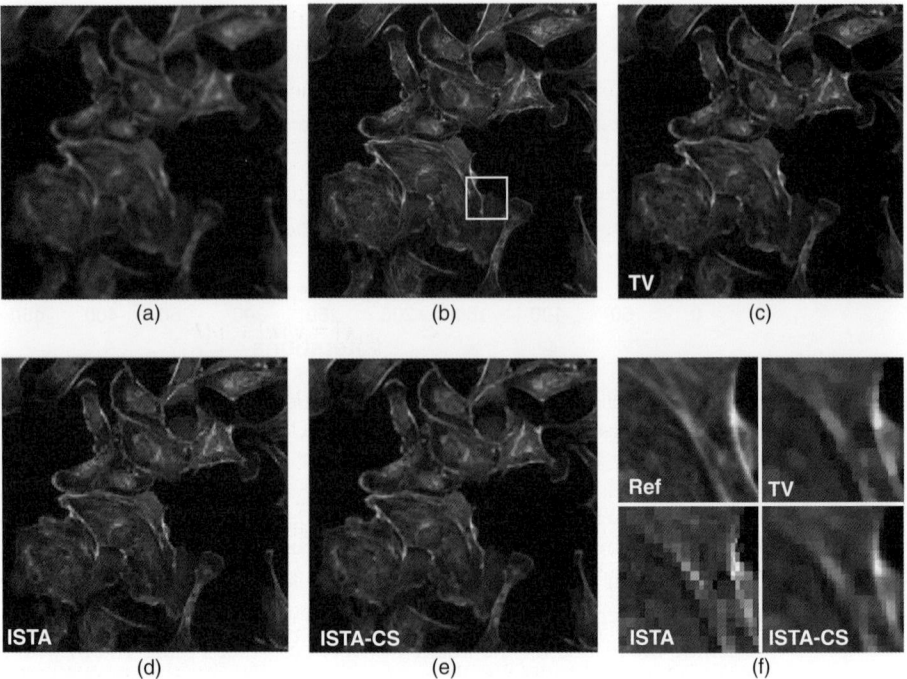

Figure 11.2 Results of deconvolution experiment: (a) Blurry and noisy input of the deconvolution algorithm (BSNR=20dB). (b) Ground truth image (nerve cells). (c) Result of MAP deconvolution with TV regularization (SNR=15.23 dB). (d) Result of wavelet-based deconvolution (SNR=12.73dB). (e) Result of wavelet-based deconvolution with cycle spinning (SNR=15.18dB). (f) Zoomed comparison of results for the region marked in (b).

minimizer of (11.10) but, rather, to some kind of average over a family of "shifted" wavelet transforms. While this description is largely empirical, there is a theoretical explanation of the phenomenon for the simpler signal-denoising problem. Specifically, in Section 11.4, we shall demonstrate that cycle spinning necessarily improves denoising performance (see Proposition 11.3) and that it can be seen as an alternative means of computing the "exact" MAP estimators of Section 10.4.3. In other words, cycle spinning somehow compensates for the inter-scale dependencies of wavelet coefficients that were neglected when writing (11.5).

The most favorable aspect of wavelet-domain processing is that it offers direct control over the reconstruction error, thanks to Parseval's relation. In particular, it allows for a more refined design of thresholding functions based on the minimum-mean-square-error (MMSE) principle. This is the reason why we shall now investigate non-iterative strategies for improving simple wavelet-domain denoising.

11.3   Study of wavelet-domain shrinkage estimators

In the remainder of the chapter, we concentrate on the problem of signal denoising with $\mathbf{H} = \mathbf{I}$ (identity) or, equivalently, $\mathbf{H}_{\text{wav}} = \mathbf{W}$, under the assumption that the transform matrix $\mathbf{W}$ is orthonormal. The latter ensures that any reduction of the quadratic error achieved in the wavelet domain is automatically transferred to the signal domain.

In this particular setting, we can address the important issue of the dependency between the wavelet-domain thresholding functions and the prior probability model. Our practical motivation is to improve the standard algorithm by identifying the solution that minimizes the mean-square estimation error. To specify the underlying scalar estimation problem, we transpose the measurement equation $\mathbf{y} = \mathbf{s} + \mathbf{n}$ into the wavelet domain as

$$\mathbf{z} = \mathbf{W}^T \mathbf{s} + \mathbf{W}^T \mathbf{n} = \mathbf{v} + \mathbf{n}'$$
$$\Leftrightarrow \quad z_i[k] = v_i[k] + n_i[k], \tag{11.18}$$

where v_i and n_i are the wavelet coefficients of the noise-free signal $\mathbf{s}$ and of the AWGN $\mathbf{n}$, respectively. Since the wavelet transform is orthonormal, the transformed noise $\mathbf{n}' = \mathbf{W}^T \mathbf{n}$ remains white, so that n_i is Gaussian i.i.d. with variance σ^2. Now, when the wavelet coefficients v_i are statistically independent as has been assumed so far, the denoising can be performed in a separable fashion by considering the wavelet coefficients individually. The estimation problem is then to recover v from the noisy coefficient $z = v + n$, where we have dropped the wavelet indices to simplify the notation. Irrespective of the statistical criterion used (MAP vs. MMSE), the estimator $\tilde{v}(z)$ will be a function of the (scalar) noisy input z, in agreement with the standard wavelet-denoising procedure.

Next, we develop the theory associated with the statistical wavelet-based estimators. The prior information is provided by the wavelet-domain pdfs p_{V_i}, which are known to be infinitely divisible (see Proposition 8.6). We then make use of those results to characterize and compare the shrinkage/thresholding functions associated with the id distributions of Table 4.1.

11.3.1 Pointwise MAP estimators for AWGN

Our baseline is the MAP solution to the denoising problem given by (11.14). For later reference, we give the scalar formulation of this estimator

$$v_{\mathrm{MAP}}(z) = \arg \min_{v \in \mathbb{R}} \left\{ \frac{1}{2} |z - v|^2 + \sigma^2 \Phi_{V_i}(v) \right\}$$
$$= \mathrm{prox}_{\Phi_{V_i}}(z; \sigma^2), \tag{11.19}$$

which involves a scale-specific proximity operator of the type investigated in Section 10.2.3. Explicit formulas and graphs of $v_{\mathrm{MAP}}(z)$ for the primary types of probability models/sparsity patterns are presented in Section 11.3.3.

11.3.2 Pointwise MMSE estimators for AWGN

From a mean-square-error point of view, performing denoising in the wavelet domain is equivalent to signal-domain processing since the ℓ_2-error is preserved. This makes the use of MMSE shrinkage functions highly relevant, even when the wavelet coefficients are only approximately independent. The MMSE estimator of v_i, given the noisy coefficient z, is provided by the posterior mean

$$v_{\text{MMSE}}(z) = \mathbb{E}\{V|Z=z\} = \int_{\mathbb{R}} v \cdot p_{V|Z}(v|z)\, dv, \tag{11.20}$$

where $p_{V|Z}(v|z) = \frac{p_{Z|V}(z|v) \cdot p_{V_i}(v)}{p_Z(z)}$ by Bayes' rule. In the present context of AWGN, we have that $p_{Z|V}(z|v) = g_\sigma(z-v)$ and $p_Z = g_\sigma * p_{V_i}$, where g_σ is a centered Gaussian distribution with standard deviation σ. Moreover, we can bypass the integration step in (11.20) by taking advantage of the Miyasawa–Stein formula for the posterior mean of a random variable corrupted by Gaussian noise [Miy61, Ste81], which states that

$$v_{\text{MMSE}}(z) = z - \sigma^2 \Phi'_Z(z), \tag{11.21}$$

where $\Phi'_Z(z) = -\frac{d}{dz}\log p_Z(z) = -\frac{p'_Z(z)}{p_Z(z)}$. This classical formula, which capitalizes on special properties of the Gaussian distribution, is established as follows:

$$\sigma^2 p'_Z(z) = \sigma^2 (g'_\sigma * p_{V_i})(z)$$

$$= \int_{\mathbb{R}} -(z-v)g_\sigma(z-v)p_{V_i}(v)\, dv$$

$$= -z \int_{\mathbb{R}} g_\sigma(z-v)p_{V_i}(v)\, dv + p_Z(z) \int_{\mathbb{R}} v \frac{g_\sigma(z-v)p_{V_i}(v)}{p_Z(z)}\, dv$$

$$= -z p_Z(z) + p_Z(z) v_{\text{MMSE}}(z).$$

This means that we can derive the explicit form of $v_{\text{MMSE}}(z)$ for any given p_{V_i} via the evaluation of the Gaussian convolution integrals

$$p_Z(z) = (g_\sigma * p_{V_i})(z) = \mathscr{F}^{-1}\left\{ e^{-\frac{\omega^2 \sigma^2}{2}} \widehat{p}_{V_i}(\omega) \right\}(z) \tag{11.22}$$

$$p'_Z(z) = (g'_\sigma * p_{V_i})(z) = \mathscr{F}^{-1}\left\{ j\omega e^{-\frac{\omega^2 \sigma^2}{2}} \widehat{p}_{V_i}(\omega) \right\}(z). \tag{11.23}$$

These can be calculated in either the time or frequency domain. The frequency-domain formulation offers more convenience for the majority of id distributions and is also directly amenable to numerical computation with the help of the FFT. Likewise, we use (11.21) to infer the general asymptotic behavior of this estimator.

THEOREM 11.1   *Let $z = v+n$, where v is infinitely divisible with symmetric pdf p_V and n is Gaussian-distributed with variance σ^2. Then, the MMSE estimator of v given z has the linear behavior around the origin given by*

$$v_{\text{MMSE}}(z) = z \left(1 - \sigma^2 \Phi''_Z(0) \right) + O(z^3), \tag{11.24}$$

where

$$\Phi''_Z(0) = \frac{\int_{\mathbb{R}} \omega^2 e^{-\frac{\omega^2 \sigma^2}{2}} \widehat{p}_{V_i}(\omega)\, d\omega}{\int_{\mathbb{R}} e^{-\frac{\omega^2 \sigma^2}{2}} \widehat{p}_{V_i}(\omega)\, d\omega} > 0. \tag{11.25}$$

If, in addition, p_{V_i} is unimodal and does not decay faster than an exponential, then

$$v_{\text{MMSE}}(z) \sim v_{\text{MAP}}(z) \sim z - \sigma^2 b_1' \ as \ z \to \infty,$$

where $b_1' = \lim_{x \to \infty} \Phi_Z'(x) = \lim_{x \to \infty} \Phi_V'(x) \geq 0.$

Proof Since the Gaussian kernel g_σ is infinitely differentiable, the same holds true for $p_Z = p_{V_i} * g_\sigma$ even if p_{V_i} is not necessarily smooth to start with (e.g., it is a compound-Poisson or Laplace distribution.) This implies that the second-order Taylor series $\Phi_Z(z) = -\log(p_Z(z)) = \Phi_Z(0) + \frac{1}{2}\Phi_Z''(0)z^2 + O(z^4)$ is well defined, which yields (11.24). The expression for $\Phi_Z''(0)$ follows from (10.20) with $\hat{p}_Z(\omega) = e^{-\omega^2\sigma^2/2}\hat{p}_{V_i}(\omega)$. We also note that the Fourier-domain moments that appear in (11.25) are positive and finite because $\hat{p}_{V_i}(\omega) = e^{\hat{f}_{\phi_i}(\omega)} \geq 0$ is tempered by the Gaussian window. Next, we recall that the Gaussian is part of the family of strongly unimodal functions which have the remarkable property of preserving the unimodality of the functions they are convolved with [Sat94, pp. 394–399]. The second part then follows from the fact that the convolution with g_σ, which decays much faster than p_{V_i}, does not modify the decay at the tail of the distribution. □

Several remarks are in order. First, the linear approximation (11.24) is exact in the Gaussian case. It actually yields the classical linear (LMMSE) estimator

$$v_{\text{LMMSE}}(z) = \frac{\sigma_i^2}{\sigma_i^2 + \sigma^2}z,$$

where σ_i^2 is the variance of the signal contribution in the ith wavelet channel. Indeed, when p_{V_i} is a Gaussian distribution, we have that $\Phi_Z(z) = \frac{z^2}{2(\sigma_i^2+\sigma^2)}$, which, upon substitution in (11.21), yields the v_{LMMSE} estimator.

Second, by applying Parseval's relation, we can express the slope of the MMSE estimator at the origin as the ratio of time-domain integrals

$$1 - \sigma^2\Phi_Z''(0) = 1 - \sigma^2\frac{\int_{\mathbb{R}}\frac{\sigma^2-x^2}{\sigma^4}e^{-\frac{x^2}{2\sigma^2}}p_{V_i}(x)\,dx}{\int_{\mathbb{R}}e^{-\frac{x^2}{2\sigma^2}}p_{V_i}(x)\,dx}$$

$$= \frac{\int_{\mathbb{R}}x^2 e^{-\frac{x^2}{2\sigma^2}}p_{V_i}(x)\,dx}{\sigma^2\int_{\mathbb{R}}e^{-\frac{x^2}{2\sigma^2}}p_{V_i}(x)\,dx}, \tag{11.26}$$

which may be simpler to evaluate for some id distributions.

11.3.3 Comparison of shrinkage functions: MAP vs. MMSE

In order to gain practical insights and to make the connection with existing methods, we now investigate solutions that are tied to specific id distributions. We consider the prior models listed in Table 4.1, which cover a broad range of sparsity behaviors. The common feature is that these pdfs are symmetric and unimodal with tails fatter than a Gaussian. Their practical relevance is that they may be used to fit the wavelet-domain

statistics of real-world signals or to derive corresponding families of parametric algorithms. Unless stated otherwise, the graphs that follow display series of comparable
estimators with a normalized signal input (SNR$_0$ = 1).

Laplace distribution

The Laplace distribution with parameter λ is defined as

$$p_{\text{Laplace}}(x; \lambda) = \frac{1}{2}\lambda e^{-\lambda|x|}.$$

Its variance is given by $\sigma_0^2 = \frac{2}{\lambda^2}$. The Lévy exponent is $f_{\text{Laplace}}(\omega; \lambda) = \log \widehat{p}_{\text{Laplace}}(\omega; \lambda)$
$= \log(\frac{\lambda^2}{\lambda^2+\omega^2})$, which is p-admissible with $p = 2$. The Laplacian potential is

$$\Phi_{\text{Laplace}}(x; \lambda) = \lambda|x| - \log(\lambda/2).$$

Since the second term of Φ_{Laplace} does not depend on x, this translates into a MAP
estimator that minimizes the ℓ_1-norm in the corresponding wavelet channel. It is well
known that the solution of this optimization problem yields the soft-threshold estimator
(see [Tib96, CDLL98, ML99])

$$\nu_{\text{MAP}}(z; \lambda) = \begin{cases} z - \lambda, & z > \lambda \\ 0, & z \in [-\lambda, \lambda] \\ z + \lambda, & z < \lambda. \end{cases}$$

By applying the time-domain versions of (11.22) and (11.23), one can also derive the
analytical form of the corresponding MMSE estimator in AWGN. For reference purposes, we give its normalized version with $\sigma^2 = 1$ as

$$\nu_{\text{MMSE}}(z; \lambda) = z - \frac{\lambda\left(\text{erf}\left(\frac{z-\lambda}{\sqrt{2}}\right) - e^{2\lambda z}\text{erfc}\left(\frac{\lambda+z}{\sqrt{2}}\right) + 1\right)}{\text{erf}\left(\frac{z-\lambda}{\sqrt{2}}\right) + e^{2\lambda z}\text{erfc}\left(\frac{\lambda+z}{\sqrt{2}}\right) + 1},$$

where erfc$(t) = 1 - \text{erf}(t)$ denotes the complementary (Gaussian) error function, which is
a result that can be traced back to [HY00, Proposition 1]. A comparison of the estimators
for the Laplace distribution with $\lambda = 2$ and unit noise variance is given in Figure 11.3b.
While the graph of the MMSE estimator has a smoother appearance than that of the
soft-thresholding function, it does also exhibit two distinct regimes that are well represented by first-order polynomials: behavior around the origin vs. behavior at $\pm\infty$. However, the transition between the two regimes is much more progressive in the MMSE
case. Asymptotically, the MAP and MMSE estimators are equivalent, as predicted by
Theorem 11.1. The key difference occurs around the origin, where the MMSE estimator
is linear (in accordance with Theorem 11.1) and quite distinct from a thresholding function. This means that the MMSE estimator will never annihilate a wavelet coefficient,
which somewhat contradicts the predominant paradigm for recovering sparse signals.

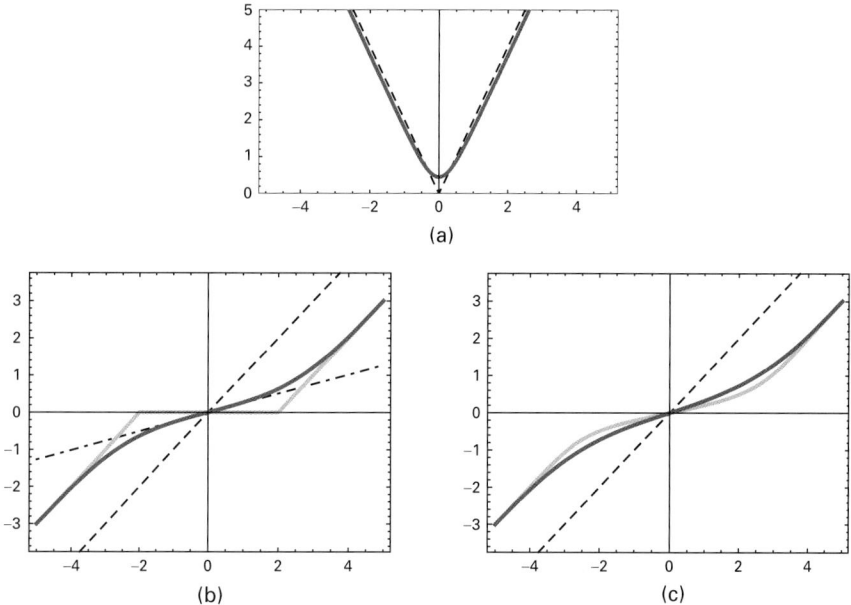

Figure 11.3 Comparison of potential functions $\Phi(z)$ and pointwise estimators $v(z)$ for signals with matched Laplace and sech distributions corrupted by AWGN with $\sigma = 1$. (a) Laplace (dashed) and sech (solid) potentials. (b) Laplace MAP estimator (light), MMSE (dark) estimator, and its first-order equivalent (dot-dashed line) for $\lambda = 2$. (c) Sech MAP (light) and MMSE (dark) estimators for $\sigma_0 = \pi/4$.

Hyperbolic secant distribution

The hyperbolic secant (reciprocal of the hyperbolic cosine) is a classical example of id distribution [Fel71]. It seems to us as interesting a candidate for regularization as the Laplace distribution. Its generic version with standard deviation σ_0 is given by

$$p_{\text{sech}}(x; \sigma_0) = \frac{\text{sech}\left(\frac{\pi x}{2\sigma_0}\right)}{2\sigma_0} = \frac{1}{\sigma_0\left(e^{-\frac{\pi x}{2\sigma_0}} + e^{\frac{\pi x}{2\sigma_0}}\right)}.$$

Remarkably, its characteristic function is part of the same class of distributions, with

$$\widehat{p}_{\text{sech}}(\omega; \sigma_0) = \text{sech}(\omega\sigma_0).$$

The hyperbolic secant potential function is

$$\Phi_{\text{sech}}(x; \sigma_0) = -\log p_{\text{sech}}(x; \sigma_0) = \log(e^{-\frac{\pi x}{2\sigma_0}} + e^{\frac{\pi x}{2\sigma_0}}) + \log \sigma_0,$$

which is convex and increasing for $x \geq 0$. Indeed, the second derivative of the potential function is

$$\Phi''_{\text{sech}}(x) = \frac{\pi^2}{4\sigma_0}\text{sech}^2\left(\frac{\pi x}{2\sigma_0}\right),$$

which is positive. Note that, for large absolute values of x, $\Phi_{\text{sech}}(x) \sim \frac{\pi}{2\sigma_0}|x| + \log \sigma_0$, suggesting that it is essentially equivalent to the ℓ_1-type Laplace potential (see Figure 11.3a). However, unlike the latter, it is infinitely differentiable everywhere, with a quadratic behavior around the origin.

The corresponding MAP and MMSE estimators, with a parameter value that is matched to the Laplace example, are shown in Figure 11.3c. An interesting observation is that the sech MAP thresholding functions are very similar to the MMSE Laplacian ones over the whole range of values. This would suggest using hyperbolic-secant-penalized least-squares regression as a practical substitute for the MMSE Laplace solution.

Symmetric Student family

We define the symmetric Student distribution with standard deviation σ_0 and algebraic decay parameter $r > 1$ as

$$p_{\text{Student}}(x; r, \sigma_0) = A_{r,\sigma_0} \left(\frac{1}{C_{r,\sigma_0} + x^2} \right)^{r+\frac{1}{2}} \tag{11.27}$$

with $C_{r,\sigma_0} = \sigma_0^2(2r - 2) > 0$ and normalizing constant $A_{r,\sigma_0} = \frac{(C_{r,\sigma_0})^r}{B(r,\frac{1}{2})}$, where $B(r, \frac{1}{2})$ is the beta function (see Appendix C.2). Despite the widespread use of this distribution in statistics, it took until the late 1970s to establish its infinite divisibility [SVH03]. The interest for signal processing is that the Student model offers fine control of the behavior of the tail, which conditions the level of sparsity of signals. The Student potential is logarithmic:

$$\Phi_{\text{Student}}(x; r, \sigma_0) = a_0 + \left(r + \frac{1}{2} \right) \log \left(C_{r,\sigma_0} + x^2 \right) \tag{11.28}$$

with $a_0 = \log A_{r,\sigma_0}$.

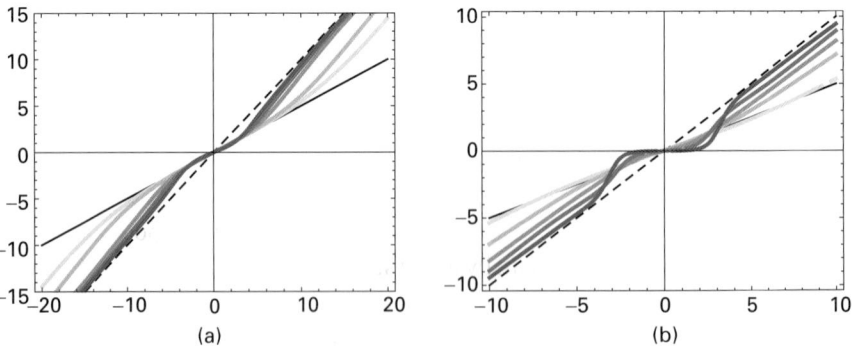

Figure 11.4 Examples of pointwise MMSE estimators $\nu_{\text{MMSE}}(z)$ for signals corrupted by AWGN with $\sigma = 1$ and the fixed SNR $= 1$. (a) Student priors with $r = 2, 4, 8, 16, 32, +\infty$ (dark to light). (b) Compound-Poisson priors with $\lambda_i = 1/8, 1/4, 1/2, 1, 2, 4, \infty$ (dark to light) and Gaussian amplitude distribution.

The Student MAP estimator is specified by a third-order polynomial equation that can be solved explicitly. This results in the thresholding functions shown in Figure 10.1b. We have also observed experimentally that the Student MAP and MMSE estimators are rather close to each other, with linear trends around the origin that become indistinguishable as r increases; this can be verified by comparing Figure 10.1b and Figure 11.4a. This finding is also consistent with the distributions becoming more Gaussian-like for larger r.

Note that Definition (11.27) remains valid in the super-sparse regimes with $r \in (0, 1]$, provided that the normalization constant $C > 0$ is no longer tied to r and σ_0. The catch is that the variance of the signal is unbounded for $r \leq 1$, which tends to flatten the shrinkage function around the origin, but maintains continuity since Φ_{Student} is infinitely differentiable.

Compound-Poisson family

We have already mentioned that the Poisson case results in pdfs that exhibit a Dirac distribution at the origin and are therefore unsuitable for MAP estimation. A compound-Poisson variable is typically generated by integration of a random sequence of Dirac impulses with some amplitude distribution p_A and a density parameter λ corresponding to the average number of impulses within the integration window. The generic form of a compound-Poisson pdf is given by (4.9). It can be written as $p_{\text{Poisson}}(x) = e^{-\lambda}\delta(x) + (1 - e^{-\lambda})p_{A,\lambda}(x)$, where the pdf $p_{A,\lambda}$ describes the distribution of the non-zero values.

The determination of the MMSE estimator from (11.21) requires the computation of $\Phi_Z'(z) = -p_Z'(z)/p_Z(z)$. The most convenient approach is to evaluate the required factors using the right-hand expressions in (11.22) and (11.23), where $\widehat{p}_{V_i}$ is specified by its Poisson parameters as in Table 4.1. This leads to

$$\widehat{p}_{V_i}(\omega) = \exp\left(\lambda_i(\widehat{p}_{A_i}(\omega) - 1)\right),$$

where $\lambda_i \in \mathbb{R}^+$ and $\widehat{p}_{A_i} : \mathbb{R} \to \mathbb{C}$ are the Poisson rate and the characteristic function of the Poisson amplitude distribution at resolution i, respectively. Moreover, due to the multiscale structure of the analysis, the wavelet-domain Poisson parameters are related to each other by

$$\lambda_i = \lambda_0 2^{id}$$
$$\widehat{p}_{A_i}(\omega) = \widehat{p}_{A_0}\left(2^{i(\gamma - d/2)}\omega\right),$$

which follows directly from (11.7). The first formula highlights the fact that the sparseness of the wavelet distributions, as measured by the proportion $e^{-\lambda_i}$ of zero coefficients, decreases substantially as the scale gets coarser. Also note that the strength of this effect increases with the number of dimensions.

Some examples of MMSE thresholding functions corresponding to a sequence of compound-Poisson signals with Gaussian amplitude distributions are shown in Figure 11.4b. Not too surprisingly, the smaller λ (dark curve), the stronger the thresholding behavior at the origin. In that experiment, we have considered a wavelet-like progression of the rate parameter λ, while keeping the signal-to-noise ratio constant to facilitate the comparison. For larger values of λ (light), the estimator converges to the LMMSE solution (thin black line), which is consistent with the fact that the distribution becomes more and more Gaussian-like.

Evolution of the estimators across wavelet scales

The increase of λ_i in Figure 11.4b is consistent with the one predicted for a wavelet-domain analysis. Nonetheless, this graph only accounts for part of the story because we enforced a constant signal-to-noise-ratio. In a realistic wavelet-domain scenario, another effect that predominates as the scale gets coarser must also be accounted for: the amplification of the quadratic signal-to-noise ratio that follows from (9.23) and that results in

$$\text{SNR}_i = \frac{\text{Var}(Z_i)}{\sigma^2} = \left(2^{2\gamma}\right)^i \text{SNR}_0,$$

where γ is the scaling order of the stochastic process. Consequently, the potential function Φ is dilated by $b_i = (2^{\gamma-d/2})^i$. The net effect is to make the estimators more identity-like as i increases, both around the origin and at infinity because of the corresponding decrease of the magnitude of $\Phi''(0)$ and $\lim_{x\to\infty} \Phi'(x)$, respectively. This

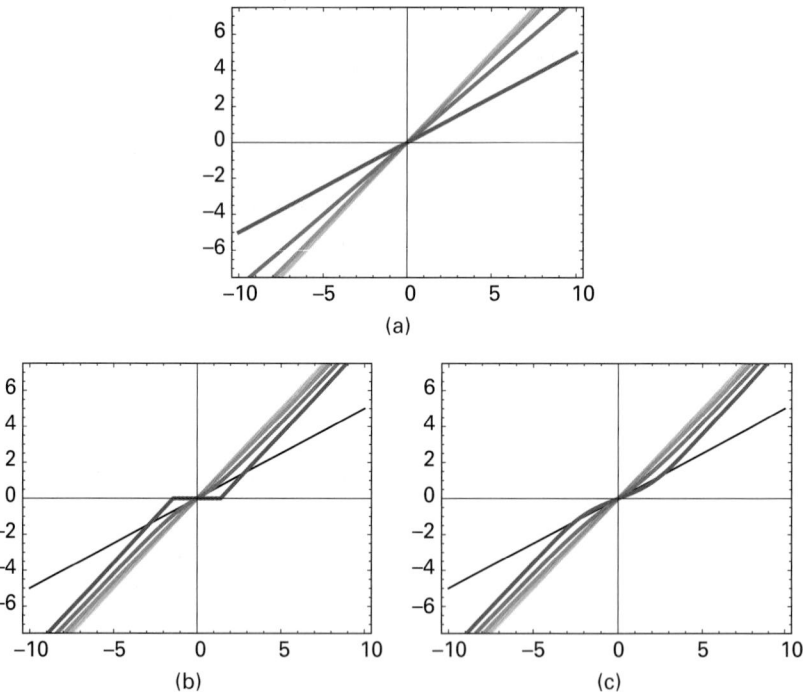

(a)

(b)   (c)

Figure 11.5 Sequence of wavelet-domain estimators $v_i(z)$ for a Laplace-type Lévy process corrupted by AWGN with $\sigma = 1$ and wavelet resolutions $i = 0$ (dark) to 4 (light). (a) LMMSE (or Brownian-motion MMSE) estimators. (b) Sym gamma MAP estimators. (c) Sym gamma MMSE estimators. The reference (fine-scale) parameters are $\sigma_0^2 = 1$ (SNR$_0 = 1$) and $r_0 = 1$ (Laplace distribution). The scale progression is dyadic.

progressive convergence to the identity map is easiest to describe for a Gaussian signal where the sequence of estimators is linear – this is illustrated in Figure 11.5a for $\gamma = 1$ and $d = 1$ (Brownian motion) so that $b_1 = 2^{1/2}$.

In the non-Gaussian case, the sequence of wavelet-domain estimators will be part of some specific family that is completely determined by p_{V_0}, the wavelet pdf at scale 0, as described in Section 11.1.2. When the variance of the signal is finite, the implication of the underlying semigroup structure and iterated convolution relations is that the MAP and MMSE estimators both converge to the Gaussian solution (LMMSE estimator) as the scale gets coarser (light curves). Thus, the global picture remains very similar to the Gaussian one, as illustrated in Figure 11.5b,c. Clearly, the most significant non-linearities can be found at the finer scale (dark curves), where the sparsity effect is prominent.

The example shown (wavelet analysis of a Lévy process in a Haar basis) was set up so that the fine-level MAP estimator is a soft-threshold. As discussed next, the wavelet-domain estimators are all part of the sym gamma family, which is the semigroup extension of the Laplace distribution. An interesting observation is that the thresholding behavior fades with a coarsening of the scale. Again, this points to the fact that the non-Gaussian effects (non-linearities) are the most significant at the finer levels of the wavelet analysis where the signal-to-noise ratio is also the least favorable.

Symmetric gamma and Meixner distributions

Several authors have proposed representing wavelet statistics using Bessel K-forms [SLG02, FB05]. The Bessel K-forms are in fact equivalent to the symmetric gamma (sym gamma) distributions specified by Table 4.1. Here, we would like to emphasize a further theoretical advantage: the semigroup structure of this family ensures compatibility across wavelet scales, provided that one properly links the distribution parameters.

The sym gamma distribution with bandwidth λ and order parameter r is best specified in terms of its characteristic function

$$\widehat{p}_{\text{gamma}}(\omega; \lambda, r) = \left(\frac{\lambda^2}{\lambda^2 + \omega^2} \right)^r.$$

The inverse Fourier transform of this expression yields a Bessel function of the second kind (a.k.a. Bessel K-form), as discussed in Appendix C. The variance of the distribution is $2r/\lambda^2$. We also note that $p_{\text{gamma}}(x; \lambda, 1)$ is equivalent to the Laplace pdf and that $p_{\text{gamma}}(x; \lambda, r)$ is the r-fold convolution of the former (semigroup property). In the case of a dyadic wavelet analysis, we invoke (11.7) to show that the evolution of the sym gamma parameters across scales is given by

$$\lambda_i = \frac{\lambda_0}{\left(2^{\gamma - d/2} \right)^i}$$
$$r_i = r_0 \left(2^d \right)^i,$$

where (λ_0, r_0) are the parameters at resolution $i = 0$. These relations underly the generation of the graphs in Figure 11.5b,c with $\gamma = 1$, $d = 1$, $\lambda_0 = \sqrt{2}$, and $r_0 = 1$.

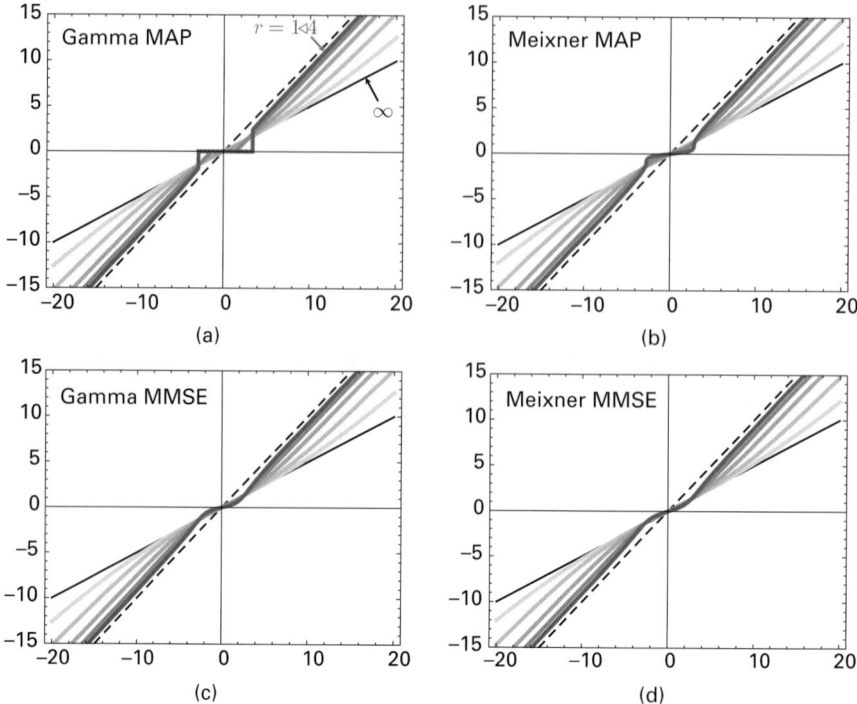

Figure 11.6 Comparison of MAP and MMSE estimators $v(z)$ for a series of sym gamma and Meixner-distributed random variables with $r = 1/4, 1, 4, 64, +\infty$ (dark to light) corrupted by white Gaussian noise of the same power as the signal (SNR=1). (a) Sym gamma MAP estimators. (b) Meixner MAP estimators. (c) Sym gamma MMSE estimators. (d) Meixner MMSE estimators.

Some further examples of sym gamma MAP and MMSE estimators over a range of orders are shown in Figure 11.6 under constant signal-to-noise ratio to highlight the differences in sparsity behavior. We observe that the MAP estimators have a hard-to-soft-threshold behavior for $r < 3/2$, which is consistent with the discontinuity of the potential at the origin. For larger values of r, the trend becomes more linear. By contrast, the MMSE estimator is much closer to the LMMSE (thin black line) around the origin. For larger signal values, both estimators result in a more or less progressive transition between the two extreme lines of the cone (identity and LMMSE) that is controlled by r – the smaller values of r correspond to the sparser scenarios with v_{MMSE} being closer to identity.

The Meixner family in Table 4.1 with order $r > 0$ and scale parameter $s_0 \in \mathbb{R}^+$ provides the same type of extension for the hyperbolic secant distribution with essentially the same functionality. Mathematically, it is closely linked to the gamma function whose relevant properties are summarized in Appendix C. As shown in Table 10.1, the Meixner potential has the same asymptotic behavior as the sym gamma potential at infinity, with the advantage of being much smoother (infinitely differentiable) at the origin. This implies that the curves of the gamma and Meixner estimators are globally quite similar. The main difference is that the Meixner MAP estimator is guaranteed to be linear around the origin, irrespective of the value of r, and in better agreement with the MMSE solution than its gamma counterpart.

Cauchy distribution

The prototypical example of a heavy-tail distribution is the symmetric Cauchy distribution with dispersion parameter s_0, which is given by

$$p_{\text{Cauchy}}(x; s_0) = \frac{s_0}{\pi \left(s_0^2 + x^2\right)}. \tag{11.29}$$

It is a special case of a $S\alpha S$ distribution (with $\alpha = 1$) as well as a symmetric Student with $r = \frac{1}{2}$.

Since the Cauchy distribution is stable, we can invoke Proposition 9.8, which ensures that the wavelet coefficients of a Cauchy process are Cauchy-distributed too. For illustration purposes, we consider the analysis of a stable Lévy process (a.k.a. Lévy flight) in an orthonormal Haar wavelet basis with $\psi = D^*\phi$, where ϕ is a triangular smoothing kernel. The corresponding wavelet-domain Cauchy parameters may be determined from (9.27) with $\gamma = 1$, $d = 1$, and $\alpha = 1$, which yields $s_i = s_0(2\sqrt{2})^i$.

While the variance of the Cauchy distribution is unbounded, an analytical characterization of the corresponding MAP estimator can be obtained by solving a cubic equation. The MMSE solution is then described by a cumbersome formula that involves exponentials and the error function erf. In particular, we can evaluate (11.24) to linearize its behavior around the origin as

$$\nu_{\text{MMSE}}(z_i; s_i) = z_i \left(1 - \sigma^2 \left(\frac{\sqrt{\frac{2}{\pi}} e^{-\frac{s_i^2}{2}} s}{\text{erfc}\left(\frac{s_i}{\sqrt{2}}\right)} - s_i^2 - 1\right)\right) + O(z_i^3). \tag{11.30}$$

The corresponding MAP and MMSE shrinkage functions with $s_0 = \frac{1}{4}$ and resolution levels $i = 0, \ldots, 4$ are shown in Figure 11.7. The difference between the two types of estimator is striking around the origin and is much more dramatic at finer scales ($i = 0$

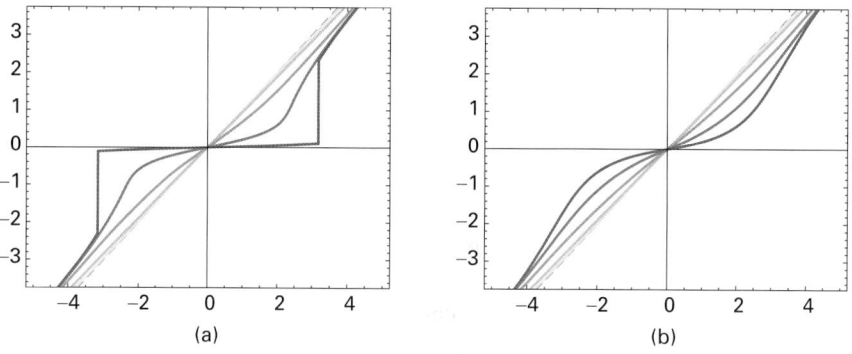

(a)   (b)

Figure 11.7 Comparison of pointwise wavelet-domain estimators $\nu(z)$ with $\{s_i\}_{i=0}^4 = \{\frac{1}{4}, \frac{1}{\sqrt{2}}, 2, 4\sqrt{2}, 16\}$ (dark to light) for a Cauchy-Lévy process corrupted by AWGN with $\sigma = 1$. (a) Cauchy MAP estimators. (b) Cauchy MMSE estimators.

(dark) and $i = 1$). As expected, all estimators converge to the identity map for large input values, due to the slow (algebraic) decay of the Cauchy distribution. We observe that the effect of processing (deviation from identity) becomes less and less significant at coarser scales (light curves). This is consistent with the relative increase of the signal contribution while the power of the noise remains constant across wavelet channels.

11.3.4   Conclusion on simple wavelet-domain shrinkage estimators

The general conclusions that can be drawn from the present statistical analysis are as follows:

- The thresholding functions should be tuned to the wavelet-domain statistics, which necessarily involve *infinitely divisible* distributions. In particular, this excludes the generalized Gaussian models with $1 < p < 2$, which have been invoked in the past to justify ℓ_p-minimization algorithms. The specification of wavelet-domain estimators can be carried out explicitly – at least, numerically – for the primary families of sparse processes characterized by their Lévy exponent $f(\omega)$ and scaling order γ.
- Pointwise MAP and MMSE estimators can differ quite substantially, especially for small input values. The MAP estimator sometimes acts as a soft-threshold, setting small wavelet coefficients to zero, while the MMSE solution is always linear around the origin where it essentially replicates the traditional Wiener solution. On the other hand, the two estimators are indistinguishable for large input values: they exhibit a *shrinkage* behavior with an offset that depends on the decay (exponential vs. algebraic) of the canonical id distribution.
- Wavelet-domain shrinkage functions must be adapted to the scale. In particular, the present statistical formulation does not support the use of a single, universal denoising function – such as the fixed soft-threshold dictated by a global ℓ_1-minimization argument – that could be applied to all coefficients in a non-discriminative manner.
- Two effects come into play as the scale gets coarser. The first is a progressive increase of the signal-to-noise ratio, which results in a shrinkage function that becomes more and more identity-like. This justifies the heuristic strategy to leave the coarser-scale coefficients untouched. The second is a Gaussianization of the wavelet-domain statistics (under the finite-variance hypothesis) due to the summation of a large number of random components (generalized version of the central-limit theorem.) Concretely, this means that the estimator ought to progressively switch to a linear regime when the scale gets coarser, which is not what is currently done in practice.
- The present analysis did not take into account the statistical dependencies of wavelet coefficients across scales. While these dependencies affect neither the performance of pointwise estimators nor the present conclusions, their existence clearly suggests that the basic application of wavelet-domain shrinkage functions is suboptimal. A possible refinement is to specify higher-order estimators (e.g., bivariate shrinkage functions.) Such designs could benefit from a tree-like structure where each wavelet coefficient is statistically linked to its parents. The other alternative is the algorithmic solution described next, which constitutes a promising mechanism for turning a suboptimal solution into an optimal one.

11.4 Improved denoising by consistent cycle spinning

A powerful strategy for improving the performance of the basic wavelet-based denoisers described in Section 11.3 is through the use of an overcomplete representation. Here, we formalize the idea of cycle spinning by expanding the signal in a wavelet frame. In essence, this is equivalent to considering a series of "shifted" orthogonal wavelet transforms in parallel. The denoising task thereby reduces to finding a consensus solution. We show that this can be done either through simple averaging or by constructing a solution that is globally consistent by way of an iterative refinement procedure.

To demonstrate the concept and the virtues of an optimized design, we concentrate on the model-based scenario of Section 10.4. The first important ingredient is the proper choice of basis functions, which is discussed in Section 11.4.1. Then, in Section 11.4.2, we switch to a redundant representation (tight wavelet frame) with a demonstration of its benefits for noise reduction. In Section 11.4.3, we introduce the idea of consistent cycle spinning, which results in an iterative variant of the basic denoising algorithm. The impact of each of these refinements, including the use of the MMSE shrinkage functions of Section 11.3, is evaluated experimentally in Section 11.4.4. The final outcome is an optimized wavelet-based algorithm that is able to replicate the MMSE results of Chapter 10.

11.4.1 First-order wavelets: design and implementation

In line with the results of Sections 8.5 and 10.4, we focus on the first-order (or Markov) processes, which lend themselves to an analytical treatment. The underlying statistical model is characterized by the first-order whitening operator $L = D - \alpha_1 \mathrm{Id}$ with $\alpha_1 \in \mathbb{R}$ and the Lévy exponent f of the innovation. We then apply the design procedure of Section 6.5 to determine the operator-like wavelet at resolution level $i = 1$, which is given by $\psi_{\alpha_1,1}(t) = L^* \varphi_{\mathrm{int}}(t-1)$ where $L^* = -D - \alpha_1 \mathrm{Id}$. Here, φ_{int} is the unique interpolant in the space of cardinal L^*L-splines which is calculated as

$$\varphi_{\mathrm{int}}(t) = \frac{1}{(\beta^\vee_{\alpha_1} * \beta_{\alpha_1})(0)} (\beta^\vee_{\alpha_1} * \beta_{\alpha_1})(t)$$

$$= \begin{cases} \frac{e^{-\alpha_1 |t|} - e^{2\alpha_1 + \alpha_1 |t|}}{1 - e^{2\alpha_1}}, & \text{for } t \in [-1,1] \text{ and } \alpha_1 \neq 0 \\ 1 - |t|, & \text{for } t \in [-1,1] \text{ and } \alpha_1 = 0 \\ 0, & \text{otherwise,} \end{cases}$$

where β_{α_1} is the first-order exponential spline defined by (6.21).

Examples of the functions $\beta_{\alpha_1} \propto \beta_{\alpha_1,0}$ (B-spline), $\varphi_{\mathrm{int}} = \phi$ (wavelet smoothing kernel), and $\psi_{\alpha_1,1}$ (operator-like wavelet) are shown in Figure 11.8. The B-spline $\beta_{\alpha_1,1}$ in Figure 11.8b is an extrapolated version of β_{α_1}; it generates the coarser-resolution space $V_1 = \mathrm{span}\{\beta_{\alpha_1,1}(\cdot - 2k)\}_{k \in \mathbb{Z}}$ which is such that $V_0 = \mathrm{span}\{\beta_{\alpha_1}(\cdot - k)\}_{k \in \mathbb{Z}} = V_1 + W_1$ with $W_1 = \mathrm{span}\{\psi_{\alpha_1,1}(\cdot - 2k)\}_{k \in \mathbb{Z}}$ and $W_1 \perp V_1$. A key property of the first-order model is that these basis functions are orthogonal and non-overlapping, as a result of the construction (B-spline of unit support).

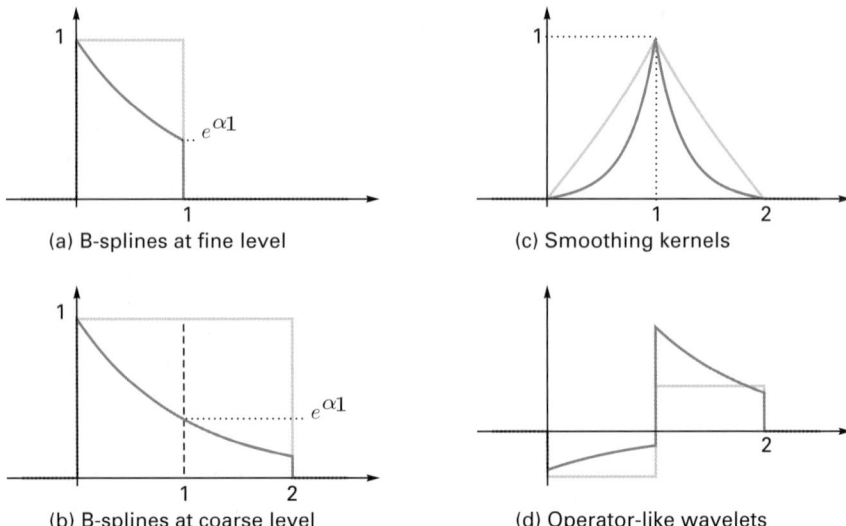

(a) B-splines at fine level

(c) Smoothing kernels

(b) B-splines at coarse level

(d) Operator-like wavelets

Figure 11.8 Operator-like wavelets and exponential B-splines for the first-order operator $L = D - \alpha_1 \text{Id}$ with $\alpha_1 = 0$ (light) and $\alpha_1 = -1$ (dark). (a) Fine-level exponential B-splines $\beta_{\alpha_1,0}(t)$. (b) Coarse-level exponential B-splines $\beta_{\alpha_1,1}(t)$. (c) Wavelet smoothing kernels $\varphi_{\text{int}}(t-1)$. (d) Operator-like wavelets $\psi_{\alpha_1,1}(t) = L^*\varphi_{\text{int}}(t-1)$.

The fundamental ingredient for the implementation of the wavelet transform is that the scaling function (exponential B-splines) and wavelets at resolution i satisfy the two-scale relation

$$
\begin{bmatrix} \beta_{\alpha_1,i}(t - 2^i k) \\ \psi_{\alpha_1,i}(t - 2^i k) \end{bmatrix} \propto \begin{bmatrix} 1 & e^{2^{(i-1)}\alpha_1} \\ -e^{2^{(i-1)}\alpha_1} & 1 \end{bmatrix} \cdot \begin{bmatrix} \beta_{\alpha_1,i-1}(t - 2^i k) \\ \beta_{\alpha_1,i-1}(t - 2^i k - 2^{i-1}) \end{bmatrix},
$$

(11.31)

which involves two filters of length 2 (row vectors of the (2×2) transition matrix) since the underlying B-splines and wavelets are non-overlapping for distinct values of k. In particular, for $i = 1$ and $k = 0$, we get that $\beta_{\alpha_1,1}(t) \propto \beta_{\alpha_1}(t) + a_1 \beta_{\alpha_1}(t - 1)$ and $\psi_{\alpha_1,1}(t) \propto -a_1 \beta_{\alpha_1}(t) + \beta_{\alpha_1}(t - 1)$ with $a_1 = e^{\alpha_1}$. These relations can be visualized in Figures 11.8b and 11.8d, respectively. Also, for $\alpha_1 = 0$, we recover the Haar system for which the underlying filters h and g (sum and difference with $e^{2^i \alpha_1} = 1 = a_1$) do not depend upon the scale i (see (6.6) and (6.7) in Section 6.1). Finally, we note that the proportionality factor in (11.31) is set by renormalizing the basis functions on both sides such that their norm is unity, which results in an *orthonormal* transform.

The corresponding fast wavelet-transform algorithm is derived by assuming that the fine-scale expansion of the input signal is $s(t) = \sum_{k \in Z} s[k]\beta_{\alpha_1,0}(t - k)$. To specify the first iteration of the algorithm, we observe that the support of $\psi_{\alpha_1,1}(\cdot - 2k)$ (resp., $\beta_{\alpha_1,1}(\cdot - 2k)$) overlaps with the fine-scale B-splines at locations $2k$ and $2k + 1$ only. Due to the orthonormality of the underlying basis functions, this results in

$$
\begin{bmatrix} s_1[k] \\ v_1[k] \end{bmatrix} = \begin{bmatrix} \langle s, \beta_{\alpha_1,1}(\cdot - 2k) \rangle \\ \langle s, \psi_{\alpha_1,1}(\cdot - 2k) \rangle \end{bmatrix}
$$

$$
= \frac{1}{\sqrt{1 + |e^{\alpha_1}|^2}} \begin{bmatrix} 1 & e^{\alpha_1} \\ -e^{\bar{\alpha}_1} & 1 \end{bmatrix} \cdot \begin{bmatrix} s[2k] \\ s[2k+1] \end{bmatrix}, \tag{11.32}
$$

which is consistent with (11.31) and $i = 1$. The reconstruction algorithm is obtained by straightforward matrix inversion:

$$
\begin{bmatrix} s[2k] \\ s[2k+1] \end{bmatrix} = \frac{1}{\sqrt{1 + |e^{\alpha_1}|^2}} \begin{bmatrix} 1 & -e^{\alpha_1} \\ e^{\bar{\alpha}_1} & 1 \end{bmatrix} \cdot \begin{bmatrix} s_1[k] \\ v_1[k] \end{bmatrix}. \tag{11.33}
$$

The final key observation is that the computation of the first level of wavelet coefficients is analogous to the determination of the discrete increment process $u[k] = s[k] - a_1 s[k-1]$ (see Section 8.5.1) in the sense that $v_1[k] \propto u[2k+1]$ is a subsampled version of the latter.

11.4.2 From wavelet bases to tight wavelet frames

From now on, we shall pool the computed wavelet and approximation coefficients of a signal $s \in \mathbb{R}^N$ in the wavelet vector $\mathbf{v}$ and formally represent the decomposition/reconstruction process $\big($Equations (11.32) and (11.33)$\big)$ by their vector-matrix counterparts $\mathbf{v} = \mathbf{W}^T \mathbf{s}$ and $\mathbf{s} = \mathbf{W}\mathbf{v}$, respectively. Moreover, since the choice of the origin of the signal is arbitrary, we shall consider a series of "m-shifted" versions of the wavelet transform matrix $\mathbf{W}_m = \mathbf{Z}^m \mathbf{W}$, where $\mathbf{Z}$ (resp., $\mathbf{Z}^m$) is the unitary matrix that circularly shifts the samples of the vector to which it is applied by one (resp., by m) to the left. With this convention, the solution to the denoising problem (11.13) in the orthogonal basis $\mathbf{W}_m$ is given by

$$
\tilde{\mathbf{v}}_m = \arg\min_{\mathbf{v}} \left\{ \frac{1}{2} \|\mathbf{v} - \mathbf{W}_m^T \mathbf{y}\|_2^2 + \tau \Phi(\mathbf{v}) \right\}
$$

$$
= \mathrm{prox}_\Phi(\mathbf{W}_m^T \mathbf{y}; \tau), \tag{11.34}
$$

which amounts to a component-wise shrinkage of the wavelet coefficients. For reference, we also give the equivalent signal-domain (or analysis) formulation of the algorithm:

$$
\tilde{\mathbf{s}}_m = \arg\min_{\mathbf{s}} \left\{ \frac{1}{2} \|\mathbf{s} - \mathbf{y}\|_2^2 + \tau \Phi(\mathbf{W}_m^T \mathbf{s}) \right\}
$$

$$
= \mathbf{W}_m \mathrm{prox}_\Phi(\mathbf{W}_m^T \mathbf{y}, \tau). \tag{11.35}
$$

Next, instead of a single orthogonal wavelet transform, we shall consider a wavelet frame expansion which is built from the concatenation of M shifted orthonormal transforms. The corresponding ($MN \times N$) transformation matrix is denoted by

$$
\mathbf{A} = \begin{bmatrix} \mathbf{W}_1^T \\ \vdots \\ \mathbf{W}_M^T \end{bmatrix} \tag{11.36}
$$

while the augmented wavelet vector is

$$
\mathbf{z} = \mathbf{As} = \begin{bmatrix} \mathbf{v}_1 \\ \vdots \\ \mathbf{v}_M \end{bmatrix},
$$

where $\mathbf{v}_m = \mathbf{W}_m^T \mathbf{s}$.

PROPOSITION 11.2   *The transformation matrix* $\mathbf{A} : \mathbb{R}^N \to \mathbb{R}^{MN}$*, which is formed from the concatenation of* M *orthonormal matrices* $\mathbf{W}_m$ *as in (11.36), defines a tight frame of* $\mathbb{R}^N$ *in the sense that*

$$
\|\mathbf{Ax}\|^2 = M\|\mathbf{x}\|^2
$$

for all $\mathbf{x} \in \mathbb{R}^N$*. Moreover, its pseudo-inverse* $\mathbf{A}^\dagger : \mathbb{R}^{MN} \to \mathbb{R}^N$ *is given by*

$$
\mathbf{A}^\dagger = \tfrac{1}{M}[\mathbf{W}_1 \ \cdots \ \mathbf{W}_M] = \tfrac{1}{M}\mathbf{A}^T,
$$

with the property that

$$
\arg\min_{\mathbf{x}\in\mathbb{R}^N} \left\{ \|\mathbf{z} - \mathbf{Ax}\|^2 \right\} = \mathbf{A}^\dagger \mathbf{z}
$$

for all $\mathbf{z} \in \mathbb{R}^{MN}$ *and* $\mathbf{A}^\dagger \mathbf{A} = \mathbf{I}$.

Proof   The frame expansion of $\mathbf{x}$ is $\mathbf{z} = \mathbf{Ax} = (\mathbf{v}_1, \dots, \mathbf{v}_M)$. The energy preservation then follows from

$$
\|\mathbf{z}\|_2^2 = \sum_{m=1}^M \|\mathbf{v}_m\|_2^2 = \sum_{m=1}^M \|\mathbf{W}_m^T\mathbf{x}\|_2^2 = M\,\|\mathbf{x}\|_2^2,
$$

where the equality on the right-hand side results from the application of Parseval's identity for each individual basis. Next, we express the quadratic error between an arbitrary vector $\mathbf{z} = (\mathbf{z}_1, \dots, \mathbf{z}_M) \in \mathbb{R}^{MN}$ and its approximation by $\mathbf{Ax}$ as

$$
\|\mathbf{z} - \mathbf{Ax}\|^2 = \sum_{m=1}^M \|\mathbf{z}_m - \mathbf{W}_m^T\mathbf{x}\|^2
$$

$$
= \sum_{m=1}^M \|\mathbf{W}_m\mathbf{z}_m - \mathbf{x}\|^2. \qquad\qquad \text{(by Parseval)}
$$

This error is minimized by setting its gradient with respect to $\mathbf{x}$ to zero; that is,

$$
\frac{\partial}{\partial\mathbf{x}} \|\mathbf{z} - \mathbf{Ax}\|^2 = -\sum_{m=1}^M (\mathbf{W}_m\mathbf{z}_m - \mathbf{x}) = \mathbf{0},
$$

which yields

$$
\mathbf{x}_{\mathrm{LS}} = \frac{1}{M}\sum_{m=1}^M \mathbf{W}_m\mathbf{z}_m = \mathbf{A}^\dagger\mathbf{z}.
$$

Finally, we check the left-inverse property,

$$\mathbf{A}^{\dagger}\mathbf{A} = \frac{1}{M}\sum_{m=1}^{M}\mathbf{W}_m\mathbf{W}_m^T = \mathbf{I},$$

which follows from the orthonormality of the matrices $\mathbf{W}_m$. □

In practice, when the wavelet expansion is performed over I resolution levels, the number of distinct shifted wavelet transforms is at most $M = 2^I$. In direct analogy with (11.13), the transposition of the wavelet-denoising problem to the context of wavelet frames is then

$$\tilde{\mathbf{z}} = \arg\min_{\mathbf{z}}\left\{\frac{1}{2}\|\underbrace{\mathbf{A}^{\dagger}\mathbf{z}}_{\mathbf{s}} - \mathbf{y}\|_2^2 + \frac{\tau}{M}\Phi(\mathbf{z})\right\} \tag{11.37}$$

$$= \arg\min_{\mathbf{z}}\left\{\frac{1}{2}\|\mathbf{z} - \mathbf{A}\mathbf{y}\|_2^2 + \tau\Phi(\mathbf{z})\right\} \qquad \text{(due to the tight-frame property)}$$

$$= \mathrm{prox}_{\Phi}(\mathbf{A}\mathbf{y};\tau) = (\tilde{\mathbf{v}}_1,\ldots,\tilde{\mathbf{v}}_M),$$

which simply amounts to performing M basic wavelet-denoising operations in parallel since the cost function is separable. We refer to (11.37) as the *synthesis-with-cycle-spinning* formulation of the wavelet-denoising problem. The corresponding signal reconstruction is given by

$$\tilde{\mathbf{s}} = \mathbf{A}^{\dagger}\tilde{\mathbf{z}} = \frac{1}{M}\sum_{m=1}^{M}\tilde{\mathbf{s}}_m, \tag{11.38}$$

which is the average of the solutions (11.34) obtained with each individual wavelet basis.

A remarkable property is that the cycle-spun version of wavelet denoising is guaranteed to improve upon the non-redundant version of the algorithm.

PROPOSITION 11.3 *Let* $\mathbf{y} = \mathbf{s} + \mathbf{n}$ *be the samples of a signal* $\mathbf{s}$ *corrupted by zero-mean i.i.d. noise* $\mathbf{n}$ *and* $\tilde{\mathbf{s}}_m$ *the corresponding signal estimates given by the wavelet-based denoising algorithm (11.35) with* $m = 1,\ldots,M$. *Then, under the assumption that the mean-square errors of the individual wavelet denoisers are equivalent, the averaged signal estimate (11.38) satisfies*

$$\mathbb{E}\{\|\tilde{\mathbf{s}} - \mathbf{s}\|^2\} \leq \mathbb{E}\{\|\tilde{\mathbf{s}}_m - \mathbf{s}\|^2\}$$

for any $m = 1,\ldots,M$.

Proof The residual noise in the orthonormal wavelet basis $\mathbf{W}_m$ is $(\tilde{\mathbf{v}}_m - \mathbb{E}\{\mathbf{v}_m\})$, where $\tilde{\mathbf{v}}_m = \mathbf{W}_m^T\tilde{\mathbf{s}}_m$ and $\mathbb{E}\{\mathbf{v}_m\} = \mathbf{W}_m^T\mathbb{E}\{\mathbf{y}\} = \mathbf{W}_m^T\mathbf{s}$, because of the assumption of zero-mean noise. This allows us to express the total noise power over the M wavelet bases as

$$\|\tilde{\mathbf{z}} - \mathbf{A}\mathbf{s}\|^2 = \sum_{m=1}^{M}\|\tilde{\mathbf{v}}_m - \mathbb{E}\{\mathbf{v}_m\}\|^2.$$

The favorable aspect of considering a redundant representation is that the inverse frame operator $\mathbf{A}^\dagger$ is an orthogonal projector onto the signal space $\mathbb{R}^N$ with the property that

$$\|\mathbf{A}^\dagger\mathbf{w}\|^2 \leq \tfrac{1}{M}\|\mathbf{w}\|^2$$

for all $\mathbf{w} \in \mathbb{R}^{MN}$. This follows from the Pythagorean relation $\|\mathbf{w}\|^2 = \|\mathbf{A}\mathbf{A}^\dagger\mathbf{w}\|^2 + \|(\mathbf{I} - \mathbf{A}\mathbf{A}^\dagger)\mathbf{w}\|^2$ (projection theorem) and the tight-frame property, which is equivalent to $\|\mathbf{A}\mathbf{A}^\dagger\mathbf{w}\|^2 = M\|\mathbf{A}^\dagger\mathbf{w}\|^2$. By applying this result to $\mathbf{w} = \tilde{\mathbf{z}} - \mathbf{A}\mathbf{s}$, we obtain

$$\|\mathbf{A}^\dagger(\tilde{\mathbf{z}} - \mathbf{A}\mathbf{s})\|^2 = \|\tilde{\mathbf{s}} - \mathbf{s}\|^2 \leq \frac{1}{M}\sum_{m=1}^{M}\|\tilde{\mathbf{v}}_m - \mathbb{E}\{\mathbf{v}_m\}\|^2. \tag{11.39}$$

Next, we take the statistical expectation of (11.39), which yields

$$\mathbb{E}\{\|\tilde{\mathbf{s}} - \mathbf{s}\|^2\} \leq \frac{1}{M}\sum_{m=1}^{M}\mathbb{E}\left\{\|\tilde{\mathbf{v}}_m - \mathbf{W}_m^T\mathbf{s}\|^2\right\}. \tag{11.40}$$

The final result then follows from Parseval's relation (norm preservation of individual wavelet transforms) and the weak stationarity hypothesis (MSE equivalence of shifted-wavelet denoisers). □

Note that this general result does not depend on the type of wavelet-domain processing – MAP vs. MMSE, or even scalar vs. vectorial – as long as the non-linear mapping $\tilde{\mathbf{v}}_m = \mathbf{f}(\mathbf{v}_m)$ is fixed and applied in a consistent fashion. The inequality in Proposition 11.3 also suggests that one can push the denoising performance further by optimizing the MSE globally in the signal domain, which is not the same as minimizing the error for each individual wavelet denoiser. The only downside of the redundant synthesis formulation (11.37) is that the underlying cost function loses its statistical interpretation (e.g., MAP criterion) because of the inherent coupling that results from considering multiple series of wavelet coefficients. The fundamental limitation there is that it is impossible to specify a proper innovation model in an overcomplete system.

11.4.3   Iterative MAP denoising

The alternative way of making use of wavelet frames is the dual *analysis* formulation of the denoising problem

$$\tilde{\mathbf{s}} = \arg\min_{\mathbf{s}}\left\{\tfrac{1}{2}\|\mathbf{s} - \mathbf{y}\|_2^2 + \tfrac{\tau}{M}\Phi(\mathbf{A}\mathbf{s})\right\}. \tag{11.41}$$

The advantage there is that the cost function is compatible with the statistical innovation-based formulation of Section 10.4.3, provided that the weights and components of the potential function are properly chosen. In light of the comment at the end of Section 11.4.1, the equivalence with MAP estimation is exact if we perform a single level of wavelet decomposition with $M = 2$ and if we do not apply any penalty to the lowpass coefficients s_1. Yet, the price to pay is that we are now facing a harder optimization problem.

The difficulty stems from the fact that we do no longer benefit from Parseval's norm equivalence between the signal and wavelet domains. A workaround is to reinstate the equivalence by insisting that the wavelet-frame expansion be consistent with the signal. This leads to the reformulation of (11.41) in synthesis form as

$$\tilde{\mathbf{z}} = \arg\min_{\mathbf{z}} \left\{ \frac{1}{2} \|\mathbf{z} - \mathbf{A}\mathbf{y}\|_2^2 + \tau \Phi(\mathbf{z}) \right\} \quad \text{s.t. } \mathbf{A}\mathbf{A}^\dagger \mathbf{z} = \mathbf{z}, \tag{11.42}$$

which is the *consistent cycle-spinning* version of denoising. Rather than attempting to solve the constrained optimization problem (11.42) directly, we shall exploit the link with conventional wavelet shrinkage. To that end, we introduce the augmented Lagrangian penalty function

$$\mathcal{L}_{\mathscr{A}}(\mathbf{z}, \mathbf{x}, \boldsymbol{\lambda}; \mu) = \frac{1}{2} \|\mathbf{z} - \mathbf{A}\mathbf{y}\|_2^2 + \tau \Phi(\mathbf{z}) + \frac{\mu}{2} \|\mathbf{z} - \mathbf{A}\mathbf{x}\|_2^2 - \boldsymbol{\lambda}^T (\mathbf{z} - \mathbf{A}\mathbf{x}) \tag{11.43}$$

with penalty parameter $\mu \in \mathbb{R}^+$ and Lagrangian multiplier vector $\boldsymbol{\lambda} \in \mathbb{R}^{MN}$. Observe that the minimization of (11.43) over $(\mathbf{z}, \mathbf{x}, \boldsymbol{\lambda})$ is equivalent to solving (11.42). Indeed, the consistency condition $\mathbf{z} = \mathbf{A}\mathbf{x}$ asserted by (11.43) is equivalent to $\mathbf{A}\mathbf{A}^\dagger \mathbf{z} = \mathbf{z}$, while the auxiliary variable $\mathbf{x} = \mathbf{A}^\dagger \mathbf{z}$ is the sought-after signal.

The standard strategy in the augmented-Lagrangian method of multipliers is to solve the problem iteratively by first minimizing $\mathcal{L}_{\mathscr{A}}(\mathbf{z}, \mathbf{x}, \boldsymbol{\lambda}; \mu)$ with respect to $(\mathbf{z}, \mathbf{x})$ while keeping μ fixed and updating $\boldsymbol{\lambda}$ according to the rule

$$\boldsymbol{\lambda}^{k+1} = \boldsymbol{\lambda}^k - \mu(\mathbf{z}^{k+1} - \mathbf{A}\mathbf{x}^{k+1}).$$

Here, the task is simplified by applying the alternating-direction method of multipliers; that is, by first minimizing $\mathcal{L}_{\mathscr{A}}(\mathbf{z}, \mathbf{x}, \boldsymbol{\lambda}; \mu)$ with respect to $\mathbf{z}$ with $\mathbf{x}$ fixed and then the other way around. The link with conventional wavelet denoising is obtained by rewriting (11.43) as

$$\mathcal{L}_{\mathscr{A}}(\mathbf{z}, \mathbf{x}, \boldsymbol{\lambda}; \mu) = \frac{1+\mu}{2} \|\mathbf{z} - \tilde{\mathbf{z}}\|_2^2 + \tau \Phi(\mathbf{z}) + C_0(\mathbf{x}, \boldsymbol{\lambda}; \mu),$$

where

$$\tilde{\mathbf{z}} = \frac{1}{1+\mu} (\mathbf{A}\mathbf{y} + \mu \mathbf{A}\mathbf{x} + \boldsymbol{\lambda})$$

and C_0 is a term that does not depend on $\mathbf{z}$. Since the underlying cost function is separable, the solution of the minimization of $\mathcal{L}_{\mathscr{A}}$ with respect to $\mathbf{z}$ is obtained by suitable shrinkage of $\tilde{\mathbf{z}}$, leading to

$$\mathbf{z}^{k+1} = \mathrm{prox}_\Phi \left(\tilde{\mathbf{z}}^{k+1}; \frac{\tau}{1+\mu} \right) \tag{11.44}$$

and involving the same kind of pointwise non-linearity as algorithm (11.34). The converse task of optimizing $\mathcal{L}_{\mathscr{A}}$ over $\mathbf{x}$ with $\mathbf{z} = \mathbf{z}^{k+1}$ fixed is a quadratic problem. The required partial derivatives are obtained as

$$\frac{\partial \mathcal{L}_{\mathscr{A}}(\mathbf{z}, \mathbf{x}, \boldsymbol{\lambda}; \mu)}{\partial \mathbf{x}} = -\mu \mathbf{A}^T (\mathbf{z} - \mathbf{A}\mathbf{x}) - \mathbf{A}^T \boldsymbol{\lambda}.$$

This leads to the closed-form solution

$$\mathbf{x}^{k+1} = \mathbf{A}^\dagger \mathbf{z}^{k+1} - \frac{1}{\mu} \mathbf{A}^\dagger \boldsymbol{\lambda}^k,$$

where we have taken advantage of the tight-frame/pseudo-inverse property $\mathbf{A}^\dagger\mathbf{A} = \mathbf{I}$ with $\mathbf{A}^\dagger = \frac{1}{M}\mathbf{A}^T$.

The complete CCS (consistent cycle spinning) denoising procedure is summarized in Algorithm 3. It is an iterative variant of wavelet shrinkage where the thresholding

---

Algorithm 3: CCS denoising solves Problem (11.41) where $\mathbf{A}$ is a tight-frame matrix

> **input:** $\mathbf{y}, \mathbf{s}^0 \in \mathbb{R}^N, \tau, \mu \in \mathbb{R}^+$
> **set:** $k = 0, \boldsymbol{\lambda}^0 = \mathbf{0}, \mathbf{u} = \mathbf{Ay};$
> **repeat**
> $\quad \mathbf{z}^{k+1} = \text{prox}_\Phi\left(\frac{1}{1+\mu}\left(\mathbf{u} + \mu\mathbf{As}^k + \boldsymbol{\lambda}^k\right); \frac{\tau}{1+\mu}\right)$
> $\quad \mathbf{s}^{k+1} = \mathbf{A}^\dagger\left(\mathbf{z}^{k+1} - \frac{1}{\mu}\boldsymbol{\lambda}^k\right)$
> $\quad \boldsymbol{\lambda}^{k+1} = \boldsymbol{\lambda}^k - \mu\left(\mathbf{z}^{k+1} - \mathbf{As}^{k+1}\right)$
> $\quad k = k + 1$
> **until** stopping criterion
> **return** $\mathbf{s} = \mathbf{s}^k$

---

function is determined by the statistics of the signal and applied in a consensus fashion. Its cost per iteration is $O(N \times M)$ operations, which is essentially that of the fast wavelet transform. This makes the method very fast. Since every step is the outcome of an exact minimization, the cost function decreases monotonically until the algorithm reaches a fixed point. The convergence to a global optimum is guaranteed when the potential function Φ is convex.

One may also observe that the CCS denoising algorithm is similar to the MAP estimation method of Section 10.2.4 since both rely on ADMM. Besides the fact that the latter can handle an arbitrary system matrix $\mathbf{H}$, the crucial difference is in the choice of the auxiliary variable $\mathbf{u} = \mathbf{Ls}$ (discrete innovation) vs. $\mathbf{z} = \mathbf{As}$ (redundant wavelet transform). While the two representations have a significant intersection, the tight wavelet frame has the advantage of resulting in a better conditioned problem (because of the norm-preservation property) and hence a faster convergence to the solution.

11.4.4   Iterative MMSE denoising

CCS denoising constitutes an attractive alternative to the more traditional iterative MAP estimators described in Chapter 10. The scheme is appealing conceptually because it bridges the gap between finite-difference and wavelet-based schemes.

The fact that the algorithm cycles through a series of orthogonal wavelet representations facilitates the Bayesian interpretation of the procedure. Fundamentally, there are two complementary mechanisms at play: the first is a wavelet-domain shrinkage whose first iteration is identical to the direct wavelet-denoising methods investigated in Section 11.3. The second is the cycling through the "shifted" transforms which results in the progressive refinement of the solution. Clearly, it is the search for a signal that is globally consistent that constitutes the major improvement upon simple

averaging. The natural idea that we shall test now is to replace the proximal operator prox_Φ by the optimal MMSE shrinkage function dictated by the theory of Section
11.3.2. This change essentially comes for free – the mere substitution in a lookup table
of $\text{prox}_\Phi(z, \tau) = \nu_{\text{MAP}}(z; \sigma)$ by $\nu_{\text{MMSE}}(z; \sigma)$ as defined by (11.20) – but it has a tre-
mendous effect on performance, to the point that the algorithm reaches the best level
achievable. While there is not yet a proof that the proposed scheme converges to the
true MMSE solution, we shall document the behavior experimentally.

In order to compare the various wavelet-based denoising techniques, we have applied
them to the same series of Lévy processes as in Section 10.4.3: Brownian motion,
Laplace motion, compound-Poisson process with a standard Gaussian distribution and
$\lambda = 0.6$, and Lévy flight with Cauchy-distributed increments. Each realization of the
signal of length $N = 100$ is corrupted by AWGN of variance σ^2. The signal is then
expanded in a Haar basis and denoised using the following algorithms:

- Soft-thresholding ($\Phi(z) \propto |z|$) in the Haar wavelet basis with optimized τ (ortho-ST)
- Model-based shrinkage in the orthogonal wavelet basis (ortho-MAP vs. ortho-MMSE)
- Model-based shrinkage in a tight frame with $M = 2$ (frame-MAP vs. frame-MMSE)
- Global MAP estimator implemented by consistent cycle spinning (CCS-MAP), as described in Section 11.4.3
- Model-based consistent cycle spinning with MMSE shrinkage function (CCS-MMSE)

To simplify the comparison, the depth of the transform is set to $I = 1$ and the lowpass
coefficients are kept untouched. The experimental conditions are exactly the same as in
Section 10.4.3, with each data point (SNR value) being the result of an average over
500 trials.

The model-based denoisers are derived from the knowledge of the pdf of the wave-
let coefficients, which is given by $p_{V_1}(x) = \sqrt{2}p_U(\sqrt{2}x)$, where p_U is the pdf of the
increments of the Lévy process. The rescaling by $\sqrt{2}$ accounts for the fact that the
(redundant) Haar wavelet coefficients are a renormalized version of the increments. For
the direct methods, we set $\tau = \sigma^2$, which corresponds to a standard wavelet-domain
MAP estimator, as described by (11.13). For the iterative CCS-MAP solution, the identification of (11.41) with the standard form (10.46) of the MAP estimator dictates the
choice $\tau = \sigma^2$ and $\Phi(z) = \Phi_U(\sqrt{2}z)$. Similarly, the setting of the shrinkage function for
the MMSE version of CCS relies on the property that the noise variance in the wavelet
domain is σ^2 even though the components are no longer independent. By exploiting
the analogy with the orthogonal scenario, one then simply replaces the proximal step in
Algorithm 3, described in Equation (11.44), by

$$\mathbf{z}^{k+1} = \nu_{\text{MMSE}}\left(\tilde{\mathbf{z}}^{k+1}; \tfrac{\sigma^2}{1+\mu}\right),$$

which amounts to the component-wise application of the theoretical MMSE estimator
that is derived from the wavelet-domain statistics $\big(\text{see (11.20) and (11.21)}\big)$.

In the case of Brownian motion for which the wavelet coefficients are Gaussian-
distributed, there is no distinction between the MAP and MMSE shrinkage functions,

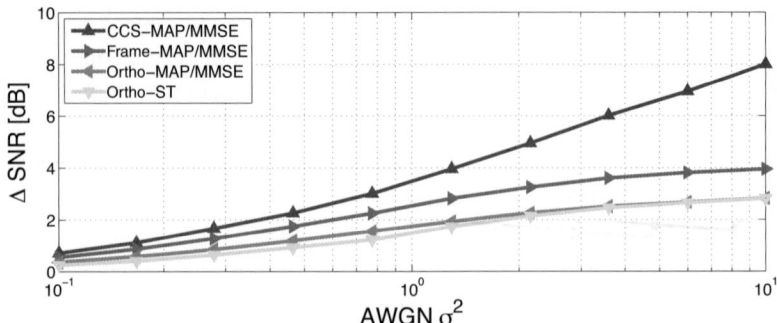

Figure 11.9 SNR improvement as a function of the level of noise for Brownian motion. The wavelet-denoising methods by reverse order of performance are: standard soft-thresholding (ortho-ST), optimal shrinkage in a wavelet basis (ortho-MAP/MMSE), shrinkage in a redundant system (frame-MAP/MMSE), and optimal shrinkage with consistent cycle spinning (CCS-MAP/MMSE).

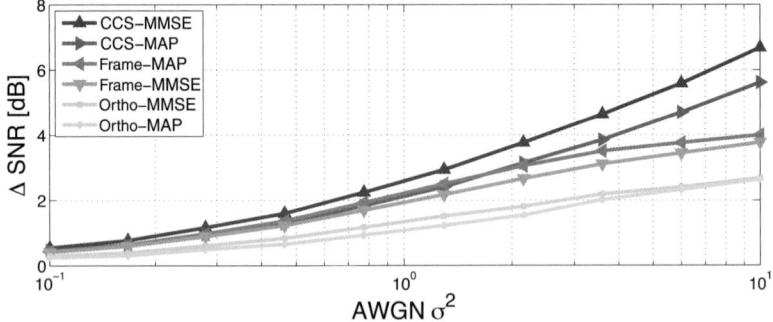

Figure 11.10 SNR improvement as a function of the level of noise for a Lévy process with Laplace-distributed increments. The wavelet-denoising methods by reverse order of performance are: ortho-MAP (equivalent to soft-thresholding with fixed τ), ortho-MMSE, frame-MMSE, frame-MAP, CCS-MAP, and CCS-MMSE. The results of CCS-MMSE are indistinguishable from the ones of the reference MMSE estimator obtained using message passing (see Figure 10.12).

which are linear. The corresponding denoising results are shown in Figure 11.9. They are consistent with our expectations: the model-based approach (ortho-MAP/MMSE) results in a (slight) improvement over basic wavelet-domain soft-thresholding while the performance gain brought by redundancy (frame-MAP/MMSE) is more substantial, in accordance with Proposition 11.3. The optimal denoising (global MMSE=MAP solution) is achieved by running the CCS version of the algorithm, which produces a linear solution that is equivalent to the Wiener filter (LMMSE estimator).

The same performance hierarchy can also be observed for the other types of signals (see Figures 11.10–11.12), confirming the relevance of the proposed series of refinements. For non-Gaussian processes, ortho-MMSE is systematically better than ortho-MAP, not to mention soft-thresholding, in agreement with the theoretical predictions of Section 11.3. Since the MAP estimator for the compound-Poisson process is useless (identical to zero), the corresponding ortho-MMSE thresholding can be compared

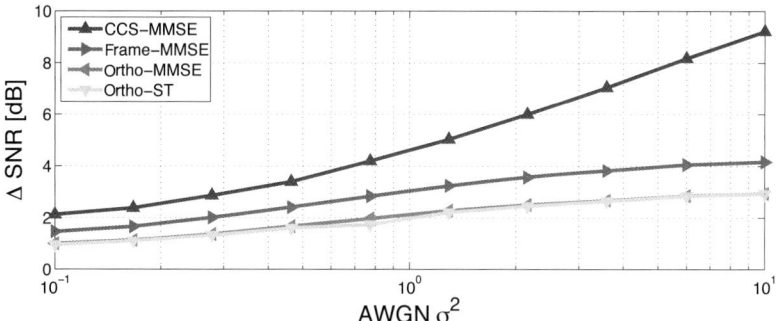

Figure 11.11 SNR improvement as a function of the level of noise for a compound-Poisson process (piecewise-constant signal). The wavelet-denoising methods by reverse order of performance are: ortho-ST, ortho-MMSE, frame-MMSE, and CCS-MMSE. The results of CCS-MMSE are indistinguishable from the ones of the reference MMSE estimator obtained using message passing (see Figure 10.10).

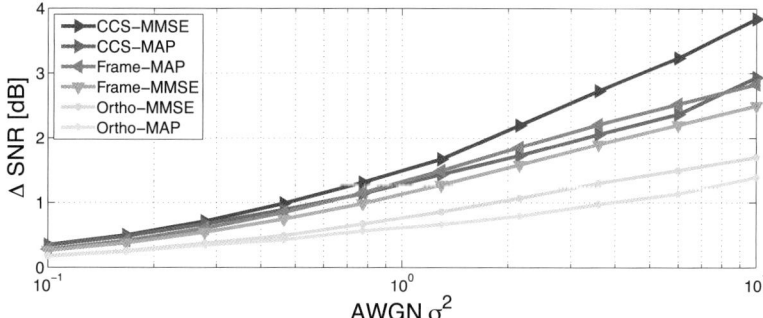

Figure 11.12 SNR improvement as a function of the level of noise for a Lévy flight with Cauchy-distributed increments. The wavelet-denoising methods by reverse order of performance are: ortho-MAP, ortho-MMSE, frame-MMSE, frame-MAP, CCS-MAP, and CCS-MMSE. The results of CCS-MMSE are indistinguishable from the ones of the reference MMSE estimator obtained using message passing (see Figure 10.11).

against the optimized soft-thresholding (ortho-ST) where τ is tuned for maximum SNR (see Figure 11.11). This is actually the scenario where this standard sparsity-promoting scheme performs best, which is not too surprising since a piecewise-constant signal is intrinsically sparse with a large proportion of its wavelet coefficients being zero. Switching to a redundant system (frame) is beneficial in all instances. The only caveat is that frame-MMSE is not necessarily the best design because the corresponding one-step modification of wavelet coefficients typically destroys Parseval's relation (lack of consistency). In fact, one can observe a degradation, with frame-MMSE being worse than frame-MAP in most cases. By contrast, the full power of the Bayesian formulation is reinstated when the denoising is performed iteratively according to the CCS strategy. Here, thanks to the consistency requirement, CCS-MMSE is always better than CCS-MAP which actually corresponds to a wavelet-based implementation of the true MAP estimator of the signal (see Section 10.4.3). In particular, CCS-MAP under the Laplace hypothesis is equivalent to the standard total-variation denoiser. Finally, the most important finding is that, in all tested scenarios, the results of CCS-MMSE are indistinguishable from those of the belief-propagation algorithm of Section 10.4.2 which implements the reference MMSE solution. This leads us to conjecture that the CCS-MMSE estimator is optimal for the class of first-order processes. The practical benefit is that CCS-MMSE is much faster than BP, which necessitates the computation of two FFTs per data point.

While we are still missing a theoretical explanation of the ability of CCS-MMSE to resolve non-Gaussian statistical interdependencies, we believe that the results presented are important conceptually, for they demonstrate the possibility of specifying iterative signal-reconstruction algorithms that minimize the reconstruction error. Moreover, it should be possible, by reverse engineering, to reformulate such schemes in terms of the minimization of a pseudo-MAP cost functional that is tied to the underlying signal model. Whether such ideas and design principles are transposable to higher-order signals and/or more general types of inverse problems is an open question that calls for further investigation.

11.5   Bibliographical notes

The earliest instance of wavelet-based denoising is a soft-thresholding algorithm that was developed for magnetic resonance imaging [WYHJC91]. The same algorithm was discovered independently by Donoho and Johnstone for the reconstruction of signals from noisy samples [DJ94]. The key contribution of these authors was to establish the statistical optimality of the procedure in a minimax sense as well as its smoothness-preservation properties [Don95, DJ95]. This series of papers triggered the interest of the statistical and signal-processing communities and got researchers working on extending the technique and applying it to a variety of image-reconstruction problems.

Sections 11.1 and 11.2
The discovery of the connection between soft-thresholding and ℓ_1-minimization [Tib96, CDLL98] was a significant breakthrough. It opened the door to a variety of novel methods for the recovery of sparse signals based on non-quadratic wavelet-domain regularization, while also providing a link with statistical estimation techniques. For instance, Figueiredo and Nowak [FN03] developed an approach to image restoration based on the maximization of a likelihood criterion that is equivalent to (11.5) with a Laplacian prior. These authors also introduced an expectation-maximization algorithm that is one of the earliest incarnations of ISTA. The algorithm was brought into prominence when Daubechies *et al.* were able to establish its convergence for general linear inverse problems [DDDM04]. This motivated researchers to improve the convergence speed of ISTA through the use of appropriate preconditioning and/or over-relaxation [BDF07, VU08, FR08, BT09b]. A favored algorithm is FISTA because of its ease of deployment and superior convergence guarantees [BT09b]. Biomedical-imaging applications of wavelet-based image reconstruction include image restoration [BDF07], 3-D deconvolution microscopy [VU09], and parallel MRI [GKHPU11]. These works

involve accelerated versions of ISTA or FISTA that capitalize on the specificities of the underlying system matrices.

Section 11.3

The signal-processing community's response to the publication of Donoho and Johnstone's work on the optimality of wavelet-domain soft-thresholding was a friendly competition to improve denoising performance. The Bayesian reformulation of the basic signal-denoising problem naturally led to the derivation of thresholding functions that are optimal in the MMSE sense [SA96, Sil99, ASS98]. Moulin and Lui presented a mathematical analysis of pointwise MAP estimators, establishing their shrinkage behavior for heavy-tailed distributions [ML99]. In this statistical view of the problem, the thresholding function is determined by the assumed prior distribution of the wavelet coefficients, the most prominent choices being the generalized Gaussian distribution [Mal89, CYV00, PP06] or a mixture of Gaussians with a peak at the origin [CKM97]. While infinite divisibility is not a property that has been emphasized in the image-processing literature, researchers have considered a number of wavelet-domain models that are compatible with the property and therefore part of the general framework investigated in Section 11.3. These include the Laplace distribution (see [HY00, Mar05] for pointwise MMSE estimator), $S\alpha S$ laws (see [ATB03] and [ABT01, BF06] for pointwise MAP and MMSE estimators, respectively), the Cauchy distribution [BAS07], as well as the sym gamma family [FB05]. The latter choice (a.k.a. Bessel K-form) is supported by a constructive model of images that is reminiscent of generalized Poisson processes [GS01]. There is also experimental evidence that this class of models is able to fit the observed transform-domain histograms well over a variety of natural images [MG01, SLG02].

The multivariate version of (11.21) can be found in [Ste81, Equation (3.3)]. This formula is also central to the derivation of Stein's unbiased risk estimator (SURE), which provides a powerful data-driven scheme for adjusting the free parameters of a statistical estimator under the AWGN hypothesis. SURE has been applied to the automatic adjustment of the thresholding parameters of wavelet-based denoising algorithms such as the SURE-shrink [DJ95] and SURELET [LBU07] approaches.

The possibility of defining bivariate shrinkage functions for exploiting inter-scale wavelet dependencies is investigated in [SS02].

Section 11.4

The concept of redundant wavelet-based denoising was introduced by Coifman under the name of cycle spinning [CD95]. The fact that this scheme always improves upon non-redundant wavelet-based denoising (see Proposition 11.3) was pointed out by Raphan and Simoncelli [RS08]. A frame is an overcomplete (and stable) generalization of a basis; see for instance [Ald95, Chr03]. For the design and implementation of the operator-like wavelets (including the first-order ones which are orthogonal), we refer to [KU06].

The concept of consistent cycle spinning was developed by Kamilov *et al.* [KBU12]. The CCS Haar-denoising algorithm was then modified appropriately to provide the MMSE estimator for Lévy processes [KKBU12].

12 Conclusion

We have presented a mathematical framework that results in the specification of the broadest possible class of linear stochastic processes. The remarkable aspect is that these continuous-domain processes are either Gaussian or sparse, as a direct consequence of the theory. While the formulation relies on advanced mathematical concepts (distribution theory, functional analysis), the underlying principles are simple and very much in line with the traditional methods of statistical signal processing. The main point is that one can achieve a whole variety of sparsity patterns by combining relatively simple building blocks: non-Gaussian white-noise excitations and suitable integral operators. This results in non-Gaussian processes whose properties are compatible with the modern sparsity-based paradigm for signal processing. Yet, the proposed class of models is also backward-compatible with the linear theory of signal processing (LMMSE = linear minimum mean square estimation) since the correlation structure of the processes remains the same as in the traditional Gaussian case – the processes are ruled by the same stochastic differential equations and it is only the driving terms (innovations) that differ in their level of sparsity. On the theoretical front, we have highlighted the crucial role of the generalized B-spline function β_L – in one-to-one relation with the whitening operator L – that provides the functional link between the continuous-domain specification of the stochastic model and the discrete-domain handling of the sample values. We have also shown that these processes admit a sparse wavelet decomposition whenever the wavelet is matched to the whitening operator.

Possible applications and directions of future research include:

- The generation of sparse stochastic processes. These may be useful for testing algorithms or for providing artificial sounds or textures.
- The development of identification procedures for estimating: (1) the whitening operator L and (2) the noise parameters (Lévy exponent). The former problem may be addressed by suitable adaptation of well-established Gaussian estimation techniques. The second task is less standard and potentially more challenging. One possibility is to estimate the noise parameters in the transformed domain (generalized increments or wavelet coefficients) using cumulants and higher-order statistical methods.
- The design of optimal denoising and restoration procedures that expand upon the first-order techniques described in Sections 10.4 and 11.4. The challenge is to develop efficient algorithms for MMSE signal estimation using the present class of sparse prior models. This should provide a principled approach for tackling sparse signal recovery problems and possibly result in higher-quality reconstructions.

- The formulation of a model-based approach for regularization and wavelet-domain processing with the derivation of optimal thresholding estimators.
- The definition and investigation of richer classes of sparse processes – especially in higher dimensions – by a mixture of elementary ones associated with individual operators L_i.
- The theory predicts that, for the proposed class of models, the transformed domain statistics should be infinitely divisible. This needs to be tested on real signals and images. One also needs to design appropriate estimators for capturing the tail behavior of the pdf, which is the most important indicator of sparsity.

Appendix A Singular integrals

In this appendix, we are concerned with integrals involving functions that are singular at a finite (or at least countable) number of isolated points. Without further loss of generality, we consider the singularities to arise at the origin.

Suppose that we are given a function f that is locally integrable in any neighborhood in $\mathbb{R}^d$ that excludes the origin, but not if the origin is included. Then, for any test function $\varphi \in \mathscr{D}(\mathbb{R}^d)$ with $0 \notin \text{support}(\varphi)$, the integral

$$\langle \varphi, f \rangle = \int_{\mathbb{R}^d} \varphi(\boldsymbol{r}) f(\boldsymbol{r}) \, \mathrm{d}\boldsymbol{r}$$

converges in the sense of Lebesgue and is continuous in φ for sequences that exclude a neighborhood of the origin. It may also converge for some other, but not all, $\varphi \in \mathscr{D}$. In general, if f grows no faster than some inverse power of $|\boldsymbol{r}|$ as $\boldsymbol{r} \to 0$, then $\langle \varphi, f \rangle$ will converge for all φ whose value as well as all derivatives up to some order k do vanish at 0. This is the situation that will be of interest to us here.

In some cases we may be able to continuously extend the bilinear form $\langle \varphi, f \rangle$ to all test functions in $\mathscr{D}(\mathbb{R}^d)$ (or even in $\mathscr{S}(\mathbb{R}^d)$). In other words, it may be possible to find a generalized function $\tilde{f}$ such that $\langle \varphi, f \rangle = \langle \varphi, \tilde{f} \rangle$ for all $\varphi \in \mathscr{D}$ for which the left-hand side converges in the sense of Lebesgue. $\tilde{f}$ is then called a *regularization* of f.

Note that regularizations of f, when they exist, are not unique, as they can differ by a generalized function that is supported at $\boldsymbol{r} = 0$. This also implies that the difference of any two regularizations of f can be written as a finite sum of the form

$$\sum_{|n| \le k} c_n \delta^{(n)}(\boldsymbol{r}),$$

where $\delta^{(n)}$ denotes the nth (partial) derivative of Dirac's δ distribution.

We proceed to present some of the standard ways to regularize singular integrals, and close the appendix with a table of standard or *canonical* regularizations of some common singular functions. The reader who is interested in a more complete account may refer to the books by Gelfand and Shilov [GS68], Hörmander [Hör05], Mikhlin and Prössdorf [MP86], and Estrada and Kanwal [EK00], among others, where all of the examples and explanations given below and more may be found.

A.1   Regularization of singular integrals by analytic continuation

The first approach to regularization we shall describe is useful for regularizing para-metric families of integrals. As an illustration, consider the family of functions $x_+^\lambda = x^\lambda \mathbb{1}_{[0,\infty)}$ in one scalar variable. For $\lambda \in U = \{\lambda \in \mathbb{C} : \mathrm{Re}(\lambda) > -1\}$, the integral

$$\langle \varphi, x_+^\lambda \rangle = \int_{\mathbb{R}} \varphi(x) x_+^\lambda \, dx$$

converges for all $\varphi \in \mathscr{D}(\mathbb{R})$. Moreover, the function

$$F_\varphi(\lambda) = \int_{\mathbb{R}} \varphi(x) x_+^\lambda \, dx$$

is (complex) analytic in λ over its (initial) domain U, as can be seen by differentiation under the integral sign. [1] Additionally, as we shall see shortly, $F_\varphi(\lambda)$ has a (necessarily unique) analytic continuation to a larger domain $\tilde{U} \supset U$ in the complex plane. Denoting the analytic continuation of F_φ by $\tilde{F}_\varphi$, we can use it to define a regularization of x_+^λ for $\lambda \in \tilde{U}\backslash U$ by the identity

$$\langle \varphi, x_+^\lambda \rangle = \tilde{F}_\varphi(\lambda) \quad \text{for } \lambda \in \tilde{U}\backslash U.$$

In the above example, one can show that the largest domain to which $F_\varphi(\lambda) = \int_{\mathbb{R}} \varphi(x) x_+^\lambda \, dx$ (integration in the sense of Lebesgue) can be extended analytically is the set

$$\tilde{U} = \mathbb{C}\backslash\{-1, -2, -3, \ldots\}.$$

Over $\tilde{U}$, the analytic continuation of $F_\varphi(\lambda)$ to $-1 \geq \mathrm{Re}(\lambda) > -2$, $\lambda \neq -1$, is found by using the formula

$$x_+^\lambda = \frac{1}{\lambda + 1} \frac{d}{dx} x_+^{\lambda+1}$$

to write

$$\langle \varphi, x_+^\lambda \rangle = \frac{1}{\lambda + 1} \langle \varphi, \frac{d}{dx} x_+^{\lambda+1} \rangle = \frac{-1}{\lambda + 1} \langle \frac{d}{dx} \varphi, x_+^{\lambda+1} \rangle,$$

where the rightmost member is well defined and where we have used duality to transfer the derivative operator to the side of the test function. Similarly, we can find the analytic continuation for $-n \geq \mathrm{Re}(\lambda) > -(n + 1)$, $\lambda \neq -n$, by successive re-application of the same approach, which leads to

$$\langle \varphi, x_+^\lambda \rangle = \frac{(-1)^n}{(\lambda + 1) \cdots (\lambda + n)} \langle \frac{d^n}{dx^n} \varphi, x_+^{\lambda+n} \rangle.$$

---

[1] We can differentiate under the integral sign with respect to λ due to the compact support of φ, whereby we obtain the integral

$$\frac{d}{d\lambda} F_\varphi(\lambda) = \int_{\mathbb{R}} \varphi(x) x_+^\lambda \log x \, dx,$$

which also converges for $\lambda \in U$.

Within the band $-n > \mathrm{Re}(\lambda) > -(n+1)$, we may also compute $\langle \varphi, x_+^\lambda \rangle$ using the formulas

$$\langle \varphi, x_+^\lambda \rangle = \int_0^\infty \left(\varphi(x) - \sum_{k \leq \lfloor -\mathrm{Re}(\lambda)-1 \rfloor} \frac{\varphi^{(k)}(0)x^k}{k!} \right) x^\lambda \, dx \tag{A.1}$$

$$= \int_1^\infty \varphi(x)x^k \, dx$$

$$+ \int_0^1 \left(\varphi(x) - \sum_{k \leq \lfloor -\mathrm{Re}(\lambda)-1 \rfloor} \frac{\varphi^{(k)}(0)x^k}{k!} \right) x^\lambda \, dx$$

$$- \sum_{k \leq \lfloor -\mathrm{Re}(\lambda)-1 \rfloor} \int_1^\infty \frac{\varphi^{(k)}(0)}{k!} x^{\lambda+k} \, dx$$

$$= \int_1^\infty \varphi(x)x^k \, dx$$

$$+ \int_0^1 \left(\varphi(x) - \sum_{k \leq \lfloor -\mathrm{Re}(\lambda)-1 \rfloor} \frac{\varphi^{(k)}(0)x^k}{k!} \right) x^\lambda \, dx$$

$$+ \sum_{k \leq \lfloor -\mathrm{Re}(\lambda)-1 \rfloor} \frac{\varphi^{(k)}(0)}{k!(\lambda+k+1)}. \tag{A.2}$$

The effect of subtracting the first n terms of the Taylor expansion of $\varphi(x)$ in (A.2) is to create a zero of sufficiently high order at 0 to make the singularity of x_+^λ integrable.

We use the above definition of x_+^λ to define $x_-^\lambda = (-x)_+^\lambda$ for $\lambda \neq -1, -2, -3, \ldots$, as well as $\mathrm{sign}(x)|x|^\lambda = x_+^\lambda - x_-^\lambda$ and $|x|^\lambda = x_+^\lambda + x_-^\lambda$. Due to the cancellation of some poles in λ, the generalized functions $\mathrm{sign}(x)|x|^\lambda$ and $|x|^\lambda$ have an extended domain of definition: they are defined for $\lambda \neq -1, -3, -5, \ldots$ and $\lambda \neq -2, -4, -6, \ldots$, respectively.

The singular function $\|r\|^\lambda$ in d dimensions is regularized by switching to (hyper)spherical coordinates and applying the definition of $x_+^{\lambda+d-1}$. Due to symmetries, the definition obtained in this way is valid for all $\lambda \in \mathbb{C}$ with the exception of the individual points $-d, -(d+2), -(d+4), -(d+6), \ldots$. As was the case in 1-D, we find formulas based on removing terms from the Taylor expansion of φ for computing $\langle \varphi, \|r\|^\lambda \rangle$ in bands of the form $-(d+2m) < \mathrm{Re}(\lambda) < -(d+2m-2)$, $m = 1, 2, 3, \ldots$, which results in

$$\langle \varphi, \|r\|^\lambda \rangle = \int_0^\infty \left(S_\varphi(r) - \sum_{n \leq \lfloor -\mathrm{Re}(\lambda)-d \rfloor} \frac{r^n}{n!} S_\varphi^{(n)}(0) \right) r^{\lambda+d-1} \, dr$$

$$= \int_{\mathbb{R}^d} \left(\varphi(r) - \sum_{|k| \leq \lfloor -\mathrm{Re}(\lambda)-d \rfloor} \frac{r^k}{k!} \varphi^{(k)}(0) \right) \|r\|^\lambda \, dr.$$

The first equality is obtained by considering (hyper)spherical coordinates; $S_\varphi(r)$ therein denotes the integral of $\varphi(\boldsymbol{r})$ over the (hyper)sphere of radius r. The second equality is obtained by rewriting the first in Cartesian coordinates. In the second formula we use standard multi-index notation $\boldsymbol{r^k} = x_1^{k_1}\cdots x_d^{k_d}$, $\boldsymbol{k}! = k_1!\cdots k_d!$, and $\varphi^{(\boldsymbol{k})} = \partial_1^{k_1}\cdots\partial_d^{k_d}\varphi$.

By normalizing $\|\boldsymbol{r}\|^\lambda$ as

$$\rho^\lambda(\boldsymbol{r}) = \frac{\|\boldsymbol{r}\|^\lambda}{2^{\frac{\lambda}{2}}\Gamma\left(\frac{\lambda+d}{2}\right)},$$

where the gamma function $\Gamma\left(\frac{\lambda+d}{2}\right)$ has poles in λ over the same set of points $-d, -(d+2), -(d+4), -(d+6), \ldots$, we obtain a generalized function that is well defined for all $\lambda \in \mathbb{C}$. Moreover, it is possible to show that, for $\lambda = -(d+2m)$, $m = 0, 1, 2, 3, \ldots$, we have

$$\rho^\lambda(\boldsymbol{r}) \equiv (-\Delta)^m \delta(\boldsymbol{r}), \tag{A.3}$$

where Δ^m is the mth iteration of the Laplacian operator. For $m = 0$, we can find the result (A.3) directly by taking the limit $\lambda \to -d$ of $\langle\varphi, \rho^\lambda\rangle$. From there, the general result is obtained by iterating the relation

$$(-\Delta)\rho^\lambda = \lambda\rho^{\lambda-2},$$

which is true for all λ.

A.2   Fourier transform of homogeneous distributions

The definition of the 1-D generalized functions of Section A.1 extends to Schwartz' space, $\mathscr{S}(\mathbb{R})$, and to $\mathscr{S}(\mathbb{R}^d)$ in multiple dimensions. We can therefore consider these families of generalized functions as members of the space $\mathscr{S}'$ of tempered distributions. In particular, this implies that the Fourier transform of any of these generalized functions will belong to (the complex version of) $\mathscr{S}'$ as well. We recall from Section 3.3.3 that, in general, the (generalized) Fourier transform $\widehat{g}$ of a tempered distribution g is defined as the tempered distribution that makes the following identity hold true for all $\varphi \in \mathscr{S}$:

$$\langle g, \widehat{\varphi}\rangle = \langle\widehat{g}, \varphi\rangle. \tag{A.4}$$

The distributions we considered above are all homogeneous, in the sense that, using g to denote any one of them, $g(a\cdot)$ for some $a > 0$ is equal to $a^\lambda g$. It then follows from the properties of the Fourier transform that $\widehat{g}$ is also homogeneous, albeit of order $-(d+\lambda)$. By invoking additional symmetry properties of these distributions, one then finds that the Fourier transform of members of each of these families belongs to the same family, with some normalization factor that can be computed; for instance, by plugging a Gaussian function φ (which is its own Fourier transform) in (A.4). We summarize the

Table A.1 Table of canonical regularizations of some singular functions, and their Fourier transforms. The one-sided power function is $r_+^\lambda = \frac{1}{2}\left(|r|^\lambda + \text{sign}(r)\,|r|^\lambda\right)$, $m = 1, 2, 3, \cdots$, and Γ denotes the gamma function. Derivatives of δ are also included for completeness.

Singular function	Canonical regularization	Fourier transform
r_+^λ, $-m-1 < \text{Re}(\lambda) < -m$	$\langle \varphi, \tilde{r}_+^\lambda \rangle$ $= \int_0^\infty r^\lambda \left(\varphi(r) - \sum_{0\le i\le m-1} \frac{r^i \varphi^{(i)}(0)}{i!} \right) dr$	$\dfrac{\Gamma(\lambda+1)}{(j\omega)^{\lambda+1}}$
r_+^n, $n = 0, 1, 2, \ldots$	N/A	$j^n \pi \delta^{(n)}(\omega) + \dfrac{n!}{(j\omega)^{n+1}}$
$|r|^\lambda$, $-2m - 2 < \text{Re}(\lambda) < -2m$	$\langle \varphi, |\tilde{r}|^\lambda \rangle$ $= \int_0^\infty r^\lambda \left(\varphi(r) + \varphi(-r) \right.$ $\left. -2 \sum_{0\le i\le m-1} \frac{r^{2i}\varphi^{(2i)}(0)}{(2i)!} \right) dr$	$-2\sin(\tfrac{\pi}{2}\lambda)\dfrac{\Gamma(\lambda+1)}{|\omega|^{\lambda+1}}$
$|r|^\lambda \text{sign}(r)$, $-2m - 1 < \text{Re}(\lambda) < -2m+1$	$\langle \varphi, |\tilde{r}|^\lambda \text{sign}(r) \rangle$ $= \int_0^\infty r^\lambda \left(\varphi(r) - \varphi(-r) \right.$ $\left. -2 \sum_{0\le i\le m-1} \frac{r^{2i+1}\varphi^{(2i+1)}(0)}{(2i+1)!} \right) dr$	$-2j\cos(\tfrac{\pi}{2}\lambda)\dfrac{\Gamma(\lambda+1)}{|\omega|^{\lambda+1}}\text{sign}(\omega)$
r^n, $n = 0, 1, 2, \ldots$	N/A	$j^n 2\pi \delta^{(n)}(\omega)$
$r^n \text{sign}(r)$, $n = 0, 1, 2, \ldots$	N/A	$2\dfrac{n!}{(j\omega)^{n+1}}$
$1/r$	$\int_0^{+\infty} \frac{\varphi(r) - \varphi(-r)}{r} dr$	$-j\pi\text{sign}(\omega)$
$\|r\|^\lambda$, $r \in \mathbb{R}^d$, $-(d + 2m) < \text{Re}(\lambda) < -(d+2m-2)$	$\int_{\mathbb{R}^d} \left(\varphi(r) - \sum_{|k|=0}^{\lfloor -\text{Re}(\lambda)-d \rfloor} \frac{r^k}{k!}\varphi^{(k)}(0) \right) \|r\|^\lambda\, dr$	$\dfrac{2^{\lambda+d}\pi^{d/2}\Gamma\left(\frac{d+\lambda}{2}\right)}{\Gamma\left(-\frac{\lambda}{2}\right)}\dfrac{1}{\|\omega\|^{\lambda+d}}$
r^n, $r \in \mathbb{R}^d$, $n \in \mathbb{N}^d$	N/A	$j^{|n|}(2\pi)^d \delta^{(n)}(\omega)$

Fourier transforms found in this way in Table A.1. The interested reader may refer to Gelfand and Shilov [GS68] for the details of their calculations.

A.3   Hadamard's finite part

A second approach to normalizing singular integrals is known as Hadamard's *finite part*, and can be considered a generalization of Cauchy's principal value. We recall that, for the singular integral $\int_{-\infty}^{\infty} \frac{\varphi(x)}{x} \, dx$, the principal value is defined as

$$\text{p.v.} \int_{-\infty}^{\infty} \frac{\varphi(x)}{x} \, dx = \lim_{\epsilon \to 0} \int_{\epsilon}^{\infty} \frac{\varphi(x)}{x} \, dx + \int_{-\infty}^{-\epsilon} \frac{\varphi(x)}{x} \, dx$$

$$= \lim_{\epsilon \to 0} \int_{\epsilon}^{\infty} \frac{\varphi(x) - \varphi(-x)}{x} \, dx$$

$$= \int_{0}^{\infty} \frac{\varphi(x) - \varphi(-x)}{x} \, dx$$

where the last integral converges in the sense of Lebesgue.

In essence, Cauchy's definition of principal value relies on the "infinite parts" of the integrals $\int_{0}^{\infty}$ and $\int_{-\infty}^{0}$ cancelling one another out. To generalize this idea, consider the integral

$$\int_{0}^{\infty} \varphi(x) f(x) \, dx,$$

where the function f is assumed to be singular at 0. Let

$$\Phi(\epsilon) = \int_{\epsilon}^{\infty} \varphi(x) f(x) \, dx$$

and suppose that, for some pre-chosen family of functions $H_k(\epsilon)$ approaching infinity at 0, we can find an $n \in \mathbb{Z}^+$ and coefficients a_k, $1 \le k \le n$, such that

$$\lim_{\epsilon \to 0+} \Phi(\epsilon) - \sum_{k=1}^{n} a_n H_k(\epsilon) = A < \infty.$$

A is then called the *finite part* (in French, *partie finie*) of the integral $\int_{0}^{\infty} \varphi(x) f(x) \, dx$ and is denoted as [EK00]

$$\text{p.f.} \int_{0}^{\infty} \varphi(x) f(x) \, dx.$$

In cases of interest to us, the family $H_k(\epsilon)$ consists of inverse integer powers of ϵ and logarithms. With this choice, the finite-part regularization of the singular integrals considered in Section A.1 can be obtained. It is found to coincide with their regularization by analytic continuation (note that all H_k are analytic in ϵ). But, in addition, we can use the finite part to regularize x_+^λ and related functions in cases where the previous method fails (namely, for $\lambda = -1, -2, -3, \ldots$). Indeed, for $\lambda = -n$, $n \in \mathbb{Z}^+$, we may write

$$\int_{\epsilon}^{\infty} \frac{\varphi(x)}{x^n} \, dx = \sum_{k=0}^{n-1} \int_{\epsilon}^{1} \frac{\varphi^{(k)}(0) x^{k-n}}{k!} \, dx$$

$$+ \int_{\epsilon}^{1} \frac{\varphi(x) - \sum_{k=0}^{n-1} \frac{\varphi^{(k)}(0) x^k}{k!}}{x^n} \, dx + \int_{1}^{\infty} \frac{\varphi(x)}{x^n} \, dx$$

$$= -\frac{\varphi^{(n-1)}(0)}{(n-1)!} \log \epsilon + \sum_{k=0}^{n-2} \frac{\varphi^{(k)}(0)}{k!} \cdot \frac{\epsilon^{-n+k+1} - 1}{n - k - 1}$$

$$+ \int_{\epsilon}^{1} \frac{\varphi(x) - \sum_{k=0}^{n-1} \frac{\varphi^{(k)}(0) x^k}{k!}}{x^n} \, dx + \int_{1}^{\infty} \frac{\varphi(x)}{x^n} \, dx.$$

From there, by discarding the logarithm and inverse powers of ϵ and taking the limit $\epsilon \to 0$ of what remains, we find

$$\text{p.f.} \int_0^\infty \frac{\varphi(x)}{x^n}\,dx = \int_1^\infty \frac{\varphi(x)}{x^n}\,dx + \int_0^1 \frac{\varphi(x) - \sum_{k=0}^{n-1}\frac{\varphi^{(k)}(0)x^k}{k!}}{x^n}\,dx - \sum_{k=0}^{n-2}\frac{\varphi^{(k)}(0)}{k!(n-k-1)},$$

where the two integrals of the right-hand side converge in the sense of Lebesgue.

Using similar calculations, for $\langle\varphi, x_+^\lambda\rangle$ with $\lambda \neq -1, -2, -3, \ldots$, we find the same regularization as the one given by (A.2).

In general, a singular function f does not define a distribution in a unique way. However, in many of the cases that are of interest to us there exists a particular regularization of f that is considered standard or *canonical*. For the parametric families discussed so far, these essentially correspond to the regularization obtained by analytic continuation. In Table A.1, we have summarized the formulas for canonical regularization as well as the Fourier transforms of the singular distributions that are of interest to us.

Finally, we point out that the approaches presented in this appendix to regularize singularities at the origin generalize in an obvious way to isolated singularities at any other point and also to a finite (or even countable) number of isolated singularities.

A.4   Some convolution integrals with singular kernels

As we noted earlier, the scaling property of the Fourier transform demands that the Fourier transform of a homogeneous distribution of order λ be homogeneous of order $-(\lambda+d)$. Thus, for $\text{Re}(\lambda) \leq -d$ where the original distribution is singular at $\mathbf{0}$, its Fourier transform is locally integrable everywhere and vice versa. Consequently, convolutions with homogeneous singular kernels are often easier to evaluate in the Fourier domain by employing the convolution-multiplication rule. Important examples of such convolutions are the Hilbert and Riesz transforms.

The Hilbert transform of a test function $\varphi \in \mathscr{S}(\mathbb{R})$ can thus be defined either by the convolution with the singular kernel $h(x) = 1/(\pi x)$ as

$$H\varphi(x) = \text{p.v.} \int_{-\infty}^{+\infty} \varphi(y)h(x-y)\,dy,$$

which involves a principal-value limit, or by the Fourier-domain formula

$$H\varphi(x) = \mathscr{F}^{-1}\{\widehat{\varphi}\widehat{h}\},$$

with $\widehat{h}(\omega) = -\mathrm{j}\,\text{sign}(\omega)$. These definitions extend beyond $\mathscr{S}(\mathbb{R})$ to $\varphi \in L_p(\mathbb{R})$ for $1 < p < \infty$ (the standard reference here is Stein and Weiss [SW71]).

Similarly, the ith component of the Riesz transform of a test function $\varphi \in \mathscr{S}(\mathbb{R}^d)$ is defined in the spatial domain by the convolution with the kernel $h_i(r) = \Gamma(d/2 + 1/2)r_i/(\pi^{d/2+1/2}|r|^{d+1})$ as

$$R_i\varphi(r) = \text{p.v.} \int_{\mathbb{R}^d} \varphi(t)h_i(r - t)\, dt,$$

which is equivalent to the Fourier integral

$$R_i\varphi(r) = \mathscr{F}^{-1}\{\widehat{\varphi}\widehat{R_i}\}$$

with $\widehat{R_i}(\omega) = -j\frac{\omega_i}{|\omega|}$. Once again, these definitions extend to $\varphi \in L_p(\mathbb{R}^d)$ for $1 < p < \infty$ (see Mikhlin's Theorem 3.6).

Appendix B Positive definiteness

Positive–definite functions play a central role in statistics, approximation theory [Mic86, Wen05], and machine learning [HSS08]. They allow for a convenient Fourier-domain specification of characteristic functions, autocorrelation functions, and inter-polation/approximation kernels (e.g., radial basis functions) with the guarantee that the underlying approximation problems are well posed, irrespective of the location of the data points. In this appendix, we provide the basic definitions of positive defini-teness and conditional positive definiteness in the multidimensional setting, together with a review of corresponding mathematical results. We distinguish between contin-uous functions on the one hand and generalized functions on the other. We also give a self-contained derivation of Gelfand and Vilenkin's characterization of condition-ally positive definite generalized functions in 1-D and discuss its connection with the celebrated Lévy–Khintchine formula of statisticians. For a historical account of the rich topic of positive definiteness, we refer to [Ste76].

B.1   Positive definiteness and Bochner's theorem

DEFINITION B.1   A continuous, complex-valued function f of the vector variable $\boldsymbol{\omega} \in \mathbb{R}^d$ is said to be *positive semidefinite* if and only if

$$\sum_{m=1}^{N}\sum_{n=1}^{N} \xi_m \overline{\xi}_n f(\boldsymbol{\omega}_m - \boldsymbol{\omega}_n) \geq 0$$

for every choice of $\boldsymbol{\omega}_1, \ldots, \boldsymbol{\omega}_N \in \mathbb{R}^d$, $\xi_1, \ldots, \xi_N \in \mathbb{C}$, and $N \in \mathbb{N}$. Such a function is called *positive definite in the strict sense* if the quadratic form is greater than 0 for all $\xi_1, \ldots, \xi_N \in \mathbb{C}\backslash\{0\}$.

In what follows, we shall abbreviate "positive semidefinite" by *positive–definite*. This property is equivalent to the requirement that the $N \times N$ matrix $\mathbf{F}$ whose elements are given by $[\mathbf{F}]_{m,n} = f(\boldsymbol{\omega}_m - \boldsymbol{\omega}_n)$ is positive semidefinite (or, equivalently, non-negative definite), for all N, no matter how the $\boldsymbol{\omega}_n$ are chosen.

The prototypical example of a positive–definite function is the Gaussian kernel $\widehat{g}(\omega) = e^{-\omega^2/2}$. To establish the property, we express this Gaussian as the Fourier transform of $g(x) = \frac{1}{\sqrt{2\pi}} e^{-\frac{x^2}{2}}$:

$$\sum_{m=1}^{N}\sum_{n=1}^{N} \xi_m \overline{\xi}_n \, \widehat{g}(\omega_m - \omega_n) = \sum_{m=1}^{N}\sum_{n=1}^{N} \xi_m \overline{\xi}_n \int_{\mathbb{R}} e^{-j(\omega_m - \omega_n)x} g(x) \, dx$$

$$= \int_{\mathbb{R}} \sum_{m=1}^{N}\sum_{n=1}^{N} \xi_m \overline{\xi}_n e^{-j(\omega_m - \omega_n)x} g(x) \, dx$$

$$= \int_{\mathbb{R}} \underbrace{\left| \sum_{m=1}^{N} \xi_m e^{-j\omega_m x} \right|^2}_{\geq 0} \underbrace{g(x)}_{>0} \, dx \geq 0,$$

where we have made use of the fact that $g(x)$, the (inverse) Fourier transform of $e^{-\omega^2/2}$, is positive. It is not hard to see that the argument above remains valid for any (multi-dimensional) function $f(\omega)$ that is the Fourier transform of some non-negative kernel $g(x) \geq 0$. The more impressive result is that the converse implication is also true.

THEOREM B.1 (Bochner's theorem) *Let f be a bounded continuous function on $\mathbb{R}^d$. Then, f is positive definite if and only if it is the (conjugate) Fourier transform of a non-negative and finite Borel measure μ:*

$$f(\omega) = \int_{\mathbb{R}^d} e^{j\langle \omega, x \rangle} \mu(dx).$$

In particular, Bochner's theorem implies that f is a valid characteristic function – that is, $f(\omega) = \mathbb{E}\{e^{j\langle \omega, x \rangle}\} = \int_{\mathbb{R}^d} e^{j\langle \omega, x \rangle} \mathscr{P}_X(dx)$ where $\mathscr{P}_X$ is some probability measure on $\mathbb{R}^d$ – if and only if f is continuous, positive definite with $f(0) = 1$ (see Section 3.4.3 and Theorem 3.7).

Bochner's theorem is also fundamental to the theory of scattered data interpolation, although it requires a very slight restriction on the Fourier transform of f to ensure positive definiteness in the strict sense [Wen05].

THEOREM B.2 *A function $f : \mathbb{R}^d \to \mathbb{C}$ that is the (inverse) Fourier transform of a non-negative, finite Borel measure μ is positive definite in the strict sense if there exists an open set $E \subseteq \mathbb{R}^d$ such that $\mu(E) \neq 0$.*

Proof Let $g(x) \geq 0$ be the (generalized) density associated with μ such that $\mu(E) = \int_E g(x) \, dx$ for any Borel set E. We then write $f(\omega) = \int_{\mathbb{R}^d} e^{-j\langle \omega, x \rangle} g(x) \, dx$ and perform the same manipulation as for the Gaussian example above, which yields

$$\sum_{m=1}^{N}\sum_{n=1}^{N} \xi_m \overline{\xi}_n f(\omega_m - \omega_n) = \int_{\mathbb{R}^d} \underbrace{\left| \sum_{m=1}^{N} \xi_m e^{-j\langle \omega_m, x \rangle} \right|^2}_{\geq 0} \underbrace{g(x)}_{\geq 0} \, dx > 0.$$

The key observation is that the zero set of the sum of exponentials $\sum_{m=1}^{N} \xi_m e^{-j\langle \omega_m, x \rangle}$ (which is an entire function) has measure zero. Since the above integral involves positive terms only, the only possibility for it to be vanishing is that g be identically zero on the complement of this zero set, which contradicts the assumption on the existence of E.   □

In particular, the latter constraint is verified whenever $f(\omega) = \mathcal{F}\{g\}(\omega)$, where g is a continuous, non-negative function with a bounded Lebesgue integral; i.e., $0 < \int_{\mathbb{R}^d} g(x)\, dx < +\infty$. This kind of result is highly relevant to approximation and learning theory: indeed, the choice of a strictly positive definite interpolation kernel (or radial basis function) ensures that the solution of the generic scattered data interpolation problem is well defined and unique, no matter how the data centers are distributed [Mic86]. Here too, the prototypical example of a valid kernel is the Gaussian, which is (strictly) positive definite.

There is also an extension of Bochner's theorem for generalized functions that is due to Laurent Schwartz. In a nutshell, the idea is to replace each finite sum $\sum_{n=1}^{N} \xi_n f(\omega - \omega_n)$ by an infinite one (integral) $\int_{\mathbb{R}^d} \varphi(\omega') f(\omega - \omega')\, d\omega' = \int_{\mathbb{R}^d} \varphi(\omega - \omega') f(\omega')\, d\omega' = \langle f, \varphi(\cdot - \omega) \rangle$, which amounts to considering appropriate linear functionals of f over Schwartz' class of test functions $\mathcal{S}(\mathbb{R}^d)$. In doing so, the double sum in Definition B.1 collapses into a scalar product between f and the autocorrelation function of the test function $\varphi \in \mathcal{S}(\mathbb{R}^d)$, the latter being written as

$$(\varphi * \overline{\varphi}^\vee)(\omega) = \int_{\mathbb{R}^d} \varphi(\omega')\overline{\varphi(\omega' - \omega)}\, d\omega'.$$

DEFINITION B.2   A generalized function $f \in \mathcal{S}'(\mathbb{R}^d)$ is said to be *positive–definite* if and only if, for all $\varphi \in \mathcal{S}(\mathbb{R}^d)$,

$$\langle f, (\varphi * \overline{\varphi}^\vee) \rangle \geq 0.$$

It can be shown that this is equivalent to Definition B.1 in the case where $f(\omega)$ is continuous.

THEOREM B.3 (Schwartz–Bochner theorem)   *A generalized function $f \in \mathcal{S}'(\mathbb{R}^d)$ is positive definite if and only if it is the generalized Fourier transform of a non-negative tempered measure μ; that is,*

$$\langle f, \widehat{\varphi} \rangle = \langle \widehat{f}, \varphi \rangle = \int_{\mathbb{R}^d} \varphi(x)\mu(dx).$$

The term "tempered measure" refers to a generic type of mildly singular generalized function that can be defined by the Lebesgue integral $\int_{\mathbb{R}^d} \varphi(x)\mu(dx) < \infty$ for all $\varphi \in \mathcal{S}(\mathbb{R}^d)$. Such measures are allowed to exhibit polynomial growth at infinity subject to the restriction that they remain finite on any compact set.

The fact that the above form implies positive definiteness can be verified by direct substitution and application of Parseval's relation, by which we obtain

$$\langle f, (\varphi * \overline{\varphi}^\vee) \rangle = \frac{1}{(2\pi)^d} \langle \widehat{f}, |\widehat{\varphi}|^2 \rangle = \frac{1}{(2\pi)^d} \int_{\mathbb{R}^d} |\widehat{\varphi}(x)|^2 \mu(dx) \geq 0,$$

where the summability property against $\mathscr{S}(\mathbb{R}^d)$ ensures that the integral is convergent (since $|\widehat{\varphi}(x)|^2$ is rapidly decreasing).

The improvement over Theorem B.1 is that $\mu(\mathbb{R}^d)$ is no longer constrained to be finite. While this extension is of no direct help for the specification of characteristic functions, it happens to be quite useful for the definition of spline-like interpolation kernels that result in well-posed data fitting/approximation problems. We also note that the above definitions and results generalize to the infinite-dimensional setting (e.g., the Minlos–Bochner theorem which involves measures over topological vector spaces).

B.2 Conditionally positive–definite functions

DEFINITION B.3 A continuous, complex-valued function f of the vector variable $\boldsymbol{\omega} \in \mathbb{R}^d$ is said to be *conditionally positive–definite* of (integer) order $k \geq 0$ if and only if

$$\sum_{m=1}^{N} \sum_{n=1}^{N} \xi_m \overline{\xi}_n f(\boldsymbol{\omega}_m - \boldsymbol{\omega}_n) \geq 0$$

under the condition

$$\sum_{n=1}^{N} \xi_n p(\boldsymbol{\omega}_n) = 0, \quad \text{for all } p \in \Pi_{k-1}(\mathbb{R}^d),$$

for all possible choices of $\boldsymbol{\omega}_1, \ldots, \boldsymbol{\omega}_N \in \mathbb{R}^d$, $\xi_1, \ldots, \xi_N \in \mathbb{C}$, and $N \in \mathbb{N}$, where $\Pi_{k-1}(\mathbb{R}^d)$ denotes the space of multidimensional polynomials of degree $(k - 1)$.

This definition is also extendable to generalized functions using the line of thought that leads to Definition B.2. To keep the presentation reasonably simple and to make the link with the definition of the Lévy exponents in Section 4.2, we now focus on the 1-D case ($d = 1$). Specifically, we consider the polynomial constraint $\sum_{n=1}^{N} \xi_n \omega_n^m = 0$, $m \in \{0, \ldots, k - 1\}$ and derive the generic form of conditionally positive definite generalized functions of order k, including the continuous ones which are of greatest interest to us.

The distributional counterpart of the kth-order constraint for $d = 1$ is the orthogonality condition $\int_{\mathbb{R}} \varphi(\omega)\omega^m \, d\omega = 0$ for $m \in \{0, \ldots, k - 1\}$. It is enforced by restricting the analysis to the class of test functions whose moments up to order $(k - 1)$ are vanishing. Without loss of generality, this is equivalent to considering some alternative test function $D^k \varphi = \varphi^{(k)}$ where D^k is the kth derivative operator.

DEFINITION B.4 A generalized function $f \in \mathscr{S}'(\mathbb{R})$ is said to be *conditionally positive definite* of order k if and only if for all $\varphi \in \mathscr{S}(\mathbb{R})$

$$\left\langle f, (\varphi^{(k)} * \overline{\varphi^{(k)}}^\vee) \right\rangle = \left\langle f, (-1)^k D^{2k}(\varphi * \overline{\varphi}^\vee) \right\rangle \geq 0.$$

This extended definition allows for the derivation of the corresponding version of Bochner's theorem which provides an explicit characterization of the family of conditionally positive–definite generalized functions, together with their generalized Fourier transform.

THEOREM B.4 (Gelfand–Villenkin)   *A generalized function $f \in \mathscr{S}'(\mathbb{R})$ is conditionally positive definite of order k if and only if it admits the following representation over $\mathscr{S}(\mathbb{R})$:*

$$\langle f, \widehat{\varphi} \rangle = \langle \widehat{f}, \varphi \rangle = \int_{\mathbb{R}\setminus\{0\}} \left(\varphi(x) - r(x) \sum_{n=0}^{2k-1} \frac{\varphi^{(n)}(0)}{n!} x^n \right) \mu(dx) + \sum_{n=0}^{2k} a_n \frac{\varphi^{(n)}(0)}{n!}, \quad (B.1)$$

where μ is a positive tempered Borel measure on $\mathbb{R}\setminus\{0\}$ satisfying

$$\int_{|x|<1} |x|^{2k} \mu(dx) < \infty.$$

Here, $r(x)$ is a function in $\mathscr{S}(\mathbb{R})$ such that $(r(x) - 1)$ has a zero of order $(2k+1)$ at $x = 0$, while the a_n are appropriate real-valued constants with the constraint that $a_{2k} \geq 0$.

Below, we provide a slightly adapted version of Gelfand and Vilenkin's proof, which is remarkably concise and quite illuminating [GV64, Theorem 1, pp. 178], at least if one compares it with the standard derivation of the Lévy–Khintchine formula, which has a much more technical flavor (see [Sat94]) and is ultimately less general.

Proof   Since $\langle f, (-1)^k D^{2k} (\varphi * \overline{\varphi}^\vee) \rangle = \langle (-1)^k D^{2k} f, (\varphi * \overline{\varphi}^\vee) \rangle$, we interpret Definition B.4 as the property that $(-1)^k D^{2k} f$ is positive definite. By the Schwartz–Bochner theorem, this is equivalent to the existence of a tempered measure ν such that

$$\langle (-1)^k D^{2k} f, \widehat{\varphi} \rangle = \langle f, (-1)^k D^{2k} \widehat{\varphi} \rangle = \langle \widehat{f}, x^{2k} \varphi \rangle = \int_{\mathbb{R}} \varphi(x) \nu(dx).$$

By defining $\phi(x) = x^{2k} \varphi(x)$, this can be rewritten as

$$\langle \widehat{f}, \phi \rangle = \int_{\mathbb{R}} \frac{\phi(x)}{x^{2k}} \nu(dx) = \langle f, \widehat{\phi} \rangle,$$

where ϕ is a test function that has a zero of order $2k$ at the origin. In particular, this implies that $\lim_{\epsilon \downarrow 0} \int_{|x|<\epsilon} \frac{\phi(x)}{x^{2k}} \nu(dx) = \frac{\phi^{(2k)}(0)}{(2k)!} a_{2k}$, where $a_{2k} \geq 0$ is the ν-measure at point $x = 0$. Introducing the new measure $\mu(dx) = \nu(dx)/x^{2k}$, we then decompose the Lebesgue integral as

$$\langle \widehat{f}, \phi \rangle = \int_{\mathbb{R}\setminus\{0\}} \phi(x) \mu(dx) + a_{2k} \frac{\phi^{(2k)}(0)}{(2k)!}, \quad (B.2)$$

which specifies f on the subset of test functions that have a $2k$th-order zero at the origin. To extend the representation to the whole space $\mathscr{S}(\mathbb{R})$, we associate with every $\varphi \in \mathscr{S}(\mathbb{R})$ the corrected function

$$\phi_c(x) = \varphi(x) - r(x) \sum_{n=0}^{2k-1} \frac{\varphi^{(n)}(0)}{n!} x^n, \quad (B.3)$$

with $r(x)$ as specified in the statement of the theorem. By construction, $\phi_c \in \mathscr{S}(\mathbb{R})$ and has the $2k$th-order zero that is required for (B.2) to be applicable. By combining (B.2) and (B.3), we find that

$$\langle \hat{f}, \varphi \rangle = \int_{\mathbb{R}\backslash\{0\}} \phi_c(x)\mu(dx) + a_{2k}\frac{\phi_c^{(2k)}(0)}{(2k)!} + \sum_{n=0}^{2k-1}\frac{\varphi^{(n)}(0)}{n!}\langle \hat{f}, r(x)x^n \rangle.$$

Next, we identify the constants $a_n = \langle \hat{f}, r(x)x^n \rangle$ and note that $\phi_c^{(2k)}(0) = \varphi^{(2k)}(0)$. The final step is to substitute these together with the expression (B.3) of ϕ_c in the above formula, which yields the desired result.

To prove the sufficiency of the representation, we apply (B.1) to evaluate the functional

$$\langle f, (\widehat{\varphi^{(k)}} * \overline{\widehat{\varphi^{(k)}}}^{\vee})\rangle = \langle \hat{f}, x^{2k}|\varphi(x)|^2 \rangle = \int_{\mathbb{R}} x^{2k}|\varphi(x)|^2\mu(dx) + a_{2k}|\varphi(0)|^2 \geq 0,$$

where we have used the property that the derivatives of $x^{2k}|\varphi(x)|^2$ are all vanishing at the origin, except the one of order $2k$, which equals $(2k)!\,|\widehat{\varphi}(0)|^2$ for $x = 0$. □

It is important to note that the choice of the function r is arbitrary as long as it fulfills the boundary condition $r(x) = 1 + O(|x|^{2k+1})$ as $x \to 0$, so as to regularize the potential kth-order singularity of μ at the origin, and that it decays sufficiently fast to temper the Taylor-series correction in (B.3) at infinity. If we compare the effect of using two different tempering functions r_1 and r_2, the modification is only in the value of the constants a_n, with $a_{n,2} - a_{n,1} = \langle \hat{f}, (r_2(x) - r_1(x))x^n \rangle$. Another way of putting this is that the corresponding distributions $\hat{f}_1$ and $\hat{f}_2$ specified by the leading integral in (B.1) will only differ by a $(2k-1)$th-order point distribution that is entirely localized at $x = 0$; that is, $\hat{f}_2(x) - \hat{f}_1(x) = \sum_{n=0}^{2k-1}\frac{a_{n,2}-a_{n,1}}{n!}\delta^{(n)}(x)$, owing to the property that a_{2k} is common to both scenarios, or, equivalently, that the difference of their inverse Fourier transforms f_1 and f_2 is a polynomial of degree $(2k-1)$.

Thanks to Theorem B.4, it is also possible to derive an integral representation that is the kth-order generalization of the Lévy–Khintchine formula. For a detailed treatment of the multidimensional version of the problem, we refer to the works of Madych, Nelson, and Sun [MN90a, Sun93].

COROLLARY B.5 *Let $f(\omega)$ be a continuous function of $\omega \in \mathbb{R}$. Then, f is conditionally positive–definite of order k if and only if it can be represented as*

$$f(\omega) = \frac{1}{2\pi}\left(\int_{\mathbb{R}\backslash\{0\}}\left(e^{j\omega x} - r(x)\sum_{n=0}^{2k-1}\frac{(j\omega x)^n}{n!}\right)\mu(dx) + \sum_{n=0}^{2k}a_n\frac{(j\omega)^n}{n!}\right),$$

where μ is a positive Borel measure on $\mathbb{R}\backslash\{0\}$ satisfying

$$\int_{\mathbb{R}}\min(|x|^{2k}, 1)\mu(dx) < \infty,$$

where $r(x)$ and a_n are as in Theorem B.4.

The result is obtained by plugging $\varphi(x) = \frac{1}{2\pi}e^{j\omega x} \longleftrightarrow \widehat{\varphi}(\cdot) = \delta(\cdot - \omega)$ into (B.1), which is justifiable using a continuity argument. The key is that the corresponding integral is bounded when μ satisfies the admissibility condition, which ensures the continuity of $f(\omega)$ (by Lebesgue's dominated-convergence theorem), and vice versa.

B.3   Lévy–Khintchine formula from the point of view of generalized functions

We now make the link with the Lévy–Khintchine theorem of statisticians (see Section 4.2.1) which is equivalent to characterizing the functions that are conditionally positive definite of order one. To that end, we rewrite the formula in Corollary B.5 for $k = 1$ under the additional constraint that $f_1(0) = 0$ (which fixes the value of a_0) as

$$f_1(\omega) = \frac{1}{2\pi}\left(a_0 + a_1 j\omega - \frac{a_2}{2}\omega^2 + \int_{\mathbb{R}\backslash\{0\}} \left(e^{j\omega x} - r(x) - r(x)j\omega x\right)\mu(dx)\right)$$

$$= b_1 j\omega - \frac{b_2}{2}\omega^2 + \int_{\mathbb{R}\backslash\{0\}} \left(e^{j\omega x} - 1 - r(x)j\omega x\right)v(x)\,dx,$$

where $v(x)\,dx = \frac{1}{2\pi}\mu(dx)$, $b_n = \frac{1}{2\pi}a_n$, $r(x) = 1 + O(|x|^3)$ as $x \to 0$, and $\lim_{x\to\pm\infty} r(x) = 0$. Clearly, the new form is equivalent to the Lévy–Khintchine formula (4.3) with the slight difference that the bias compensation is achieved by using a bell-shaped, infinitely differentiable function r instead of the rectangular window $\mathbb{1}_{|x|<1}(x)$.

Likewise, we are able to transcribe the generalized Fourier-transform-pair relation (B.1) for the Lévy–Khintchine representation (4.3), which yields

$$\langle \widehat{f}_{L-K}, \varphi\rangle = \langle f_{L-K}, \widehat{\varphi}\rangle$$

$$= \int_{\mathbb{R}\backslash\{0\}} \left(\varphi(x) - \varphi(0) - x\,\mathbb{1}_{|x|<1}(x)\varphi^{(1)}(0)\right)v(x)\,dx + b_1'\varphi^{(1)}(0) + \frac{b_2}{2}\varphi^{(2)}(0).$$

$$(B.4)$$

The interest of (B.4) is that it uniquely specifies the generalized Fourier transform of a Lévy exponent f_{L-K} as a linear functional of φ.

We can also give a "time-domain" (or pointwise) interpretation of this result by characterizing the generalized function

$$g(x) = \frac{1}{2\pi}\mathcal{F}\{f_{K-L}\}(x) = G\{\delta\}(x),$$

which also represents the impulse response of the infinitesimal semigroup generator G investigated in Section 9.7. This is achieved by distinguishing between three cases:

(1) Lebesgue-integrable Lévy density

In this simpler scenario, we are able to split the leading integral in (B.4) into its subparts, which results in

$$g(x) = v(x) - \delta(x)\left(\int_{\mathbb{R}} v(a)\,da\right) - \delta'(x)\left(b_1' - \int_{|a|<1} av(a)\,da\right) + \delta''(x)\frac{b_2}{2}.$$

The underlying principle is that the so-defined generalized function will result in the same measurements as (B.4) when applied to the test function φ. In particular, the values of $\varphi^{(n)}$ at the origin are sampled using the Dirac distribution and its derivatives in accordance with the rule $\langle \varphi, \delta^{(n)} \rangle = (-1)^n \varphi^{(n)}(0)$.

(2) Non-integrable Lévy density with finite absolute moment

To ensure that the integral in (B.4) is convergent, we need to retain the zero-order correction. Yet, when $\int_{\mathbb{R}} |a|v(a)\, da < \infty$, we can still pull out the third term, which results in the interpretation

$$g(x) = \text{p.f.}(v) - \delta'(x)\left(b_1' - \int_{|a|<1} av(a)\, da\right) + \delta''(x)\frac{b_2}{2},$$

where p.f. stands for the finite-part operator that implicitly implements the Taylor-series adjustment that stabilizes the scalar-product integral $\langle v, \varphi \rangle$ (see Appendix A).

(3) Non-integrable Lévy density and unbounded absolute moment

Here, we cannot split the integral anymore. However, when $\int_{|a|>1} |a|v(a)\, da < \infty$, we can stabilize the integral by applying a full first-order Taylor-series correction. This leads to the finite-part interpretation

$$g(x) - \text{p.f.}(v) - b_1'\delta'(x) + \frac{b_2}{2}\delta''(x),$$

which is the direct counterpart of (4.5). For $\int_{|a|>1} |a|v(a)\, da = \infty$, the proper pointwise interpretation becomes more delicate and it is safer to stick to the distributional definition (B.4).

Appendix C Special functions

C.1 Modified Bessel functions

The modified Bessel function of the second kind with order parameter $\alpha \in \mathbb{R}$ admits the Fourier-based representation [AS72]

$$K_\alpha(\omega) = \int_{\mathbb{R}} \frac{e^{-j\omega x}}{(1+x^2)^{|\alpha|}}\, dx.$$

It has the property that $K_\alpha(x) = K_{-\alpha}(x)$. A special case of interest is $K_{\frac{1}{2}}(x) = \left(\frac{\pi}{2x}\right)^{\frac{1}{2}} e^{-x}$.

The small-scale behavior of $K_\alpha(x)$ is $K_\alpha(x) \sim \frac{\Gamma(\alpha)}{2}\left(\frac{2}{x}\right)^\alpha$ as $x \to 0$. In order to determine the form of the variance-gamma distribution around the origin, we can rely on the following expansion which includes a few more terms:

$$K_\alpha(x) = x^{-\alpha}\left(2^{\alpha-1}\Gamma(\alpha) - \frac{2^{\alpha-3}\Gamma(\alpha)x^2}{\alpha-1} + O\left(x^4\right)\right)$$

$$+ x^\alpha\left(2^{-\alpha-1}\Gamma(-\alpha) + \frac{2^{-\alpha-3}\Gamma(-\alpha)x^2}{\alpha+1} + O\left(x^4\right)\right).$$

At the other end of the scale, its asymptotic behavior is

$$K_\alpha(x) \sim \sqrt{\frac{\pi}{2x}} e^{-x} \text{ as } x \to +\infty.$$

C.2 Gamma function

Euler's gamma function constitutes an analytic extension of the factorial function $n! = \Gamma(n+1)$ to the complex plane. It is defined by the integral

$$\Gamma(z) = \int_0^{+\infty} t^{z-1}e^{-t}\, dt,$$

which is convergent for $\mathrm{Re}(z) > 0$. Specific values are $\Gamma(1) = 1$ and $\Gamma(1/2) = \sqrt{\pi}$. The gamma function satisfies the functional equation

$$\Gamma(z+1) = z\Gamma(z), \tag{C.1}$$

which is compatible with the recursive definition of the factorial $n! = n(n-1)!$. Another useful result is Euler's reflection formula,

$$\Gamma(1-z)\Gamma(z) = \frac{\pi}{\sin(\pi z)}.$$

By combining the above with (C.1), we obtain

$$\text{sinc}(z) = \frac{\sin(\pi z)}{\pi z} = \frac{1}{\Gamma(1-z)\,\Gamma(1+z)}, \tag{C.2}$$

which makes an intriguing connection with the *sinus cardinalis* function. There is a similar link with Euler's beta function,

$$B(z_1, z_2) = \int_0^1 t^{z_1-1}(1-t)^{z_2-1}\,dt \tag{C.3}$$

$$= \frac{\Gamma(z_1)\Gamma(z_2)}{\Gamma(z_1 + z_2)}$$

with $\text{Re}(z_1), \text{Re}(z_2) > 0$.

$\Gamma(z)$ also admits the well-known product decomposition

$$\Gamma(z) = \frac{e^{-\gamma_0 z}}{z} \prod_{n=1}^{\infty} \left(1 + \frac{z}{n}\right)^{-1} e^{z/n}, \tag{C.4}$$

where γ_0 is the Euler–Mascheroni constant. The above allows us to derive the expansion

$$-\log |\Gamma(z)|^2 = 2\gamma_0 \text{Re}(z) + \log |z|^2 + \sum_{n=1}^{\infty} \left(\log\left|1 + \frac{z}{n}\right|^2 - 2\frac{\text{Re}(z)}{n}\right),$$

which is directly applicable to the likelihood function associated with the Meixner distribution. Also relevant to that context is the integral relation

$$\int_{\mathbb{R}} \left|\Gamma(\frac{r}{2} + jx)\right|^2 e^{jzx}\,dx = 2\pi\Gamma(r)\left(\frac{1}{2\cosh\frac{z}{2}}\right)^r$$

for $r > 0$ and $z \in \mathbb{C}$, which can be interpreted as a Fourier transform by setting $z = -j\omega$.

Euler's digamma function is defined as

$$\psi(z) = \frac{d}{dz}\log\Gamma(z) = \frac{\Gamma'(z)}{\Gamma(z)}, \tag{C.5}$$

while its mth-order derivative

$$\psi^{(m)}(z) = \frac{d^{m+1}}{dz^{m+1}}\log\Gamma(z) \tag{C.6}$$

is called the *polygamma function of order m*.

C.3   Symmetric-alpha-stable distributions

The SαS pdf of degree $\alpha \in (0, 2]$ and scale parameter s_0 is best defined via its characteristic function,

$$p(x; \alpha, s_0) = \int_{\mathbb{R}} e^{-|s_0\omega|^\alpha} e^{j\omega x} \frac{d\omega}{2\pi}.$$

Alpha-stable distributions do not admit closed-form expressions, except for the special cases $\alpha = 1$ (Cauchy) and 2 (Gauss distribution). Moreover, their absolute moments of order p, $\mathbb{E}\{|X|^p\}$, are unbounded for $p > \alpha$, which is characteristic of heavy-tailed distributions. We can relate the (symmetric) γth-order moments of their characteristic function to the gamma function by performing the change of variable $t = (s_0\omega)^\alpha$, which leads to

$$\int_{\mathbb{R}} |\omega|^\gamma e^{-|s_0\omega|^\alpha} d\omega = 2 \int_0^\infty \frac{s_0^{-\gamma-1}}{\alpha} t^{\frac{\gamma-\alpha+1}{\alpha}} e^{-t} dt = 2 \frac{s_0^{-\gamma-1} \Gamma\left(\frac{\gamma+1}{\alpha}\right)}{\alpha}. \tag{C.7}$$

By using the correspondence between Fourier-domain moments and time-domain derivatives, we use this result to write the Taylor series of $p(x; \alpha, s_0)$ around $x = 0$ as

$$p(x; \alpha, s_0) = \sum_{k=0}^\infty \frac{s_0^{-2k-1}}{\pi\alpha} \Gamma\left(\frac{2k+1}{\alpha}\right) (-1)^k \frac{|x|^{2k}}{(2k)!}, \tag{C.8}$$

which involves even terms only (because of symmetry). The moment formula (C.7) also yields a simple expression for the slope of the score at the origin, which is given by

$$\Phi_X''(0) = -\frac{p_X''(0)}{p_X(0)} = \frac{\Gamma\left(\frac{3}{\alpha}\right)}{s_0^2 \Gamma\left(\frac{1}{\alpha}\right)}.$$

Similar techniques are applicable to obtain the asymptotic form of $p(x; \alpha, s_0)$ as x tends to infinity [Ber52, TN95]. To characterize the tail behavior, it is sufficient to consider the first term of the asymptotic expansion

$$p(x; \alpha, s_0) \sim \frac{1}{\pi} \Gamma(\alpha + 1) \sin\left(\frac{\pi\alpha}{2}\right) s_0^\alpha \frac{1}{|x|^{\alpha+1}} \quad \text{as } x \to \pm\infty, \tag{C.9}$$

which emphasizes the algebraic decay of order $(\alpha + 1)$ at infinity.

References

[ABDF11] M. V. Afonso, J. M. Bioucas-Dias, and M. A. T. Figueiredo, An augmented Lagrangian approach to the constrained optimization formulation of imaging inverse problems, *IEEE Transactions on Image Processing* **20** (2011), no. 3, 681–695.

[ABT01] A. Achim, A. Bezerianos, and P. Tsakalides, Novel Bayesian multiscale method for speckle removal in medical ultrasound images, *IEEE Transactions on Medical Imaging* **20** (2001), no. 8, 772–783.

[AG01] A. Aldroubi and K. Gröchenig, Nonuniform sampling and reconstruction in shift-invariant spaces, *SIAM Review* **43** (2001), 585–620.

[Ahm74] N. Ahmed, Discrete cosine transform, *IEEE Transactions on Communications* **23** (1974), no. 1, 90 93.

[AKBU13] A. Amini, U. S. Kamilov, E. Bostan, and M. Unser, Bayesian estimation for continuous-time sparse stochastic processes, *IEEE Transactions on Signal Processing* **61** (2013), no. 4, 907–920.

[Ald95] A. Aldroubi, Portraits of frames, *Proceedings of the American Mathematical Society* **123** (1995), no. 6, 1661–1668.

[App09] D. Appelbaum, *Lévy Processes and Stochastic Calculus*, 2nd edn., Cambridge University Press, 2009.

[AS72] M. Abramowitz and I. A. Stegun, *Handbook of Mathematical Functions*, National Bureau of Standards, 1972.

[AS83] D. A. Agard and J. W. Sedat, Three-dimensional architecture of a polytene nucleus, *Nature* **302** (1983), no. 5910, 676–681.

[ASS98] F. Abramovich, T. Sapatinas, and B. W. Silverman, Wavelet thresholding via a Bayesian approach, *Journal of the Royal Statistical Society: Series B (Statistical Methodology)* **60** (1998), no. 4, 725–749.

[ATB03] A. Achim, P. Tsakalides, and A. Bezerianos, SAR image denoising via Bayesian wavelet shrinkage based on heavy-tailed modeling, *IEEE Transactions on Geoscience and Remote Sensing* **41** (2003), no. 8, 1773–1784.

[AU94] A. Aldroubi and M. Unser, Sampling procedures in function spaces and asymptotic equivalence with Shannon's sampling theory, *Numerical Functional Analysis and Optimization* **15** (1994), no. 1–2, 1–21.

[AU14] A. Amini and M. Unser, Sparsity and infinite divisibility, *IEEE Transactions on Information Theory* **60** (2014), no. 4, 2346–2358.

[AUM11] A. Amini, M. Unser, and F. Marvasti, Compressibility of deterministic and random infinite sequences, *IEEE Transactions on Signal Processing* **59** (2011), no. 11, 5193–5201.

[Bar61]   M. S. Bartlett, *An Introduction to Stochastic Processes, with Special Reference to Methods and Applications*, Cambridge Universty Press, 1961.

[BAS07]   M. I. H. Bhuiyan, M. O. Ahmad, and M. N. S. Swamy, Spatially adaptive wavelet-based method using the Cauchy prior for denoising the SAR images, *IEEE Transactions on Circuits and Systems for Video Technology* **17** (2007), no. 4, 500–507.

[BDE09]   A. M. Bruckstein, D. L. Donoho, and M. Elad, From sparse solutions of systems of equations to sparse modeling of signals and images, *SIAM Review* **51** (2009), no. 1, 34–81.

[BDF07]   J. M. Bioucas-Dias and M. A. T. Figueiredo, A new twist: Two-step iterative shrinkage/thresholding algorithms for image restoration, *IEEE Transactions on Image Processing* **16** (2007), no. 12, 2992–3004.

[BDR02]   A. Bose, A. Dasgupta, and H. Rubin, A contemporary review and bibliography of infinitely divisible distributions and processes, *Sankhya: The Indian Journal of Statistics: Series A* **64** (2002), no. 3, 763–819.

[Ber52]   H. Bergström, On some expansions of stable distributions, *Arkiv für Mathematik* **2** (1952), no. 18, 375–378.

[BF06]   L. Boubchir and J. M. Fadili, A closed-form nonparametric Bayesian estimator in the wavelet domain of images using an approximate alpha-stable prior, *Pattern Recognition Letters* **27** (2006), no. 12, 1370–1382.

[BH10]   P. J. Brockwell and J. Hannig, CARMA(p,q) generalized random processes, *Journal of Statistical Planning and Inference* **140** (2010), no. 12, 3613–3618.

[BK51]   M. S. Bartlett and D. G. Kendall, On the use of the characteristic functional in the analysis of some stochastic processes occurring in physics and biology, *Mathematical Proceedings of the Cambridge Philosophical Society* **47** (1951), 65–76.

[BKNU13]   E. Bostan, U. S. Kamilov, M. Nilchian, and M. Unser, Sparse stochastic processes and discretization of linear inverse problems, *IEEE Transactions on Image Processing* **22** (2013), no. 7, 2699–2710.

[BM02]   P. Brémaud and L. Massoulié, Power spectra of general shot noises and Hawkes point processes with a random excitation, *Advances in Applied Probability* **34** (2002), no. 1, 205–222.

[BNS01]   O. E. Barndorff-Nielsen and N. Shephard, Non-Gaussian Ornstein-Uhlenbeck-based models and some of their uses in financial economics, *Journal of the Royal Statistical Society: Series B (Statistical Methodology)* **63** (2001), no. 2, 167–241.

[Boc32]   S. Bochner, *Vorlesungen über Fouriersche Integrale*, Akademische Verlagsgesellschaft, 1932.

[Boc47]   Stochastic processes, *Annals of Mathematics* **48** (1947), no. 4, 1014–1061.

[Bog07]   V. I. Bogachev, *Measure Theory, Vol. I, II*, Springer, 2007.

[BPCE10]   S. Boyd, N. Parikh, E. Chu, and J. Eckstein, Distributed optimization and statistical learning via the alternating direction method of multipliers, *Information Systems Journal* **3** (2010), no. 1, 1–122.

[Bro01]   P. J. Brockwell, Lévy-driven CARMA processes, *Annals of the Institute of Statistical Mathematics* **53** (2001), 113–124.

[BS50]   H. W. Bode and C. E. Shannon, A simplified derivation of linear least square smoothing and prediction theory, *Proceedings of the IRE* **38** (1950), no. 4, 417–425.

[BS93]   C. Bouman and K. Sauer, A generalized Gaussian image model for edge-preserving MAP estimation, *IEEE Transactions on Image Processing* **2** (1993), no. 3, 296–310.

[BT09a]   A. Beck and M. Teboulle, Fast gradient-based algorithms for constrained total variation image denoising and deblurring problems, *IEEE Transactions on Image Processing* **18** (2009), no. 11, 2419–2434.

[BT09b]   A fast iterative shrinkage-thresholding algorithm for linear inverse problems, *SIAM Journal on Imaging Sciences* **2** (2009), no. 1, 183–202.

[BTU01]   T. Blu, P. Thévenaz, and M. Unser, MOMS: Maximal-order interpolation of minimal support, *IEEE Transactions on Image Processing* **10** (2001), no. 7, 1069–1080.

[BU03]   T. Blu and M. Unser, A complete family of scaling functions: The (α, τ)-fractional splines, *Proceedings of the IEEE International Conference on Acoustics, Speech, and Signal Processing (ICASSP'03)* (Hong Kong SAR, People's Republic of China, April 6–10, 2003), vol. VI, IEEE, pp. 421–424.

[BU07]   Self-similarity. Part II: Optimal estimation of fractal processes, *IEEE Transactions on Signal Processing* **55** (2007), no. 4, 1364–1378.

[BUF07]   K. T. Block, M. Uecker, and J. Frahm, Undersampled radial MRI with multiple coils. Iterative image reconstruction using a total variation constraint, *Magnetic Resonance in Medicine* **57** (2007), no. 6, 1086–1098.

[BY02]   B. Bru and M. Yor, Comments on the life and mathematical legacy of Wolfgang Döblin, *Finance and Stochastics* **6** (2002), no. 1, 3–47.

[Car99]   J. F. Cardoso and D. L. Donoho, Some experiments on independent component analysis of non-Gaussian processes, *Proceedings of the IEEE Signal Processing Workshop on Higher-Order Statistics (SPW-HOS'99)* (Caesarea, Istael, June 14–16, 1999), 1999, pp. 74–77.

[CBFAB97]   P. Charbonnier, L. Blanc-Féraud, G. Aubert, and M. Barlaud, Deterministic edge-preserving regularization in computed imaging, *IEEE Transactions on Image Processing* **6** (1997), no. 2, 298–311.

[CD95]   R. R. Coifman and D. L. Donoho, Translation-invariant de-noising, *Wavelets and statistics*, (A. Antoniadis and G, Oppeheim, eds.), Lecture Notes in Statistics, vol. 103, Springer, 1995, pp. 125–150.

[CDLL98]   A. Chambolle, R. A. DeVore, N.-Y. Lee, and B. J. Lucier, Nonlinear wavelet image processing: Variational problems, compression, and noise removal through wavelet shrinkage, *IEEE Transactions on Image Processing* **7** (1998), no. 33, 319–335.

[Chr03]   O. Christensen, *An Introduction to Frames and Riesz Bases*, Birkhäuser, 2003.

[CKM97]   H. A. Chipman, E. D. Kolaczyk, and R. E. McCulloch, Adaptive Bayesian wavelet shrinkage, *Journal of the American Statistical Association* **92** (1997), no. 440, 1413–1421.

[CLM$^+$95]   W. A. Carrington, R. M. Lynch, E. D. Moore, G. Isenberg, K. E. Fogarty, and F. S. Fay, Superresolution three-dimensional images of fluorescence in cells with minimal light exposure, *Science* **268** (1995), no. 5216, 1483–1487.

[CNB98]   M. S. Crouse, R. D. Nowak, and R. G. Baraniuk, Wavelet-based statistical signal processing using hidden Markov models, *IEEE Transactions on Signal Processing* **46** (1998), no. 4, 886–902.

[Com94]   P. Comon, Independent component analysis: A new concept, *Signal Processing* **36** (1994), no. 3, 287–314.

[CP11]   P. L. Combettes and J.-C. Pesquet, Proximal splitting methods in signal proces-sing, *Fixed-Point Algorithms for Inverse Problems in Science and Engineering* (H. H. Bauschke, R. S. Burachik, P. L. Combettes, V. Elser, D. R. Luke, and H. Wolkowicz, eds.), vol. 49, Springer New York, 2011, pp. 185–212.

[Cra40]   H. Cramér, On the theory of stationary random processes, *The Annals of Mathematics* **41** (1940), no. 1, 215–230.

[CSE00]   C. Christopoulos, A. S. Skodras, and T. Ebrahimi, The JPEG2000 still image coding system: An overview, *IEEE Transactions on Consumer Electronics* **16** (2000), no. 4, 1103–1127.

[CT04]   R. Cont and P. Tankov, *Financial Modelling with Jump Processes*, Chapman & Hall, 2004.

[CU10]   K. N. Chaudhury and M. Unser, On the shiftability of dual-tree complex wavelet transforms, *IEEE Transactions on Signal Processing* **58** (2010), no. 1, 221–232.

[CW91]   C. K. Chui and J.-Z. Wang, A cardinal spline approach to wavelets, *Proceedings of the American Mathematical Society* **113** (1991), no. 3, 785–793.

[CW08]   E. J. Candès and M. B. Wakin, An introduction to compressive sampling, *IEEE Signal Processing Magazine* **25** (2008), no. 2, 21–30.

[CYV00]   S. G. Chang, B. Yu, and M. Vetterli, Spatially adaptive wavelet thresholding with context modeling for image denoising, *IEEE Transactions on Image Pro-cessing* **9** (2000), no. 9, 1522 –1531.

[Dau88]   I. Daubechies, Orthogonal bases of compactly supported wavelets, *Communi-cations on Pure and Applied Mathematics* **41** (1988), 909–996.

[Dau92]   ———— *Ten Lectures on Wavelets*, Society for Industrial and Applied Mathematics, 1992.

[dB78]   C. de Boor, *A Practical Guide to Splines*, Springer, 1978.

[dB87]   ———— The polynomials in the linear span of integer translates of a compactly sup-ported function, *Constructive Approximation* **3** (1987), 199–208.

[dBDR93]   C. de Boor, R. A. DeVore, and A. Ron, On the construction of multivariate (pre) wavelets, *Constructive Approximation* **9** (1993), 123–123.

[DBFZ$^+$06] N. Dey, L. Blanc-Féraud, C. Zimmer, P. Roux, Z. Kam, J.-C. Olivo-Marin, and J. Zerubia, Richardson-Lucy algorithm with total variation regularization for 3D confocal microscope deconvolution, *Microscopy Research and Technique* **69** (2006), no. 4, 260–266.

[dBH82]   C. de Boor and K. Höllig, B-splines from parallelepipeds, *Journal d'Analyse Mathématique* **42** (1982), no. 1, 99–115.

[dBHR93]   C. de Boor, K. Höllig, and S. Riemenschneider, *Box Splines*, Springer, 1993.

[DDDM04]   I. Daubechies, M. Defrise, and C. De Mol, An iterative thresholding algorithm for linear inverse problems with a sparsity constraint, *Communications on Pure and Applied Mathematics* **57** (2004), no. 11, 1413–1457.

[DJ94]   D. L. Donoho and I. M. Johnstone, Ideal spatial adaptation via wavelet shrin-kage, *Biometrika* **81** (1994), 425–455.

[DJ95]   ———— Adapting to unknown smoothness via wavelet shrinkage, *Journal of the Ame-rican Statistical Association* **90** (1995), no. 432, 1200–1224.

[Don95]   D. L. Donoho, De-noising by soft-thresholding, *IEEE Transactions on Infor-mation Theory* **41** (1995), no. 3, 613–627.

[Don06]   ———— Compressed sensing, *IEEE Transactions on Information Theory* **52** (2006), no. 4, 1289–1306.

[Doo37]   J. L. Doob, Stochastic processes depending on a continuous parameter, *Transactions of the American Mathematical Society* **42** (1937), no. 1, 107–140.

[Doo90]   ——— *Stochastic Processes*, John Wiley & Sons, 1990.

[Duc77]   J. Duchon, Splines minimizing rotation-invariant semi-norms in Sobolev spaces, *Constructive Theory of Functions of Several Variables* (W. Schempp and K. Zeller, eds.), Springer, 1977, pp. 85–100.

[Ehr54]   L. Ehrenpreis, Solutions of some problems of division I, *American Journal of Mathematics* **76** (1954), 883–903.

[EK00]   R. Estrada and R. P. Kanwal, *Singular Integral Equations*, Birkhäuser, 2000.

[ENU12]   A. Entezari, M. Nilchian, and M. Unser, A box spline calculus for the discretization of computed tomography reconstruction problems, *IEEE Transactions on Medical Imaging* **31** (2012), no. 8, 1532–1541.

[Fag14]   J. Fageot, A. Amini and M. Unser, On the continuity of characteristic functionals and sparse stochastic modeling, Preprint (2014), arXiv:1401.6850(math.PR).

[FB05]   J. M. Fadili and L. Boubchir, Analytical form for a Bayesian wavelet estimator of images using the Bessel K form densities, *IEEE Transactions on Image Processing* **14** (2005), no. 2, 231–240.

[FBD10]   M. A. T. Figueiredo and J. Bioucas-Dias, Restoration of Poissonian images using alternating direction optimization, *IEEE Transactions on Image Processing* **19** (2010), no. 12, 3133–3145.

[Fel71]   W. Feller, *An Introduction to Probability Theory and its Applications, vol. 2*, John Wiley & Sons, 1971.

[Fer67]   X. Fernique, Lois indéfiniment divisibles sur l'espace des distributions, *Inventiones Mathematicae* **3** (1967), no. 4, 282–292.

[Fie87]   D. J. Field, Relations between the statistics of natural images and the response properties of cortical cells, *Journal of the Optical Society of America A: Optics Image Science and Vision* **4** (1987), no. 12, 2379–2394.

[Fla89]   P. Flandrin, On the spectrum of fractional Brownian motions, *IEEE Transactions on Information Theory* **35** (1989), no. 1, 197–199.

[Fla92]   ——— Wavelet analysis and synthesis of fractional Brownian motion, *IEEE Transactions on Information Theory* **38** (1992), no. 2, 910–917.

[FN03]   M. A. T. Figueiredo and R. D. Nowak, An EM algorithm for wavelet-based image restoration, *IEEE Transactions on Image Processing* **12** (2003), no. 8, 906–916.

[Fou77]   J. Fournier, Sharpness in Young's inequality for convolution, *Pacific Journal of Mathematics* **72** (1977), no. 2, 383–397.

[FR08]   M. Fornasier and H. Rauhut, Iterative thresholding algorithms, *Applied and Computational Harmonic Analysis* **25** (2008), no. 2, 187–208.

[Fra92]   J. Frank, *Electron Tomography: Three-Dimensional Imaging with the Transmission Electron Microscope*, Springer, 1992.

[Fre00]   D. H. Fremlin, *Measure Theory, Vol. 1*, Torres Fremlin, 2000.

[Fre01]   ——— *Measure Theory, Vol. 2*, Torres Fremlin, 2001.

[Fre02]   ——— *Measure Theory, Vol. 3*, Torres Fremlin, 2002.

[Fre03]   ——— *Measure Theory, Vol. 4*, Torres Fremlin, 2003.

[Fre08]   ——— *Measure Theory, Vol. 5*, Torres Fremlin, 2008.

[GBH70]   R. Gordon, R. Bender, and G. T. Herman, Algebraic reconstruction techniques (ART) for three-dimensional electron microscopy and X-ray photography, *Journal of Theoretical Biology* **29** (1970), no. 3, 471–481.

[Gel55]   I. M. Gelfand, Generalized random processes, *Doklady Akademii Nauk SSSR* **100** (1955), no. 5, 853–856, in Russian.

[GG84]   S. Geman and D. Geman, Stochastic relaxation, Gibbs distributions, and the Bayesian restoration of images, *IEEE Transactions on Pattern Analysis and Machine Intelligence* **6** (1984), no. 6, 721–741.

[GK68]   B. V. Gnedenko and A. N. Kolmogorov, *Limit Distributions for Sums of Independent Random Variables*, Addison-Wesley, 1968.

[GKHPU11]   M. Guerquin-Kern, M. Häberlin, K. P. Pruessmann, and M. Unser, A fast wavelet-based reconstruction method for magnetic resonance imaging, *IEEE Transactions on Medical Imaging* **30** (2011), no. 9, 1649–1660.

[GR92]   D. Geman and G. Reynolds, Constrained restoration and the recovery of discontinuities, *IEEE Transactions on Pattern Analysis and Machine Intelligence* **14** (1992), no. 3, 367–383.

[Gra08]   L. Grafakos, *Classical Fourier Analysis*, Springer, 2008.

[Gri11]   R. Gribonval, Should penalized least squares regression be interpreted as maximum a posteriori estimation?, *IEEE Transactions on Signal Processing* **59** (2011), no. 5, 2405–2410.

[Gro55]   A. Grothendieck, Produits tensoriels topologiques et espaces nucléaires, *Memoirs of the American Mathematical Society* **16** (1955).

[GS64]   I. M. Gelfand and G. Shilov, *Generalized Functions, Vol. 1: Properties and Operations*, Academic Press, 1964.

[GS68]   *Generalized Functions, Vol. 2: Spaces of Fundamental and Generalized Functions*, Academic Press, New York, 1968.

[GS01]   U. Grenander and A. Srivastava, Probability models for clutter in natural images, *IEEE Transactions on Pattern Analysis and Machine Intelligence* **23** (2001), no. 4, 424–429.

[GV64]   I. M. Gelfand and N. Ya. Vilenkin, *Generalized Functions, Vol. 4: Applications of Harmonic Analysis*, Academic Press, New York, 1964.

[Haa10]   A. Haar, Zur Theorie der orthogonalen Funktionensysteme, *Mathematische Annalen* **69** (1910), no. 3, 331–371.

[Hea71]   O. Heaviside, *Electromagnetic Theory: Including an Account of Heaviside's Unpublished Notes for a Fourth Volume*, Chelsea Publishing, 1971.

[HHLL79]   G. T. Herman, H. Hurwitz, A. Lent, and H. P. Lung, On the Bayesian approach to image reconstruction, *Information and Control* **42** (1979), no. 1, 60–71.

[HKPS93]   T. Hida, H.-H. Kuo, J. Potthoff, and L. Streit, *White Noise: An Infinite Dimensional Calculus*, Kluver, 1993.

[HK70]   A. E. Hoerl and R. W. Kennard, Ridge regression: Biased estimation for nonorthogonal problems, *Technometrics* **12** (1970), no. 1, 55–67.

[HL76]   G. T. Herman and A. Lent, Quadratic optimization for image reconstruction. I, *Computer Graphics and Image Processing* **5** (1976), no. 3, 319–332.

[HO00]   A. Hyvärinen and E. Oja, Independent component analysis: Algorithms and applications, Neural Networks **13** (2000), no. 4, 411–430.

[Hör80]   L. Hörmander, *The Analysis of Linear Differential Operators I: Distribution Theory and Fourier Analysis*, 2nd edn., Springer, 1980.

[Hör05]   *The Analysis of Linear Partial Differential Operators II. Differential Operators with Constant Coefficients*. Classics in mathematics, Springer, 2005.

[HP76]   M. Hamidi and J. Pearl, Comparison of the cosine and Fourier transforms of Markov-1 signals, *IEEE Transactions on Acoustics, Speech and Signal Processing* **24** (1976), no. 5, 428–429.

[HS04]   T. Hida and S. Si, *An Innovation Approach to Random Fields: Application of White Noise Theory*, World Scientific, 2004.

[HS08]   *Lectures on White Noise Functionals*, World Scientific, 2008.

[HSS08]   T. Hofmann, B. Schölkopf, and A. J. Smola, Kernel methods in machine learning, *Annals of Statistics* **36** (2008), no. 3, 1171–1220.

[HT02]   D. W. Holdsworth and M. M. Thornton, Micro-CT in small animal and specimen imaging, *Trends in Biotechnology* **20** (2002), no. 8, S34–S39.

[Hun77]   B. R. Hunt, Bayesian methods in nonlinear digital image restoration, *IEEE Transactions on Computers* **C-26** (1977), no. 3, 219–229.

[HW85]   K. M. Hanson and G. W. Wecksung, Local basis-function approach to computed tomography, *Applied Optics* **24** (1985), no. 23, 4028–4039.

[HY00]   M. Hansen and B. Yu, Wavelet thresholding via MDL for natural images, *IEEE Transactions on Information Theory* **46** (2000), no. 5, 1778–1788.

[Itô54]   K. Itô, Stationary random distributions, *Kyoto Journal of Mathematics* **28** (1954), no. 3, 209–223.

[Itô84]   *Foundations of Stochastic Differential Equations in Infinite-Dimensional Spaces*, CBMS-NSF Regional Conference Series in Applied Mathematics, vol. 47, Society for Industrial and Applied Mathematics (SIAM), 1984.

[Jac01]   N. Jacob, *Pseudo Differential Operators & Markov Processes, Vol. 1: Fourier Analysis and Semigroups*, World Scientific, 2001.

[Jai79]   A. K. Jain, A sinusoidal family of unitary transforms, *IEEE Transactions on Pattern Analysis and Machine Intelligence* **1** (1979), no. 4, 356–365.

[Jai89]   *Fundamentals of Digital Image Processing*, Prentice-Hall, 1989.

[JMR01]   S. Jaffard, Y. Meyer, and R. D. Ryan, *Wavelets: Tools for Science and Technology*, SIAM, 2001.

[JN84]   N. S. Jayant and P. Noll, *Digital Coding of Waveforms: Principles and Application to Speech and Video Coding*, Prentice-Hall, 1984.

[Joh66]   S. Johansen, An application of extreme point methods to the representation of infinitely divisible distributions, *Probability Theory and Related Fields* **5** (1966), 304–316.

[Kai70]   T. Kailath, The innovations approach to detection and estimation theory, *Proceedings of the IEEE* **58** (1970), no. 5, 680–695.

[Kal06]   W. A. Kalender, X-ray computed tomography, *Physics in Medicine and Biology* **51** (2006), no. 13, R29.

[Kap62]   W. Kaplan, *Operational Methods for Linear Systems*, Addison-Wesley, 1962.

[KBU12]   U. Kamilov, E. Bostan, and M. Unser, Wavelet shrinkage with consistent cycle spinning generalizes total variation denoising, *IEEE Signal Processing Letters* **19** (2012), no. 4, 187–190.

[KFL01]   F. R. Kschischang, B. J. Frey, and H.-A. Loeliger, Factor graphs and the sum-product algorithm, *IEEE Transactions on Information Theory* **47** (2001), no. 2, 498–519.

[Khi34]   A. Khintchine, Korrelationstheorie der stationären stochastischen Prozesse, *Mathematische Annalen* **109** (1934), no. 1, 604–615.

[Khi37a]   A new derivation of one formula by P. Lévy, *Bulletin of Moscow State University*, **I** (1937), no. 1, 1–5.

[Khi37b]   Zur Theorie der unbeschränkt teilbaren Verteilungsgesetze, *Recueil Mathé-matique (Matematiceskij Sbornik)* **44** (1937), no. 2, 79–119.

[KKBU12]   A. Kazerouni, U. S. Kamilov, E. Bostan, and M. Unser, Bayesian denoising: From MAP to MMSE using consistent cycle spinning, *IEEE Signal Processing Letters* **20** (2012), no. 3, 249–252.

[KMU11]   H. Kirshner, S. Maggio, and M. Unser, A sampling theory approach for continuous ARMA identification, *IEEE Transactions on Signal Processing* **59** (2011), no. 10, 4620–4634.

[Kol35]   A. N. Kolmogoroff, La transformation de Laplace dans les espaces linéaires, *Comptes Rendus de l'Académie des Sciences* **200** (1935), 1717–1718, Note de A. N. Kolmogoroff, présentée par Jacques Hadamard.

[Kol40]   Wienersche Spiralen und einige andere interessante Kurven im Hilbertschen Raum, *Comptes Rendus (Doklady) de l'Académie des Sciences de l'URSS* **26** (1940), no. 2, 115–118.

[Kol41]   A. N. Kolmogorov, Stationary sequences in Hilbert space, *Vestnik Moskovskogo Universiteta, Seriya 1: Matematika, Mekhanika* **2** (1941), no. 6, 1–40.

[Kol56]   *Foundations of the Theory of Probability*, 2nd English edn., Chelsea Publishing, 1956.

[Kol59]   A note on the papers of R. A. Minlos and V. Sazonov, *Theory of Probability and Its Applications* **4** (1959), no. 2, 221–223.

[KPAU13]   U. S. Kamilov, P. Pad, A. Amini, and M. Unser, MMSE estimation of sparse Lévy processes, *IEEE Transactions on Signal Processing* **61** (2013), no. 1, 137–147.

[Kru70]   V. M. Kruglov, A note on infinitely divisible distributions, *Theory of Probability and Its Applications* **15** (1970), no. 2, 319–324.

[KU06]   I. Khalidov and M. Unser, From differential equations to the construction of new wavelet-like bases, *IEEE Transactions on Signal Processing* **54** (2006), no. 4, 1256–1267.

[KUW13]   I. Khalidov, M. Unser, and J. P. Ward, Operator-like bases of $L_2(\mathbb{R}^d)$, *Journal of Fourier Analysis and Applications* **19**, (2013), no. 6, 1294–1322.

[KV90]   N. B. Karayiannis and A. N. Venetsanopoulos, Regularization theory in image restoration: The stabilizing functional approach, *IEEE Transactions on Acoustics, Speech and Signal Processing* **38** (1990), no. 7, 1155–1179.

[KW70]   G. Kimeldorf and G. Wahba, A correspondence between Bayesian estimation on stochastic processes and smoothing by splines, *Annals of Mathematical Statistics* **41** (1970), no. 2, 495–502.

[Lat98]   B. P. Lathy, *Signal Processing and Linear Systems*, Berkeley-Cambridge Press, 1998.

[LBU07]   F. Luisier, T. Blu, and M. Unser, A new SURE approach to image denoising: Interscale orthonormal wavelet thresholding, *IEEE Transactions on Image Processing* **16** (2007), no. 3, 593–606.

[LC47]   L. Le Cam, Un instrument d'étude des fonctions aléatoires: la fonctionnelle caractéristique, *Comptes Rendus de l'Académie des Sciences* **224** (1947), no. 3, 710–711.

[LDP07]   M. Lustig, D. L. Donoho, and J. M. Pauly, Sparse MRI: The application of compressed sensing for rapid MR imaging, *Magnetic Resonance in Medicine* **58** (2007), no. 6, 1182–1195.

[Lév25]   P. Lévy, *Calcul des Probabilités*, Gauthier-Villars, 1925.

[Lév34]   Sur les intégrales dont les éléments sont des variables aléatoires indépendantes, *Annali della Scuola Normale Superiore di Pisa: Classe di Scienze* **3** (1934), no. 3–4, 337–366.

[Lév54]   *Le Mouvement Brownien*, Mémorial des Sciences Mathématiques, vol. CXXVI, Gauthier–Villars, 1954.

[Lév65]   *Processus Stochastiques et Mouvement Brownien*, 2nd edn., Gauthier-Villars, 1965.

[Lew90]   R. M. Lewitt, Multidimensional digital image representations using generalized Kaiser Bessel window functions, *Journal of the Optical Society of America A* **7** (1990), no. 10, 1834–1846.

[LFB05]   V. Lučić, F. Förster, and W. Baumeister, Structural studies by electron tomography: From cells to molecules, *Annual Review of Biochemistry* **74** (2005), 833–865.

[Loè73]   M. Loève, Paul Lévy, 1886-1971, *Annals of Probability* **1** (1973), no. 1, 1–8.

[Loe04]   H.-A. Loeliger, An introduction to factor graphs, *IEEE Signal Processing Magazine* **21** (2004), no. 1, 28–41.

[LR78]   A. V. Lazo and P. Rathie, On the entropy of continuous probability distributions, *IEEE Transactions on Information Theory* **24** (1978), no. 1, 120–122.

[Luc74]   L. B. Lucy, An iterative technique for the rectification of observed distributions, *The Astronomical Journal* **6** (1974), no. 6, 745–754.

[Mal56]   B. Malgrange, Existence et approximation des solutions des équations aux dérivées partielles et des équations de convolution, *Annales de l'Institut Fourier* **6** (1956), 271–355.

[Mal89]   S. G. Mallat, A theory of multiresolution signal decomposition: The wavelet representation, *IEEE Transactions on Pattern Analysis and Machine Intelligence* **11** (1989), no. 7, 674–693.

[Mal98]   S. Mallat, *A Wavelet Tour of Signal Processing*, Academic Press, 1998.

[Mal09]   *A Wavelet Tour of Signal Processing: The Sparse Way*, 3rd edn., Academic Press, 2009.

[Man63]   B. B. Mandelbrot, The variation of certain speculative prices, *The Journal of Business* **36** (1963), no. 4, 393–413.

[Man82]   *The Fractal Geometry of Nature*, Freeman, 1982.

[Man01]   *Gaussian Self-Affinity and Fractals*, Springer, 2001.

[Mar05]   R. Martin, Speech enhancement based on minimum mean-square error estimation and supergaussian priors, *IEEE Transactions on Speech and Audio Processing* **13** (2005), no. 5, 845–856.

[Mat63]   G. Matheron, Principles of geostatistics, *Economic Geology* **58** (1963), no. 8, 1246–1266.

[Mey90]   Y. Meyer, *Ondelettes et Opérateurs I: Ondelettes*, Hermann, 1990.

[MG01]   D. Mumford and B. Gidas, Stochastic models for generic images, *Quarterly of Applied Mathematics* **59** (2001), no. 1, 85–112.

[Mic76]   C. Micchelli, Cardinal L-splines, *Studies in Spline Functions and Approximation Theory* (S. Karlin, C. Micchelli, A. Pinkus, and I. Schoenberg, eds.), Academic Press, 1976, pp. 203–250.

[Mic86]   Interpolation of scattered data: Distance matrices and conditionally positive definite functions, *Constructive Approximation* **2** (1986), no. 1, 11–22.

[Min63]   R. A. Minlos, *Generalized Random Processes and Their Extension to a Measure*, Selected Translations in Mathematical Statististics and Probability, vol. 3, American Mathematical Society, 1963, pp. 291–313.

[Miy61]   K. Miyasawa, An empirical Bayes estimator of the mean of a normal population, *Bulletin de l'Institut International de Statistique* **38** (1961), no. 4, 181–188.

[MKCC99]   J. G. McNally, T. Karpova, J. Cooper, and J. A. Conchello, Three-dimensional imaging by deconvolution microscopy, *Methods* **19** (1999), no. 3, 373–385.

[ML99]   P. Moulin and J. Liu, Analysis of multiresolution image denoising schemes using generalized Gaussian and complexity priors, *IEEE Transactions on Information Theory* **45** (1999), no. 3, 909–919.

[MN90a]   W. R. Madych and S. A. Nelson, Multivariate interpolation and conditionally positive definite functions. II, *Mathematics of Computation* **54** (1990), no. 189, 211–230.

[MN90b]   Polyharmonic cardinal splines, *Journal of Approximation Theory* **60** (1990), no. 2, 141–156.

[MP86]   S. G. Mikhlin and S. Prössdorf, *Singular Integral Operators*, Springer, 1986.

[MR06]   F. Mainardi and S. Rogosin, The origin of infinitely divisible distributions: From de Finetti's problem to Lévy-Khintchine formula, *Mathematical Methods in Economics and Finance* **1** (2006), no. 1, 7–55.

[MVN68]   B. B. Mandelbrot and J. W. Van Ness, Fractional Brownian motions, fractional noises and applications, *SIAM Review* **10** (1968), no. 4, 422–437.

[Mye92]   D. E. Myers, Kriging, cokriging, radial basis functions and the role of positive definiteness, *Computers and Mathematics with Applications* **24** (1992), no. 12, 139–148.

[Nat84]   F. Natterer, *The Mathematics of Computed Tomography*, John Wiley & Sons, 1984.

[New75]   D. J. Newman, A simple proof of Wiener's $1/f$ theorem, *Proceedings of the American Mathematical Society* **48** (1975), no. 1, 264–265.

[Nik07]   M. Nikolova, Model distortions in Bayesian MAP reconstruction, *Inverse Problems and Imaging* **1** (2007), no. 2, 399–422.

[OF96a]   B. A. Olshausen and D. J. Field, Emergence of simple-cell receptive field properties by learning a sparse code for natural images, *Nature* **381** (1996), no. 6583, 607–609.

[OF96b]   Natural image statistics and efficient coding, *Network: Computation in Neural Systems* **7** (1996), no. 2, 333–339.

[Øks07]   B. Øksendal, *Stochastic Differential Equations*, 6th edn., Springer, 2007.

[OSB99]   A. V. Oppenheim, R. W. Schafer, and J. R. Buck, *Discrete-Time Signal Processing*, Prentice Hall, 1999.

[PAP72]   J. Pearl, H. C. Andrews, and W. Pratt, Performance measures for transform data coding, *IEEE Transactions on Communications* **20** (1972), no. 3, 411–415.

[Pap91]   A. Papoulis, *Probability, Random Variables, and Stochastic Processes*, McGraw-Hill, 1991.

[PBE01]   M. Persson, D. Bone, and H. Elmqvist, Total variation norm for three-dimensional iterative reconstruction in limited view angle tomography, *Physics in Medicine and Biology* **46** (2001), no. 3, 853.

[Pet05]   M. Petit, *L'équation de Kolmogoroff: Vie et Mort de Wolfgang Doeblin, un Génie dans la Tourmente Nazie*, Gallimard, 2005.

[PHBJ⁺01]   E. Perrin, R. Harba, C. Berzin-Joseph, I. Iribarren, and A. Bonami, nth-order fractional Brownian motion and fractional Gaussian noises, *IEEE Transactions on Signal Processing* **49** (2001), no. 5, 1049–1059.

[Pie72]   A. Pietsch, *Nuclear Locally Convex Spaces*, Springer, 1972.

[PM62]   D. P. Petersen and D. Middleton, *Sampling and reconstruction of wave-number-limited functions in n-dimensional Euclidean spaces*, Information and Control **5** (1962), no. 4, 279–323.

[PMC93]   C. Preza, M. I. Miller, and J.-A. Conchello, Image reconstruction for 3D light microscopy with a regularized linear method incorporating a smoothness prior, *Proceedings of the IS&T/SPIE Symposium on Electronic Imaging Science and Technology* (San Jose, CA, January 1993), vol. 1905, SPIE 1993, pp. 129–139.

[PP06]   A. Pizurica and W. Philips, Estimating the probability of the presence of a signal of interest in multiresolution single- and multiband image denoising, *IEEE Transactions on Image Processing* **15** (2006), no. 3, 654–665.

[PPLV02]   B. Pesquet-Popescu and J. Lévy Véhel, Stochastic fractal models for image processing, *IEEE Signal Processing Magazine* **19** (2002), no. 5, 48–62.

[PR70]   B. L. S. Prakasa Rao, Infinitely divisible characteristic functionals on locally convex topological vector spaces, *Pacific Journal of Mathematics* **35** (1970), no. 1, 221–225.

[Pra91]   W. K. Pratt, *Digital Image Processing*, John Wiley & Sons, 1991.

[Pro04]   P. Protter, *Stochastic Integration and Differential Equations*, Springer, 2004.

[Pru06]   K. P. Pruessmann, Encoding and reconstruction in parallel MRI, *NMR in Biomedicine* **19** (2006), no. 3, 288–299.

[PSV09]   X. Pan, E. Y. Sidky, and M. Vannier, Why do commercial CT scanners still employ traditional, filtered back-projection for image reconstruction?, *Inverse Problems* **25** (2009), no. 12, 123009.

[PSWS03]   J. Portilla, V. Strela, M. J. Wainwright, and E. P. Simoncelli, Image denoising using scale mixtures of Gaussians in the wavelet domain, *IEEE Transactions on Image Processing* **12** (2003), no. 11, 1338–1351.

[PU13]   P. Pad and M. Unser, On the optimality of operator-like wavelets for sparse AR(1) processes, *Proceedings of the 2013 IEEE International Conference on Acoustics, Speech, and Signal Processing (ICASSP'13)* (Vancouver, BC, Canada, May 26–31, 2013), IEEE, 2013, pp. 5598–5602.

[PWSB99]   K. P. Pruessmann, M. Weiger, M. B. Scheidegger, and P. Boesiger, SENSE: Sensitivity encoding for fast MRI, *Magnetic Resonance in Medicine* **42** (1999), no. 5, 952–962.

[Rab92a]   C. Rabut, Elementary m-harmonic cardinal B-splines, *Numerical Algorithms* **2** (1992), no. 2, 39–61.

[Rab92b]   High level m-harmonic cardinal B-splines, *Numerical Algorithms* **2** (1992), 63–84.

[Ram69]   B. Ramachandran, On characteristic functions and moments, *Sankhya: The Indian Journal of Statistics, Series A* **31** (1969), no. 1, 1–12.

[RB94]   D. L. Ruderman and W. Bialek, Statistics of natural images: Scaling in the woods, *Physical Review Letters* **73** (1994), no. 6, 814–818.

[RBFK11]   L. Ritschl, F. Bergner, C. Fleischmann, and M. Kachelrieß, Improved total variation-based CT image reconstruction applied to clinical data, *Physics in Medicine and Biology* **56** (2011), no. 6, 1545.

[RF11]   S. Ramani and J. A. Fessler, Parallel MR image reconstruction using augmented Lagrangian methods, *IEEE Transactions on Medical Imaging* **30** (2011), no. 3, 694–706.

[Ric72]   W. Richardson, Bayesian-based iterative method of image restoration, *Journal of the Optical Society of America* **62** (1972), no. 1, 55–59.

[Ric77]   J. Rice, On generalized shot noise, *Advances in Applied Probability* **9** (1977), no. 3, 553–565.

[RL71]   G. N. Ramachandran and A. V. Lakshminarayanan, Three-dimensional reconstruction from radiographs and electron micrographs: Application of convolutions instead of Fourier transforms, *Proceedings of the National Academy of Sciences* **68** (1971), no. 9, 2236–2240.

[ROF92]   L. I. Rudin, S. Osher, and E. Fatemi, Nonlinear total variation based noise removal algorithms, *Physica D* **60** (1992), no. 1–4, 259–268.

[Ron88]   A. Ron, Exponential box splines, *Constructive Approximation* **4** (1988), no. 1, 357–378.

[RS08]   M. Raphan and E. P. Simoncelli, Optimal denoising in redundant representations, *IEEE Transactions on Image Processing* **17** (2008), no. 8, 1342–1352.

[RU06]   S. Ramani and M. Unser, Matérn B-splines and the optimal reconstruction of signals, *IEEE Signal Processing Letters* **13** (2006), no. 7, 437–440.

[Rud73]   W. Rudin, *Functional analysis*, McGraw-Hill Series in Higher Mathematics, McGraw-Hill, 1973.

[Rud97]   D. L. Ruderman, Origins of scaling in natural images, *Vision Research* **37** (1997), no. 23, 3385–3398.

[RVDVBU08] S. Ramani, D. Van De Ville, T. Blu, and M. Unser, Nonideal sampling and regularization theory, *IEEE Transactions on Signal Processing* **56** (2008), no. 3, 1055–1070.

[SA96]   E. P. Simoncelli and E. H. Adelson, Noise removal via Bayesian wavelet coring, *Proceedings of the IEEE International Conference on Image Processing (ICIP2010)*, (Lausanne, Switzerland September 16–19, 1996), vol. 1, IEEE, 1996, pp. 379–382.

[Sat94]   K.-I. Sato, *Lévy Processes and Infinitely Divisible Distributions*, Chapman & Hall, 1994.

[Sch38]   I. J. Schoenberg, Metric spaces and positive definite functions, *Transactions of the American Mathematical Society* **44** (1938), no. 3, 522–536.

[Sch46]   Contribution to the problem of approximation of equidistant data by analytic functions, *Quarterly of Applied Mathematics* **4** (1946), 45–99, 112–141.

[Sch66]   L. Schwartz, *Théorie des Distributions*, Hermann, 1966.

[Sch73a]   I. J. Schoenberg, *Cardinal Spline Interpolation*, Society of Industrial and Applied Mathematics, 1973.

[Sch73b]   L. Schwartz, *Radon Measures on Arbitrary Topological Spaces and Cylindrical Measures*, Studies in Mathematics, vol. 6, Tata Institute of Fundamental Research, Bombay Oxford University Press, 1973.

[Sch81a]   L. L. Schumaker, *Spline Functions: Basic Theory*, John Wiley & Sons, 1981.

[Sch81b]   L. Schwartz, *Geometry and Probability in Banach Spaces*, Lecture Notes in Mathematics, vol. 852, Springer, 1981.

[Sch88]   I. J. Schoenberg, *A Brief Account of My Life and Work*, Selected papers (C. de Boor, ed.), vol. 1, Birkhäuser, 1988, pp. 1–10.

[Sch99]   H. H. Schaefer, *Topological Vector Spaces*, 2nd edn., Graduate Texts in Mathematics, vol. 3, Springer, 1999.

[SDFR13]   J.-L. Starck, D. L. Donoho, M. J. Fadili, and A. Rassat, Sparsity and the Bayesian perspective, *Astronomy and Astrophysics* **552** (2013), A133.

[SF71]   G. Strang and G. Fix, A Fourier analysis of the finite element variational method, *Constructive Aspects of Functional Analysis*, Edizioni Cremonese, 1971, pp. 793–840.

[Sha93]   J. Shapiro, Embedded image coding using zerotrees of wavelet coefficients, *IEEE Transactions on Acoustics, Speech and Signal Processing* **41** (1993), no. 12, 3445–3462.

[Sil99]   B. W. Silverman, Wavelets in statistics: Beyond the standard assumptions, *Philosophical Transactions: Mathematical, Physical and Engineering Sciences* **357** (1999), no. 1760, 2459–2473.

[SLG02]   A. Srivastava, X. Liu, and U. Grenander, Universal analytical forms for modeling image probabilities, *IEEE Transactions on Pattern Analysis and Machine Intelligence* **24** (2002), no. 9, 1200–1214.

[SLSZ03]   A. Srivastava, A. B. Lee, E. P. Simoncelli, and S.-C. Zhu, On advances in statistical modeling of natural images, *Journal of Mathematical Imaging and Vision* **18** (2003), 17–33.

[SO01]   E. P. Simoncelli and B. A. Olshausen, Natural image statistics and neural representation, *Annual Review of Neuroscience* **24** (2001), 1193–1216.

[Sob36]   S. Soboleff, Méthode nouvelle à résoudre le problème de Cauchy pour les équations linéaires hyperboliques normales, *Recueil Mathématique (Matematiceskij Sbornik)* **1(43)** (1936), no. 1, 39–72.

[SP08]   E. Y. Sidky and X. Pan, Image reconstruction in circular cone-beam computed tomography by constrained, total-variation minimization, *Physics in Medicine and Biology* **53** (2008), no. 17, 4777.

[SPM02]   J.-L. Starck, E. Pantin, and F. Murtagh, Deconvolution in astronomy: A review, *Publications of the Astronomical Society of the Pacific* **114** (2002), no. 800, 1051–1069.

[SS02]   L. Sendur and I. W. Selesnick, Bivariate shrinkage functions for wavelet-based denoising exploiting interscale dependency, *IEEE Transactions on Signal Processing* **50** (2002), no. 11, 2744–2756.

[ST94]   G. Samorodnitsky and M. S. Taqqu, *Stable Non-Gaussian Random Processes: Stochastic Models with Infinite Variance*, Chapman & Hall, 1994.

[Ste76]   J. Stewart, Positive definite functions and generalizations: An historical survey, *Rocky Mountain Journal of Mathematics* **6** (1976), no. 3, 409–434.

[Ste81]   C. Stein, Estimation of the mean of a multivariate normal distribution, *Annals of Statistics* **9** (1981), no. 6, 1135–1151.

[SU12]   Q. Sun and M. Unser, Left inverses of fractional Laplacian and sparse stochastic processes, *Advances in Computational Mathematics* **36** (2012), no. 3, 399–441.

[Sun93]   X. Sun, Conditionally positive definite functions and their application to multivariate interpolations, *Journal of Approximation Theory* **74** (1993), no. 2, 159–180.

[Sun07]   Q. Sun, Wiener's lemma for infinite matrices, *Transactions of the American Mathematical Society* **359** (2007), no. 7, 3099–3123.

[SV67]   M. H. Schultz and R. S. Varga, L-splines, *Numerische Mathematik* **10** (1967), no. 4, 345–369.

[SVH03]   F. W. Steutel and K. Van Harn, *Infinite Divisibility of Probability Distributions on the Real Line*, Marcel Dekker, 2003.

[SW71]   E. M. Stein and G. Weiss, *Introduction to Fourier Analysis on Euclidean Spaces*, Princeton University Press, 1971.

[TA77]   A. N. Tikhonov and V. Y. Arsenin, *Solution of Ill-Posed Problems*, Winston & Sons, 1977.

[Taf11]   P. D. Tafti, Self-similar vector fields, unpublished Ph.D. thesis, École Polytechnique Fédérate de Lausanne (2011).

[Tay75]   S. J. Taylor, Paul Lévy, *Bulletin of the London Mathematical Society* **7** (1975), no. 3, 300–320.

[TBU00]   P. Thévenaz, T. Blu, and M. Unser, Image interpolation and resampling, *Handbook of Medical Imaging, Processing and Analysis* (I. N. Bankman, ed.), Academic Press, 2000, pp. 393–420.

[TH79]   H. J. Trussell and B. R. Hunt, Improved methods of maximum a posteriori restoration, *IEEE Transactions on Computers* **100** (1979), no. 1, 57–62.

[Tib96]   R. Tibshirani, Regression shrinkage and selection via the Lasso, *Journal of the Royal Statistical Society, Series B* **58** (1996), no. 1, 265–288.

[Tik63]   A. N. Tikhonov, Solution of incorrectly formulated problems and the regularization method, *Soviet Mathematics* **4** (1963), 1035–1038.

[TM09]   J. Trzasko and A. Manduca, Highly undersampled magnetic resonance image reconstruction via homotopic ℓ_0-minimization, *IEEE Transactions on Medical Imaging* **28** (2009), no. 1, 106–121.

[TN95]   G. A. Tsihrintzis and C. L. Nikias, Performance of optimum and suboptimum receivers in the presence of impulsive noise modeled as an alpha-stable process, *IEEE Transactions on Communications* **43** (1995), no. 234, 904 –914.

[TVDVU09]   P. D. Tafti, D. Van De Ville, and M. Unser, Invariances, Laplacian-like wavelet bases, and the whitening of fractal processes, *IEEE Transactions on Image Processing* **18** (2009), no. 4, 689–702.

[UAE92]   M. Unser, A. Aldroubi, and M. Eden, On the asymptotic convergence of B-spline wavelets to Gabor functions, *IEEE Transactions on Information Theory* **38** (1992), no. 2, 864–872.

[UAE93]   A family of polynomial spline wavelet transforms, *Signal Processing* **30** (1993), no. 2, 141–162.

[UB00]   M. Unser and T. Blu, Fractional splines and wavelets, *SIAM Review* **42** (2000), no. 1, 43–67.

[UB03]   Wavelet theory demystified, *IEEE Transactions on Signal Processing* **51** (2003), no. 2, 470–483.

[UB05a]   Cardinal exponential splines, Part I: Theory and filtering algorithms, *IEEE Transactions on Signal Processing* **53** (2005), no. 4, 1425–1449.

[UB05b]   Generalized smoothing splines and the optimal discretization of the Wiener filter, *IEEE Transactions on Signal Processing* **53** (2005), no. 6, 2146–2159.

[UB07]   Self-similarity, Part I: Splines and operators, *IEEE Transactions on Signal Processing* **55** (2007), no. 4, 1352–1363.

[Uns84]   M. Unser, On the approximation of the discrete Karhunen-Loève transform for stationary processes, *Signal Processing* **7** (1984), no. 3, 231–249.

[Uns93]   On the optimality of ideal filters for pyramid and wavelet signal approxima-
               tion, *IEEE Transactions on Signal Processing* **41** (1993), no. 12, 3591–3596.

[Uns99]   Splines: A perfect fit for signal and image processing, *IEEE Signal Proces-
               sing Magazine* **16** (1999), no. 6, 22–38.

[Uns00]   Sampling: 50 years after Shannon, *Proceedings of the IEEE* **88** (2000), no. 4,
               569–587.

[Uns05]   Cardinal exponential splines, Part II: Think analog, act digital, *IEEE Tran-
               sactions on Signal Processing* **53** (2005), no. 4, 1439–1449.

[UT11]   M. Unser and P. D. Tafti, Stochastic models for sparse and piecewise-smooth
               signals, *IEEE Transactions on Signal Processing* **59** (2011), no. 3, 989–1005.

[UTAK14]   M. Unser, P. D. Tafti, A. Amini, and H. Kirshner, A unified formulation of
               Gaussian vs. sparse stochastic processes, Part II: Discrete-domain theory, *IEEE
               Transactions on Information Theory* **60** (2014), no. 5, 3036–3051.

[UTS14]   M. Unser, P. D. Tafti, and Q. Sun, A unified formulation of Gaussian vs. sparse
               stochastic processes, Part I: Continuous-domain theory, *IEEE Transactions on
               Information Theory* **60** (2014), no. 3, 1361–1376.

[VAVU06]   C. Vonesch, F. Aguet, J.-L. Vonesch, and M. Unser, The colored revolution of
               bioimaging, *IEEE Signal Processing Magazine* **23** (2006), no. 3, 20–31.

[VBB+02]   G. M. Viswanathan, F. Bartumeus, S. V. Buldyrev, J. Catalan, U. L. Fulco,
               S. Havlin, M. G. E da Luz, M. L. Lyra, E. P. Raposo, and H. E. Stanley, Lévy
               flight random searches in biological phenomena, *Physica A: Statistical Mecha-
               nics and Its Applications* **314** (2002), no. 1–4, 208–213.

[VBU07]   C. Vonesch, T. Blu, and M. Unser, Generalized Daubechies wavelet families,
               IEEE Transactions on Signal Processing **55** (2007), no. 9, 4415–4429.

[VDVBU05]   D. Van De Ville, T. Blu, and M. Unser, Isotropic polyharmonic B-splines: Sca-
               ling functions and wavelets, *IEEE Transactions on Image Processing* **14** (2005),
               no. 11, 1798–1813.

[VDVFHUB10]   D. Van De Ville, B. Forster-Heinlein, M. Unser, and T. Blu, Analytical foot-
               prints: Compact representation of elementary singularities in wavelet bases,
               IEEE Transactions on Signal Processing **58** (2010), no. 12, 6105–6118.

[vKvVVvdV97]   G. M. P. van Kempen, L. J. van Vliet, P. J. Verveer, and H. T. M. van der
               Voort, A quantitative comparison of image restoration methods for confocal
               microscopy, *Journal of Microscopy* **185** (1997), no. 3, 345–365.

[VMB02]   M. Vetterli, P. Marziliano, and T. Blu, Sampling signals with finite rate
               of innovation, *IEEE Transactions on Signal Processing* **50** (2002), no. 6,
               1417–1428.

[VU08]   C. Vonesch and M. Unser, A fast thresholded Landweber algorithm for wavelet-
               regularized multidimensional deconvolution, *IEEE Transactions on Image Pro-
               cessing* **17** (2008), no. 4, 539–549.

[VU09]   A fast multilevel algorithm for wavelet-regularized image restoration, *IEEE
               Transactions on Image Processing* **18** (2009), no. 3, 509–523.

[Wag09]   P. Wagner, A new constructive proof of the Malgrange-Ehrenpreis theorem,
               American Mathematical Monthly **116** (2009), no. 5, 457–462.

[Wah90]   G. Wahba, *Spline Models for Observational Data*, Society for Industrial and
               Applied Mathematics, 1990.

[Wen05]   H. Wendland, *Scattered Data Approximations*, Cambridge University Press,
               2005.

[Wie30]   N. Wiener, Generalized harmonic analysis, *Acta Mathematica* **55** (1930), no. 1, 117–258.

[Wie32]   Tauberian theorems, *Annals of Mathematics* **33** (1932), no. 1, 1–100.

[Wie64]   *Extrapolation, Interpolation and Smoothing of Stationary Time Series with Engineering Applications*, MIT Press, 1964.

[WLLL06]   J. Wang, T. Li, H. Lu, and Z. Liang, Penalized weighted least-squares approach to sinogram noise reduction and image reconstruction for low-dose X-ray computed tomography, *IEEE Transactions on Medical Imaging* **25** (2006), no. 10, 1272–1283.

[WM57]   N. Wiener and P. Masani, The prediction theory of multivariate stochastic processes, *Acta Mathematica* **98** (1957), no. 1, 111–150.

[WN10]   D. Wipf and S. Nagarajan, Iterative reweighted and methods for finding sparse solutions, *IEEE Journal of Selected Topics in Signal Processing* **4** (2010), no. 2, 317–329.

[Wol71]   S. J. Wolfe, On moments of infinitely divisible distribution functions, *The Annals of Mathematical Statistics* **42** (1971), no. 6, 2036–2043.

[Wol78]   On the unimodality of infinitely divisible distribution functions, *Probability Theory and Related Fields* **45** (1978), no. 4, 329–335.

[WYHJC91]   J. B. Weaver, X. Yansun, D. M. Healy Jr., and L. D. Cromwell, Filtering noise from images with wavelet transforms, *Magnetic Resonance in Medicine* **21** (1991), no. 2, 288–295.

[WYYZ08]   Y. L. Wang, J. F. Yang, W. T. Yin, and Y. Zhang, A new alternating minimization algorithm for total variation image reconstruction, *SIAM Journal on Imaging Sciences* **1** (2008), no. 3, 248–272.

[XQJ05]   X. Q. Zhang and F. Jacques, Constrained total variation minimization and application in computerized tomography, *Energy Minimization Methods in Computer Vision and Pattern Recognition*, Lecture Notes in Computer Science, vol. 3757, Springer, 2005, pp. 456–472.

[Yag86]   A. M. Yaglom, *Correlation Theory of Stationary and Related Random Functions I: Basic Results*, Springer, 1986.

[Zem10]   A. H. Zemanian, *Distribution Theory and Transform Analysis: An Introduction to Generalized Functions, with Applications*, Dover, 2010.

Index